Photocatalysis
Applications

RSC Energy and Environment Series

Editor-in-Chief:
Professor Laurence M. Peter, *University of Bath, UK*

Series Editors:
Dr Heinz Frei, *Lawrence Berkeley National Laboratory, USA*
Dr Roberto Rinaldi, *Max Planck Institute for Coal Research, Germany*
Professor Tim S. Zhao, *The Hong Kong University of Science and Technology, Hong Kong, China*

Titles in the Series:
1: Thermochemical Conversion of Biomass to Liquid Fuels and Chemicals
2: Innovations in Fuel Cell Technologies
3: Energy Crops
4: Chemical and Biochemical Catalysis for Next Generation Biofuels
5: Molecular Solar Fuels
6: Catalysts for Alcohol-Fuelled Direct Oxidation Fuel Cells
7: Solid Oxide Fuel Cells: From Materials to System Modeling
8: Solar Energy Conversion: Dynamics of Interfacial Electron and Excitation Transfer
9: Photoelectrochemical Water Splitting: Materials, Processes and Architectures
10: Biological Conversion of Biomass for Fuels and Chemicals: Explorations from Natural Utilization Systems
11: Advanced Concepts in Photovoltaics
12: Materials Challenges: Inorganic Photovoltaic Solar Energy
13: Catalytic Hydrogenation for Biomass Valorization
14: Photocatalysis: Fundamentals and Perspectives
15: Photocatalysis: Applications

How to obtain future titles on publication:
A standing order plan is available for this series. A standing order will bring delivery of each new volume immediately on publication.

For further information please contact:
Book Sales Department, Royal Society of Chemistry, Thomas Graham House, Science Park, Milton Road, Cambridge, CB4 0WF, UK
Telephone: +44 (0)1223 420066, Fax: +44 (0)1223 420247
Email: booksales@rsc.org
Visit our website at www.rsc.org/books

Photocatalysis
Applications

Edited by

Dionysios D Dionysiou
University of Cincinnati, Cincinnati, Ohio, USA
Email: dionysios.d.dionysiou@uc.edu

Gianluca Li Puma
Loughborough University, Loughborough, UK
Email: g.lipuma@lboro.ac.uk

Jinhua Ye
National Institute for Materials Science (NIMS), Tsukuba, Japan
Email: Jinhua.YE@nims.go.jp

Jenny Schneider
Leibniz University of Hannover, Hannover, Germany
Email: schneid@iftc.uni-hannover.de

Detlef Bahnemann
Leibniz University of Hannover, Hannover, Germany
Email: bahnemann@iftc.uni-hannover.de

ROYAL SOCIETY OF CHEMISTRY

THE QUEEN'S AWARDS FOR ENTERPRISE: INTERNATIONAL TRADE 2013

RSC Energy and Environment Series No. 15

Print ISBN: 978-1-78262-709-8
PDF eISBN: 978-1-78262-710-4
EPUB eISBN: 978-1-78262-798-2
Two-volume set print ISBN: 978-1-78262-714-2
ISSN: 2044-0774

A catalogue record for this book is available from the British Library

Published by The Royal Society of Chemistry,
Thomas Graham House, Science Park, Milton Road,
Cambridge CB4 0WF, UK

Registered Charity Number 207890

For further information see our web site at www.rsc.org

Printed in the United Kingdom by CPI Group (UK) Ltd, Croydon, CR0 4YY, UK

Preface

The field of photocatalysis has experienced numerous applications for the treatment of water, air, soil and other applications in the field of environmental remediation, self-cleaning surfaces, green chemistry and green engineering, production of energy and alternative fuels, food packaging and biomedical and medical applications. The first chapter reviews certain aspects on the photocatalytic degradation of organic contaminants in water such as insecticides, chlorinated organic compounds, antibiotics, cyanotoxins and taste and odor compounds. Particular emphasis is placed on the role of water quality parameters on degradation kinetics as well as on degradation pathways. Chapter 2 discusses the photocatalytic removal of metallic and other inorganic pollutants such as chromium, mercury, lead, uranium, arsenic, and nitrate. The chapter discusses mechanistic aspects of energetics and nano-interfacial phenomena in such process as well as real world applications. Chapter 3 focuses on solar photocatalytic disinfection of water. The chapter covers mechanistic aspects of solar disinfection, the role of reactive oxygen species in the disinfection mechanisms, visible light active materials and reactors for solar disinfection, and emerging applications of the photocatalytic inactivation of antibiotic resistant bacteria. Chapter 4 deals with fundamentals, reactor design and applications of solar photocatalysis. The chapter includes integration of solar photocatalytic processes with biological treatments or membrane processes. Chapter 5 presents development, characterization and evaluation of combined photocatalysis-separation processes for water treatment using hybrid photocatalytic membrane processes. Chapter 6 presents concepts for the effective integration and coupling of photocatalytic processes with other treatment processes for the purification of various types of wastewater. Chapter 7 provides an overview of the photocatalytic purification and disinfection of air. The chapter discusses mechanisms

RSC Energy and Environment Series No. 15
Photocatalysis: Applications
Edited by Dionysios D Dionysiou, Gianluca Li Puma, Jinhua Ye, Jenny Schneider, and Detlef Bahnemann

Published by the Royal Society of Chemistry, www.rsc.org

of removal of organic contaminants and microbiological airborne objects in indoor air, the role of various parameters on the mechanisms and kinetics of removal, commercial and prototype photocatalytic reactors for air purification, and present challenges in the field. Chapter 8 reviews development, mechanisms, applications, progress and challenges in the field of self-cleaning photocatalytic materials. Chapter 9 discusses the development, underlying photocatalysis and photoelectrocatalysis mechanisms, and performance evaluation of PhotoFuelCells for energy generation. Chapter 10 is focused on photocatalytic hydrogen generation. The chapter presents the mechanism and reviews applications in the field as well as insights on approaches for process optimization for higher efficiency. Chapter 11 presents examples of applications of various photocatalysts in reactions relevant to synthesis of chemicals. The chapter discusses the role of catalyst type, reaction yields and selectivities, and concepts of green chemistry and green engineering. The final chapter (Chapter 12) presents an overview of the potential applications of TiO_2 photocatalysis in food packaging, osseointegration and disinfection of medical implants, and controlled delivery of chemical compounds.

Contents

RSC Energy and Environment Series No. 15
Photocatalysis: Applications
Edited by Dionysios D Dionysiou, Gianluca Li Puma, Jinhua Ye, Jenny Schneider, and Detlef Bahnemann

Published by the Royal Society of Chemistry, www.rsc.org

CHAPTER 1

Photocatalytic Degradation of Organic Contaminants in Water: Process Optimization and Degradation Pathways†

MARIA G. ANTONIOU*[a], CEN ZHAO[b], KEVIN E. O'SHEA[b], GESHAN ZHANG[c,d], DIONYSIOS D. DIONYSIOU[c], CHUN ZHAO[e], CHANGSEOK HAN[f], MALLIKARJUNA N. NADAGOUDA[g], HYEOK CHOI[h], THEODORA FOTIOU[i], THEODOROS M. TRIANTIS[i], AND ANASTASIA HISKIA[i]

[a]Department of Environmental Science and Technology, Cyprus University of Technology, 3036 Lemesos, Cyprus; [b]Department of Chemistry and Biochemistry, Florida International University, Miami, FL 33199, USA; [c]Environmental Engineering and Science Program, Department of Biomedical, Chemical and Environmental Engineering (DBCEE), University of Cincinnati, Cincinnati, OH 45221-0012, USA; [d]Department of Chemical Engineering, Zhejiang University of Technology, Hangzhou, Zhejiang, 310014, PR China; [e]Key Laboratory of the Three Gorges Reservoir

†The U.S. Environmental Protection Agency, through its Office of Research and Development, funded and managed, or partially funded and collaborated in, the research described herein. It has been subjected to the Agency's administrative review and has been approved for external publication. Any opinions expressed in this chapter are those of the author(s) and do not necessarily reflect the views of the Agency, therefore, no official endorsement should be inferred. Any mention of trade names or commercial products does not constitute endorsement or recommendation for use.

RSC Energy and Environment Series No. 15
Photocatalysis: Applications
Edited by Dionysios D Dionysiou, Gianluca Li Puma, Jinhua Ye, Jenny Schneider, and Detlef Bahnemann

Published by the Royal Society of Chemistry, www.rsc.org

Region's Eco-Environment, Ministry of Education, Chongqing University, Chongqing 400045, PR China; [f]ORISE Post-doctoral Fellow, Clean Processes Branch, Sustainable Technology Division, The United States Environmental Protection Agency, Cincinnati, OH 45268, USA; [g]Water Quality Management Branch, Water Supply and Water Resources Division, National Risk Management Research Laboratory, The United States Environmental Protection Agency, Cincinnati, OH 45268, USA; [h]Department of Civil Engineering, The University of Texas at Arlington, 416 Yates Street, Arlington, Texas 76019-0308, USA; [i]Catalytic-Photocatalytic Processes and Environmental Analysis Lab., Institute of Nanoscience and Nanotechnology, NCSR "DEMOKRITOS", Neapoleos 25, Agia Paraskevi, 15341, Athens, Greece
*E-mail: maria.antoniou@cut.ac.cy

1.1 Introduction

Ever since Fujishima and Honda (1972) successfully applied titanium dioxide (TiO_2) for the photoelectrochemical splitting of water, TiO_2 has been widely utilized as a remediation technology for the removal of pollutants from the air, drinking water, and wastewater.[1] TiO_2 is considered an attractive alternative to other applied technologies because of its superior photo-activity, low toxicity, chemical and biological inertness, low cost, and corrosion resistance.[2,3] TiO_2 exists in three different polymorphic forms: anatase, rutile and brookite with band gaps of 3.2, 3.0, and ~3.2 eV, respectively, activated in the UV range. Anatase and rutile are the most common polymorphs, with the anatase phase possessing higher photocatalytic activity than rutile and brookite.[4] Higher reactivity is associated with the number and type of reactive oxygen or free radicals species (ROS) formed that react and transform the targeted pollutants first into smaller molecular weight products and ultimately mineralize them into simpler and non-toxic products H_2O, CO_2, and mineral acids. ROS such as hydroxyl radical ($HO^{\bullet}$), superoxide anion radical ($O_2^{-\bullet}$), hydroperoxyl radical ($HO_2^{\bullet}$), and singlet oxygen ($^1O_2$) are mainly produced during UV TiO_2 photocatalysis.[5–7] Hydroxyl radicals, generated on the surface of the catalyst following oxidation of water from the positive holes of TiO_2, are non-selective oxidizing species with strong oxidation potential (+2.80 V) that rapidly react with most organic compounds with rate constants of the order of 10^6–10^{10} M^{-1} s^{-1}.[8–10] While several ROS are formed during UV TiO_2 photocatalysis, the effectiveness of the degradation process is generally dependent on the production and subsequent reactions of $HO^{\bullet}$ at or very near the TiO_2 surface.[11,12] The surface adsorption and reactivity of the target compound with $HO^{\bullet}$ are key factors in the UV promoted TiO_2 photocatalytic degradation of organic compounds. Several heterogeneous kinetic models have been used to determine apparent kinetic parameters and effectively model the observed degradation processes.[5] Detailed product studies as well as structure–reactivity relationships provide convincing evidence that generated $HO^\bullet$ adsorbed at the surface of TiO_2 has similar reactivity to $HO^\bullet$ in a homogenous solution.[13] Radiolysis methods are commonly used to study the $HO^\bullet$ induced oxidative degradation of problematic pollutants and toxins. Radiolytic and UV TiO_2 photocatalytic degradation of organic compounds can follow similar degradation pathways.[14,15] Under photocatalytic conditions the adsorption of a target compound at a higher surface concentration than in the bulk solution can enhance the rate of $HO^\bullet$ induced degradation since both the oxidant and target are localized at the surface. For the purpose of this chapter the differences in reactivities of free and adsorbed $HO^\bullet$ will not be discussed.

Hydroxyl radicals generally react with organic compounds *via* three pathways: electrophilic addition (eqn (1.1)), hydrogen atom abstraction (eqn (1.2)) and electron transfer (eqn (1.3)). The hydroxyl radical additions to carbon–carbon double bonds and electron-rich aromatic systems leading to hydroxylated adducts are generally faster than H-abstraction and electron transfer pathways. Electron transfer with hydroxyl radical is limited to very electron rich systems.[8] The electron transfer pathway generates a radical cation which in aqueous media is hydrolyzed, often generating the same hydroxylated adducts observed by direct addition of $HO^\bullet$. The H-abstraction pathways result in oxidation of a sp^3 hybridized carbon. The hydroxyl radical induced oxidation of primary (1°), secondary (2°), and tertiary (3°) aliphatic carbons is initiated by hydrogen atom abstraction. The hydroxyl radical mediated oxidation of a 1° alcohol *via* the H-atom abstraction pathways are given in eqn (1.4). The presence of a heteroatom attached to the carbon significantly enhances the rate of hydrogen atom abstraction and leads to a stabilized carbon-centered radical. Such stabilized carbon-centered radicals can add a molecular oxygen and subsequently eliminate a hydroperoxyl radical to yield the corresponding oxidation product.

$$RHC{=}CR_2 + HO^\bullet \rightarrow RHC(OH){-}C(^\bullet)R_2 \tag{1.1}$$

$$R_3C{-}H + HO^\bullet \rightarrow R_3C(^\bullet) + H_2O \tag{1.2}$$

$$RH + HO^\bullet \rightarrow R^\bullet + H_2O \tag{1.3}$$

$$RCH_2OH + HO^\bullet \rightarrow RHC(^\bullet)OH + O_2 \rightarrow RHC(OO^\bullet)OH \rightarrow RCH({=}O) + HO_2^\bullet \tag{1.4}$$

UV/TiO_2 photocatalytic and radiolytic oxidations of benzene and related aromatic compounds have been extensively studied.[16,17] A primary product of UV/TiO_2 photocatalytic degradation of benzene is phenol, which results from the reaction of $HO^\bullet$ addition to the benzene ring. The TiO_2 photocatalysis of benzene involves $HO^\bullet$ addition to form hydroxylated and polyhydroxylated adducts, which upon further oxidation can be converted into quinone type structures. Subsequent oxidation can also lead to ring open products, affording low molecular weight carboxylic acids and ultimately CO_2. ^{13}C isotope labeling was used to establish different pathways involved in the formation of reaction products, muconaldehyde and phenol[17,18] (Figure 1.1). Muconaldehyde can be readily oxidized to the corresponding carboxylic acids, with extensive oxidation leading to CO_2 as final product. The reaction pathways

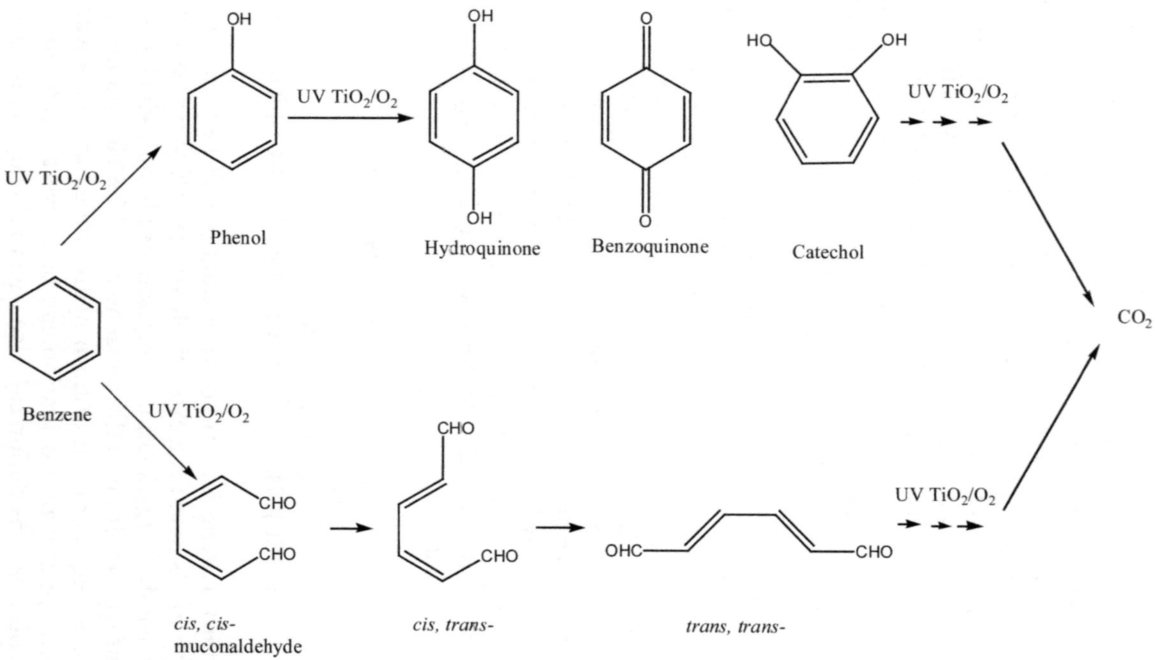

Figure 1.1 Two pathways for TiO_2 photocatalysis of benzene. (Reproduced with permission from ref. 17. Copyright 2011 Elsevier.)

leading to muconaldehyde account for 60–70% and pathways *via* phenol account for 30–40% of benzene consumption.[17]

Halogenated organic compounds are among the most problematic water pollutants. The presence of chlorine, fluorine, bromine, and iodine atoms can dramatically affect the toxicity and treatability of organic compounds. Halo-organics have been extensively used as herbicides, pesticides, and in several personal care products.[10,19,20] Because of the negative health environmental consequences associated with many halogenated organic compounds extensive studies on TiO_2 photocatalytic degradation of halogenated compounds have been reported.[19,20] As an example, the degradation mechanisms of 1,2-dichlorobenzene are initiated with $HO^•$ addition to the 3-position of 1,2-dichlorobenzene to form 2,3-dichlorophenol (1A) or substitution of chloro by $HO^•$ to generate 2-chlorophenol (1B) (Figure 1.2).[19] Intermediates produced along pathway A are 2,3-dichlorophenol (1A), 2,3-dihydroxychlorobenzene (2A) and 1,2,3-trihydrobenzene (3A) and further oxidation to CO_2. Analogous substitution in pathway B can form 1,2-dihydroxybenzene (4B) and *o*-benzoquinone (5B).

Phosphorus, sulfur and nitrogen containing organic substrates are also a target of TiO_2 photocatalytic degradation due to their wide usage for insect controls in agricultural crops, pharmaceutical, and personal care products.[2,21] These heteroatom containing compounds have widespread presence in groundwater and sediment. While there are numerous successful cases of remediation of such compounds, we choose to highlight examples of UV/TiO_2 photocatalytic degradation of select phosphorus, sulfur, and nitrogen containing compounds. The TiO_2 photocatalysis of organophosphorus insecticides (dimethyl 2,2-dichlorovinyl phosphate and dimethyl 2,2,2-trichloro-1-hydroxyethyl phosphate) has been investigated and the final degradation products are mineralized to Cl^-, PO_4^{3-}, H^+, and CO_2.[22] The photocatalytic degradation of sulfur and nitrogen containing molinate under simulated solar irradiation has been studied using a suspension of TiO_2 in aqueous media.[21] Different reaction pathways are involved in the degradation of molinate by the positive holes and $HO^•$. Ring fragmentation at the amine group is initiated *via* hole-mediated oxidation to form an immonium radical cation which undergoes hydrolysis to yield amine and carboxyl derivatives.[23] An H-abstraction pathway leads to the alkyl radical, which undergoes addition of molecular oxygen with the subsequent elimination of peroxyl radical to yield keto-molinate. The holes can also mediate oxidation through the sulfur atom by an electron transfer process leading to sulfoxide derivatives and de-alkylated products.[21] Extensive treatment can lead to complete mineralization with the formation of NH_3, CO_2, SO_4^{2-}, HX, and H_2O.

In the following sections, the degradation kinetics of TiO_2 photocatalysis with various pollutants will be extensively discussed along with the effect of water quality parameters such as pH, natural organic matter (NOM), and alkalinity. Finally, a thorough discussion of the transformation products formed that lead to the degradation pathways of select compounds is given.

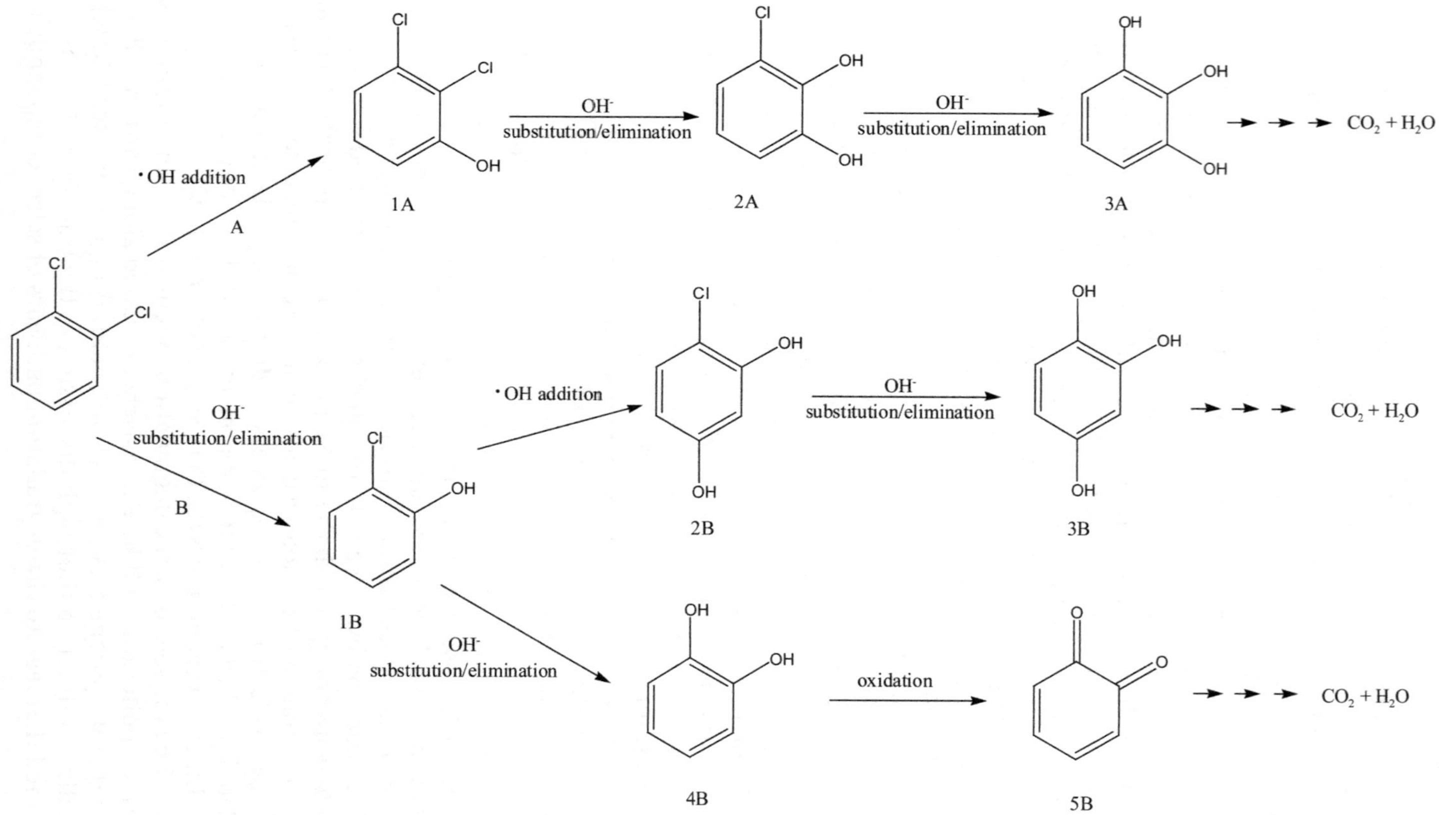

Figure 1.2 Proposed pathways for TiO_2 photocatalysis of 1,2-dichlorobenzene. (Reproduced with permission from ref. 19. Copyright 2002 Elsevier.)

1.2 Degradation Efficiency-Kinetics of Emerging Contaminants

1.2.1 Photocatalytic Degradation of Cyanotoxins

Photocatalysis has become one of the most effective approaches to degrade highly toxic naturally produced compounds, the cyanotoxins, that cannot be easily removed through conventional treatment processes. A lot of effort has been made to investigate the TiO_2 photocatalytic degradation of various cyanotoxins, such as microcystins (MCs) and cylindrospermopsin (CYN), under UV or solar irradiation. Shephard *et al.* reported the high degradation efficiency of microcystin-LR (MC-LR) and MC-RR using immobilized TiO_2 on a fiberglass sheet under UVC irradiation with the observed first rate constants of 0.255 ± 0.017 and 0.199 ± 0.016 min^{-1}, respectively.[24] The photocatalytic degradation of [D-Leu]-MC-LR using Degussa P25 TiO_2 film under simulated solar light was investigated by Vilela *et al.*[25] They reported a treatment time of 150 min could decrease the toxin concentration from 10 to 1 $\mu g\ L^{-1}$ and remove 90% initial total organic carbon (TOC) under the experimental condition. Su *et al.* investigated the high degradation efficiency of MC-LR with a much higher concentration (5 $mg\ L^{-1}$) using TiO_2 nanotubes as the photocatalyst under the irradiation of UV and natural solar light.[26] This study confirmed a better performance of TiO_2 nanotubes than TiO_2 nanofilm. Senogles *et al.* examined the effectiveness of two commercial TiO_2 nanoparticles under UV light for the removal of CYN and found Degussa P-25 was more efficient than Hombikat UV-100.[27] In addition, a polymorphic TiO_2 nanoparticle investigated by Zhang *et al.*,[28] containing anatase, brookite and rutile phases, showed good performance for CYN photocatalytic removal.

To extend the light response of TiO_2 into the visible region, many studies focused on the modification of TiO_2 by doping with metals (Fe, Co, Ag, or Ni)[29] or non-metals (N, F, S, or C).[30–33] This causes narrowing of the bandgap of TiO_2 or introduces mid-gap energy states. The study of Choi *et al.*[34] reported the efficient removal of MC-LR using a self-synthesized nitrogen-doped TiO_2 (N–TiO_2) under visible light. In addition, the N–TiO_2 was found to be 3–4 times more efficient than control TiO_2 for the degradation of MC-LR under UVA irradiation. Another study compared the removal efficiency of MC-LR, -LA, -YR, -RR, and CYN using N and F doped TiO_2 films under visible or UV-vis irradiation.[35] In a mixture of MCs, the reaction rate of an individual toxin under UV-vis irradiation was shown in this study as follow: MC-LA $\geq$ MC-LR $\geq$ MC-YR $>$ MC-RR, which was attributed to the extent of adsorption of an individual toxin.

1.2.2 Photocatalytic Degradation and Detoxification of Oxytetracycline

Oxytetracycline (OTC) belongs to tetracyclines, one of the most common antibiotics group for disease treatment and prevention, and growth promotion in livestock worldwide. Thus, it is one of the most frequently detected antibiotic chemicals in water bodies and sediments around the world, and it was also found to be potentially toxic for the ecosystem and humans.[36–39] The occurrence of OTC in the environment is due to the discharge of pharmaceutical, agriculture, and livestock waste. In the past few years, several studies have demonstrated the presence of OTC in surface water at concentrations ranging from a few μg L^{-1} to mg L^{-1}.[40,41] The detoxification of water by removal of OTC has become a pressing environmental problem because of the high toxicity of OTC and the inability of conventional treatment processes to remove it due to its stable naphthol ring structure and the high toxicity to the microorganisms used in biological treatment. Herein, the most important studies related to the application of TiO_2 based photocatalysts for the treatment of OTC will be presented.

OTC is degradable by both hydrolysis and photolysis. Xuan *et al.*[42] reported the effects of pH and temperature on the decomposition of OTC. Solution pH and temperature were found to have important effects on OTC hydrolysis and photolysis. The hydrolysis was much faster in acidic and alkaline solutions than in neutral. With the enhancement of temperature from 4 ± 0.8 to 60 ± 1 °C, the half-life of OTC reduced from 1.2×10^2 d to 0.15 d. In addition, the photolytic degradation of OTC was found to be fast with a degradation rate constant of 3.61 ± 0.06 per day at pH 5.85. The degradation rate is highly pH dependent and drastically increased in alkaline solutions. OTC has four species at different pH ranges according to the three ionization equilibrium constants with pK_a values of 3.57, 7.49, and 9.88. The four protonation states can be represented by H_3OTC^+, H_2OTC, $HOTC^-$, and OTC^{2-}. The adsorption spectra of OTC at different pH values exhibited a redshift to visible light, paralleled to the change of the protonation states of OTC. Zhao *et al.*[43] revealed a self-photosensitization pathway with evidence of singlet oxygen generation for $HOTC^-$ and OTC^{2-} during photolysis under solar light irradiation. Meanwhile, the three-dimensional fluorescence spectra of OTC at different pH values also demonstrate that only $HOTC^-$ and OTC^{2-} exhibit significant emission spectra, but not H_3OTC^+ and H_2OTC. Therefore, the change of the internal electrostatic force of OTC molecule as a function of solution pH values and species was proposed as a significant factor influencing the energy states and observed reaction pathways of OTC species under light irradiation. Nevertheless, only 13.5% reduction of TOC occurred despite a rapid photolytic degradation of 90% of OTC at 20 mg L^{-1} after 240 min of irradiation by a 500 W medium mercury lamp (light intensity 5.25×10^{-4} W cm^{-2} at 365 nm).[44] Moreover, the photolytic degradation byproducts (maintained the naphthol ring) have been proved to be even more toxic than the parent when tested with luminescent bacterium with the inhibition rate increasing from 21% to 47% on *P. phosphoreum*.[44]

The photocatalytic removal of OTC with TiO_2 was also investigated under UV, visible, and solar light irradiation. Pereira *et al.*[45] reported the effects of TiO_2 load, solution pH, and inorganic ions on the degradation, mineralization, and detoxification of OTC using simulated solar light and natural solar radiation. The solution pH was critical in decomposing OTC.

The highest photocatalytic activity of TiO_2 was observed at pH 4.4 and 0.5 g L^{-1} TiO_2 for the removal of both OTC (100% after 40 min of irradiation; 7.5 kJ L^{-1} of UV dose) and TOC (>90% after 180 min of irradiation; 38.3 kJ L^{-1} of UV dose), while 100% OTC and ~80% TOC were removed by 1.8 and 11.3 kJ L^{-1}, respectively, by natural solar energy. In addition, the BOD_5 (biological oxygen demand)/COD (chemical oxygen demand) ratio rose from 0 to nearly 0.5, showing a remarkable improvement in biodegradability, while the inhibition percentage of bioluminescence of *Vibrio fischeri* decreased significantly from 35% to 7%. The remaining degradation by-products were mainly low-molecular-weight carboxylate anions without any antibacterial activity. The presence of phosphates hindered the removal of OTC, whereas the presence of chlorides, sulfates, nitrates, ammonium, and bicarbonates did not alter the degradation kinetics.[46] Han *et al.*[47] synthesized Ag-decorated, monodisperse TiO_2 aggregates for the degradation of OTC, and found enhanced photocatalytic activity with a low leaching of Ag in the solution. Zhao *et al.*[48] also reported the photocatalytic degradation, mineralization, and detoxification of OTC by 5A and 13X zeolite particles loaded with nano-TiO_2 under UV irradiation. The best loading rate of composite photocatalysts was 15 wt% TiO_2/5A and 10 wt% TiO_2/13X for OTC degradation. The composite photocatalysts showed a better anti-interference capability of radical scavengers, and humic acid, at a low concentration than unsupported TiO_2 during the removal of OTC in aqueous solutions.[49] Meanwhile, the 10 wt% TiO_2/13X kept the inhibition percentage of bioluminescence of *Vibrio qinghaiensis* sp.-Q67 below 35% during the process due to its excellent adsorption capability of OTC and its degradation byproducts.[48] Moreover, the degradation of OTC was investigated with nitrogen and fluorine doped TiO_2 film at different solution pH under visible and solar light in the presence of a series of scavengers. Five pathways, including direct photolytic degradation, UV/vis light-induced photocatalytic oxidation and reduction, and visible light-induced self-photosensitized oxidation and reduction, were proposed and verified as the mechanism for OTC photochemical degradation with TiO_2.[43]

1.3 Effects of Water Quality Parameters on the Photocatalytic Degradation of Emerging Contaminants

1.3.1 Effects of General Water Quality Parameters

The application of TiO_2 photocatalysis for the efficient removal of contaminants of emerging concern (CEC) is dependent on many different parameters associated with both water quality and the physical chemical parameters of the contaminants. The main characteristics of the influent (*i.e.*, pH, alkalinity, dissolved oxygen (DO), and the presence of NOM), the physical chemical characteristics of the contaminant and the treatment goal will determine the final design of photocatalytic water treatment systems.[50–53] In this section we review the main publications that investigated the effects of water quality on the efficiency of TiO_2 photocatalysis to remove an array of contaminants.

Pelaez *et al.*[50] reported the effect of water quality parameters on the degradation of the cyanotoxin MC-LR when treated with visible light-active nitrogen and fluorine co-doped TiO_2. Four water parameters (solution pH, alkalinity, NOM, and DO) were chosen to be studied. The solution pH significantly affected the decomposition of MC-LR since it is directly associated with the surface charge of TiO_2 and the contaminant (in this case MC-LR). The optimal pH range to achieve the highest efficiency for MC-LR decomposition was 3–4 since TiO_2 is positively charged[30] and MC-LR is negatively charged at those pH values,[54] which allows the toxin to better interact with the surface of the catalyst and better utilize the surface formed radicals. The effect of pH was also observed with other CEC such as carbamazepine (CBZ) when treated with different types of photocatalysts[51,52] such as nitrogen doped TiO_2, conventional TiO_2 or zinc oxide (ZnO). For the degradation of CBZ, the highest degradation was achieved at basic conditions around pH values of 8–9 using nitrogen doped TiO_2.[52] At basic conditions, available hydroxyl ions on the surface of the catalyst increased to form hydroxyl radicals ($HO^•$) due to changes in the catalyst's surface ionization states.[55,56] Consequently, increased formation of $HO^•$ at these conditions enhanced the photocatalysis for the decomposition of CBZ. However, when conventional TiO_2 and ZnO were employed to decompose CBZ, the highest degradation was observed at pH 6.4 when compared to pH 3 and pH 11.[51] Therefore, the effect of pH on the degradation of CEC was found to be dependent on both the targeted CEC and type of photocatalysts used.

In addition to the effect of pH on the photocatalytic degradation of CEC, the effect of alkalinity was studied. For the degradation of MC-LR by visible light-induced photocatalysis, three alkalinity values at 50, 100, and 150 mg L^{-1} of sodium carbonate were tested at pH 7.1 using a phosphate buffer.[50] Increasing alkalinity significantly decreased the degradation of MC-LR because carbonates and bicarbonates scavenged ROS. Since ROS attack and decompose CEC in photocatalytic degradation processes, MC-LR degradation decreased as alkalinity increased. Avisar *et al.*[52] also reported the effect of alkalinity on the degradation of CBZ under simulated solar light-induced photocatalysis. In their study, five alkalinity values at 25, 50, 100, 200, and 300 mg L^{-1} of calcium carbonate ($CaCO_3$) were tested at pH 9.0–9.5. The degradation efficiency of CBZ was also found to be inversely proportional to the concentration of alkalinity due to the scavenging effect of carbonates and bicarbonates. Above 200 mg L^{-1} of $CaCO_3$, the effect of alkalinity on the degradation of CBZ decreased compared to lower alkalinity. It was reported that the reaction of carbonate and bicarbonate anions with $HO^•$ led to the formation of carbonate radicals, which have a lower redox potential (1.59 V)[57] compared to hydroxyl radicals (2.8 V).[58] The reduced reactivity of these radicals resulted in a decrease in the efficiency of the photocatalytic process.

Additionally, the effect of DO on the removal of CEC was investigated because oxygen has the ability to trap electrons and form additional ROS such as the superoxide anion ($O_2^{\bullet-}$), therefore enhancing photocatalytic processes and CEC removal.[50,59] The enhancement of photocatalytic degradation of MC-LR in the presence of DO was also reported.[50] Nitrogen or oxygen was supplied into each solution for 30 min in order to study the effect of DO on the degradation of MC-LR. The MC-LR degradation rate in the reactor purged with oxygen was higher than that of the nitrogen purged reactor. The degradation rate was also proportional to the concentration of DO. Since electron–hole pairs generated during photocatalysis recombined fast in low concentrations of oxygen, the formation of ROS reduced, and the photocatalytic degradation of CEC decreased.[60]

Finally, the effect of NOM on CEC degradation was investigated since natural water contains various forms of NOM including humic and fulvic acids, which can either inhibit or enhance photocatalytic degradation of CEC.[50,53,61,62] For the degradation of MC-LR by visible light-induced photocatalysis, the degradation rates were inversely proportional to the concentration of NOM at pH values equal to 3.0, 5.7, and 7.1. Other studies also reported the inhibition of NOM for the degradation of CEC due to their scavenging effects of ROS.[61,62] Although enhanced photolysis of CEC under simulated solar irradiation has been reported due to photosensitization of NOM,[63,64] the degradation of CEC was inhibited in the presence of photocatalysts such as TiO_2 and ZnO. Additional information on the effect of NOM is discussed in Section 1.3.2.

1.3.2 Effects of Natural Organic Matter (NOM) and Selectivity

The presence of co-existing organic chemicals is problematic in decomposing target chemicals of concern in water. Since hydroxyl radicals produced from TiO_2 non-selectively attack organic chemicals, the decomposition rates of many organic chemicals seem very similar.[55] The non-selective nature looks attractive because TiO_2 photocatalysts do not need special modification when a target chemical to decompose changes. The photocatalytic process does not differentiate one chemical from another in decomposition. However, this also represents a huge problem when only target chemicals of concern should be decomposed in a mixture with non-target chemicals.

In particular, typical source water is contaminated with highly toxic organic substances at low concentrations (*e.g.*, anthropogenic chemicals such as pharmaceuticals) and less toxic naturally occurring organic substances at high concentrations (*e.g.*, NOM). A photocatalytic process is required to focus on the decomposition of the toxic target chemicals rather than splitting its reactivity to both toxic targets and non-toxic NOM. However, target chemicals have to compete with abundant NOM. Competing chemicals take catalytic sites and consume $HO^{\bullet}$. Moreover, they foul the surface of TiO_2 materials. As demonstrated in Figure 1.3(a), the presence of humic acid (as representative NOM) significantly retarded the decomposition of a target chemical, ibuprofen (IBP, a well-known analgesic pharmaceutical found in water resources)

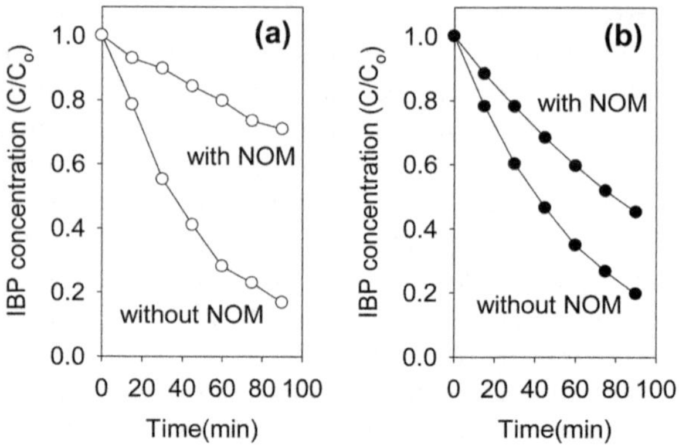

Figure 1.3 Ibuprofen (IBP) decomposition in water by (a) conventional TiO_2 (control) and (b) porous TiO_2 in the absence and presence of humic acid (NOM). The result clearly shows the impact of the presence of NOM and the significance of the surface modification of TiO_2. (Reproduced with permission from ref. 65. Copyright 2013 Elsevier.)

on conventional TiO_2.[65] IBP decomposition was significantly decreased to around 30% in the presence of NOM from 80% in the absence of NOM.

Consequently, nonselective sorption followed by oxidation of co-existing chemicals at TiO_2 surface should be prevented. Operational parameters such as reaction pH and temperature can change decomposition rates among organic chemicals while TiO_2 materials with controlled properties (*e.g.*, crystal phase) can improve selectivity.[66] An extrinsic material exhibiting specific affinity for a target chemical can also be introduced onto TiO_2. Coating TiO_2 with molecules with hydrophobic moieties can enhance sorption and oxidation of hydrophobic chemicals.[67] Specially designed organic or inorganic domains as molecular recognition sites are integrated onto TiO_2 surface to preferentially attract targets which then diffuse to the photocatalytic sites for chemical oxidation.[68] Cavities of target chemicals can be imprinted onto TiO_2 surfaces.[69] Such chemical modifications exhibit ultimate selectivity only to target chemicals, meaning that a chemical marker should be replaced upon change in a target chemical.

Considering the heterogeneous reaction, size-based selection onto well-defined porous materials has also been investigated. TiO_2 particles are embedded into clay structure and thus molecules smaller than the distance between sheet silicate layers can reach the embedded TiO_2 particles.[70] Similarly, they can also be coated with a SiO_2 shell with well-developed porous structure.[71] However, the approaches discussed above to improve selectivity commonly require introduction of a secondary material, which generally decreases overall reactivity due to covered TiO_2 surface, reduced UV penetration, and limited mass transport. A recent study delivered an important message that even a pure TiO_2 material, when properly designed, can work for size exclusion (Figure 1.3b).[65] TiO_2 with well-controlled pores slightly smaller than NOM and larger than IBP showed improved reactivity towards IBP in the presence of NOM as compared to nonporous control TiO_2. However, studies on preferential photocatalytic degradation in a real mixed stream have still been rarely reported.

1.4 Transformation Mechanistic Pathways and Reaction Intermediates

Photocatalysis with TiO_2 is considered to be an alternative to conventional methods for environmental detoxification. Its efficiency is based on the formation of ROS that are able to decompose a great variety of environmental pollutants of different origin (both manmade and naturally found compounds),[72–75] functionality (such as pesticides, pharmaceuticals, chlorophenols, and azo dyes),[76–79] and chemical structure (organic compounds with heteroatoms such as N, P, and S).[80] Although photocatalysis in the presence of TiO_2 under UV-A proceeds to mineralization of the organic substrates, the degradation pathways are quite complicated through the formation and decay of numerous intermediates.[81] The elucidation of the reaction mechanism is a difficult but necessary task in TiO_2 photocatalysis not only for understanding the process in detail but also for detoxification purposes due to toxicity and persistency of many intermediates.

Consequently, it is important to monitor together with the destruction of the parent compounds the formation and decay of the intermediates formed in order to control all transformation steps, to identify hazardous intermediates, and to clarify the reaction mechanism.[82–84] On some occasions, toxicity studies are contacted on the treated mixture and/or the identified intermediates to prove water detoxification following treatment.[81]

Structural identification of transformation products (TPs) requires advanced analytical methods that combine high separation efficiency with increased structural identification capabilities. Due to their high selectivity and sensitivity, gas and liquid chromatography-mass spectrometry techniques are frequently used methods of choice for the structural elucidation of TPs.[85]

Determination of TPs based on mass spectrometry requires three working steps. Following the exposure of the contaminant of interest to the relevant treatment, the sample is analyzed using a chromatographic and mass spectrometric method optimized for the parent compound. Then the mass spectrometer scans a range of masses that account for potential additions on the contaminant or its fragments. The second step consists of careful examination of the TP mass spectrum and comparison with the mass spectrum of the parent compound (contaminant). The final third step consists of additional analysis by tandem mass spectrometry which insures proper structural identification of TPs. Following that, TPs can be organized in degradation routes and pathways.

This section gives an overview of the different intermediates and their mechanisms of formation upon TiO_2 photocatalysis with UV light. For that reason selected compounds (cyanotoxins *i.e.* MC-LR, CYN and water taste and odor compounds, *i.e.* geosmin (GSM) and 2-methylisoboeneol (MIB)) will be discussed to cover different mechanisms of degradation based on the functional groups treated with the various ROS that TiO_2 photocatalysis forms. This target group of compounds are emerging contaminants of biological origin, the photocatalytic degradation of which was recently studied in detail.[6,86–88]

1.4.1 Transformation Products of Compounds Resulting from the Reaction of Carbon Bonds

Light activation of titania results in the formation of an array of ROS including hydroxyl radicals $HO^{\bullet}$ (main species), superoxide radical anion ($O_2^{\bullet-}$), and perhydroxyl radical ($HO_2^{\bullet-}$) and the conduction band electron (e^-_{CB}).[89,90] More importantly, the positive holes formed in the valence band following electron photoexcitation can oxidize any compound (organic and/or inorganic) found adsorbed on the surface of the catalyst, also contributing to the routes of oxidation.

Since it is not possible to go over all the previously mentioned studies in this section, select groups of contaminants that incorporate in their structure basic functional groups (such as aromatic bonds, simple carbon bonds) and heteroatoms (N) will be discussed as well as their TPs and mechanisms of formation with TiO_2 photocatalysis. The selected group of contaminants chosen is characterized by great structural variability and concerns a family of naturally occurring toxins, the cyanotoxins. Cyanotoxins are formed from the harmful strains of cyanobacterial algal blooms and greatly affect the ecosystem and human health.[81,91] They can also be grouped based on structure: cyclic peptides (microcystins, nodularins–hepatotoxins), alkaloids (saxitoxin–neurotoxins), and lipopolysaccharides (LPSs). Cylindrospermopsin though structurally grouped as an alkaloid is not a neurotoxin but rather affects the liver and kidneys. Herein, the intermediates of MC-LR, CYN, and two taste and odor compounds with TiO_2 photocatalysis ($HO^{\bullet}$ attack) will be discussed.

In general, the main mechanistic steps that organic compounds undergo following $HO^{\bullet}$ attack are hydroxylation *via* substitution or addition with possible simultaneous isomerization, oxidation, and oxidative bond cleavage. Unsaturated bonds such as aromatic rings and carbon double bonds can undergo hydroxyl substitution of a hydrogen to form hydroxylated intermediates such as *m*/*z* 1011.5 (Table 1.1). The necessary mechanistic steps include the addition of a $HO^{\bullet}$ to one of the double bonds and formation of a carbon-centered radical, which rapidly reacts with oxygen, to form a peroxy radical. The release of a perhydroxyl radical results eventually in the substitution of the hydrogen with a hydroxyl group.[92] If another $HO^{\bullet}$ reacts with the carbon centered radical, then hydroxyl addition occurs which changes the carbon hybridization from sp^2 to sp^3, and therefore the overall three-dimensional structure of the TP compared to the parent compound. As will be seen from the studied cases

Table 1.1 Structures of reaction intermediates of microcystin-LR with TiO_2 based photocatalysts.

No.	Structure	*m/z*	Compound	Technology[a]	Reference
MC-LR	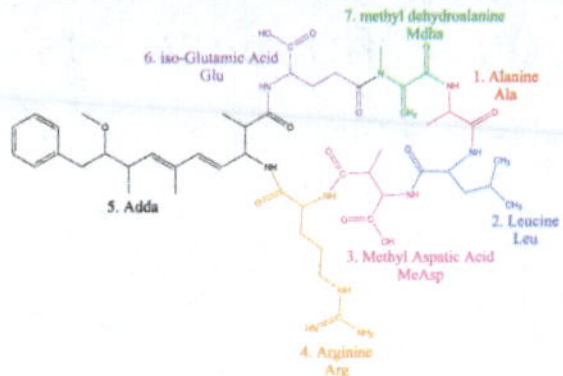	995.5	$C_{49}H_{74}N_{10}O_{12}$	—[b]	—[b]
1A	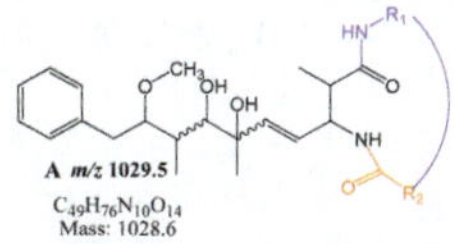	1029.5	$C_{49}H_{76}N_{10}O_{14}$	TiO_2_p/UVA TiO_2_f/UVA NTiO_2_p/λ > 420 nm NTiO_2_p/UVA NFTiO_2_f/λ > 420 nm	15 95[c] 34 15 103[d]
1B	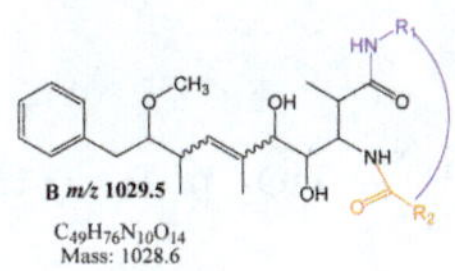				
1C	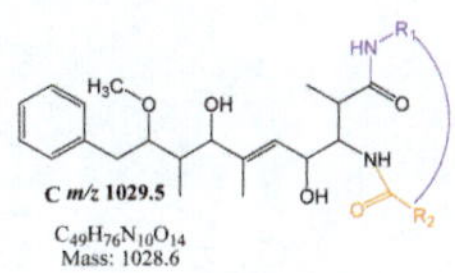				
1D	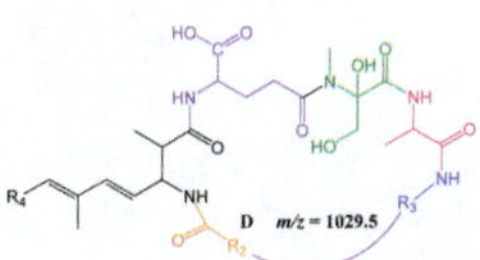				
1E	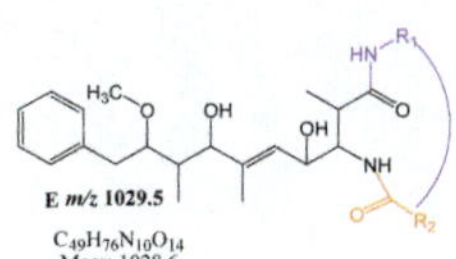				
2	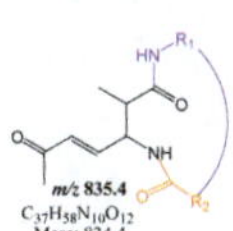	835.5	$C_{37}H_{58}N_{10}O_{12}$	TiO_2_p/UVA TiO_2_f/UVA NTiO_2_p/UVA NFTiO_2_f/λ > 420 nm	15 95 34 103
3	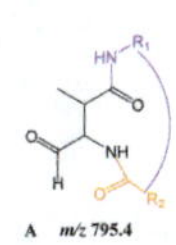	795.4	$C_{34}H_{54}N_{10}O_{12}$	TiO_2_p/UVA TiO_2_f/UVA NTiO_2_p/UVA NFTiO_2_f/λ > 420 nm	15 95 34 103

(*continued*)

Table 1.1 (*continued*)

No.	Structure	*m/z*	Compound	Technology	Reference
4	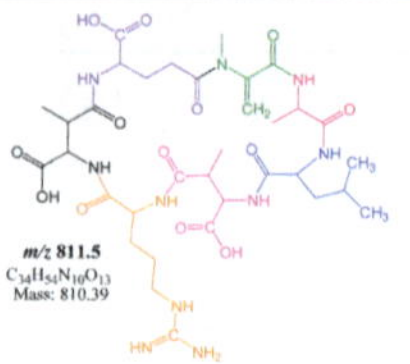	811.5	$C_{34}H_{54}N_{10}O_{13}$	TiO_2_p/UVA	15
5	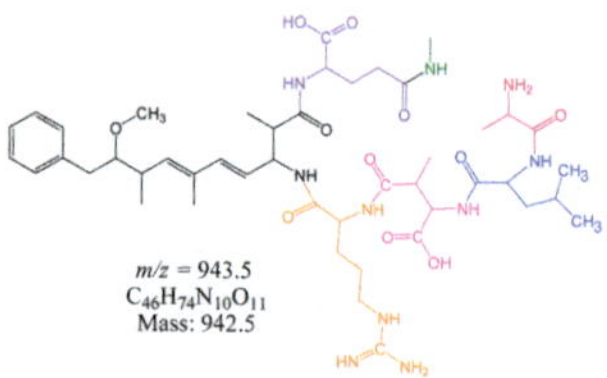	965 943	$C_{46}H_{74}N_{10}O_{11}Na$ $C_{46}H_{74}N_{10}O_{11}$	TiO_2_p/UVA	15
6	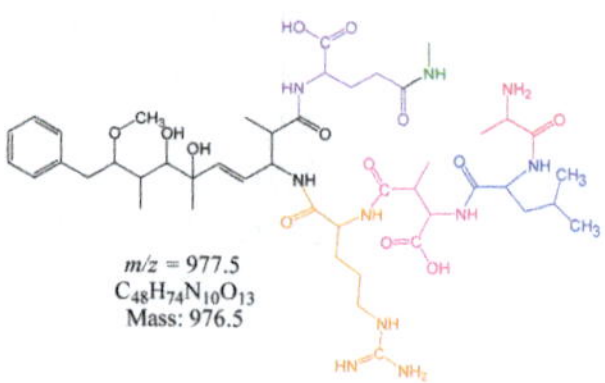	999 977	$C_{46}H_{75}N_{10}O_{13}Na$ $C_{46}H_{76}N_{10}O_{13}$	TiO_2_p/UVA	15
7A	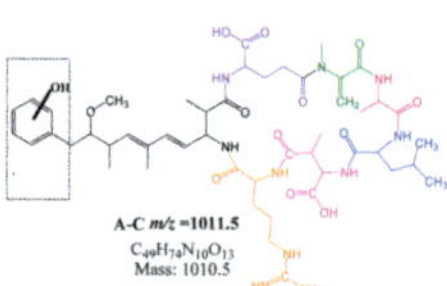	1011.5	$C_{49}H_{74}N_{10}O_{13}$	TiO_2_p/UVA	15
7B	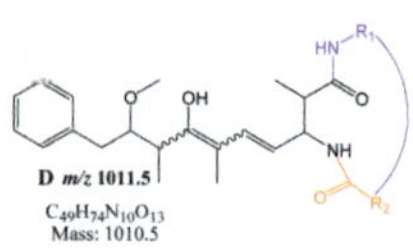				
7C	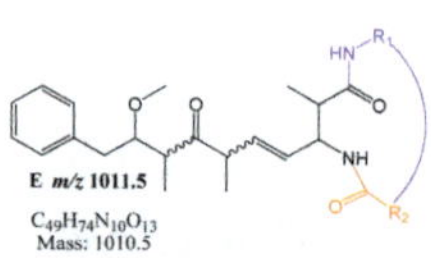				
7D	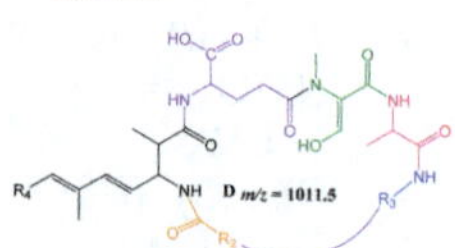				
7E	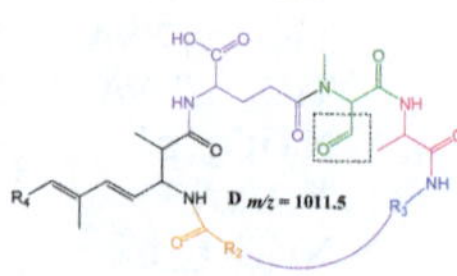				
8	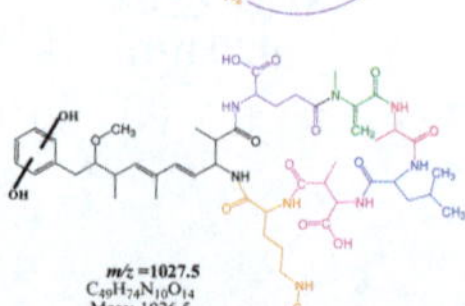	1027.5	$C_{49}H_{74}N_{10}O_{14}$	TiO_2_f/UVA $NFTiO_2$_f/$\lambda >$ 420 nm	95 103

9	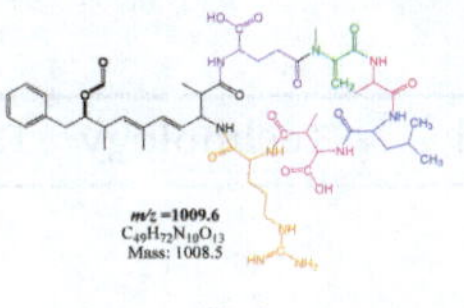	1009.6	$C_{49}H_{72}N_{10}O_{13}$	TiO_2_f/UVA	95
10A	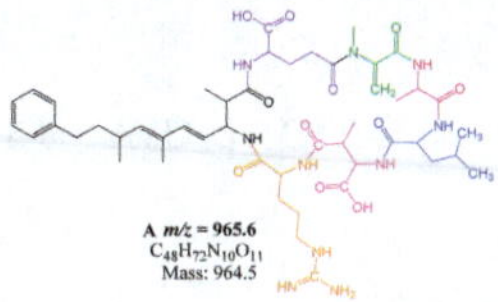	965.6	$C_{48}H_{72}N_{10}O_{11}$	TiO_2_f/UVA	95
11A	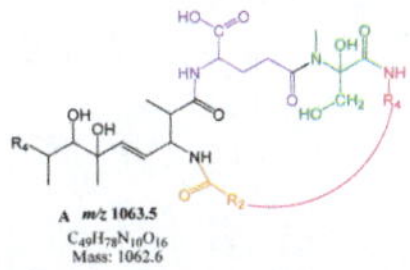	1063.5	$C_{49}H_{78}N_{10}O_{14}$	TiO_2_f/UVA	95
11B	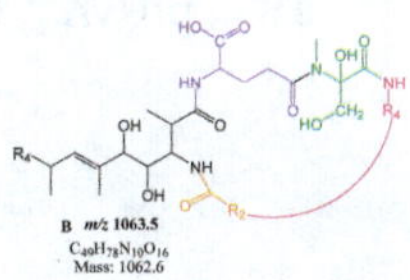				
11C	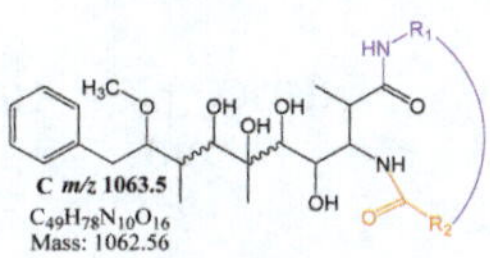				
12A	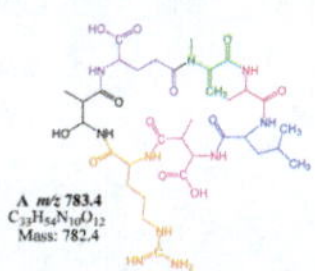	783.4	$C_{33}H_{54}N_{10}O_{12}$	TiO_2_f/UVA NTiO_2_p/UVA	95 34
12B	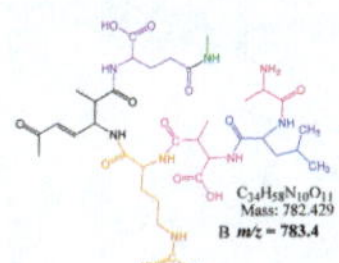	783.4	$C_{34}H_{58}N_{10}O_{11}$	TiO_2_p/UVA TiO_2_f/UVA	15 95
13	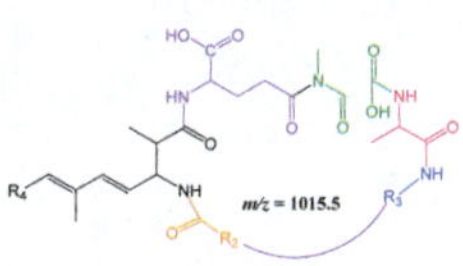	1015.5	$C_{48}H_{74}N_{10}O_{14}$	TiO_2_f/UVA	95
14	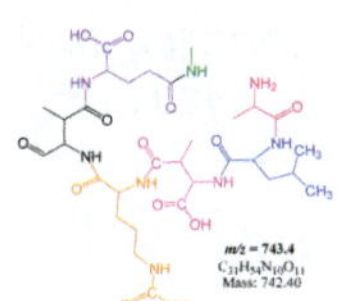	743.4	$C_{31}H_{54}N_{10}O_{11}$	TiO_2_p/UVA	15
15	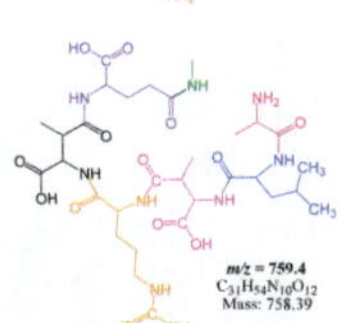	759.4	$C_{31}H_{54}N_{10}O_{12}$	TiO_2_p/UVA	15

(*continued*)

Table 1.1 (*continued*)

No.	Structure	m/z	Compound	Technology	Reference
16	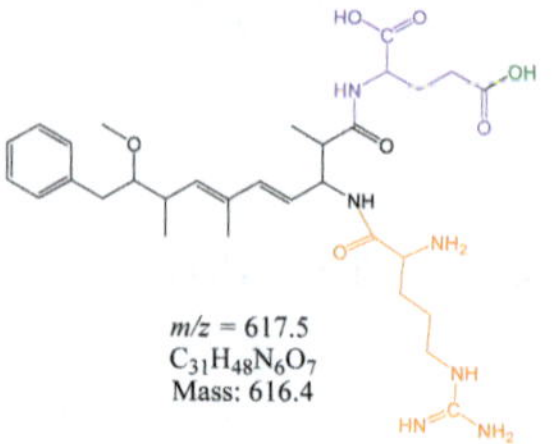	617	$C_{31}H_{48}N_6O_7$	TiO_2_p/UVA	15
17	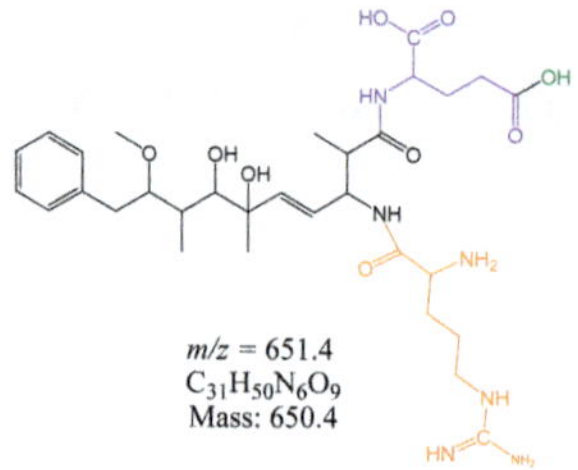	651	$C_{31}H_{50}N_6O_9$	TiO_2_p/UVA	15
18A	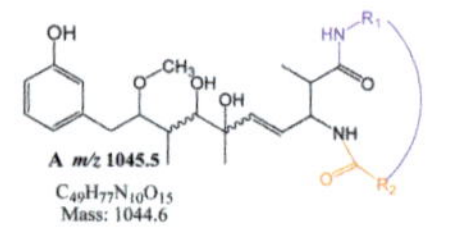	1045.5	$C_{49}H_{77}N_{10}O_{15}$	NFTiO_2_p/$\lambda >$ 420 nm	103
18B	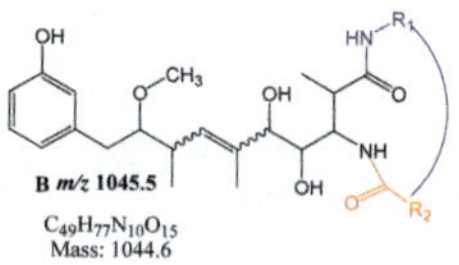				
18C	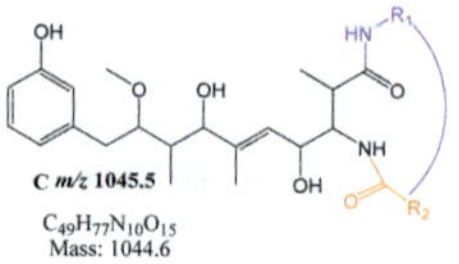				
18D	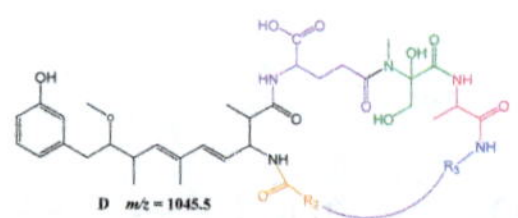				
19	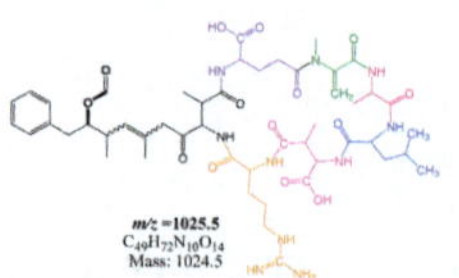	1025.5	$C_{49}H_{72}N_{10}O_{14}$	NFTiO_2_f/$\lambda >$ 420 nm	103
NOD	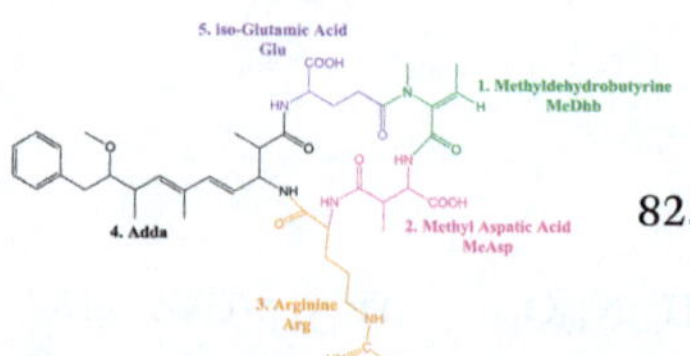	825.5	$C_{41}H_{60}N_8O_{10}$	—	—

20A	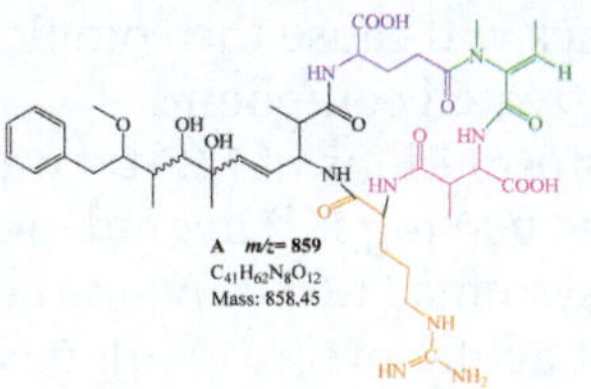	859	$C_{41}H_{62}N_8O_{12}$	TiO_2_p/UVA	94
20B	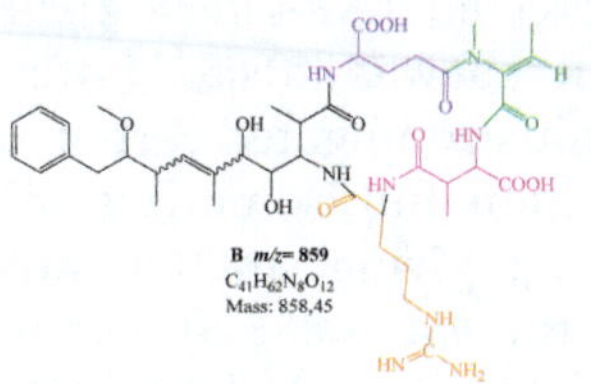				
21	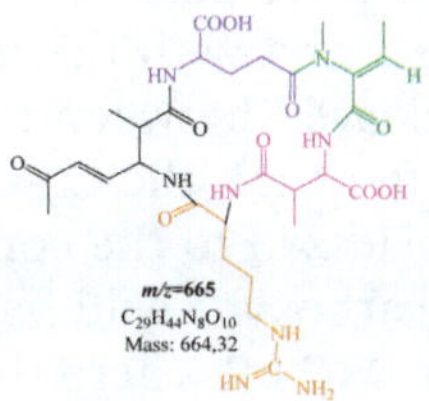	665	$C_{29}H_{44}N_8O_{10}$	TiO_2_p/UVA	94
22	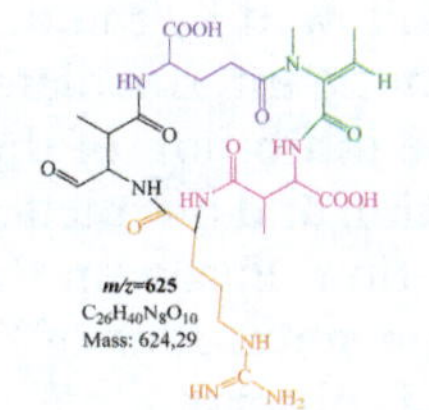	625	$C_{26}H_{40}N_8O_{10}$	TiO_2_p/UVA	94
23	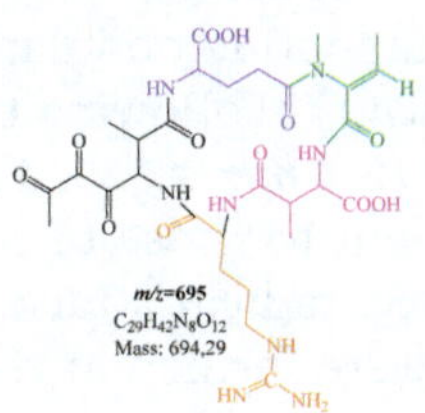	695	$C_{29}H_{42}N_8O_{12}$	TiO_2_p/UVA	94
24	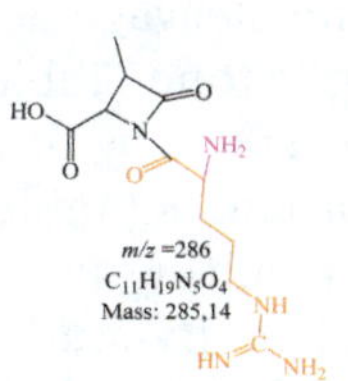	286	$C_{11}H_{19}N_5O_4$	TiO_2_p/UVA	94
25	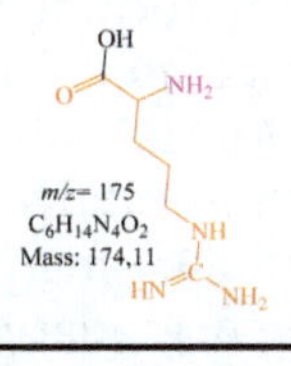	175	$C_6H_{14}N_4O_2$	TiO_2_p/UVA	94

[a]TiO_2_p = TiO_2 particles. TiO_2_f = TiO_2 films.
[b]— Not available or not referred at the publication. UVA stands for UV(300 < λ < 400 nm).
[c]Displayed structures A–D were observed in Ref. 95.
[d]Structure 1E was only reported by Ref. 103.

extensively analyzed below, further radical attack will cause the complete oxidation of carbon and the fragmentation of the treated compound.

The first study on TiO_2 photocatalytic transformations of MC-LR was conducted by Liu and coworkers.[15,93] MC-LR (C_o = 1000 mg L^{-1}) degradation was performed with TiO_2 nanoparticles in slurry systems (1.0% w/v) in the presence of hydrogen peroxide (42.8 mM H_2O_2) at acidic pH (pH 4.0). Based on these experimental conditions, ten TP were detected ($651 \leq m/z \leq 1029$) leading to three oxidation routes. The side of MC-LR that was affected initially was the chain of the Adda amino acid, which contains conjugated carbon double bonds that underwent simultaneous hydroxylation and isomerization (m/z = 1029.5 products, Table 1.1), and bond cleavage giving smaller molecular weight aldehydes (m/z = 795.5) and ketones (the m/z = 835.5). Further oxidation of m/z = 795.5, produced the carboxylated intermediate with m/z = 811.5. The next proposed pathway is similar to the first with the difference that the cyclic structure of MC-LR was initially cleaved between Mdha and Ala to produce a linear TP with [M + Na = 965]. After that, the C_6–C_7 double bond of Adda was hydroxylated (m/z = 977) and oxidized to the corresponding ketone (m/z = 783), aldehyde (m/z = 743), and carboxylic acid (m/z = 759). The last pathway involved a highly oxidized linear MC-LR where the Mdha–Ala–MeAsp moieties were removed (m/z = 617), followed by the dihydroxylation of the Adda chain (m/z = 651). In addition to the intermediate studies, the authors performed toxicity tests based on the inhibition of the protein phosphatases 1 (PP1) enzymes of the treated solution and on fractions of the treated solution collected following HPLC separation. Based on the inhibition data, though the fractions contained high concentrations of linearized TPs no significant toxicity against PP1 enzymes was observed.[15,93]

The same research group also studied the pentapeptide nodularin (NOD) by utilizing TiO_2 nanoparticles in a slurry (0.1% w/v).[94] Following treatment eleven TP were identified with m/z in the range $175 \leq m/z \leq 859$. The degradation pathways that NOD followed were equivalent to those of MC-LR[15,93] since the same types of bonds underwent hydroxyl radical attack. Initially, dihydroxylated TPs of different stereochemistry were formed at the carbon diene bonds (Table 1.1, m/z = 859), followed by bond cleavage at C_4–C_5 and/or C_6–C_7 bonds of Adda to form the corresponding ketone (Table 1.1, m/z = 665) and aldehyde products (Table 1.1, m/z = 625). These by-products were then oxidized to the corresponding peroxidated products (Table 1.1, m/z = 695), followed by hydrolysis of the peptide bonds resulting in small amino acid fragments (Table 1.1, m/z = 286; m/z = 175). Toxicity studies with the PP1 enzyme showed again loss of toxicity due to the photocatalytic oxidation (PCO), as well as the lack of formation of toxic intermediates.

Another study investigated the degradation of MC-LR with TiO_2/UVA, Antoniou *et al.* (2008),[95] but utilized two different photocatalytic films with different thickness (mass of catalyst was 1.4 mg and 50.4 mg per thin and thick film, respectively) instead of nanoparticles. The degradation of MC-LR (C_o = 20 mg L^{-1}) occurred at neutral pH and 21 types of TPs were detected as m/z. Differences in the type and number of intermediates isolated between the two studies[15,93] can be attributed to differences in the experimental conditions, frequency of sampling, addition of oxidants (which enhance degradation), and the pH of the treated solution (which affects the interaction between pollutant and catalyst). In their study, Antoniou and coworkers conducted their experiments in the absence of additional oxidants and at pH of milli-Q water which reduced interactions between MC-LR and the TiO_2 and allowed the $HO^{\bullet}$ to interact with more sites of the toxin.[95]

Tandem mass spectrometry analysis led to the structural characterization of the formed TPs and revealed four sites of MC-LR where degradation was initiated: three of them are on the Adda amino acid (aromatic ring, methoxy group, and conjugated double bonds) and one is on the cyclic structure (Mdha amino acid). Overall, the most susceptible group for hydroxyl radical attack was the unsaturated bonds of the Adda amino acid (side chain of the toxin) possibly because of their location and vulnerability to oxidation by $HO^{\bullet}$ (k_{OH} ~10^{9-10} M^{-1} s^{-1}).[96] The first oxidation step was the hydroxylation of the aromatic ring through substitution of an aromatic hydrogen to form the *m/z* 1011.5 intermediates at *o*-, *p*- and at a lesser extend the *m*-substitution, followed by a second hydroxylation (*m/z* 1027.5, Table 1.1). Detection of *m/z* 1027.5 confirmed the first substitution, because hydroxyl groups increase the electron density of the aromatic ring and, thus, the kinetic rates of electrophilic reactions increase.[97,98]

Hydroxyl radical attack not only causes hydrogen abstraction but also removal of small functional groups such as methyl and methoxy groups. Antoniou *et. al.* (2008)[95] were the first to observe the demethoxylation of the Adda chain (*m/z* 965.6, DmADDA) through the formation of the formic acid ester (*m/z* 1009.6, Table 1.1). Another mechanist step observed was hydroxyl addition, especially at the diene carbon bonds. The *m/z* 1029.5 and *m/z* 1063.5 were formed following double and quadruple hydroxyl addition. It has been observed that when conjugated dienes undergo electrophilic reactions, mixtures of TPs are formed (additional to the stereoisomers) because of 1,2- (*m/z* = 1029.5, 1A and 1B) or 1,4-additions (*m/z* = 1029, 1C) (Table 1.1). A separate oxidation pathway initiated at the diene bonds resulted in the complete removal of the Adda chain.[15,95] Once the enol-MC-LR (*m/z* 1011.5, 7D) was formed it was isomerized to the more stable tautomer of ketone-MC-LR (*m/z* = 1011.5, 7E). Following consequent oxidative bond cleavage steps, the ketone tautomer *m/z* = 1011.5 (7E) transformed into *m/z* 835.4, then into the aldehyde-derivative with *m/z* = 795.4, and eventually to a hydroxyl-derivative with *m/z* 783.4 (product 12A, Table 1.1).

The last site where Antoniou *et. al.* (2008)[95] reported initiation of MC-LR's oxidation was in the cyclic structure at the Mdha amino acid. The cyclic structure proved to be more resilient to ROS attack, possibly due to shielding from the functional groups of the other amino acids, as well as competition with the three other sites for ROS utilization. TPs formed following consecutive oxidation, including double hydroxylation of the double bond of the Mdha (*m/z* 1029.5, 1D), its oxidation to aldehyde (*m/z* 1011.5, 7D), and cleavage of the R_2C–COR bond (*m/z* 1015.5), were reported. Finally, TPs where degradation occurred simultaneously at the Adda chain and cyclic structure, such as *m*/*z* 783.4 (product 12B, Table 1.1), were observed.

Following TPs' identification, the authors performed toxicity studies measuring the activity of the PP1 enzymes in the treated solution,[99] though it was already proven that some of the identified TPs did not possess toxic properties [(*Z*)-MC-LR and the demethoxylated-MC-LR (DmADDA)].[95,100,101] The PP1 inhibition studies showed that as photocatalytic degradation progressed, the enzyme's activity was recovered, and towards the end of the treatment it was completely restored. Treatment with the photocatalytic films impaired MC-LR's structure so much that its toxic properties disappeared with limited carbon mineralization. This proves that water detoxification can be achieved without the complete mineralization of a compound, but instead through targeted oxidation.

Aligned with the current trend of PCO for utilizing sustainable light sources for catalyst activation such as sunlight, a few studies reported intermediates with light activated modified TiO_2 photocatalysts. Sunlight contains only 5% of UV radiation; however, incorporation of heteroatoms (such as N and F) into the TiO_2 structure allows visible light activation due to narrowing of the band gap energy.[30,34,102] In this case, $HO^•$ radicals are not the primary oxidation species produced by the positive holes of the conduction band and the oxidation is speculated to arise from $O_2^{•-}$ and $HO_2^{•-}$ radicals. The latter react in a similar way to $HO^•$ and can also produce hydroxylated TPs. So far, two studies have identified the reaction intermediates of MC-LR under visible light activation of N–TiO_2 nanoparticles[34] and NF–TiO_2 photocatalytic films[103] respectively. When N–TiO_2 nanoparticles where illuminated with visible light, only the *m*/*z* = 1029.5 intermediate was isolated in multiple peaks. When UVA radiation was used instead for N–TiO_2 activation, besides *m*/*z* = 1029.5 the oxidative cleavage of the Adda chain through the formation of a ketone (*m*/*z* 835.5), an aldehyde (*m*/*z* 795.5), and a hydroxyl derivative (*m*/*z* 783.5, product 12A, Table 1.1) were observed as well. In Andersen's study, seven different TPs were formed following 8 h of treatment with *m*/*z* = 795.4; 1011.5 (4 peaks); 1027.5; 1029.5; 1045.5; 835.4; and 1025.5 (Table 1.1). The latter one was a unique intermediate that was never before reported with PCO, and was formed after the hydroxyl substitution of the Adda amino acid double bond at C_4 (enol ↔ ketone), followed the oxidation of the methoxy group to an aldehyde. The reported pathways of the remaining of TPs were the same as those reported by Antoniou *et al.* (2008).[95]

Microcystins are cyanobacterial metabolites with relatively large and complex structures and, therefore, unveiling the corresponding TPs structures following free radical attack can be tedious work. TPs have also been identified for simpler secondary metabolites of cyanobacterial blooms such as taste and odor causing compounds GSM and MIB.[104] Removing these metabolites from water is necessary for its general usage and consumption.

Two studies have worked on TPs formed during the photocatalytic degradation of GSM with Degussa P-25 nanoparticles.[87,105] During the first study only a few intermediates were identified which involved oxidation of GSM *via*

HO$^•$ and by the reactive holes on the surface of the catalyst. Results in this study pointed that GSM undergoes rapid ring opening giving aliphatic saturated and unsaturated compounds.[105] In a more elaborate study by Fotiou *et al.*,[87] many degradation products were detected. The majority of the TPs formed were oxygenated, suggesting that the primer degradation mechanism was due to HO$^•$ oxidation through electrophilic substitutions. Bond cleavages also occurred at multiple sites, producing cyclic ketones that with further bond cleavages gave open chain saturated and unsaturated products (Figure 1.4).[87] Mechanisms concerning the formation of the main intermediate products are α-hydrogen abstraction from GSM tertiary carbon, hydroxylation from HO$^•$ attack and ketone formation with β-scission abstraction.

In the same study, degradation of MIB using of TiO_2 under UV-A light revealed the formation of several products, prior to their decay and final decomposition to CO_2.[87] Alcohol-, ketone- and diketone-derivatives of MIB, oxygen containing cyclic compounds and open chain aliphatic compounds, were the identified intermediates (Figure 1.5). D-Camphor was the primary intermediate formed during the photocatalytic degradation of MIB which was formed with a β-scission reaction mechanism on the methyl group of

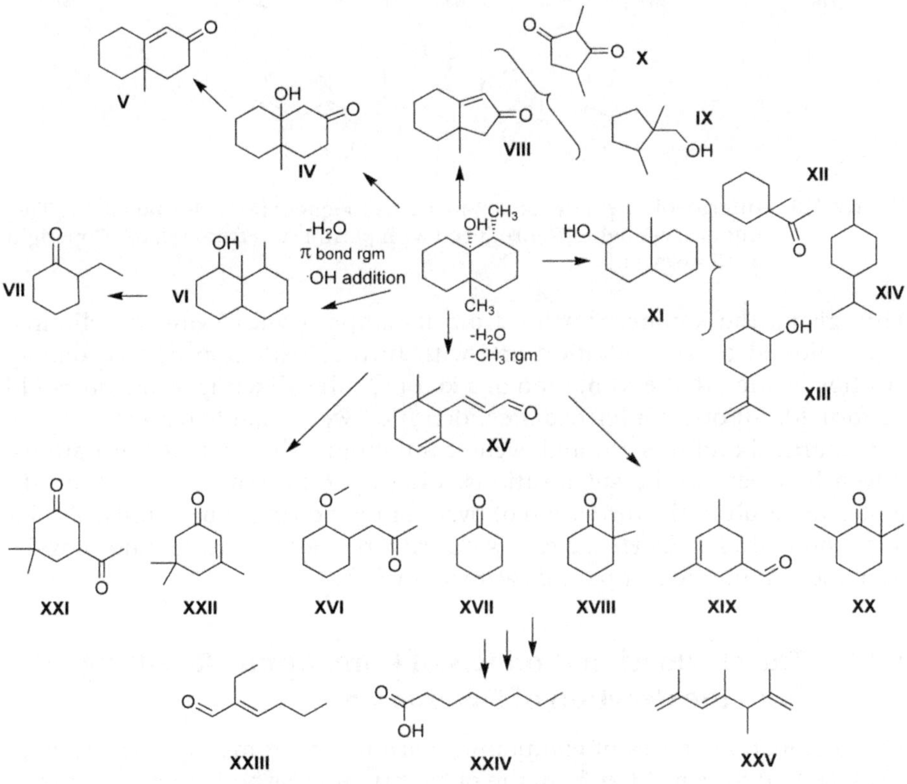

Figure 1.4 Intermediates products of geosmin formed using TiO_2 under UV-A light. (Reproduced with permission from ref. 87. Copyright 2015 Elsevier.)

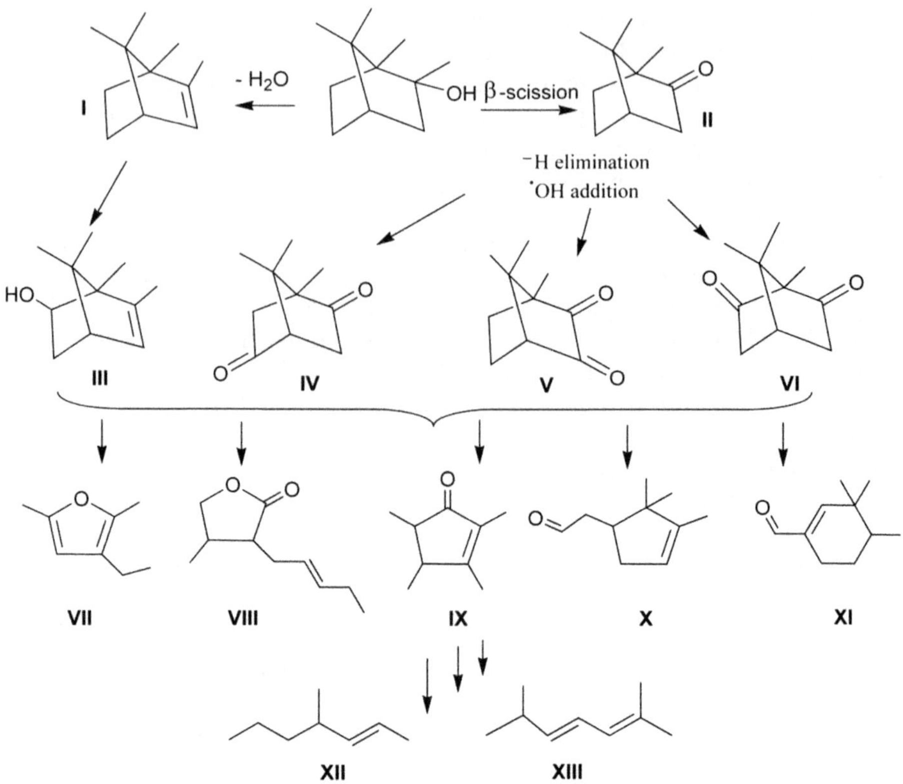

Figure 1.5 Intermediate products of 2-methylisoboeneol (MIB) formed using TiO_2 under UV-A light. (Reproduced with permission from ref. 87. Copyright 2015 Elsevier.)

MIB, giving the ketone. Starting from D-camphor with hydrogen elimination followed by HO˙ addition and with further oxidation of a secondary alcohol, leading to the formation of a ketone, three diketone products could be formed. All other intermediates identified were mainly oxygen containing saturated and unsaturated cyclic compounds, through HO˙ oxidations, driven by electrophilic substitutions. Ring opening reactions on the MIB molecule result in the formation of five-membered ring compounds. Similar to previous studies by Hiskia *et al.*,[106] during the later stages of photo-oxidation linear aliphatic compounds were identified.

1.4.2 Transformation Products of Compounds Resulting from the Reaction of Heteroatoms

The degradation routes of compounds with heteroatoms in their structure (such as N, P, S) may differ from the ones with only carbon atoms because of differences in the kinetic rates but also the reaction mechanisms that need to take place for stable intermediates to form. Herein, the TPs of CYN and taste and odor compounds will be reported as characteristic examples of this group of compounds.

CYN is a uracil derivative with a tricyclic guanidine and a sulfate group. Only a few studies have dealt with the degradation mechanism of CYN using TiO_2 photocatalysis,[6,86] where it has been stated that the main reaction pathways were through hydroxyl radical attack, with hydrogen abstractions and addition to double bonds. Table 1.2 gives the structures of reaction intermediates identified upon TiO_2 photocatalysis under UV light.

Table 1.2 Structures of reaction intermediates of cylindrospermopsin with TiO_2 photocatalysis.

No.	Structure	*m*/*z*	Compound	Reference
CYN	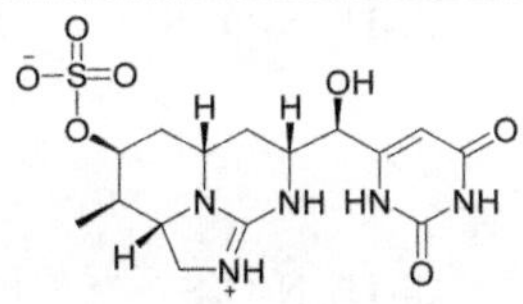	416	$C_{15}H_{21}N_5O_7S$	
1	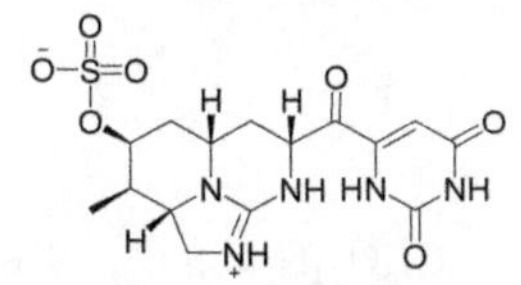	414	$C_{15}H_{19}N_5O_7S$	6,87
2A	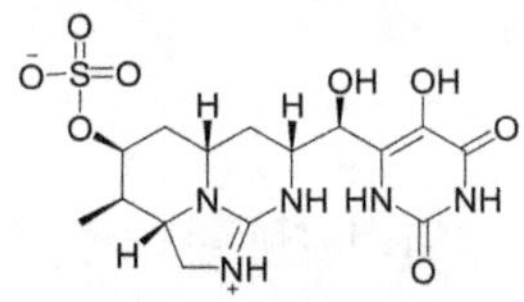	432	$C_{15}H_{21}N_5O_8S$	6,87
2B	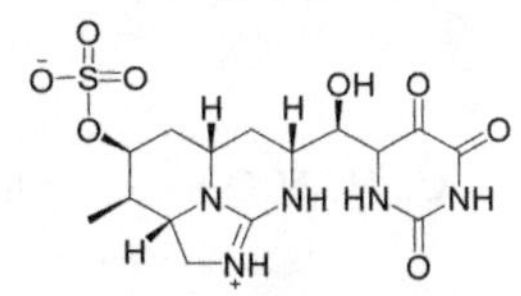			6,87
2C	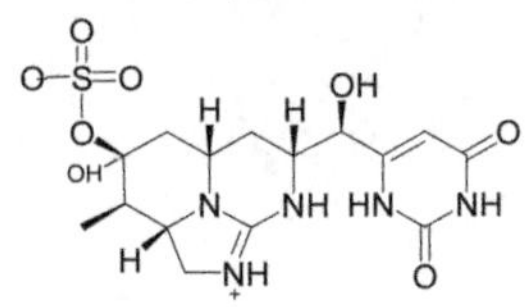		$C_{15}H_{21}N_5O_8S$	6
3	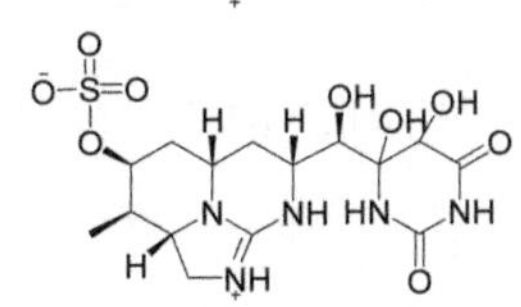	450	$C_{15}H_{23}N_5O_9S$	87
4A	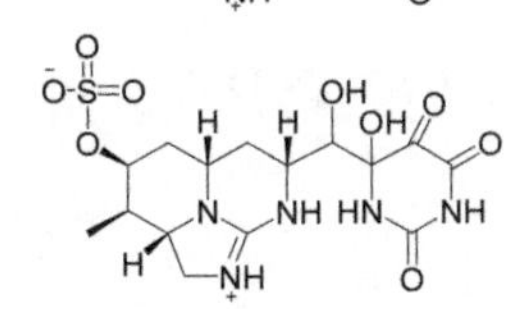	448	$C_{15}H_{21}N_5O_9S$	6,87

(*continued*)

Table 1.2 (*continued*)

No.	Structure	m/z	Compound	Reference
4B	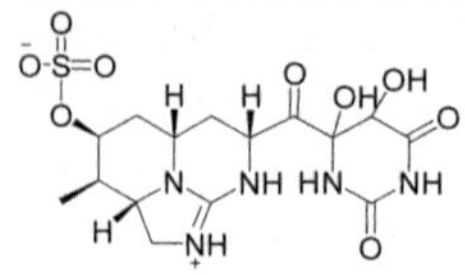	448	$C_{15}H_{21}N_5O_9S$	6,87
5A	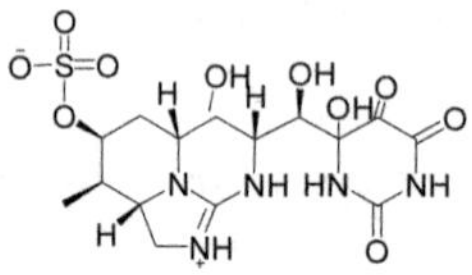	464	$C_{15}H_{21}N_5O_{10}S$	6
5B	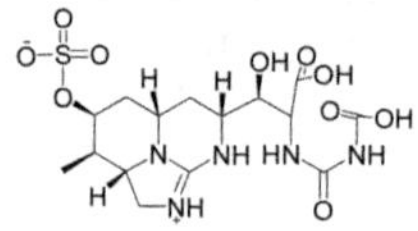	464	$C_{15}H_{23}N_5O_{10}S$	6
6	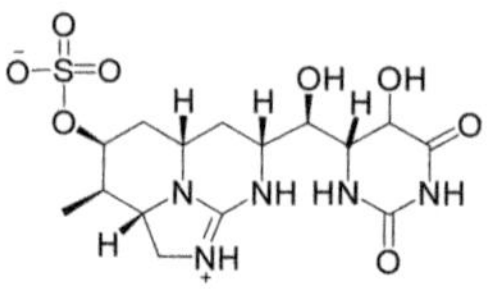	434	$C_{15}H_{23}N_5O_8S$	87
7A	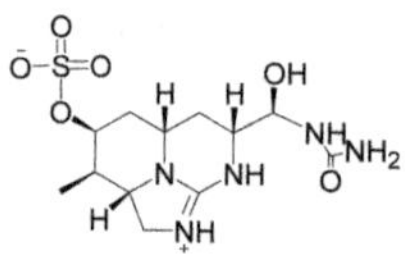	392	$C_{13}H_{21}N_5O_7S$	6
7B	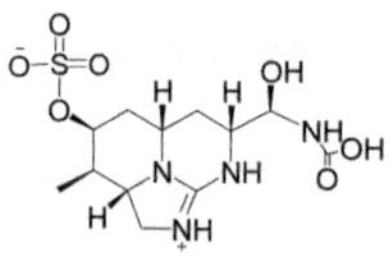	393	$C_{13}H_{20}N_4O_8S$	6
8	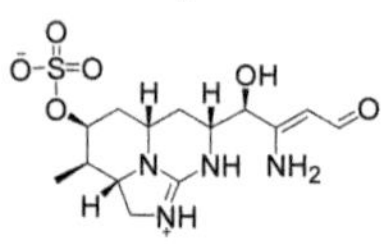	375	$C_{14}H_{22}N_4O_6S$	87
9	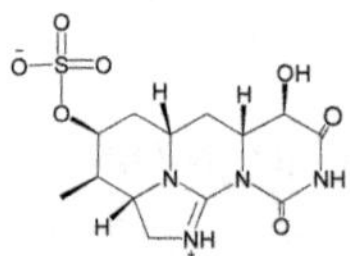	375	$C_{15}H_{19}N_4O_7S$	6
10	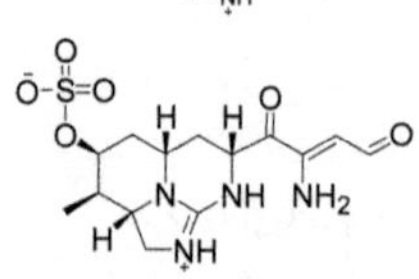	373	$C_{14}H_{20}N_4O_6S$	87
11	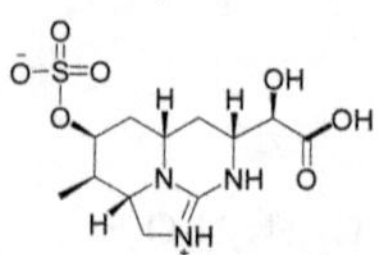	350	$C_{12}H_{19}N_3O_7S$	87

12	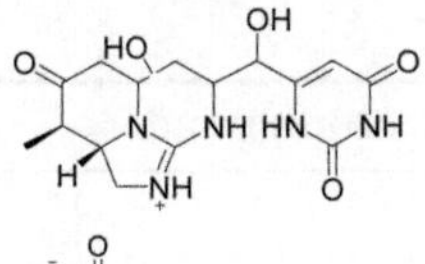	350	$C_{15}H_{19}N_5O_5$	6
13	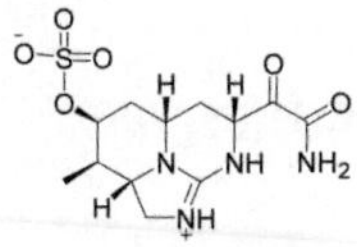	347	$C_{12}H_{18}N_4O_6S$	87
14	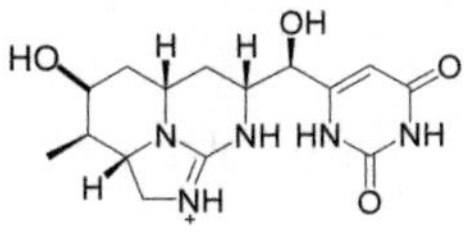	338	$C_{15}H_{23}N_5O_4^+$	87
15	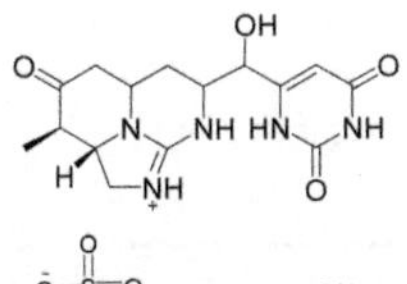	334	$C_{15}H_{19}N_5O_4$	6
16	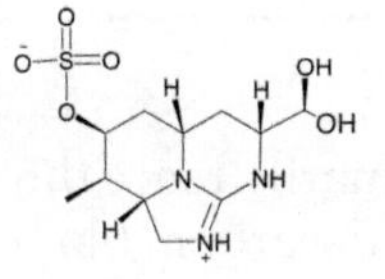	322	$C_{11}H_{19}N_3O_6S$	6
17	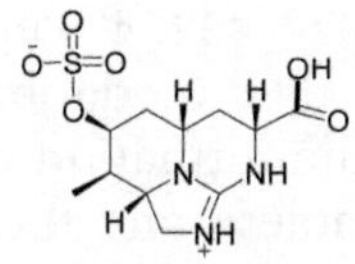	320	$C_{11}H_{17}N_3O_6S$	87
18	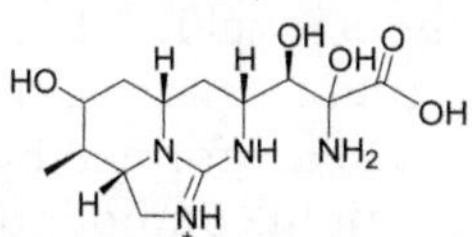	316	$C_{13}H_{23}N_4O_5^+$	87
19A	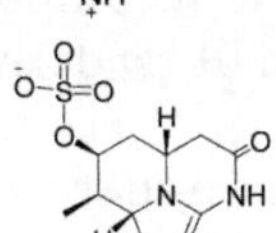	290	$C_{10}H_{15}N_3O_5S$	6
19B	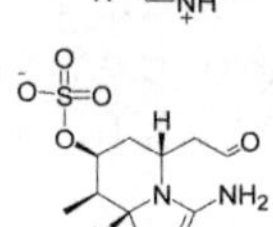	292	$C_{10}H_{17}N_3O_5S$	6
19C	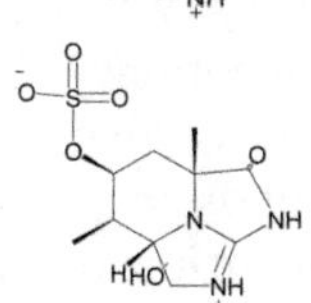	292	$C_9H_{13}N_3O_6S$	6
20	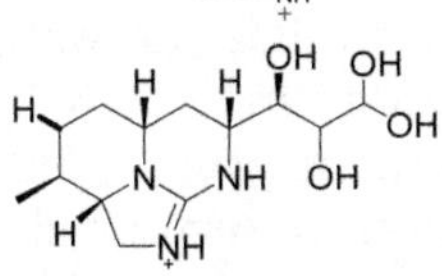	287	$C_{13}H_{24}N_3O_4^+$	87

(*continued*)

Table 1.2 (*continued*)

No.	Structure	m/z	Compound	Reference
21	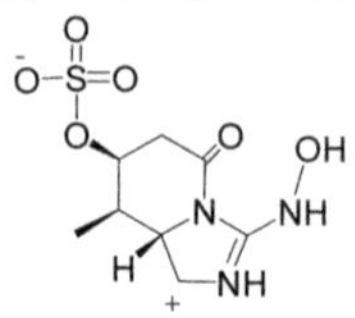	280	$C_8H_{13}N_3O_6S$	87
22	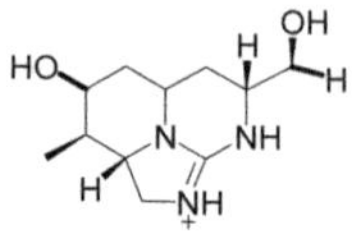	227	$C_{11}H_{20}N_3O_2^+$	87
23	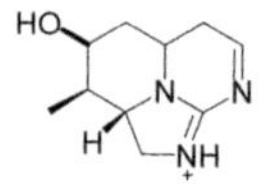	195	$C_{10}H_{16}N_3O^+$	87

As previously mentioned, HO• radicals primarily react through hydrogen abstraction and addition on the unsaturated carbon bond on the uracil group, resulting in the formation of product m/z = 432.[6,86] Further hydroxylation results in the formation of product m/z = 450.[1] Oxidation on the uracil group of product m/z = 432 can lead first to the formation of product m/z = 375, upon ring opening at the urea group moiety, and then by a further oxidation to product m/z = 350 (cylindrospermopsic acid).[86] From m/z = 375, product m/z = 373 can also be formed. From there onwards, products with m/z = 347 and the acid-like product with m/z = 320 are formed by further oxidations. HO• radical attack on m/z = 347 can result in compound m/z = 280.[86] A product with m/z = 392 was detected, which can likely derive from hydroxyl addition on product m/z = 432.[6]

Elimination of the sulfate group can also occur under free radical attack. Product m/z = 334 resulted from hydroxylation on the tricyclic guanidine moiety followed by elimination of the sulfuric acid group. With further hydroxylation, product m/z = 350 was produced,[6] from which further hydroxylation resulted to give product m/z = 366 and m/z 382.[6]

Starting from product m/z = 450, TPs with m/z = 287 and m/z = 448 are formed. Oxidation of m/z = 450 on either of the secondary alcohols, through hydrogen abstraction, can lead to the formation of two carbonyl isomers with m/z = 448,[86] while further hydroxylation leads to product m/z = 464.[6] In addition, m/z = 287 is formed, through substitution of the sulfate group followed by oxidation on the uracil group. By a similar pathway, the carbonyl product with m/z = 414 corresponds to the oxidation of CYN by HO• radical attack on the secondary alcohol (bridging methane group), followed by substitution of the sulfate group with HO• addition to form product m/z = 338, from which a compound with m/z = 316 can be produced with oxidation on the uracil group.[86]

Overall, it was found that the mechanism of CYN degradation proceeds through $HO^{\bullet}$ attack on different sites of CYN: (a) hydroxylation on the uracil group, with uracil ring opening and further oxidation, (b) attack on the tricyclic alkaloid group, (c) attack on the tricyclic alkaloid group with ring opening, and (d) attack on the sulfate group, leading to various oxidation products (Table 1.2). After prolonged irradiation under UV-A light, complete mineralization of CYN to CO_2, SO_4^{2-}, NO_3^-, and NH_4^+ takes place.[86]

1.5 Conclusions

UV-based TiO_2 photocatalysis is effective for the degradation of a wide variety of organic compounds with different functional groups. Complex organic compounds can be broken down into individual functional groups. The different functional groups have different reactivity and susceptibility to photocatalytic degradation. Given the large number of studies reported on the UV/TiO_2 degradation of an extensive number of organic pollutants and toxins, it is possible to predict the relative reactivity of different reaction sites and reaction pathway partitioning in complex substrates. Fundamental mechanistic understanding is critical for the development and application of UV/TiO_2 photocatalysis for real water treatment of problematic pollutant and mixed waste. Therefore, the influent water quality parameters must be optimized to achieve the highest efficiency of TiO_2 photocatalysis. In addition, physical and chemical modification of TiO_2 materials can improve selectivity towards target chemicals, thus achieving higher removal efficiencies than conventional with UV/TiO_2. Future research will also focus on unveiling experimental conditions under which the photocatalytic processes applied in water treatment results in detoxified water by focusing on the contaminant's sites that initiate its toxic behavior.

Acknowledgements

M. G. Antoniou is thankful to the Cyprus University of Technology for a start-up package (EX-90). H. Choi is grateful to the Texas Higher Education Coordinating Board for the Norman Hackerman Advanced Research Program fund (THECB13311). D. D. Dionysiou also acknowledges support from the University of Cincinnati through a UNESCO co-Chair Professor position on "Water Access and Sustainability". C. Han was supported by the Postgraduate Research Program at the National Risk Management Research Laboratory administered by the Oak Ridge Institute for Science and Education through an interagency agreement between the U.S. Department of Energy and the U.S. Environmental Protection Agency. T. Fotiou, T. M. Triantis and A. Hiskia acknowledge financial support from the European Social Fund and Greek national funds through ARISTEIA operational research program "Cyanotoxins in Fresh Waters, Advances in Analysis, Occurrence and Treatment CYANOWATER" (Grant N.: 2455).

References

1. A. Fujishima and K. Honda, *Nature*, 1972, **238**, 37.
2. M. A. Fox and M. T. Dulay, *Chem. Rev.*, 1993, **93**, 341.
3. S. Banerjee, S. C. Pillai, P. Falaras, K. E. O'Shea, J. A. Byrne and D. D. Dionysiou, *J. Phys. Chem. Lett.*, 2014, **5**, 2543.
4. M. Pelaez, N. T. Nolan, S. C. Pillai, M. K. Seery, P. Falaras, A. G. Kontos, P. S. M. Dunlop, J. W. J. Hamilton, J. A. Byrne, K. O'Shea, M. H. Entezari and D. D. Dionysiou, *Appl. Catal., B*, 2012, **125**, 331.
5. C. Zhao, M. Pelaez, D. D. Dionysiou, S. C. Pillai, J. A. Byrne and K. E. O'Shea, *Catal. Today*, 2014, **224**, 70.
6. G. Zhang, E. M. Wurtzler, X. He, M. N. Nadagouda, K. O'Shea, S. M. El-Sheikh, A. A. Ismail, D. Wendell and D. D. Dionysiou, *Appl. Catal., B*, 2015, **163**, 591.
7. S. Zheng, W. Jiang, Y. Cai, D. D. Dionysiou and K. E. O'Shea, *Catal. Today*, 2014, **224**, 83.
8. G. V. Buxton, C. L. Greenstock, W. P. Helman and A. B. Ross, *J. Phys. Chem. Ref. Data*, 1988, **17**, 513.
9. R. Andreozzi, V. Caprio, A. Insola and R. Marotta, *Catal. Today*, 1999, **53**, 51.
10. C. Zhao, L. E. Arroyo-Mora, A. P. DeCaprio, V. K. Sharma, D. D. Dionysiou and K. E. O'Shea, *Water Res.*, 2014, **67**, 144.
11. C. S. Turchi and D. F. Ollis, *J. Catal.*, 1990, **122**, 178.
12. D. Lawless, N. Serpone and D. Meisel, *J. Phys. Chem.*, 1991, **95**, 5166.
13. K. E. O'Shea and C. Cardona, *J. Org. Chem.*, 1994, **59**, 5005.
14. W. Song, T. Xu, W. J. Cooper, D. D. Dionysiou, A. A. de la Cruz and K. E. O'Shea, *Environ. Sci. Technol.*, 2009, **43**, 1487.
15. I. Liu, L. A. Lawton and P. K. J. Robertson, *Environ. Sci. Technol.*, 2003, **37**, 3214.
16. O. d'Hennezel, P. Pchat and D. F. Ollis, *J. Photochem. Photobiol., A*, 1998, **118**, 197.
17. T. D. Bui, A. Kimura, S. Higashida, S. Ikeda and M. Matsumura, *Appl. Catal., B*, 2011, **107**, 119.
18. X. Li and W. S. Jenks, *J. Am. Chem. Soc.*, 2000, **122**, 11864.
19. H. F. Lin, R. Ravikrishna and K. T. Valsaraj, *Sep. Purif. Technol.*, 2002, **28**, 87.
20. H. Hidaka, H. Honjou, T. Koike, Y. Mitsutsuka, T. Oyama and N. Serpone, *J. Photochem. Photobiol., A*, 2008, **197**, 115.
21. I. K. Konstantinou, V. A. Sakkas and T. A. Albanis, *Appl. Catal., B*, 2001, **34**, 227.
22. K. Harada, T. Hisanaga and K. Tanaka, *Water Res.*, 1990, **24**, 1415.
23. A. Vidal, Z. Dinya and F. Mogyorodi, *Appl. Catal., B*, 1999, **21**, 259.
24. G. S. Shephard, S. Stockenstrom, D. de Villiers, W. J. Engelbrecht and G. F. S. Wessels, *Water Res.*, 2002, **36**, 140.
25. W. F. D. Vilela, A. Minillo, O. Rocha, E. M. Vieira and E. B. Azevedo, *Sol. Energy*, 2012, **86**, 2746.

26. Y. L. Su, Y. R. Deng, L. Zhao and Y. X. Du, *Chin. Sci. Bull.*, 2013, **58**, 1156.
27. P. J. Senogles, J. A. Scott, G. Shaw and H. Stratton, *Water Res.*, 2001, **35**, 1245.
28. G. S. Zhang, M. N. Nadagouda, K. O'Shea, S. M. El-Sheikh, A. A. Ismail, V. Likodimos, P. Falaras and D. D. Dionysiou, *Catal. Today*, 2014, **224**, 49.
29. J. Choi, H. Park and M. R. Hoffmann, *J. Phys. Chem. C*, 2010, **114**, 783.
30. M. Pelaez, A. A. de la Cruz, E. Stathatos, P. Falaras and D. D. Dionysiou, *Catal. Today*, 2009, **144**, 19.
31. G. L. Liu, C. Han, M. Pelaez, D. W. Zhu, S. J. Liao, V. Likodimos, N. Ioannidis, A. G. Kontos, P. Falaras, P. S. M. Dunlop, J. A. Byrne and D. D. Dionysiou, *Nanotechnology*, 2012, **23**, 29.
32. C. Han, M. Pelaez, V. Likodimos, A. G. Kontos, P. Falaras, K. O'Shea and D. D. Dionysiou, *Appl. Catal., B*, 2011, **107**, 77.
33. Y. C. Zhang, M. Yang, G. Zhang and D. D. Dionysiou, *Appl. Catal., B*, 2013, **142–143**, 249.
34. H. Choi, M. G. Antoniou, M. Pelaez, A. A. De la Cruz, J. A. Shoemaker and D. D. Dionysiou, *Environ. Sci. Technol.*, 2007, **41**, 7530.
35. M. Pelaez, P. Falaras, A. G. Kontos, A. A. de la Cruz, K. O'Shea, P. S. M. Dunlop, J. A. Byrne and D. D. Dionysiou, *Appl. Catal., B*, 2012, **121–122**, 30.
36. L. Kiszczak and J. Tropilo, *Rocz. Panstw. Zakl. Hig.*, 1988, **39**, 314.
37. N. Milic, M. Milanovic, N. G. Letic, M. T. Sekulic, J. Radonic, I. Mihajlovic and M. V. Miloradov, *Int. J. Environ. Health Res.*, 2013, **23**, 296.
38. R. Wei, F. Ge, S. Huang, M. Chen and R. Wang, *Chemosphere*, 2011, **82**, 1408.
39. J. M. Brausch and G. M. Rand, *Chemosphere*, 2011, **82**, 1518.
40. R. Hirsch, T. Ternes, K. Haberer and K. L. Kratz, *Sci. Total Environ.*, 1999, **225**, 109.
41. K. Li, A. Yediler, M. Yang, S. Schulte-Hostede and M. H. Wong, *Chemosphere*, 2008, **72**, 473.
42. R. Xuan, L. Arisi, Q. Wang, S. R. Yates and K. C. Biswas, *J. Environ. Sci. Health, Part B*, 2010, **45**, 73.
43. C. Zhao, M. Pelaez, X. Duan, H. Deng, K. O'Shea, D. Fatta-Kassinos and D. D. Dionysiou, *Appl. Catal., B*, 2013, **134–135**, 83.
44. S. Jiao, S. Zheng, D. Yin, L. Wang and L. Chen, *J. Environ. Sci*, 2008, **20**, 806.
45. J. H. O. S. Pereira, V. J. P. Vilar, M. T. Borges, O. González, S. Esplugas and R. A. R. Boaventura, *Sol. Energy*, 2011, **85**, 2732.
46. J. H. O. S. Pereira, A. C. Reis, D. Queirós, O. C. Nunes, M. T. Borges, V. J. P. Vilar and R. A. R. Boaventura, *Sci. Total Environ.*, 2013, **463–464**, 274.
47. C. Han, V. Likodimos, J. A. Khan, M. N. Nadagouda, J. Andersen, P. Falaras, P. Rosales-Lombardi and D. D. Dionysiou, *Environ. Sci. Pollut. Res.*, 2014, **21**, 11781.
48. C. Zhao, H. Deng, Y. Li and Z. Liu, *J. Hazard. Mater.*, 2010, **176**, 884.
49. C. Zhao, Y. Zhou, D. J. D. Ridder, J. Zhai, Y. Wei and H. Deng, *Chem. Eng. J.*, 2014, **248**, 280.

50. M. Pelaez, A. A. de la Cruz, K. O'Shea, P. Falaras and D. D. Dionysiou, *Water Res.*, 2011, **45**, 3787.
51. L. Haroune, M. Salaun, A. Ménard, C. Y. Legault and J.-P. Bellenger, *Sci. Total Environ.*, 2014, **475**, 16.
52. D. Avisar, I. Horovitz, L. Lozzi, F. Ruggieri, M. Baker, M.-L. Abel and H. Mamane, *J. Hazard. Mater.*, 2013, **244–245**, 463.
53. M. N. Chong, B. Jin, C. W. K. Chow and C. Saint, *Water Res.*, 2010, **44**, 2997.
54. L. A. Lawton, P. K. J. Robertson, B. J. P. A. Cornish, I. L. Marr and M. Jaspars, *J. Catal.*, 2003, **213**, 109.
55. R. W. Mathews, *Water Res.*, 1986, **20**, 569.
56. C. Shifu and C. Gengyu, *Sol. Energy*, 2005, **79**, 1.
57. V. Shafirvich, A. Dourandin, W. Huang and N. E. Geacintov, *J. Biol. Chem.*, 2001, **276**, 24621.
58. D. Zhao, X. Liao, X. Yan, S. G. Huling, T. Chai and H. Tao, *J. Hazard. Mater.*, 2013, **254–255**, 228.
59. M. R. Assalin, S. G. De Moraes, S. C. N. Queiroz, V. L. Ferracini and N. Duran, *J. Environ. Sci. Health, Part B*, 2010, **45**, 89.
60. M. Cho, H. Chung, W. Choi and J. Yoon, *Water Res.*, 2004, **38**, 1069.
61. R. Andreozzi, R. Marotta, G. Pinto and A. Pollio, *Water Res.*, 2002, **36**, 2869.
62. T. E. Doll and F. H. Frimmel, *Catal. Today*, 2005, **101**, 195.
63. V. A. Sakkas, D. A. Lambropoulou and T. A. Albanis, *Chemosphere*, 2002, **48**, 939.
64. J. R. Grbin, D. M. Milori, M. L. Simões, W. T. da Silva and L. M. Neto, *Chemosphere*, 2007, **66**, 1692.
65. A. Zakersalehi, M. Nadagouda and H. Choi, *Catal. Comm.*, 2013, **41**, 79.
66. Y. Paz, *C. R. Chim.*, 2006, **9**, 774.
67. K. Inumaru, M. Murashima, T. Kasahara and S. Yamanaka, *Appl. Catal., B*, 2004, **52**, 275.
68. Y. Paz, *Solid State Phenom.*, 2010, **162**, 135.
69. O. Ramstörm and R. J. Ansell, *Chirality*, 1998, **10**, 195.
70. K. I. Shimizu, T. Kaneko, T. Fujishima, T. Kodama, H. Yoshida and Y. Kitayama, *Appl. Catal., A*, 2002, **225**, 185.
71. R. Kato, N. Shimura and M. Ogawa, *Chem. Lett.*, 2008, **37**, 76.
72. M. R. Hoffmann, S. T. Martin, W. Choi and D. W. Bahnemann, *Chem. Rev.*, 1995, **95**, 69.
73. S. Malato, P. Fernández-Ibáñez, M. I. Maldonado, J. Blanco and W. Gernjak, *Catal. Today*, 2009, **147**, 1.
74. T. M. Triantis, T. Fotiou, T. Kaloudis, A. G. Kontos, P. Falaras, D. D. Dionysiou, M. Pelaez and A. Hiskia, *J. Hazard. Mater.*, 2012, **211–212**, 196.
75. M. G. Antoniou, J. A. Shoemaker, A. A. De la Cruz and D. D. Dionysiou, *Toxicon*, 2008, **51**, 1103.
76. K. Vinodgopal, I. Bedja, S. Hotchandani and P. V. Kamat, *Langmuir*, 1994, **10**, 1767.
77. E. Pelizzetti, V. Maurino, C. Minero, V. Carlin, M. L. Tosato, E. Pramauro and O. Zerbinati, *Environ. Sci. Technol.*, 1990, **24**, 1559.

78. J. C. D'Oliveira, C. Minero, E. Pelizzetti and P. Pichat, *J. Photochem. Photobiol., A*, 1993, **72**, 261.
79. E. Hapeshi, A. Achilleos, M. I. Vasquez, C. Michael, N. P. Xekoukoulotakis, D. Mantzavinos and D. Kassinos, *Water Res.*, 2010, **44**, 1737.
80. P. Calza, E. Pelizzetti and C. Minero, *J. Appl. Electrochem.*, 2005, **35**, 665.
81. A. Hiskia, M. T. Triantis, M. G. Antoniou, A. A. De la Cruz, K. O'Shea, W. Song, T. Fotiou, T. Kaloudis, X. He, J. Andersen and D. D. Dionysiou, in *Transformation Products of Emerging Contaminants in the Environment: Analysis, Processes, Occurrence, Effects and Risks*, ed. L. Nollet and D. Lambropoulou, John Wiley & Sons, Inc., 2013, pp. 687–720.
82. P. Kormali, D. Dimoticali, D. Tsipi, A. Hiskia and E. Papaconstantinou, *Appl. Catal., B*, 2004, **48**, 175.
83. V. Kouloumbos, D. Tsipi, A. Hiskia, D. Nikolic and R. van Breemen, *J. Am. Soc. Mass Spectrom.*, 2003, **14**, 803.
84. P. Calza, C. Massolino and E. Pelizzetti, *J. Photochem. Photobiol., A*, 2008, **200**, 356.
85. I. K. Konstantinou, in *Mass Spectrometry for Analysis of Pesticide Residues and their Metabolites*, ed. D. Tsipi, H. Botitsi and A. Economou, John Wiley & Sons, 2015.
86. T. Fotiou, T. Triantis, T. Kaloudis and A. Hiskia, *Chemosphere*, 2015, **119**, 89.
87. T. Fotiou, T. M. Triantis, T. Kaloudis, E. Papaconstantinou and A. Hiskia, *J. Photochem. Photobiol., A*, 2014, **286**, 1.
88. M. G. Antoniou, A. A. De la Cruz and D. D. Dionysiou, *Environ. Sci. Technol.*, 2010, **44**, 7238.
89. M. R. Hoffmann, S. T. Martin, W. Choi and D. W. Bahnemann, *Chem. Rev.*, 1995, **95**, 69.
90. P. Pichat, *Water Sci. Technol.*, 2007, **55**, 167.
91. V. K. Sharma, T. M. Triantis, M. G. Antoniou, X. He, M. Pelaez, C. Han, W. Song, K. E. O'Shea, A. A. De La Cruz, T. Kaloudis, A. Hiskia and D. D. Dionysiou, *Sep. Purif. Technol.*, 2012, **91**, 3.
92. L. Cermenati, P. Pichat, C. Guillard and A. Albini, *J. Phys. Chem. B*, 1997, **101**, 2650.
93. I. Liu, L. A. Lawton, B. Cornish and P. K. J. Robertson, *J. Photochem. Photobiol., A*, 2002, **148**, 349.
94. I. Liu, L. A. Lawton, D. W. Bahnemann and P. K. J. Robertson, *Appl. Catal., B*, 2005, **60**, 245.
95. M. G. Antoniou, J. A. Shoemaker, A. A. De la Cruz and D. D. Dionysiou, *Environ. Sci Technol.*, 2008, **42**, 8877.
96. G. V. Buxton, C. L. Greenstock, W. P. Helman and A. B. Ross, *J. Phys. Chem. Ref. Data*, 1988, **17**, 513.
97. W. Song, A. A. De La Cruz, K. Rein and K. E. O'Shea, *Environ. Sci. Technol.*, 2006, **40**, 3941.
98. J. McMurry, *Organic Chemistry*, Brooks Cole Pub Co, 1995.
99. M. G. Antoniou, P. A. Nicolaou, J. A. Shoemaker, A. A. De la Cruz and D. D. Dionysiou, *Appl. Catal., B*, 2009, **91**, 165.
100. K. Tsuji, T. Watanuki, F. Kondo, M. F. Watanabe, S. Suzuki, H. Nakazawa, M. Suzuki, H. Uchida and K. Harada, *Toxicon*, 1995, **33**, 1619.

101. S. Takenaka and Y. Tanaka, *Chemosphere*, 1995, **31**, 3635.
102. M. Pelaez, P. Falaras, V. Likodimos, A. G. Kontos, A. A. De la Cruz, K. O'shea and D. D. Dionysiou, *Appl. Catal., B*, 2010, **99**, 378.
103. J. Andersen, C. Han, K. O'Shea and D. D. Dionysiou, *Appl. Catal., B*, 2014, **154–155**, 259.
104. G. A. Burlingame, R. M. Dann and G. L. Brock, *Am. Water Works Assoc.*, 1986, **78**, 56.
105. E. E. Bamuza Pemu and E. M. Chirwa, *Chem. Eng. Trans.*, 2011, **24**, 91.
106. A. Hiskia, E. Androulaki, A. Mylonas, S. Boyatzis, D. Dimoticali, C. Minero, E. Pelizzetti and E. Papaconstantinou, *Res. Chem. Intermed.*, 2000, **26**, 235.

CHAPTER 2

Photocatalytic Removal of Metallic and Other Inorganic Pollutants

MARTA I. LITTER*[a,b,c], NATALIA QUICI[a,b], J. MARTÍN MEICHTRY[a,b], AND ALEJANDRO M. SENN[a,b]

[a]Comisión Nacional de Energía Atómica, Centro Atómico Constituyentes, Avenida General Paz 1499, 1650 San Martín, Prov. de Bs. As., Argentina; [b]Consejo Nacional de Investigaciones Científicas y Técnicas (CONICET), Godoy Cruz 2290, 1425 Ciudad Autónoma de Buenos Aires, Argentina; [c]Instituto de Investigación e Ingeniería Ambiental, Universidad Nacional de General San Martín, Peatonal Belgrano 3563, 1° piso, 1650 San Martín, Provincia de Buenos Aires, Argentina
*E-mail: marta.litter@gmail.com

2.1 Introduction

The presence of metals, metalloids and inorganic pollutants such as nitrate in wastewaters and drinking water represents nowadays one of the most important environmental problems because the annual total toxicity of these mobilized species is greater than the total toxicity of all the anthropogenic radioactive and organic wastes.[1] In addition to natural sources, anthropic activities introduce a huge amount of metals and nitrate in the terrestrial ambient. The accumulation of metals, metalloids and other inorganic

RSC Energy and Environment Series No. 15
Photocatalysis: Applications
Edited by Dionysios D Dionysiou, Gianluca Li Puma, Jinhua Ye, Jenny Schneider, and Detlef Bahnemann

Published by the Royal Society of Chemistry, www.rsc.org

species in effluents and in industrial wastes represents significant losses in raw materials and causes perturbation of the ecological equilibrium. Precious and common metals enter waters through washing, rinsing, pickling and surface treatment procedures of industrial processes such as hydrometallurgy, plating or photography. Inorganic species such as chromium, mercury, uranium, arsenic and nitrate are considered priority pollutants by several environmental agencies worldwide, and the allowed amount of these species in drinking and wastewaters is becoming increasingly stringent.[2–8]

Although the removal of organic and microbiological pollutants from waters has been thoroughly studied, less attention has been paid to the processes leading to the transformation of inorganic species into species of lower toxicity or that are more easily isolated. Metals in their various oxidation states have infinite lifetimes, and chemical or biological treatments present severe restrictions or are economically prohibitive. Removal of these species is carried out, generally, by precipitation, electrolysis, chemical oxidation, adsorption, or chelation, all of them presenting drawbacks or high economic costs. Conventional treatment methodologies such as biological processes, ozonation, activated carbon, *etc.* do not always allow us to decrease the concentration of certain pollutants to desirable levels, as in the case of recalcitrant compounds or when the processes are slow or very expensive.[9]

Advanced oxidative or reductive processes (AOPs/ARPs) can be simple and low-cost procedures for removal of this type of pollutants. These processes are based on the use of some form of energy to produce highly reactive species, which, in turn, are able to promote oxidations or reductions on the target pollutants; the main oxidative species is the hydroxyl radical, $HO^{\bullet}$, which has the ability to attack most organic and inorganic chemical species.[10–15] These processes have been generally tested for treatment of microbiological and organic pollutants (phenols, dyes, pesticides, *etc.*). The transformation of pollutants such as nitrate, metal or metalloids using PAOs/PARs has been less studied, but this study is at present very relevant due to its practical consequences and possible applications.

Heterogeneous photocatalysis (HP) is a very well-known AOP/ARP, valuable for purification and remediation of water and air. Several excellent reviews exist on the subject, with different approaches.[11,16–18] As is very well established, in HP, after irradiation of a semiconductor with light of enough energy for excitation, electrons are promoted from the valence band (VB) to the conduction band (CB), leaving holes behind; thermodynamically, an acceptor A can be photocatalytically reduced by CB electrons (e^-_{CB}) if its redox potential is more positive than that of the e^-_{CB}, and a donor D can be oxidized by VB holes (h^+_{VB}) if its redox potential is more negative than that of the h^+_{VB}.[16] Titanium dioxide (TiO_2) is the most widely investigated semiconductor because of its high stability and photocatalytic efficiency, favourable chemical properties and low cost.[19] For the most popular model photocatalyst, Degussa P-25 (now Evonik P25), the values of the edges of CB and VB at pH 0 have been calculated as −0.3 and +2.9 V *versus* SHE, respectively.[20] Importantly, the potential of both e^-_{CB} and h^+_{VB} goes to more negative values with the increase of pH.[21–23] The processes can be depicted by the following simplified equations (for a more detailed description see, for example, ref. 16 and 24):

$$TiO_2 + h\nu \rightarrow e^-_{CB} + h^+_{VB} \tag{2.1}$$
$$e^-_{CB} + A \rightarrow A^{\bullet -} \tag{2.2}$$
$$h^+_{VB} + D \rightarrow D^{\bullet +} \tag{2.3}$$

In particular, water adsorbed to TiO_2 can be oxidized by h^+_{VB}, generating $HO^\bullet$:

$$h^+_{VB} + H_2O_{ads} \rightarrow HO^\bullet + H^+ \tag{2.4}$$

On the other hand, adsorbed O_2 can be reduced by e^-_{CB}, generating superoxide radicals, $O_2^{\bullet -}$, in a thermodynamically feasible but rather slow electron transfer reaction.[21] As the following set of equations indicates, this cathodic pathway is an additional source of $HO^\bullet$:

$$O_{2ads} + e^-_{CB}\,(+H^+) \rightarrow O_2^{\bullet -}\,(HO_2^\bullet) \tag{2.5}$$
$$O_2^{\bullet -} + H_2O \rightarrow HO_2^\bullet + OH^- \tag{2.6}$$
$$2HO_2^\bullet \rightarrow O_2 + H_2O_2 \tag{2.7}$$
$$H_2O_2 + e^-_{CB}\,(O_2^{\bullet -}) \rightarrow HO^- + HO^\bullet\,(+O_2) \tag{2.8}$$

The photocatalytic activity of TiO_2 depends on several factors such as the surface area, crystallinity and nature of the surface of the material, pH, the presence of doping metals, *etc.* As photocatalytic reactions occur at the surface of TiO_2, the surface area and affinities of reactants are considered among the most important factors determining the photocatalytic activity, together with the survival of photogenerated charge-carriers before recombination.[25]

In anoxic conditions, the cathodic pathway can be the reduction of protons to adsorbed hydrogen atoms, which would combine to produce hydrogen molecules:[26]

$$H^+ + e^-_{CB} \rightarrow H_{ads}^\bullet \tag{2.9}$$
$$2H_{ads}^\bullet \rightarrow H_2 \tag{2.10}$$

However, Reaction (2.9) has a very negative redox potential, at least in homogeneous solutions (−2.3 eV),[27,†] rendering this reaction not feasible (see, however, footnote). Only if a metal (*e.g.* Pt or Pd) is present on the TiO_2 surface would H_2 be produced according to what follows. Doping TiO_2 with noble metals can have two main consequences. First, as their Fermi levels are lower than that of TiO_2, photoexcited electrons can be transferred from the CB of the semiconductor to the metal, while photogenerated VB holes remain on TiO_2, thereby reducing electron/hole recombination and increasing the lifetime of the charge carriers; thus, the system becomes more efficient.[28,29] In addition,

---

†All redox potentials given in this work are standard values *versus* SHE. Values correspond to those in homogeneous solutions, although reactions at the interface can be somewhat different.

the low hydrogen overpotential (HOV) of, for example, Pt and Pd makes more feasible the photocatalytic production of H_2 (Reactions (2.9) and (2.10)).

On the other hand, an inorganic chemical species $M^{n+‡}$ in the photocatalytic system can be reduced by e^-_{CB} to a lower oxidation state provided it has a suitable redox potential to make the reaction thermodynamically possible:

$$M^{n+} + e^-_{CB} \rightarrow M^{(n-1)+} \tag{2.11}$$

Alternatively, M can be oxidized by h^+_{VB} or $HO^•$:

$$M^{n+} + h^+_{VB}/HO^• \rightarrow M^{(n+1)+} \tag{2.12}$$

By these pathways, photocatalytic treatments can convert the pollutant into a less toxic and/or soluble species, or transform them into their metallic solid forms with deposition over the semiconductor surface. This last process would allow recovery of the metal by mechanical or chemical procedures, with an important economical return. From the beginning of the development of HP, transformation and deposition of metals – principally the most noble, expensive and toxic ones – was visualized as one of the potential applications of this technology in view of the involved economic and environmental aspects. Various semiconductors have been applied to the photocatalytic transformation or deposition of metals such as chromium, gold, silver, platinum, palladium, rhodium, mercury, lead, manganese, thallium and copper, among others.[30] The technology does not require expensive reagents or equipment.

In view of the enormous literature published on the subject, only the cases of chromium, mercury, lead, uranium, arsenic and nitrate will be reviewed here, and TiO_2 or its modifications will be the only photocatalysts considered. Previous reviews by our research group have been published in which the early literature and other cases are reported.[31–35] In this chapter, the most important mechanistic features will be reinforced together with an update of recent information and brief reference to the most important advances. Other relevant pollutants, such as nitrite or cyanide, deserve to be reviewed, but they will be not treated here due to the lack of space.

2.2 Thermodynamical Considerations and Mechanistic Pathways

The redox level of the couples related to the levels of the CB and VB can be considered as the most important parameter to predict the feasibility of photocatalytic transformations, and it has to be emphasized that under the working conditions of ordinary photocatalytic reactions, *i.e.* under non-intense

‡M^{n+} represents here a metal or a metalloid species and, by extension, it will be also applied to nitrate.

irradiation, multi-electronic reactions are rather unlikely, considering the frequency of photon absorption, the lifetime of the charge-carriers and the size of photocatalyst particle.[36] Recent experiments of our research group (see below) led to the conclusion that most photocatalytic processes on metal ions and other species occur through successive one-electron pathways that produce unstable intermediates until the most stable species is formed.[37–43]

Therefore, three types of mechanisms can be considered for the photocatalytic removal of inorganic species, with reactions taking place mainly at the solid–liquid interface: (a) direct reduction by photogenerated e^-_{CB}; (b) and (b′) indirect reduction by intermediates generated by h^+_{VB} or $HO^•$ oxidation of electron donors present in the media; (c) and (c′) oxidative removal by h^+_{VB} or $HO^•$,[44] all of which are represented in Figure 2.1.

In the direct reduction (a), the initial electron transfer step, Reaction (2.11), is usually considered as the rate determining one.[45] The feasibility of the transformation is governed by the reduction potential of the first step related to the energy of the CB. The conjugate anodic reaction of Reaction (2.11) is oxidation of water by h^+_{VB}, initiated by Reaction (2.4) and ending in protons and oxygen:

$$4h^+_{VB} + 2H_2O \rightarrow O_2 + 4H^+ \quad (2.13)$$

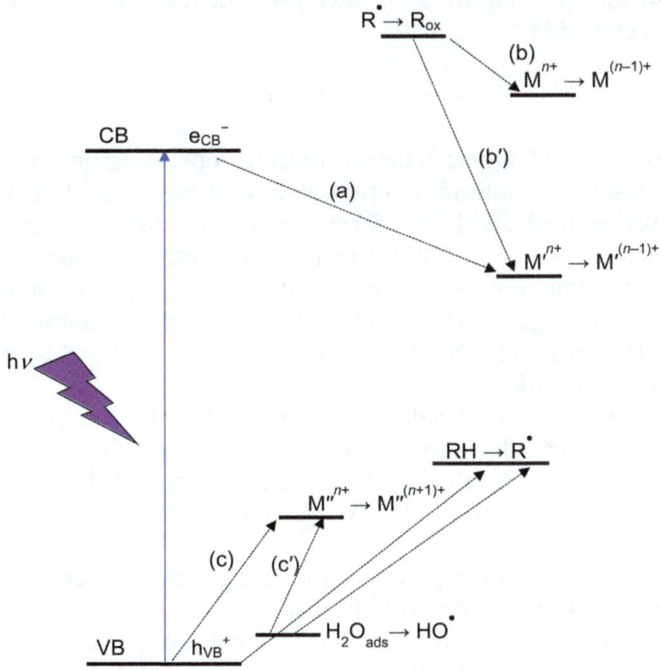

Figure 2.1 Schematic diagram for the photocatalytic transformation of metal ions on TiO_2. Different pathways (a), (b), (c) and (c′) are indicated (see text). The diagram of energy levels is only qualitative.

The photogenerated h^+_{VB} or $HO^•$ radicals can reoxidize also the species to the original one, causing a non-productive short-circuiting of the overall process:

$$M^{(n-1)+} + h^+_{VB}/HO^• \rightarrow M^{n+} \quad (2.14)$$

As the water oxidation Reaction (2.13) is a very slow four-electron process, reductive transformations can be generally improved by addition of sacrificial agents.[46,47] The process can be improved even more if a strong reducing species is formed from the sacrificial agent, generally organic compounds such as alcohols or carboxylic acids (RH), giving rise to an indirect pathway (type (b) reaction), as we will see later. These organic compounds are oxidized by h^+_{VB} or $HO^•$ through irreversible reactions, avoiding recombination of e^-_{CB} and h^+_{VB}, and enhancing reductive processes:

$$RH + h^+_{VB}/HO^• \rightarrow R^• + H^+/H_2O \quad (2.15)$$

Concurrently, Reaction (2.14) will be hindered if RH is added at relatively high concentrations because Reaction (2.15) will predominate.

Indirect reduction (mechanisms (b) and (b′)) is driven by the intermediates generated in Reaction (2.15), highly energetic species able to reduce several inorganic species (see Figure 2.1). Once produced, $R^•$ would be the effective reducing species of M^{n+}:

$$R^• + M^{n+} \rightarrow R_{ox} + M^{(n-1)+} \quad (2.16)$$

where R_{ox} can be an aldehyde, a ketone or CO_2, depending on the compound.

For the cases of methanol, ethanol and 2-propanol (2-PrOH), $R^•$ is 1-hydroxyalkyl radicals, while for formic or oxalic acids the strong reducing $CO_2^{•-}$ radical ($E^0_{(CO_2/CO_2^{•-})} \approx -2.0$ V)[48] is formed, with other carboxylic acids generating similar reducing species. In the absence of oxygen, the conjugated cathodic reaction might be H^+ reduction by e^-_{CB}, with H_2 formation (Reactions (2.9) and (2.10)), although this process will take place only in the presence of a metal acting as a sink of electrons.

After this first stage, other reducing steps driven by $R^•$ or e^-_{CB} may take place until a stable species is formed. It has been pointed out that direct reduction of metal ions to the zero-valent state by reducing radicals is rather slow, but once some metal nuclei are formed they serve as a cathodic site facilitating further reduction.[49]

Importantly, synergistic organic donors are generally present in polluted waters together with the species of interest in this chapter and the photocatalytic process can allow the simultaneous treatment of pollutants of different chemical structure and properties.

In mechanisms (c) and (c′), oxidative transformation of the M species takes place by h^+_{VB} or $HO^•$ attack (Figure 2.1). Other reactive oxygen species, ROS, formed in the system, such as $O_2^{•-}$, $HO_2^•$, H_2O_2, HO_2^-, *etc.*, can also perform oxidations depending on the redox potentials of the present species.[16] This occurs according to Reaction (2.12) when the oxidation of the metal or metalloid to a higher oxidation state is thermodynamically possible (cases of Pb(II), Mn(II), Tl(I) or As(III)).

Dyes can be used to modify the photocatalytic TiO_2 system to be activated by visible light irradiation. The mechanism of these systems is different from that taking place under UV light because the dye molecule acts as a sensitizer after being excited by light and injects an electron into the CB of the semiconductor:[50]

$$\text{Dye} + h\nu\ (\text{visible}) \rightarrow \text{Dye}^* \quad (2.17)$$

$$\text{Dye}^* + TiO_2 \rightarrow \text{Dye}^{\bullet +} + TiO_2(e^-_{CB}) \quad (2.18)$$

In some cases, the injected electron can reduce directly a chemical species, *e.g.* a metal of convenient redox potential such as Cr(VI), while the radical cation of the dye can oxidize a substrate;[51] as the oxidizing power of the radical cation is usually not very high, this allows performing selective oxidations.[52]

2.3 Chromium

Chromium is a metal with multiple industrial and technological applications including metallurgy, electroplating, textile industry, leather tanning and wood preservation. As a consequence of its uses, Cr is a frequent contaminant in wastewaters, mainly in the Cr(III) and Cr(VI) forms. Hexavalent chromium presents a high environmental threat due to its toxicity for biological organisms together with a high solubility and mobility; it is a well-known human carcinogen and may have dramatic effects on human health.[53] The World Health Organization (WHO) has established the maximum contaminant level of Cr(VI) in drinking water to be 0.05 mg L^{-1},[2] while total Cr is regulated to be discharged below 0.1 mg L^{-1} according to the US EPA.[3] Chromium(III) is considered nontoxic and its mobility is lower than that of Cr(VI).[54–56]

Several methods have been reported for Cr(VI) removal from water, including adsorption, membrane filtration, ion exchange, electrolysis and chemical and photochemical reductions.[53,57] The fact that Cr(III) is less toxic and almost insoluble at neutral pH opens up an alternative way to remove Cr(VI) from polluted waters through reduction and precipitation. The conventional treatment for reductive Cr(VI) removal involves the use of sodium thiosulfate, ferrous sulfate, sodium metabisulfite, sulfur dioxide or other chemicals, with the subsequent economical costs and generation of residues.[57] A review on conventional and advanced processes for Cr(VI) removal has been published recently.[58]

Chromium(VI) reduction is enhanced in acid solutions, and the net reaction can be written as:

$$2Cr_2O_7^{2-} + 16H^+ \rightarrow 4Cr^{3+} + 8H_2O + 3O_2\ (\Delta G^\circ_{298} = -115.8\ \text{kJ}) \quad (2.19)$$

Although the process is thermodynamically feasible, $Cr_2O_7^{2-}$ is stable at room temperature in acidic solutions due to the high overpotential of the conjugate reaction of water oxidation. Therefore, strategies to improve the reduction reaction have to be implemented.

Heterogeneous photocatalysis has been proposed as a convenient low-cost and efficient technology for Cr(VI) reduction, and numerous studies have been published employing pure and modified TiO_2 (Fe/TiO_2, Pt/TiO_2), and other semiconductors (ZnO, CdS, ZnS, WO_3). Thermodynamically, TiO_2 e^-_{CB}s have the appropriate potential to directly reduce Cr(VI), Cr(V) and Cr(IV), according to the redox potentials for chromium species shown in Figure 2.2, but further reduction is not allowed.[59]

According to several pieces of experimental evidence, the HP reductive mechanism of Cr(VI) to Cr(III) has been proposed to occur through three mono-electronic steps:[37–39]

$$\text{Cr(VI)/(V)/(IV)} + e^-_{CB} \rightarrow \text{Cr(V)/(IV)/(III)} \tag{2.20}$$

In the absence of organic electron donors, water acts as the hole scavenger, according to Reaction (2.4); however, detrimental reoxidation of reduced Cr species by h^+_{VB} or $HO^\bullet$ is possible:

$$\text{Cr(V)/(IV)/(III)} + h^+_{VB}\ (HO^\bullet) \rightarrow \text{Cr(VI)/(V)/(IV)} \tag{2.21}$$

Chromium(VI) reduction can be accelerated by the addition of organic compounds that act as h^+_{VB} or $HO^\bullet$ scavengers, either by reducing the probability of recombination or through the indirect reduction driven by radicals formed by $h^+_{VB}/HO^\bullet$ attack (Reaction (2.15)). In the latter case, the so-formed radicals, at high concentrations, hinder Reaction (2.21) and contribute to the reduction of chromium species through mechanism (b′) indicated in Figure 2.1.

The direct photocatalytic mechanism was confirmed by TRMC (time-resolved microwave conductivity) studies, recently published by our group.[60,61] These are the first studies performed with this technique on a HP reaction of a metal system. The TRMC technique offers a relevant and easy way to analyse

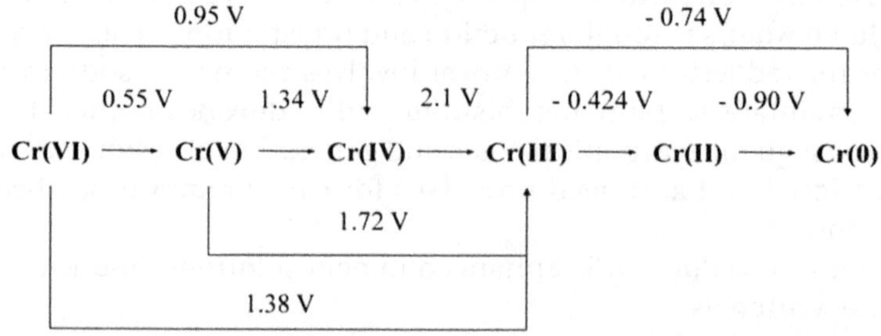

Figure 2.2 Latimer diagram connecting the different Cr species at acid pH.[59]

the behaviour of photogenerated charge carriers at very short times of reaction, and the resultant TRMC parameters can be related to the photocatalytic activity of the particular reaction and photocatalyst. A strong decrease of the TRMC signal of the electrons generated in TiO_2 in the presence of Cr(VI) was seen in all experiments with different commercial TiO_2 samples,[60] in agreement with the fact that Cr(VI) HP reduction proceeds by e^-_{CB} attack. Additionally, it was proved that the deposition of Cr(III) species over the photocatalyst inhibits the Cr(VI) decay in the absence of electron donor. However, when the reaction was conducted in the presence of EDTA, no inhibition occurred, and it was suggested that the organic compound avoids the poisoning of the photocatalyst due to formation of soluble Cr(III)–EDTA complexes.[61]

On the other hand, it has been concluded that molecular oxygen, which could compete with chromium species for e^-_{CB} through Reaction (2.5), does not affect Cr(VI) HP reduction, at least at acid pH, neither in the absence of organic donors nor in the presence of EDTA, oxalic or citric acid.[37–41] Such behaviour of this system is unique. Although it was (and is still) the object of controversies in the literature,[31–35,62–66] it can be ascribed to a thermodynamically preferential one-electron Cr(VI) reduction by e^-_{CB} over O_2 reduction ($E^0 = -0.33$ V for $O_2/O_2^{\bullet -}$)[67] and by a strong association between Cr(VI) and TiO_2 due to the formation of a charge transfer complex, which causes the fast capture of e^-_{CB} by adsorbed Cr(VI). This charge transfer complex has been identified by an absorption band at 380 nm.[68] The Cr(VI) HP independence from oxygen has been supported by experiments on platinized photocatalysts, as the same temporal evolution of Cr(VI) concentration either in the presence or in the absence of O_2 has been obtained using Pt/TiO_2 or bare TiO_2.[41,69] As Pt decreases the overpotential for electron transfer from the CB to O_2, the presence of O_2 when using Pt/TiO_2 should have affected the Cr(VI) decay in two opposite ways: (a) either enhancing the reaction rate, with $O_2^{\bullet -}$ being a mediator for Cr(VI) reduction, or (b) decreasing the rate by competition with e^-_{CB} (Reaction (2.5)).

Many examples of the HP Cr(VI) system have already been described in previous reviews,[31–35] including technological applications and use of visible light. Most of the papers cited there use the HP Cr(VI) reduction reaction as a model system for testing new photocatalysts. The following paragraphs presents an update that includes the latest literature, not mentioned in the previous reviews. However, this revision is not exhaustive as 56 papers on the subject have been published since 2013, among them ref. 70–73.

Other papers refer to different photocatalysts selected to improve Cr(VI) HP under UV irradiation, such as, for example, mesoporous anatase TiO_2,[74] TiO_2 nanotubes,[75] and N,[76–79] N–I,[66] and N–F[77] modified TiO_2. Fluorinated TiO_2 nanosheets were also tested,[80] as well as TiO_2 prepared by $TiCl_4$ hydrolysis and immobilized on serpentinite[81] and diatomite.[82] Titanate nanotubes (TNT) were mixed with P25 and successfully used for the one-step simultaneous removal of Cr(VI) and Cr(III).[83] Other useful modified TiO_2 samples for use under UV, UV-visible or sunlight irradiation were ZnO–TiO_2–reduced graphene oxide composites,[84] graphene–CNTs–TiO_2 composites,[85] micro/nanostructured nanoparticulate TiO_2 microspheres prepared by a microwave-assisted hydrothermal method,[86] a $CuAl_2O_4/TiO_2$ photocatalyst,[87] a silicate supported TiO_2–ZnO nanocomposite,[88] a ZnO/TiO_2 sample,[89] a self-rotating reactor with TiO_2 nanotubes supported on a Ti mesh,[90] carbon nanotubes (CNTs) coupled with TiO_2,[91] TiO_2 nanoparticles used in a titania float,[92] TiO_2 nanoparticles deposited over polysulfone membranes,[93] TiO_2 supported on alginate beads,[94] Mg and Ag co-impregnated TiO_2 nanoparticles,[95] sea-urchin-like rutile TiO_2 with ultrathin nanorods,[96] *etc.*

Some studies found that the efficiency of Cr(VI) reduction by photoelectrocatalysis (PEC) over TiO_2 under UV light was markedly higher than that of the normal HP reduction.[97–100] Song *et al.*[101] indicated that PEC can only take place when the Fermi potential of the cathode is more negative than the chemical potential of Cr(VI), independently of the applied voltage of the direct current. Hsu *et al.*[64,102] combined PEC with a cationic exchange membrane to enhance Cr(VI) HP reduction efficiency by preventing Cr(III) reoxidation.

Chromium(VI) and Ni(II) (another very toxic metal ion) simultaneous photocatalytic reduction with P25 was studied by Shirzad-Siboni *et al.*[103] The authors found that pH 3 was needed for preferential Cr(VI) reduction, pH 11 for Ni(II) reduction and pH 7 was the optimal pH value for the simultaneous removal of both pollutants.

Caprioli Machado *et al.*[104] performed tests of Cr(VI) reduction in wastewater from the electroplating industry, adding ethanol to improve the efficiency, and obtained almost complete Cr(VI) removal under the optimal conditions.

Most recent papers on Cr(VI) HP reduction refer to findings using visible light. As said before, dyes can sensitize HP reactions to extend the activity to the visible range according to Reactions (2.17) and (2.18). Examples have been already given in ref. 33–35 and a scheme of the reactions is displayed in Figure 2.3, in accordance with one-electron consecutive steps proposed before by our group.[51]

As examples of this sensitization under visible light, we can cite Cr(VI) reduction in the presence of cationic, zwitterionic and anionic dyes,[105] acid orange 7 (AO7),[72] and N719 dye.[65] In this last case, a TiO_2 thin film composed of a dye-sensitized zone and a cathodic catalysis zone was prepared. Charge separation was accomplished with electron transfer to the catalysis zone, where Cr(VI) was reduced, while H_2O oxidation took place at the Pt anode. The role of Cr(VI) as a visible light sensitizer was proposed in the degradation of simazine in the presence of TiO_2 and H_2O_2,[106] where no Cr(VI) reduction was observed. The mechanism indicated by the authors is, however, not clear.

Park *et al.*[107] used Cr(VI) reduction by visible-light HP as a probe for interparticle charge transfer. The authors found that bare TiO_2 nanoparticles facilitate the charge separation and the electron transfer from a dye-modified TiO_2 to Pt-modified TiO_2 by purposely decoupling the light absorbing part (dye/TiO_2) from the active catalytic centre (Pt/TiO_2). The phenomenon could

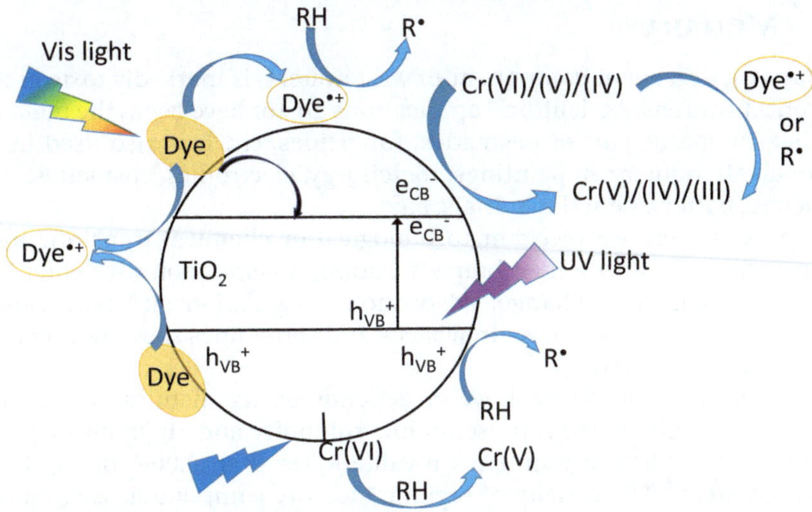

Figure 2.3 Scheme of the photocatalytic reduction of Cr(VI) in the presence of dyes under UV and vis irradiation. (Adapted from ref. 51 and 68.)

be rationalized in terms of an interparticle Fermi level gradient, which drives the charge separation.

Modification of TiO_2 with non-metals and metals was also used as a strategy to extend the active range of light for the Cr(VI) HP reduction. For example, the N-modified TiO_2 samples cited before[76] and La and Nd doped TiO_2 samples[108] showed a higher efficiency under visible irradiation ($\lambda > 420$ nm) compared with unmodified P25, which shows scarce photo-activity at those wavelengths.

Coupling of TiO_2 with smaller band-gap semiconductors was also used as another alternative to enhance the visible-light photocatalytic activity for Cr(VI) reduction. Examples are: CdS quantum dots deposited over TiO_2 films,[109] TiO_2 nanofibres/In_2S_3 nanosheets composites,[110] a graphitic carbon nitride TiO_2 (g-C_3N_4–TiO_2) composite,[111] TiO_2 nanoparticles over graphene oxide,[112] Bi_2O_3/TiO_2,[113] mesoporous yolk–shell SnS_2–TiO_2,[114] $SnIn_4S_8/TiO_2$ nanocomposites,[115] noble-metal nanoparticles (Ag, Pt, and Pd) photodeposited over TiO_2,[116] Au/TiO_2 and Au/TiO_2–Pt samples,[117] N and Pt–N modified TiO_2 samples already mentioned,[79] poly(fluorene-*co*-thiophene) coupled with Al_2O_3, ZnO and TiO_2,[118] and a triple-shelled Ag@Fe_3O_4@SiO_2@TiO_2,[119] among others.

Another sensitization procedure to extend the use of TiO_2 to longer wavelengths is the LMCT (ligand to metal charge transfer) surface complexation of TiO_2 with a suitable adsorbate, where electrons are photoexcited directly from the HOMO level of the adsorbate to the TiO_2 CB. In this line, a very recent work reports the formation of a LMCT complex between the photocatalyst and glucose, which absorbs visible light significantly up to 600 nm and exhibited good visible light activity for the photoconversion of Cr(VI) at $\lambda > 420$ nm.[120]

2.4 Mercury

Mercury(II), a frequent component of wastewaters, is markedly toxic at very low concentrations. Agricultural applications are (or have been) the main use of this element, as part of pesticides, fungicides, *etc.* It is also used in the chloro-alkali industry, in paintings, metallurgy, electronics, cosmetics, thermometers, battery, dental materials, *etc.*[8]

Mercury species are resistant to biological or chemical transformation. Common treatments are membrane filtration, precipitation with chemicals (*e.g.* as sulfide), ion exchange, adsorption, coagulation and reduction.[121] However, these methods have drawbacks and sometimes they are expensive and/or not very efficient.

The presence of mercury in water depends on its chemical state: while Hg(II) is very soluble, Hg(I) presents low solubility and Hg(0) and HgS are insoluble. The WHO recommends a value lower than 0.006 mg Hg L^{-1} in drinking water.[2,8] The toxicity of organic mercury compounds, *e.g.*, methyl- or phenylmercury, is considerably higher than that of the inorganic species. For example, the massive case of poisoning in Japan – the Minamata Bay incident – was attributed to industrial discharge of organomercurials, and declining bird populations in Sweden was blamed on the use of phenyl- and methylmercurial pesticides as seed dressings.[122] Phenylmercury chloride (PMC) and phenylmercury acetate (PMA) were commonly used as pesticides; in particular, PMA was widely used in Argentina until 1971, since when its use has been forbidden.

Mercury can also be found in flue gases as elemental mercury vapour, oxidized mercury species and particulate-bound mercury.

HP can be an alternative for treatment of Hg(II) in water, according to the Latimer diagram indicated in Figure 2.4.[123] It has been studied with the use of UV, visible and solar irradiation.[22,124,125] Several citations have been reported in previous reviews[31–34] and the main conclusions with the most important references will be described in what follows.

Mono-electronic steps have been proposed by our group for the mechanism for Hg(II) HP direct reduction according to Reaction (2.11), forming Hg(I) and then Hg(0), particularly for inorganic salts like $HgCl_2$, whose

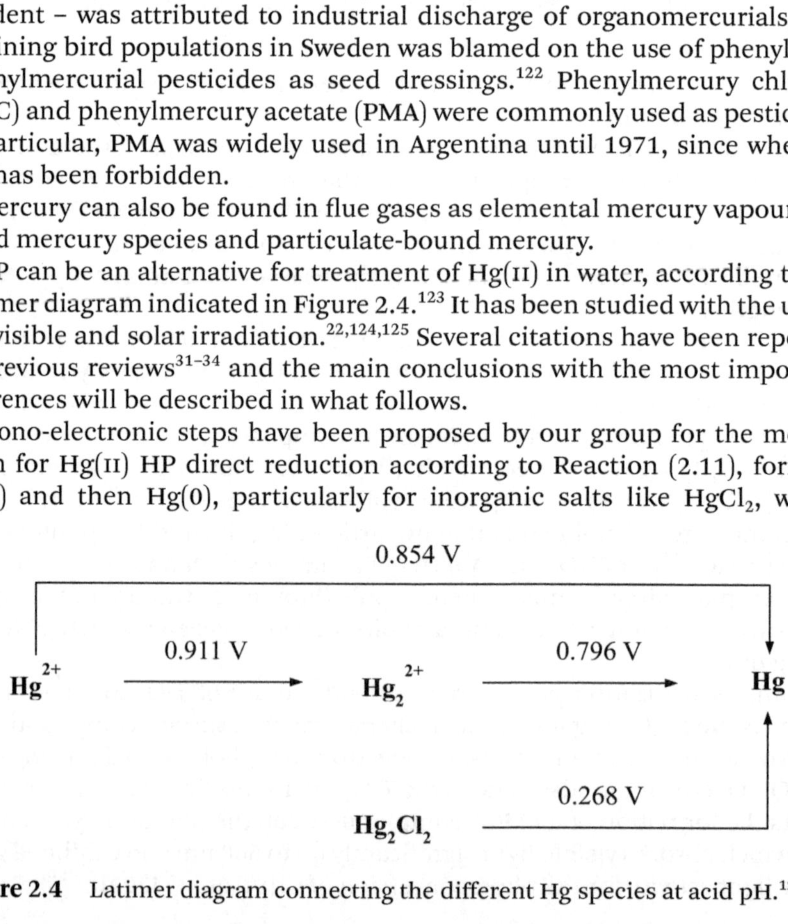

Figure 2.4 Latimer diagram connecting the different Hg species at acid pH.[123]

photocatalysis led to calomel (Hg_2Cl_2) and Hg(0) formation.[121] Regarding Hg(I) species, Fujita *et al.*[126] reported radiolytic reduction of Hg(II) in water by hydrated electrons or H radicals and distinguished between stable and unstable mercurous ions by spectroscopic studies at 220–300 nm. Interestingly, the authors indicate the involvement of the transient monovalent mercury (Hg^+) in the production of the stable mercurous ion (Hg_2^{2+}) by bimolecular reaction of zero-valent mercury with mercuric ion after Hg^+ disproportionation. A value for the reduction of $HgCl_2$ to HgCl of −0.47 V (at $[Cl^-]$ = 0.05 M) was measured later.[127] This would indicate that the direct HP Hg(II) reduction by e_{CB}^- can be favourable for $HgCl_2$ at pH 3; the formation of the insoluble calomel would help the process. From the value for the chloride species, the one-electron reduction potential of uncomplexed Hg^{2+} to Hg^+ was estimated to be lower than −2.0 V, which would indicate a much more difficult HP transformation in the absence of chloride. In acidic conditions, the reaction is slow and Hg(0) reoxidation occurs, as found many years ago by Serpone *et al.*[22] and later by us.[121] Higher conversions can be obtained at basic pH due to the shift in the potential of the e_{CB}^- with increasing pH.[21–23] Actually, our group[121] and that of López-Muñoz *et al.*[128] found an efficient removal of Hg(II) at pH 10 and 11 starting from $HgCl_2$ in water. However, much should be still investigated about the direct HP reduction of Hg(II) salts in the absence of organic compounds, taking into account the different hydrolytic and complex forms of mercury and the pH of the medium, which may vary the redox potentials and reactivity.

As expected, addition of sacrificial organic molecules (methanol, EDTA, formic or oxalic acid) significantly improves aqueous Hg(II) removal, especially under acidic conditions.[128,129] Electron donors can react with $h_{VB}^+/HO^\bullet$, preventing reoxidation of Hg(0) and Hg(I) through Reaction (2.12), but also they can transform Hg(II) into Hg(I) more easily by the indirect mechanism (Reactions (2.15) and (2.16)).

The presence of oxygen inhibits mercury reduction in all cases,[121,130] due to the competition with O_2 according to eqn (2.5).

Concerning the very toxic organomercurials, metallic mercury has been reported as the product of the HP treatment of methylmercury in the presence of methanol.[22] Miranda *et al.*[131] found a faster CH_3HgCl reduction under N_2 than with bubbling O_2, obtaining different products from the organic moiety for both conditions. Our group studied the HP of phenylmercury salts obtaining mercury removal at a large extent together with the mineralization of the organic portion.[132] In the case of phenylmercury acetate, a solid Hg(0) deposit was formed, whereas Hg_2Cl_2 was formed when starting from phenylmercury chloride. These findings support the two successive one-electron transfer reactions hypothesis, as in the case of inorganic salts. It has been proposed that, in this case, the organic moiety acts itself as electron donor and reduces Hg(II) indirectly.

An important application of HP reduction of mercury has been in the field of analytical chemistry. In this respect we described in one of our previous reviews a method for mercury determination in environmental and biological samples.[34]

A not previously reviewed subject is the use of photocatalytic systems for the treatment of gaseous mercury forms; an important description will be made in what follows. It was indicated that hydroxyl radicals oxidize Hg(0) and generate HgO on the surface of TiO_2, which can be retained on the semiconductor surface, enhancing, in parallel, Hg(0) adsorption, which was otherwise insignificant.[133] An early paper describes the use of TiO_2 and UV irradiation, with elementary mercury first adsorbed on the catalyst surface and then photo-oxidized.[134] Three different sorbent materials (Ti-, Si- and Ca-based) were compared for their mercury (Hg(0)) capture efficiencies in an entrained flow reactor. While titania particles under UV irradiation were the most effective (>98%), CaO particles had a capture efficiency of 33% (with or without irradiation), and SiO_2 was completely ineffective.[135] Lee *et al.*[136] compared the efficiency of three TiO_2 photocatalysts, pure anatase and rutile samples and P25, in the removal of Hg(0) from a simulated exhaust gas under UV and visible light, and obtained more than 85% Hg removal in all cases. XRD results showed that mercury was chemisorbed as HgO, in agreement with the proposed mechanism.

Some authors,[137–141] using different pure and modified TiO_2 materials, studied the role of moisture on Hg(0) capture (adsorption and/or photocatalytic oxidation). The results of Li and Wu[138] indicated that Hg(0) capture on a SiO_2–TiO_2 nanocomposite under humid conditions with UV irradiation is controlled by four mechanisms: adsorption, photocatalytic oxidation, desorption and photocatalytic reduction. High water vapour levels lead to competition for available adsorption sites and participate in the photocatalytic reduction of HgO to Hg(0), promoting re-emission. Later,[139] the same authors were able to describe the kinetic profiles for Hg(0) HP oxidation using a Langmuir–Hinshelwood model.

Chen *et al.*[142] found that N modification of P25, besides allowing the use of visible light irradiation, inhibited the re-emission of adsorbed Hg(0) caused by H_2O competition, possibly by strong bonding between Hg, N and O groups. Other authors[143] have used Cu-doped TiO_2, and reported appreciable Hg(0) removal under visible-light irradiation, also avoiding Hg(0) re-emission.

Recently, pilot-scale TiO_2 photocatalytic systems have been tested at real combustion facilities to control Hg(0) emissions; there, TiO_2 coated glass beads were used in three tubular photocatalytic reactors in parallel or in series and illuminated with 3–12 fluorescent UV lamps. A removal efficiency of at least 90% was obtained when the conditions were varied, and it has been shown that acid components of the flue gas, such as NO and H_2O, exhibited significant negative effects on the Hg(0) removal efficiency.[133]

2.5 Lead

Lead pollution is mainly anthropogenic, coming from industrial effluents. Lead effects on the central nervous system can be serious, and a guideline value of 0.01 mg L^{-1} in drinking water is maintained but designated as provisional by the WHO.[4]

The treatment of lead by heterogeneous photocatalysis has received scarce attention. In our previous reviews,[31–34] we cited several early papers on the subject, but the information continues to be rather scant. In what follows, some of the most important results will be cited again to review the state of the art.

Lead species can be oxidized or reduced in HP reactions, according to the Latimer diagram indicated in Figure 2.5.[144]

The oxidative HP of Pb(II) forms Pb(IV), involving two consecutive one-electron steps through h^+_{VB} or $HO^\bullet$ attack (mechanisms (c) and (c′), Figure 2.1), although redox potentials of Pb(IV)/Pb(III) and Pb(III)/Pb(II) couples are not reported:

$$\mathrm{Pb(II)} \xrightarrow{h^+_{VB}/HO^\bullet} \mathrm{Pb(III)} \xrightarrow{h^+_{VB}/HO^\bullet/ROS} \mathrm{Pb(IV)} \quad (2.22)$$

Disproportionation of Pb(III) could also take place. The participation of Pb(III) species in the electrochemical oxidation of Pb(II) in water has been indicated.[145] Under oxic conditions, photocatalytic oxidation is poor with bare TiO_2, but almost complete removal can be obtained with Pt–TiO_2, with formation of a brown PbO_2 deposit on the photocatalyst.[146] HP oxidation of Pb(II) can be promoted by the addition of ozone due to the photochemical formation of H_2O_2 and other ROS, including the ozonide radical ($HO_3^\bullet/O_3^{\bullet-}$);[147–151] ozonide reacts with Pb(II), with the subsequent formation of $PbO_{1.37}$, oxide found together with PbO_2 by XRD as deposits on the photocatalyst.[146]

In the case of Pb(II) photocatalytic reduction, the direct reductive route is not possible as the one-electron reduction potential of the Pb(II)/Pb(I) couple is very negative ($E^0 = -1.0$ V),[152] supported by experimental results that show scarce conversion even when employing Pt/TiO_2 under N_2.[146,153] On the other hand, direct reduction of Pb(II) to Pb(0) by a bi-electronic process has been reported under laser irradiation, where, due to the high photonic frequency, the authors propose that accumulation of electrons in the semiconductor particles may allow multi-electronic injection.[151,154] Nevertheless, the indirect route in the presence of electron donors (Reactions (2.15) and (2.16)) allows the photocatalytic reduction of Pb(II). In this sense, the addition of organic compounds promotes reductive HP of lead under N_2 and over pure TiO_2[155–157] and Ag/TiO_2.[158] Murruni *et al.*,[156] in experiments of Pb(II) photocatalytic removal under nitrogen over P25 in the presence of 2-PrOH and formic acid

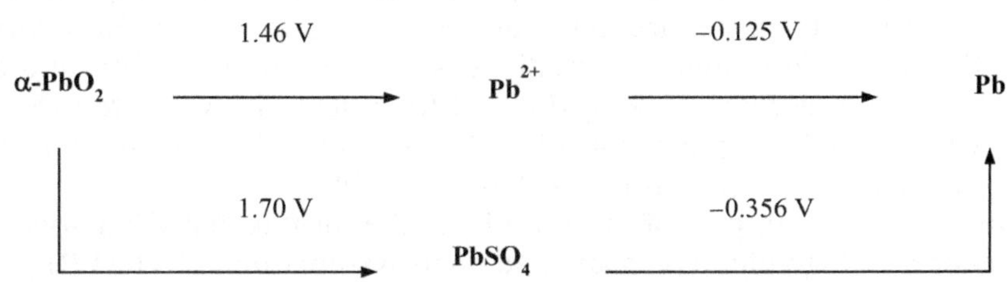

Figure 2.5 Latimer diagram connecting the different Pb species at acid pH.[155]

(HCOOH) as sacrificial donors, proposed two successive mono-electronic steps for the production of colloidal Pb(0).

Li *et al.*[158] indicated that the photocatalytic Pb(II) reduction over TiO_2 and Ag/TiO_2 in the presence of HCOOH was enhanced at pH 3.5 compared with more acid pH values, due to a stronger adsorption of formate at pH 3.5 on the photocatalyst surface, favouring the formation of $CO_2^{\bullet-}$. Prior to this paper, other authors,[159] working with nanocrystalline TiO_2 coatings and adding organic compounds, found the following order of efficiency for the photocatalytic Pb(II) reduction: HCOOH > $H_2C_2O_4$ > CH_3OH > C_2H_5OH.

The only recent papers found on photocatalytic reduction of Pb(II) deal with the use of TiO_2 nanotubes loaded with Cu^{2+} ions.[160–162] Sreekantan *et al.*[160] showed that the incorporation of an optimum Cu load (1.3%) using wet impregnation leads to a higher Pb(II) removal, and the best condition was found in alkaline solutions. Another study of the group[161] showed that a post-annealing treatment of the TiO_2/Cu nanotubes succeeded in diffusing Cu into the lattice of TiO_2; these treated nanotubes were very effective for methyl orange degradation and Pb(II) reduction. The authors indicated that the appropriate amount of Cu into the TiO_2 lattice markedly enhanced the charge-carrier separation and transport.[160–162]

Although the Pb photocatalytic system has been extensively studied, just a few works have been published in recent years, and much effort should be made to transfer the knowledge gained, making advances in the search for technological solutions.

2.6 Uranium

Uranium is a radioactive heavy metal whose prevalent isotopes in the environment are the naturally occurring ^{238}U, ^{235}U and ^{234}U. The first isotope is the most abundant (>99%).[163] In nature, uranium can be found in granites and various other minerals, and it is present in the environment as a result of leaching from natural deposits, release in mill tailings, emissions from the nuclear industry, combustion of coal and other fuels and use of uranium-containing phosphate fertilizers.[5,163,164]

In aqueous media, uranium exists mostly in its hexavalent form, with the uranyl ion (UO_2^{2+}) as the dominant species below pH 3.5, while at higher pH and/or in the presence of different anions other U(VI) complexes can be formed.[165] The tetravalent state (U(IV)), the other stable U species, is far less soluble.[163] Other U species are the +3 and +5 valence states, but these forms are rather unstable in aqueous media, easily oxidized by O_2[166,167] and are unstable even under anoxic conditions: U(III) reacts with water to produce H_2,[166] while U(V) disproportionates into U(VI) and U(IV).[167] The existence of uranium(II) in aqueous solution is still unclear.[166,168]

The main health impact of uranium is its chemical toxicity,[169] associated with nephritis, high blood pressure and bone dysfunction.[5,170] The WHO has established a guideline value of 0.015 mg L^{-1} in drinking water.[5]

Several methods are available for U(VI) removal from water, such as ion exchange, reverse osmosis and ultrafiltration, adsorption, bio-reduction and zero-valent iron;[164] references to these methods can be found in our previous reviews.[32–34]

HP can then be considered a promising alternative for U removal, by promoting U(VI) reduction to U(IV), followed by deposition over the photocatalyst.[171] However, the literature regarding U(VI) HP treatment is scarce, with only nine published works. In our previous reviews,[31–34] a few cases have been reported; due to their importance, they will be briefly described again. In recent years there have been only two scientific publications on the uranium HP system, both from our group.[42,43]

According to the reduction potentials shown in Figure 2.6,[166,172,173] TiO_2 e^-_{CB} can reduce U(VI) up to U(IV);[171] however, HP reduction may continue by the indirect pathway up to U(III) if reducing radicals are generated in the system (Reactions (2.15) and (2.16)). Reported values for the U(III)/U(II) redox potential are −1.65 V[173] and −2.53 V.[166] Despite the discrepancy between these values, HP reduction of U(III) to U(II) would be very difficult, even with reducing radicals.

The first report on U(VI) removal by HP was made by Amadelli *et al.*,[165] who studied the effect of different hole scavengers (2-PrOH, acetate or formate) at near neutral pH using P25, either in suspension or supported on electrodes. They found that formate was the most efficient electron donor, followed by acetate. U_3O_8 was detected as the final reduction product. Chen *et al.*[171] studied HP removal with pure and platinized P25 with EDTA as electron donor. The authors found that the absence of oxygen was necessary to efficiently reduce U(VI), that platinization had little effect on U(VI) removal, and that U(IV) formed over the TiO_2 could be easily reoxidized by exposure to air. Evans *et al.*[174] also reported an easy reoxidation by air exposure of U(IV) formed in a photocatalytic system with P25 at pH 2, using methanol as donor. No U(VI) removal took place in the absence of donor. Eliet *et al.*[175] studied the U(VI) photocatalytic reduction in TiO_2 aqueous suspensions by time-resolved laser induced fluorescence (TRLIF), proposing a reaction mechanism in the absence of electron donors where only adsorbed U(VI) species can be reduced to U(V), followed by disproportionation to U(VI) and U(IV) and subsequent U(IV) reoxidation by $h^+_{VB}/HO^•$. In a further study, Selli *et al.*[176] observed an

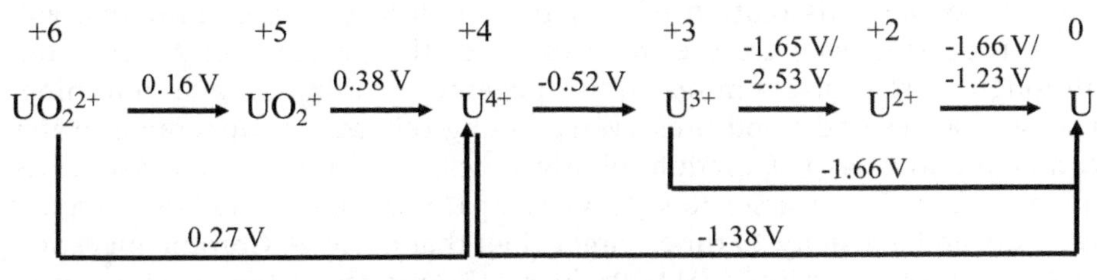

Figure 2.6 Latimer diagram connecting the different U species.[166,167]

enhancement of this photoreduction by the presence of humic acid (HA). Boxall *et al.*[177] explored the use of HP in actinide valence state control as a possible application of waste minimization in nuclear fuel processing, with separation of Np, Pu and U. Bonato *et al.*[178] studied the thermal and photocatalytic U(VI) reduction on the surface of TiO_2 anatase nanotubes; XPS results of the solids after the reaction showed the presence of adsorbed U(VI) and U(IV) species, with a higher fraction of U(IV) in the samples submitted to light irradiation.

A recent contribution of Odoh *et al.*[179] studied the photocatalytic reduction of U(VI) from a theoretical point of view, using a model rutile surface for U(VI) adsorption; proposing a $Ti_{38}O_{76}$–$UO_2(H_2O)_3^{2+}$ cluster, they calculated that the empty U-5f orbitals are between 0.0 and 1.0 eV below the Ti-3d (trapped e^-_{CB}) virtual orbitals, in good correlation with the experimental values observed between the U(VI)/U(V) couple in solution and e^-_{CB} (0.48 V), and proposed that the charge transfer mechanism for U(VI) on TiO_2 surfaces is O-2p → Ti-3d → U-5f, *i.e.* the incident light must be able to excite the semiconductor.

A study from our group[42] analysed the effect of different counter-ions (perchlorate, nitrate and acetate) on the U(VI) HP reduction at pH 3 with P25, using 2-PrOH as donor. The removal efficiency was almost identical for the nitrate and perchlorate uranyl salts, but acetate was detrimental because of the formation of a complex with U(IV), which keeps this species in solution. The spectroscopic detection of intermediate U(V) species in the uranyl HP system was reported for the first time. Another recent work[43] uses HCOOH as donor, and it is the first report of U(III) formation in HP systems. The formation of U(V) was also observed again. As U(III) was not formed with 2-PrOH as donor, it was concluded that the slow reaction between U species and organic radicals would make Reaction (2.16) only possible when $CO_2^{\bullet-}$ is formed, while, for the radical formed from 2-PrOH, electron injection into the TiO_2 CB would be the preferential pathway.

The use of U(VI)-modified composites to oxidize organic compounds using HP has been an active field in the last decade, registering many publications, including a recent review.[180] However, this subject will be not treated here.

2.7 Arsenic

Natural sources (dissolution of minerals in surface or groundwaters, volcanic processes, *etc.*) are the main cause of the presence of As in water, although there is also a minor anthropogenic contribution due to mining, use of biocides and wood preservers, among others.[181,182] Arsenic in water can be found as As(III), predominantly as arsenite (H_3AsO_3), and As(V), frequently occurring as arsenates ($H_2AsO_4^-$ or $HAsO_4^{2-}$). Arsenite has a higher mobility and a toxicity 20 times larger than that of As(V). Chronic ingestion of As produces arsenicosis (HACRE in South America), which causes severe skin lesions, ending possibly in cancer.[6] The maximum concentration of As in water recommended by the WHO is 10 μg L^{-1}.[6]

Several conventional water treatments have been used to remove As, such as oxidation/coagulation/adsorption processes on iron or aluminium hydroxides, ion exchange, activated alumina, lime softening, electrodialysis and reverse osmosis.[183–185] These applied technologies have different drawbacks and limitations, such as the need for additional removal processes, high amount of coagulants, regeneration of adsorbents and high operation and capital costs.[186] More efficient emerging technologies should be investigated to offer low-cost solutions to the arsenic problem, especially for low-income populations.[181,183,184]

Both oxidation and reduction processes can transform arsenic species by heterogeneous photocatalysis, according to the values of the redox potentials indicated in the Latimer diagram of Figure 2.7.[187]

Conventional ion exchange and adsorption techniques remove As(V) rather easily, but As(III) removal is more difficult because H_3AsO_3 is neutral at pH < 9. A first oxidation step from As(III) to As(V) is often needed; oxidative TiO_2 photocatalysis is a good alternative for this purpose, and it has been extensively used. Many examples were included in our previous reviews.[33–35] The mechanism of As(III) oxidation has been explained as a succession of mono-electronic steps involving ROS with formation of As(IV)[188–190] (Figure 2.8).

The reduction potential of the As(IV)/As(III) couple has been reported to be around +2.4 V,[191] allowing the reactions depicted in Figure 2.8. The determination of the major oxidant in the photocatalytic As(III) oxidation has been a matter of discussion and has been extensively described before;[33–35] some authors claimed that $O_2^{\bullet-}$ is the major oxidant, while others argued that the

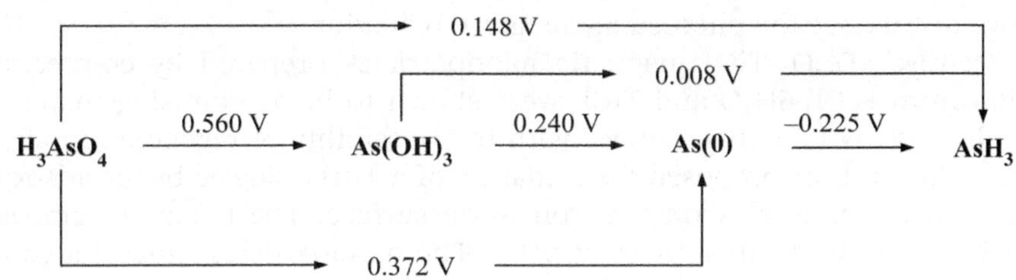

Figure 2.7 Latimer diagram connecting different As species.[187]

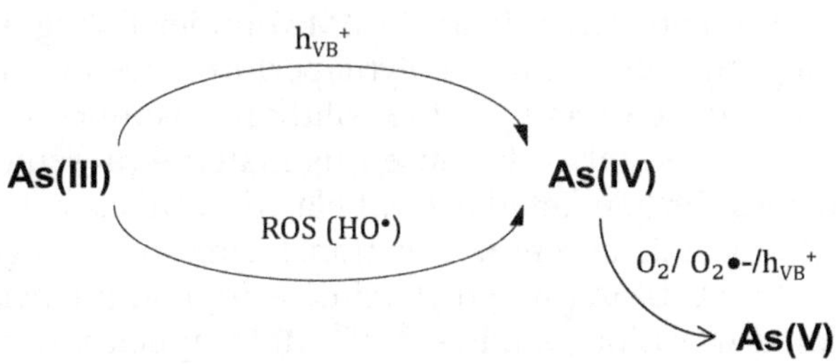

Figure 2.8 Mechanism of As(III) photocatalytic oxidation.[188–190]

major role is played by $HO^{\bullet}$ or h^{+}_{VB}. It is our opinion that superoxide radical cannot carry out oxidation of As(III) to As(IV) because the reduction potential of the $O_2^{\bullet-}/H_2O_2$ couple is around 1.7 V.[48] Regardless of the oxidant, the photocatalytic oxidation of As(III) is very efficient, but the real problem is, however, how to remove dissolved As(V) from the solution.

Improvement of the HP technology by combination with adsorbents has been proposed[192] to achieve higher photocatalytic abilities and adsorption efficiencies. Studies on the effect of TiO_2 crystalline size on As(III) and As(V) adsorption and photocatalytic oxidation[193] showed that the adsorption capacity was linearly dependent on the BET surface area, and that the photo-oxidation rate was similar for nanoparticles 6.6 and 14.8 nm of diameter but decreased when using particles of 30.1 nm. However, the potential release of arsenic from the used TiO_2 to groundwater in reducing environments should be taken into account.

The influence of coexisting solutes (silicate, phosphate, fluoride, carbonate oxyanions and HA) on HP of arsenical systems should not be overlooked, as these species can compete for the available adsorption sites of TiO_2 (see ref. 192 and references therein). HA and bicarbonate seem to be the most relevant.[192] In particular, HA has been reported as increasing[194] or decreasing[195] the oxidation rate of As(III). Lee and Choi[194] also found that the addition of Fe^{3+} enhanced the photocatalytic oxidation of As(III), acting probably as e^{-}_{CB} acceptor.

On the other hand, hybrid systems like photocatalysis with TiO_2 combined with nanoscale zero-valent iron[196] showed enhanced As(III) removal, with a five-fold reduction of the required TiO_2 mass compared with that of TiO_2 alone. Graphene oxide hybridized with TiO_2 was proposed as a low-cost alternative to enhance the photocatalytic As(III) oxidation.[197]

Recently, γ-Fe_2O_3–TiO_2 magnetic nanoparticles prepared by co-precipitation from $FeCl_3{\cdot}6H_2O$ and $TiCl_4$ were shown to be a promising material for photocatalytic As(III) removal with the possibility of regeneration after use.[198] The authors proposed the oxidation of As(III) followed by the adsorption of As(V) preferably on the iron oxide surface. The maximal removal capability for As(III) took place at pH 7. The nanoparticles showed a good stability after several reuse cycles, and could be efficiently separated by a magnetic method. In another recent paper,[199] the use of a ZnO photocatalyst synthesized by the sol–gel technique was compared with P25 for the removal of As(III) and monomethylarsonic acid (MMA) under UV light. Even when P25 showed a slightly better photocatalytic performance for the removal of both species, the release of As(V) to the solution was much higher for P25 than for the ZnO photocatalyst, because this material presented a stronger capability to adsorb the generated As(V). Related to this last work, it is relevant that the photocatalytic removal or transformation of organoarsenical species, such as MMA, DMA (dimethylarsinic acid) and phenylarsonic acid, has been explored in various studies.[199–202] MMA appears as the DMA primary degradation product and the arsenic moiety is oxidized to As(V).[199]

In all cases, $HO^•$ was found to play a major role in the oxidation. Another recent paper mentions the adsorption and photocatalytic oxidation of roxarsone (3-nitro-4-hydroxyphenylarsonic acid),[203] but in this case there was no change in the arsenic oxidation state. The authors propose that arsenate was produced by cleavage of the As–C bond due to $HO^•$ attack, with concomitant *o*-nitrophenol formation.

The use of dyes to sensitize TiO_2 or the doping of the photocatalyst with nitrogen to extend the activity to the visible was described above.[34] Recently, Li and Leng[204] studied the mechanism of sensitization of TiO_2 with N719 for the photocatalytic oxidation of As(III) using PEC measurements; the authors proposed two pathways: a direct two-electron oxidation by the dye radical cation ($E^0_{(S+/S*)} = 1.10$ V *versus* $E^0_{(As(V)/As(III))} = -0.67$ V) and an indirect one-electron oxidation carried out by ROS to generate As(IV), in accordance with Figure 2.8.

On the other hand, it should be emphasized that very few works have studied the photocatalytic reduction of arsenical species. As already indicated in ref. 34 and 35 and first studied by Yang *et al.*,[205] and later by our group,[206–209] As(V) cannot be directly reduced by e^-_{CB} because the reduction potential of the As(V)/As(IV) couple is highly negative ($E^0 \approx -1.2$ V).[191] Arsenic(V) can only be reduced *via* the indirect pathway in the presence of powerful reducing radicals (Reactions (2.15) and (2.16)). In contrast, our group determined for the first time that it is possible to reduce As(III) directly by e^-_{CB} either in the presence or also in the absence of donors (like MeOH).[206–209] This suggests that the still unknown value for the As(III)/As(II) mono-electronic couple should be located below the CB level. In both direct and indirect reduction of As(V) and As(III), and after possible As(II) formation, further steps lead to the formation of As(0) and AsH_3, which were clearly identified in the system. The detailed mechanism of this reaction can be seen in the original papers[206–209] and in the previous reviews.[34,35] Obtaining As(0) after the treatment is a promising result for real scale applications. However, careful examination of the system should be carried out to remove AsH_3 after the treatment. An alternative can be to couple the treatment in the water phase with a photocatalytic system oxidizing the gas evolved in the first step, which our group is working on at present.

In a recent paper, it was proved that As(V) can be reduced in the dark by exploiting accumulated electrons stored in TiO_2 nanoparticles previously activated under UV irradiation.[210,211] The accumulated electrons were detected by EPR as Ti(III) species; it was proposed that the reaction takes place by modification of the Fermi level of the electrons of the semiconductor by the accumulated electrons.

Low-cost applications have been reviewed in ref. 33 and 34 and refer to As removal from natural waters of rural areas with the use of PET bottles impregnated with TiO_2 under UV irradiation. The system was coupled to iron addition to complete the removal, and concentrations below the WHO limits were attained.

2.8 Nitrate

The main source of nitrate in water comes from its use in agriculture, especially as a component of fertilizers, but there are additional contributions of sewage systems, manure, effluents of septic systems, *etc.* A small proportion is produced naturally.

Nitrate ingestion causes high damage to health, as methemoglobinemia or stomach cancer; babies are the population with the highest risk of acquiring the "blue baby syndrome" or cyanosis. Thus, the WHO recommends a value of 50 mg L^{-1} as the maximum allowable concentration of NO_3^- in drinking water.[7] The most used treatments for nitrate removal are catalytic reduction,[212,213] including iron-containing nanomaterials,[214–220] electrochemical reduction,[221–228] electrodialysis, distillation, conversion into nitrogen gas by biological nitrification/denitrification,[229] ion exchange or reverse osmosis (see, for example, ref. 230), *etc.* However, more economical alternatives and fundamental studies are imperative. Nitrate reduction in water can lead to several reduced forms, such as nitrite, nitrogen oxides, hydroxylamine (NH_2OH), hydrazine (N_2H_4), nitrogen (N_2) or ammonium (NH_4^+). The production of nontoxic nitrogenated forms should be preferred, such as, for example, transformation to the innocuous molecular nitrogen. Ammonia (NH_4^+ + NH_3) is not of direct importance for human health in the concentrations to be expected in drinking-water and a health-based guideline has therefore not been derived.[231] However, the presence of ammonia in treated waters should be taken into account because of indirect interference, *e.g.* in disinfection.[232] Nitrite, which is more toxic than nitrate, should not be present at the end of treatment, having a WHO guideline in drinking water of 3 mg L^{-1}.[7]

Heterogeneous photocatalysis constitutes a potential alternative for removal of nitrate from water[28] or brines produced by other technologies.[233] Figure 2.9 shows the Latimer diagram for different nitrogen species.[234]

Although several photocatalysts have been considered, titania-based materials are the most studied.[28]

If mono-electronic pathways are proposed in HP reactions, direct nitrate photocatalytic reduction by TiO_2 e^-_{CB} is not possible because the redox potential for nitrate to the nitrate radical anion ($E^0_{(NO_3^-/NO_3^{2-})} = -0.89$ V)[235,236] hinders e^-_{CB} attack. In close relationship with the primary photocatalytic steps, Gao *et al.*[237] studied the reaction of excess electrons in TiO_2 nanocrystallites produced by radiolysis with NO_3^-, observing clearly the decay of the TiO_2 electrons. This paper was the basis of a recent stopped-flow study on kinetic and mechanistic investigations of transfer reactions induced by stored electrons

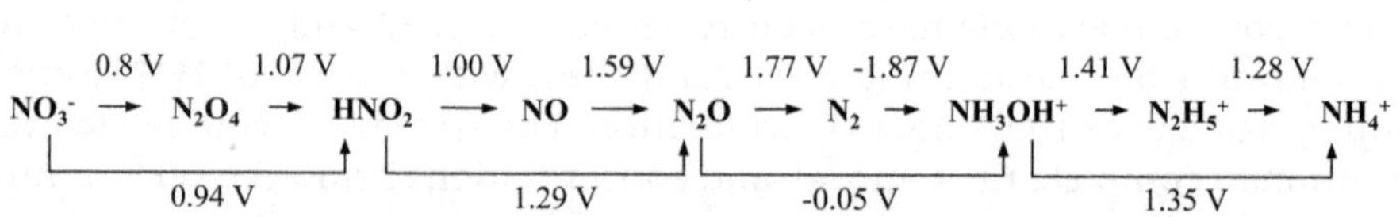

Figure 2.9 Latimer diagram connecting different nitrogenated species.[234]

in TiO_2 nanoparticles, where the decay of TiO_2 electrons in the presence of NO_3^- was also observed.[238]

Actually, poor reactivity for nitrate reduction or no reaction at all has been observed using bare TiO_2 in pure water.[28,239–251] Therefore, an electron donor has to be added to allow indirect reduction through Reactions (2.15) and (2.16).[252] As early as in 1988, Ohtani *et al.*,[253] studying the photocatalytic reduction of nitrate in acidic TiO_2 suspensions in the presence of 2-PrOH, indicated that NO_3^- itself had no ability to react with the electron of the CB and that the reduction proceeded by the unpaired electron produced as a result of hole trapping by the surface-adsorbed alcohol. Since then, a wide range of electron donors have been studied and compared,[29] such as short carboxylic acids including formic,[254,255] acetic[250] and oxalic,[235,246,256] alcohols such as 2-PrOH,[253] ethanol[257] and methanol,[258] benzene[259] and HA.[244] The electron donor influences both the efficiency and the selectivity of the reaction towards different nitrogen compounds;[28] in addition, the photocatalytic efficiency will be enhanced by a hole scavenger with fast oxidation kinetics. Comparative studies have shown that, in general, HCOOH is the best electron donor because it promotes a highly efficient reaction and an acceptable selectivity towards N_2 formation.[28,29,252,254]

According to Doudrick *et al.*,[252] commercially available TiO_2 such as Evonik P25 and Evonik P90 reduce nitrate mainly to gaseous nitrogen forms and to a lesser degree to nitrite or ammonium when using HCOOH (pH 2.57), with no NO_3^- removal observed when oxalic acid (pH 1.71) or sodium formate (pH 6.43) was used. Sachtleben Hombikat UV100 presented a poorer activity.

Another way of enhancing nitrate HP reduction is the modification of TiO_2 by deposition of metals or by doping with heteroatoms, *e.g.*, Pt, Pd, Rh, Pt–Cu, Cu, Fe and Bi. The first works were those of Kudo *et al.*,[239,240] who described the results of reduction of NO_3^- to NH_4^+ using Pt/TiO_2, Pd/TiO_2, $SrTiO_3$, NiO–TiO_2, $K_4Nb_6O_{17}$ and combinations, with different yields. Later, Ranjit *et al.*[241,242] analysed the reduction of NO_3^- to NH_4^+ using metal-modified TiO_2, finding that the photocatalytic activity depended on the metal added as the modifier, and followed the order Ru > Pt > Pd > Rh. In further papers, good efficiency for nitrate reduction was obtained with TiO_2 doped with Pd,[247,256] Ag[260,261] or Au.[262] As indicated in Section 2.1, photocatalytic production of H_2 can take place with these metal modified TiO_2 materials and compete with NO_3^- reduction, the selectivity being determined by the HOV (hydrogen overpotential) of the loaded metal working as the reduction centre.[256,263,264] For example, with Ag–TiO_2 and Cu–TiO_2 in the presence of oxalate at pH 8, reduction of H^+ to H_2 was completely inhibited due to the large HOV of Ag and Cu; the metals acted there as active sites for NO_3^- reduction, and large rates for this reaction were observed.[260,261,263] In contrast, with Pd and Pt, most of NO_3^- was recovered unchanged and a large amount of H_2 was formed, indicating that photogenerated electrons were captured preferentially by H^+.[239,247,256] With Au, the rate of NO_3^- reduction decreased, suggesting that Au particles acted more as recombination centres rather than as reduction ones.[262,263] On the other hand, the generated H_2 can also provide an additional means of NO_3^- reduction, but this will depend on the nature of the metal.[251,257,258]

Non-noble metals as TiO_2 dopants such as Fe, Cr, Co, Mg, Ni, Zn, Cu and Bi,[255,256,265–267] also promote good activity for nitrate reduction with different results in the products, depending on the metal. Combination of metals has also been tested, *e.g.* Pd–Cu/TiO_2, Pt–Cu/TiO_2 and Ni–Cu/TiO_2.[29,256,267] Pt/TiO_2 combined with SnPd/Al_2O_3 was also assayed; here, NO_3^- reduction was promoted in parallel with generation of H_2.[251] Soares *et al.*[29] indicate that when using Pd–Cu/TiO_2 with methanol or ethanol the selectivity was shifted towards N_2 in detriment of NH_4^+, but when using HA, only NH_4^+ was formed.

When TiO_2 is used as support for the metal and H_2 is externally added, both photocatalytic and non-photocatalytic reductive mechanisms can take place.[29,249,256] Wehbe *et al.*[249] and Soares *et al.*[29] used Pd–Cu/TiO_2 to treat nitrate in the presence of HCOOH and observed that UV irradiation significantly improved the catalytic reaction when H_2 was added. It was proposed that NO_2^- was essentially formed on Cu and then migrated to Pd sites, where it was further reduced to N_2 and NH_4^+.

As a general rule and as expected, O_2 should be eliminated from the system to obtain a good efficiency of nitrate reduction[252,266] because O_2 can compete with NO_3^- for the reducing agents present in the system.

Solution pH is very important for the efficiency and selectivity of nitrate reduction by HP. According to some of the global reactions, reduction is more effective in acid solutions, *e.g.*:[252]

$$NO_3^- + H_2O + 2H^+ \rightarrow NH_4^+ + 2O_2 \quad (2.23)$$
$$4NO_3^- + 4H^+ \rightarrow 2N_2 + 5O_2 + 2H_2O \quad (2.24)$$

The surface charge of TiO_2 is also affected by pH: at values below the point of zero charge (pzc, 6.4 for P25),[252] the semiconductor surface is positively charged and facilitates the adsorption of NO_3^- and, consequently, its reduction. However, the lack of photocatalytic activity on NO_3^- systems with oxalic acid as electron donor (pH around 2) is better explained by a lower rate of donor oxidation than by pH, as indicated above.[252] On the other hand, the reductive power can be increased at higher pH values due to the Nernstian shift of the band edge position.[21–23] Therefore, lowering the pH diminishes the photocatalytic reductive power but favours Reactions (2.23) or (2.24), and an optimum pH for the process is required.

Very recently, Hirayama *et al.*[268] reported the only example of photocatalytic reduction of nitrate under visible light irradiation ($\lambda > 420$ nm) using Pt/$SrTiO_3$:Rh and SnPd/Al_2O_3 as co-catalysts, finding a very good efficiency for transformation into N_2. From analysis of the reaction mechanism, it was concluded that H_2 formed by photoreduction of water over Pt/$SrTiO_3$:Rh acted as the reductant for a non-photocatalytic NO_3^- conversion reaction over SnPd/Al_2O_3.

Concerning applications, good nitrate removal in a real effluent produced by a slaughterhouse in Brazil was found using ZnO/TiO_2, and the limits of the national regulations for nitrate (10 mg N L^{-1}) and nitrite (1.0 mg N L^{-1}) were reached.[269] Although the ammonia concentration was not reduced, it remained below the allowed Brazilian limit (1.235 mg N L^{-1}). Another case was the selective decomposition of nitrate to 83% N_2 in a real groundwater at $\lambda > 300$ nm by using Pt/TiO_2 and SnPd/Al_2O_3 in the presence of glucose as hole scavenger.[251] However, pretreatment by photo-oxidation with Pt/TiO_2 was necessary to remove first the organic compounds present in the groundwater, which poisoned the catalysts. The lower decomposition rate in the groundwater compared with that of pure KNO_3 was attributed to the interference of sulfate, silicate and organic compounds. Sulfate and silicate adsorbed on the Pt/TiO_2 surface, deactivating Pt and TiO_2 sites, respectively, caused a decrease in the H_2 evolution.

Another interesting application was the photocatalytic nitrate removal from an ion exchange brine used in drinking water processes.[233] P90 was used as the photocatalyst and HCOOH as the hole scavenger, producing 83% N gases and 17% NH_4^+. Sulfate present in the system had to be removed first through barium sulfate precipitation as it competed with nitrate for active surface sites on the photocatalyst and induced aggregation.

2.9 Conclusions

Heterogeneous photocatalytic treatment of metals, metalloids and other inorganic species such as Cr, Pb, U, Hg, As and nitrate, among others, can be a valuable option for removal of these toxic and not easily removable species from water. From the point of view of application, the technology is cost-effective and does not require expensive reagents or equipment. Photocatalytic treatments can transform the pollutants into less toxic soluble species or, in the case of metals, deposit them into solid forms over the semiconductor surface.

In regular photocatalytic reactions under normal illumination, redox reactions take place usually through thermodynamically allowed successive mono-electronic transfer steps. Three types of mechanisms can be postulated: (a) direct reduction by photogenerated electrons; (b) indirect reduction by intermediates generated by hole or hydroxyl radical oxidation of electron donors (reducing radicals); and (c) oxidative removal by holes or (c′) hydroxyl radicals. Organic donors, generally present in polluted waters together with the pollutants, are good synergetic agents of the treatment; in these cases, the photocatalytic process allows the simultaneous treatment of species of different chemical structure and properties.

Hexavalent chromium is a frequent and highly toxic contaminant in wastewaters: it is a well-known human carcinogen and may have dramatic effects on human health. The Cr(VI) HP system has been extensively studied at the laboratory scale, either under UV and/or visible light. The Cr(VI) photocatalytic reduction is unique because it is not inhibited by oxygen, at least at acid pH, which represents an advantage for technological applications. Although the fundamental studies indicate advantages for a successful application, reports regarding the scale-up of the process are still scarce. The use of Cr(VI) as a sensitizer for visible light photocatalytic activity will surely be more deeply investigated in the next years.

Mercury(II) is a frequent component of wastewaters, markedly toxic at very low concentrations and resistant to biological or chemical transformation. Organic species such as methyl- or phenylmercury are much more toxic than the inorganic ones. Mercury can also be found in the gas state as elemental mercury vapour, oxidized mercury species and particulate-bound mercury. Both fundamental and applied studies on the water phase are still required, especially in the case of organomercurial species. HP treatments in the gas phase can be a very good and simple alternative for detoxification of gaseous mercury systems, which have to be considered and investigated further.

Lead pollution is mainly anthropogenic, coming from industrial effluents. Lead effects on the central nervous system can be serious. Lead(II) can be oxidized to Pb(IV) or reduced to Pb(0) in photocatalytic systems. Although the Pb photocatalytic system has been profoundly studied, few studies have been made in recent years, and much effort should be made to transfer the knowledge to technological solutions.

Uranium is a radioactive heavy metal present in nature and as a result of mining, nuclear or other applications. Uranium is toxic and has strong impacts on human health. The photocatalytic treatment of the uranyl ion (UO_2^{2+}), the most important form in water, has been explored and the reductive mechanism is well understood, with the possibility of formation of U(V), U(IV) and, recently, of U(III). The application of photocatalysis could solve not only environmental problems but also those related to actinide valence state control as a possible application of waste minimization in nuclear fuel processing.

Natural sources are the main cause of the presence of As (As(III) and As(V)), in water, with some anthropogenic contribution. Chronic ingestion of As produces arsenicosis (HACRE), which causes severe skin lesions, ending possibly in cancer. Heterogeneous photocatalysis of arsenic species is very promising, and both oxidative and reductive mechanisms may lead to less toxic species (As(V) compared with As(III)) or solid phases (As(0)). Although the oxidative system has been studied in detail, the reductive pathway is a challenging alternative that should be taken into account and improved. Organic arsenical species have also been studied.

The main source of nitrate in water is its use in agriculture, and some is produced naturally. Ingestion of nitrate causes high damage to health. The photocatalysis of nitrate to nitrogen reduced forms has been rather well explored. It was concluded that it only takes place in the presence of an electron donor or with metal doped TiO_2 photocatalysts. Although ammonia is not a concern in drinking water, its presence in the final system should be taken into account because of indirect interference. The best strategy relies mainly on the formation of N_2, the harmless form of inorganic nitrogen.

References

1. J. O. Nriagu, *Environ. Pollut.*, 1988, **50**, 139.
2. WHO, *Guidelines for Drinking-water Quality, third edition incorporating the first and second addenda*, Geneva, 2008, vol. 1, recommendations.
3. A. Baral and R. D. Engelken, *Environ. Sci. Policy*, 2002, **5**, 121.
4. WHO, *Lead in Drinking-water, Background document for development of WHO Guidelines for Drinking-water Quality*, 2011.
5. WHO, *Uranium in Drinking-water, Background document for development of WHO Guidelines for Drinking Water Quality*, WHO/SDE/WSH/03.04/118, 2004.
6. WHO, *Arsenic in drinking-water, background document for development of WHO-Guidelines for drinking-water quality*, WHO/SDE/WSH/03.04/75/Rev/1, 2011.
7. WHO, *Nitrate and nitrite in drinking-water. Background document for development of WHO Guidelines for Drinking-water Quality Nitrate and Nitrite in Drinking*, WHO/SDE/WSH/07.01/16/Rev/1, 2011.
8. D. Quality, *Mercury in Drinking-water, Background document for development of, WHO Guidelines for Drinking-water Quality*, 2005, vol. WHO/SDE/WS.
9. Degrémont, *Water Treatment Handbook*, Editorial Lavoisier Publishing, Paris, 6th edn, 1991.
10. M. I. Litter, in *The Handbook of Environmental Chemistry*, ed. P. Boule, D. W. Bahnemann and P. K. J. Robertson, Springer-Verlag, Berlin, Heidelberg, 2005, vol. 2, part M, pp. 325–366.
11. O. Legrini, E. Oliveros and A. M. Braun, *Chem. Rev.*, 1993, **93**, 671.
12. C. P. Huang, Ch. Dong and Z. Tang, *Waste Manage.*, 1993, **13**, 361.
13. US/EPA, *Handbook of Advanced Photochemical Oxidation Processes*, EPA/625/R-98/004, December 1998.
14. *The AOT Handbook*, Calgon Carbon Oxidation Technologies, Ontario, 1996.
15. X. Domènech, W. Jardim and M. Litter, in *Eliminación de contaminantes por fotocatálisis heterogénea*, ed. M. A. Blesa and B. Sánchez Cabrero, Ediciones CIEMAT, Madrid, Spain, 2004, pp. 7–34.
16. M. R. Hoffmann, S. T. Martin, W. Choi and D. Bahnemann, *Chem. Rev.*, 1995, **95**, 69.
17. A. L. Linsebigler, G. Lu and J. T. Yates Jr, *Chem. Rev.*, 1995, **95**, 735.
18. K. Nakata and A. Fujishima, *J. Photochem. Photobiol., C*, 2012, **13**, 169.
19. O. Carp, C. L. Huisman and A. Reller, *Prog. Solid State Chem.*, 2004, **32**, 33.
20. S. T. Martin, H. Herrmann and M. R. Hoffmann, *J. Chem. Soc., Faraday Trans.*, 1994, **90**, 3323.
21. H. Gerischer and A. Heller, *J. Phys. Chem.*, 1991, **95**, 5261.
22. N. Serpone, Y. K. Ah-You, T. P. Tran, R. Harris, E. Pelizzetti and H. Hidaka, *Sol. Energy*, 1987, **39**, 491.
23. A. Fujishima, X. Zhang and D. Tryk, *Surf. Sci. Rep.*, 2008, **63**, 515.

24. C. Minero, G. Mariella, V. Maurino and E. Pelizzetti, *Langmuir*, 2000, **16**, 2632.
25. H. Kominami, S. Murakami, J. Kato, Y. Kera and B. Ohtani, *J. Phys. Chem. B*, 2002, **106**, 10501.
26. D. W. Bahnemann, J. Mönig and R. Chapman, *J. Phys. Chem.*, 1987, **91**, 3782.
27. M. Breitenkamp, A. Henglein and J. Lilie, *Ber. Bunsen-Ges.*, 1976, **80**, 973.
28. M. Shand and J. A. Anderson, *Catal. Sci. Technol.*, 2013, **3**, 879.
29. O. S. G. P. Soares, M. F. R. Pereira, J. J. M. Órfão, J. L. Faria and C. G. Silva, *Chem. Eng. J.*, 2014, **251**, 123.
30. N. Serpone, E. Borgarello and E. Pelizzetti, *Photocatalysis and Environment*, ed. M. Schiavello, Kluwer Academic Publishers, Dordrecht, The Netherlands, 1988, pp. 527–565.
31. M. I. Litter, *Appl. Catal., B*, 1999, **23**, 89.
32. M. Litter, X. Domènech and H. Mansilla, in *Eliminación de contaminantes por fotocatálisis heterogénea*, ed. M. A. Blesa and B. Sánchez Cabrero, Ediciones CIEMAT, Madrid, Spain, 2004, pp. 163–187.
33. M. I. Litter, *Adv. Chem. Eng.*, 2009, **36**, 37.
34. M. I. Litter and N. Quici, in *Nanomaterials for Environmental Protection*, ed. B. I. Kharisov, O. V. Kharissova and H. V. Rasika Dias, John Wiley & Sons, 29 August 2014, ch. 9, pp. 143–167, DOI: 10.1002/9781118845530.
35. M. I. Litter, *Pure Appl. Chem.*, 2015, DOI: 10.1515/pac-2014-0710.
36. M. A. Grela and A. J. Colussi, *J. Phys. Chem.*, 1996, **100**, 18214.
37. J. J. Testa, M. A. Grela and M. I. Litter, *Langmuir*, 2001, **17**, 3515.
38. J. J. Testa, M. A. Grela and M. I. Litter, *Environ. Sci. Technol.*, 2004, **38**, 1589.
39. J. M. Meichtry, M. Brusa, G. Mailhot, M. A. Grela and M. I. Litter, *Appl. Catal., B*, 2007, **71**, 101.
40. J. M. Meichtry, Doctoral thesis, Universidad de Buenos Aires, 2011.
41. J. J. Testa, Doctoral thesis, Universidad de Buenos Aires, 2012.
42. V. N. Salomone, J. M. Meichtry, G. Zampieri and M. I. Litter, *Chem. Eng. J.*, 2014, **261**, 27.
43. V. N. Salomone, J. M. Meichtry and M. I. Litter, *Chem. Eng. J.*, 2015, in press.
44. W.-Y. Lin and K. Rajeshwar, *J. Electrochem. Soc.*, 1997, **144**, 2751.
45. A. Mills and M. A. Valenzuela, *Rev. Mex. Fis.*, 2004, **50**, 287.
46. M. R. Prairie, L. R. Evans, B. M. Stange and S. L. Martínez, *Environ. Sci. Technol.*, 1993, **27**, 1776.
47. M. R. Prairie, B. M. Stange and L. R. Evans, in *Photocatalytic Purification and Treatment of Water and Air*, ed. D. F. Ollis and H. Al-Ekabi, Elsevier Science Publishers B.V., Amsterdam, 1993, pp. 353–363.
48. P. Wardman, *J. Phys. Chem. Ref. Data*, 1989, **18**, 1637.
49. R. Baba, R. Konda, A. Fujishima and K. Honda, *Chem. Lett.*, 1986, **8**, 1307.

50. K. Vinodgopal, D. E. Wynkoop and P. V. Kamat, *Environ. Sci. Technol.*, 1996, **30**, 1660.
51. J. M. Meichtry, V. Rivera, Y. Di Iorio, E. San Román, M. A. Grela and M. I. Litter, *Photochem. Photobiol. Sci.*, 2009, **8**, 604.
52. J. Hodak, C. Quinteros, M. I. Litter and E. San Román, *J. Chem. Soc., Faraday Trans.*, 1996, **92**, 5081.
53. M. Owlad, M. K. Aroua, W. A. W. Daud and S. Baroutian, *Water, Air, Soil Pollut.*, 2009, **200**, 59.
54. J. Gauglhofer and V. Bianchi, *Metals and their compounds in the environment*, VCH, Weinheim, 1991.
55. J. W. Hamilton and K. E. Wheterhan, in *Handbook of Toxicity of Inorganic Compounds*, ed. H. G. Seiler and H. Sigel, Marcel Dekker, Inc., New York, 1988.
56. P. S. Silva, *Guidance for public water systems on enhanced monitoring for chromium-6 (hexavalent chromium) in drinking water*, EPA Guidance Memo, Washington DC, 2011.
57. J. C. Seaman, P. M. Bertsch and L. Schwallie, *Environ. Sci. Technol.*, 1999, **33**, 938.
58. C. E. Barrera-Díaz, V. Lugo-Lugo and B. Bilyeu, *J. Hazard. Mater.*, 2012, **223–224**, 1.
59. K. Niki, in *Standard potentials in aqueous solution*, ed. A. J. Bard, R. Parsons and J. Jordan, Marcel Dekker, Inc., New York, 1985, pp. 453–461.
60. J. M. Meichtry, C. Colbeau-Justin, G. Custo and M. I. Litter, *App. Catal. B*, 2014, **144**, 189.
61. J. M. Meichtry, C. Colbeau-Justin, G. Custo and M. I. Litter, *Catal. Today*, 2014, **224**, 236.
62. J. K. Yang, S. M. Lee, M. Farrokhi, O. Giahi and M. Shirzad-Siboni, *Desalin. Water Treat.*, 2012, **46**, 375.
63. Y.-S. Wang, J.-H. Shen and J.-J. Horng, *J. Hazard. Mater.*, 2014, **274**, 420.
64. H.-T. Hsu, S.-S. Chen, W.-S. Chang and C.-W. Li, *Desalin. Water Treat.*, 2013, **51**, 495.
65. Q. Wu, J. Zhao, G. Qin, C. Wang, X. Tong and S. Xue, *Appl. Catal., B*, 2013, **142–143**, 142.
66. A. E. Giannakas, M. Antonopoulou, Y. Deligiannakis and I. Konstantinou, *Appl. Catal., B*, 2013, **140–141**, 636.
67. D. W. Bahnemann, M. Hilgendorff and R. Memming, *J. Phys. Chem. B*, 1997, **101**, 4265.
68. Y. Di Iorio, E. San Román, M. I. Litter and M. A. Grela, *J. Phys. Chem. C*, 2008, **112**, 16532.
69. U. Siemon, D. Bahnemann, J. J. Testa, D. Rodríguez, N. Bruno and M. I. Litter, *J. Photochem. Photobiol., A*, 2002, **148**, 247.
70. J. K. Yang, S. M. Lee and M. Shirzad-Siboni, *Environ. Technol.*, 2012, **33**, 2027.
71. N. Huang, Y. Zhang, Q. Zhang, C. K. Chen and C. T. Chang, *J. Nanosci. Nanotechnol.*, 2014, **14**, 2718.
72. Y.-S. Wang, J.-H. Shen and J.-J. Horng, *J. Hazard. Mater.*, 2014, **274**, 420.

73. R. Djellabi and M. F. Ghorab, *Desalin. Water Treat.*, 2014, **25**, 1.
74. V. Loryuenyong, N. Jarunsak, T. Chuangchai and A. Buasri, *Adv. Mater. Sci. Eng.*, 2014, DOI: 10.1155/2014/348427.
75. X. Q. Li, Z. Hong, S. Z. Kang, L. X. Qin, G. G. Li and J. Mu, *Adv. Mater. Res.*, 2014, **864–867**, 715.
76. Y. C. Zhang, M. Yang, G. Zhang and D. D. Dionysiou, *Appl. Catal., B*, 2013, **142–143**, 249.
77. A. E. Giannakas, E. Seristatidou, Y. Deligiannakis and I. Konstantinou, *Appl. Catal., B*, 2013, **132–133**, 460.
78. A. Daya Mani, B. Rama Raju, N. Xanthopoulos, P. Ghosal, B. Sreedhar and Ch. Subrahmanyam, *Chem. Eng. J.*, 2013, **228**, 545.
79. W. Kim, T. Tachikawa, H. Kim, N. Lakshminarasimhan, P. Murugan, H. Park, T. Majima and W. Choi, *Appl. Catal., B*, 2014, **147**, 642.
80. Z. He, Q. Cai, M. Wu, Y. Shi, H. Fang, L. Li, J. Chen, J. Chen and S. Song, *Ind. Eng. Chem. Res.*, 2013, **52**, 9556.
81. Z. Sun, L. Zheng, S. Zheng and R. L. Frost, *J. Colloid Interface Sci.*, 2013, **404**, 102.
82. Q. Sun, H. Li, S. Zheng and Z. Sun, *Appl. Surf. Sci.*, 2014, **311**, 369.
83. W. Liu, J. Ni and X. Yin, *Water Res.*, 2014, **53**, 12.
84. X. Liu, L. Pan, T. Lv and Z. Sun, *J. Coll. Int. Sci.*, 2014, **394**, 441.
85. C. Wang, M. Cao, P. Wang, Y. Ao, J. Hou and J. Qian, *Appl. Catal., A*, 2014, **473**, 83.
86. Y. Yang, G. Wang, Q. Deng, D. H. L. Ng and H. Zhao, *ACS Appl. Mater. Interfaces*, 2014, **6**, 3008.
87. R. Gherbi, M. Trari and N. Nasrallah, *J. Environ. Chem. Eng.*, 2013, **1**, 1275.
88. R. Dhanalakshmi, A. Pandikumar and R. Ramaraj, *Mater. Sci. Forum*, 2013, **764**, 255.
89. M. Naeimi-Joubani, M. Shirzad-Siboni, J. K. Yang, M. Gholami and M. Farzadkia, *J. Ind. Eng. Chem.*, 2014, in press.
90. Y. Kim, H. Joo, N. Her, Y. Yoon, C.-H. Lee and J. Yoon, *Chem. Eng. J.*, 2013, **229**, 66.
91. N. Shaham-Waldmann and Y. Paz, *Chem. Eng. J.*, 2013, **231**, 49.
92. L. P. D'Souza, S. Shree and G. R. Balakrishna, *Ind. Eng. Chem. Res.*, 2013, **52**, 16162.
93. M. S. Jyothi, M. Padaki, R. G. Balakrishna and R. K. Pai, *J. Mater. Res.*, 2014, **29**, 1537.
94. Y.-g. Liu, X.-j. Hu, H. Wang, A.-w. Chen, S.-m. Liu, Y.-m. Guo, Y. He, X. Hu, J. Li, S.-h. Liu, Y.-q. Wang and L. Zhou, *Chem. Eng. J.*, 2013, **226**, 131.
95. H. Eskandarloo, A. Badiei, M. A. Behnajady and G. M. Ziarani, *RSC Adv.*, 2014, **4**, 28587.
96. S. Xu, Y. Zhang, S. Wang, J. Xu, H. Ding and G. Li, *Eur. J. Inorg. Chem.*, 2013, **2013**, 2601.
97. H. J. Wang, X. J. Wu, Y. L. Wang, Z. B. Jiao, S. W. Yan and L. H. Huang, *Chin. J. Catal.*, 2011, **32**, 637.
98. Q. Wang, J. Shang and H. Song, *Chin. J. Catal.*, 2011, **32**, 1525.

99. Q. Wang, J. Shang, T. Zhu and F. W. Zhao, *J. Mol. Catal. A: Chem.*, 2011, **335**, 242.
100. F. M. Monteiro Paschoal, M. A. Anderson, M. Valnice and B. Zanoni, *J. Hazard. Mater.*, 2009, **166**, 531.
101. H. Song, J. Shang, J. Ye and Q. Li, *Thin Solid Films*, 2014, **551**, 158.
102. H.-T. Hsu, S.-S. Chen, Y.-F. Tang and H.-C. His, *J. Hazard. Mater.*, 2013, **248–249**, 97.
103. M. Shirzad-Siboni, M. T. Samadi, J. K. Yang and S. M. Lee, *Desalin. Water Treat.*, 2012, **40**, 77.
104. T. Caprioli Machado, M. A. Lansarin and N. Matte, *Water Sci. Technol.*, 2014, **70**, 55.
105. Q. Wang, X. Chen, K. Yu, Y. Zhang and Y. Cong, *J. Hazard. Mater.*, 2013, **246–247**, 135.
106. Y.-F. Rao and W. Chu, *Ind. Eng. Chem. Res.*, 2013, **52**, 13580.
107. Y. Park, W. Kim, D. Monllor-Satoca, T. Tachikawa, T. Majima and W. Choi, *J. Phys. Chem. Lett.*, 2013, **4**, 189.
108. Y. Wang, L. Jiang and C. Feng, *Desalin. Water Treat.*, 2014, **52**, 4802.
109. X. Liu, L. Pan, T. Lv and Z. Sun, *J. Alloys Compd.*, 2013, **260**, 390.
110. X. Zhang, X. Li, C. Shao, J. Li, M. Zhang, P. Zhang, K. Wang, N. Lu and Y. Liu, *J. Hazard. Mater.*, 2013, **583**, 260.
111. K. Sridharan, E. Jang and T. J. Park, *Appl. Catal., B*, 2013, **142–143**, 718.
112. Y. Zhao, D. Zhao, C. Chen and X. Wang, *J. Coll. Int. Sci.*, 2013, **405**, 211.
113. J. Yang, J. Dai and J. Li, *Environ. Sci. Pollut. Res.*, 2013, **20**, 2435.
114. J. Wang, X. Li, X. Li, J. Zhu and H. Li, *Nanoscale*, 2013, **5**, 1876.
115. T. Wang, Y. Zhang and T. Ding, *Mater. Lett.*, 2014, **123**, 153.
116. X. Pan, Y.-J. Xu and T. J. Park, *J. Phys. Chem. C*, 2013, **117**, 17996.
117. A. Tanaka, K. Nakanishi, R. Hamada, K. Hashimoto and H. Kominami, *ACS Catal.*, 2013, **3**, 1886.
118. R. Qiu, D. Zhang, Z. Diao, X. Huang, C. He, J.-L. Morel and Y. Xiong, *Water Res.*, 2012, **46**, 2299.
119. J. Su, Y. Zhang, S. Xu, S. Wang, H. Ding, S. Pan, G. Wang, G. Li and H. Zhao, *Nanoscale*, 2014, **6**, 5181.
120. G. Kim, S.-H. Lee and W. Choi, *Appl. Catal., B*, 2015, **162**, 463.
121. S. G. Botta, D. J. Rodríguez, A. G. Leyva and M. I. Litter, *Catal. Today*, 2002, **76**, 247–258.
122. G. L. Baughman, J. A. Gordon, N. Lee Wolfe and R. G. Zepp, *Chemistry of Organomercurials in Aquatic Systems*, EPA-660/3-73-012, 1973.
123. J. Balej, in *Standard potentials in aqueous solution*, ed. A. J. Bard, R. Parsons and J. Jordan, Marcel Dekker, Inc., New York, 1985, pp. 265–286.
124. K. Tennakone and K. G. Wijayantha, *J. Photochem. Photobiol., A*, 1998, **113**, 89.
125. J. Hegyi and O. Horváth, in *From Colloids to Nanotechnology*, ed. M. Zrínyi and Z. D. Hórvölgyi, Springer Berlin Heidelberg, 2004, vol. 125, Progress in Colloid and Polymer Science, pp. 10–16.
126. S. Fujita, H. Horii and S. Taniguchi, *J. Phys. Chem.*, 1973, **77**, 2868.
127. K. Gårdfeldt and M. J. Jonsson, *J. Phys. Chem. A*, 2003, **107**, 4478.

128. M. J. López-Muñoz, J. Aguado, A. Arencibia and R. Pascual, *Appl. Catal., B*, 2011, **104**, 220.
129. G. Custo, M. I. Litter, D. Rodríguez and C. Vázquez, *Spectrochim. Acta, Part B*, 2006, **61**, 1119.
130. K. Tanaka, K. Harada and S. Murata, *Sol. Energy*, 1986, **36**, 159.
131. C. Miranda, J. Yáñez, D. Contreras, R. Garcia, W. F. Jardim and H. D. Mansilla, *Appl. Catal., B*, 2009, **90**, 115–119.
132. E. M. de la Fournière, A. G. Leyva, E. A. Gautier and M. I. Litter, *Chemosphere*, 2007, **69**, 682.
133. J. H. Cho, Y. Eom, S. H. Jeon and T. G. Lee, *J. Ind. Eng. Chem.*, 2013, **19**, 144.
134. C.-Y. Wu, T. G. Lee, G. Tyree, E. Arar and P. Biswas, *Environ. Eng. Sci.*, 1998, **15**, 138–148.
135. T. G. Lee, P. Biswas and E. Hedrick, *AIChE J.*, 2001, **47**, 954.
136. Y. G. Lee, J.-W. Park, J.-H. Kim, B. R. Min, J. Jurng, J. Kim and T. G. Lee, *Chem. Lett.*, 2004, **33**, 36.
137. E. Pitoniak, C.-Y. Wu, D. Londeree, D. Mazyck, J.-C. Bonzongo, K. Powers and W. Sigmund, *J. Nanopart. Res.*, 2003, **5**, 281.
138. Y. Li and C.-Y. Wu, *Environ. Sci. Technol.*, 2006, **40**, 6444.
139. Y. Li and C. Y. Wu, *Environ. Eng. Sci.*, 2007, **24**, 3.
140. S. Rodríguez, C. Almquist, T. G. Lee, M. Furuuchi, E. Hedrick and P. Biswas, *J. Air Waste Manage. Assoc.*, 2004, **54**, 149.
141. H.-C. His and C.-Y. Tsai, *Chem. Eng. J.*, 2012, **191**, 378.
142. S.-S. Chen, H.-C. Hsi, S.-H. Nian and C.-H. Chiu, *Appl. Catal., B*, 2014, **160–161**, 558.
143. C.-Y. Tsai, *Aerosol Air Qual. Res.*, 2013, **2**, 639.
144. Z. Galus, in *Carbon, silicon, germanium, tin and lead, in Standard potentials in aqueous solutions*, ed. A. J. Bard, R. Parsons and J. Jordan, Marcel Dekker, Inc., New York, 1985, pp. 189–236.
145. H. Liu, A. M. Kuznetsov, A. N. Masliy, J. F. Ferguson and G. V. Korshin, *Environ. Sci. Technol.*, 2012, **46**, 1430.
146. L. Murruni, G. Leyva and M. I. Litter, *Catal. Today*, 2007, **129**, 127.
147. E. Brillas, *Appl. Catal., B*, 2003, **46**, 381.
148. B. Kasprzyk-Hordern, M. Ziółek and J. Nawrocki, *Appl. Catal., B*, 2003, **46**, 639.
149. K. Tanaka, K. Abe and T. Hisanaga, *J. Photochem. Photobiol., A*, 1996, **101**, 85.
150. J. Torres and S. Cervera-March, *Chem. Eng. Sci.*, 1992, **47**, 3857.
151. M. Thurnauer and T. Rajh, *Acta Chem. Scand.*, 1997, **51**, 610.
152. M. Breitenkamp, A. Henglein and J. Lilie, *Phys. Chem. Chem. Phys.*, 1976, **80**, 973.
153. C. R. Chenthamarakshan, H. Yang, C. R. Savage and K. Rajeshwar, *Res. Chem. Intermed.*, 1999, **25**, 861.
154. T. Rajh, D. M. Tiede and M. C. Thurnauer, *J. Non-Cryst. Solids*, 1996, **205–207**, 815.
155. D. Lawless, A. Res, R. Harris, N. Serpone, C. Minero, E. Pelizzetti and H. Hidaka, *Chim. Ind.*, 1990, **72**, 139.

156. L. Murruni, F. Conde, G. Leyva and M. I. Litter, *Appl. Catal., B*, 2008, **84**, 563.
157. C. R. Chenthamarakshan, K. Rajeshwar and E. J. Wolfrum, *Langmuir*, 2000, **16**, 2715.
158. L. Li, F. Jiang, J. Liu, H. Wan, Y. Wan and S. Zheng, *J. Environ. Sci. Health, Part A*, 2012, **47**, 327.
159. Z.-p. Yang and C.-j. Zhang, *Thin Solid Films*, 2010, **518**, 6006.
160. S. Sreekantan, C. W. Lai and S. M. Zaki, *Int. J. Photoenergy*, 2014, 839106.
161. S. Sreekantan, S. M. Zaki, C. W. Lai and T. W. Tzu, *Eur. Phys. J.: Appl. Phys.*, 2014, **67**, 10404.
162. S. Sreekantan, S. M. Zaki, C. W. Lai and T. W. Tzu, *Mater. Sci. Semicond. Process.*, 2014, **26**, 620.
163. A. Meinrath, P. Schneider and G. Meinrath, *J. Environ. Radioact.*, 2003, **64**, 175.
164. P. D. Bhalara, D. Punetha and K. Balasubramanian, *J. Environ. Chem. Eng.*, 2014, **2**, 1621.
165. R. Amadelli, A. Maldotti, S. Sostero and V. Carassiti, *J. Chem. Soc., Faraday Trans.*, 1991, **87**, 3267.
166. S. A. Kulyukhin, N. B. Mikheev, A. N. Kamenskaya, N. A. Konovalova and I. A. Rumer, *Radiochemistry*, 2006, **48**, 535.
167. J. Selbin and J. D. Ortego, *Chem. Rev.*, 1969, **69**, 657.
168. I. Grenthe, J. Drożdżyński, T. Fujino, E. C. Buck, T. E. Albrecht-Schmitt and S. F. Wolf, Uranium, in *The Chemistry of the Actinide and Transactinide Elements*, ed. L. R. Morss, N. M. Edelstein, J. Fuger and J. J. Katz, Springer, Dordrecht, 4th edn, 2011.
169. S. A. Katz, *Toxics*, 2014, **2**, 50.
170. J. Nriagu, D. H. Nam, T. A. Ayanwola, H. Dinh, E. Erdenechimeg, C. Ochir and T. A. Bolormaa, *Sci. Total Environ.*, 2012, **414**, 722.
171. J. Chen, D. F. Ollis, W. H. Rulkens and H. Bruning, *Colloids Surf., A*, 1999, **151**, 339.
172. L. Martinot and J. Fuger, in *Standard potentials in aqueous solution*, ed. A. J. Bard and R. Parsons, Marcel Dekker, Inc., New York, USA, 1985, pp. 631–674.
173. M. Duflo-Plissonnier and K. Samhoun, *Radiochem. Radioanal. Lett.*, 1972, **12**, 131.
174. C. J. Evans, G. P. Nicholson, D. A. Faith and M. J. Kan, *Green Chem.*, 2004, **6**, 196.
175. V. Eliet and G. Bidoglio, *Environ. Sci. Technol.*, 1998, **32**, 3155.
176. E. Selli, V. Eliet, M. R. Spini and G. Bidoglio, *Environ. Sci. Technol.*, 2000, **34**, 3742.
177. C. Boxall, G. Le Gurun, R. J. Taylor and S. Xiao, in *The Handbook of Environmental Chemistry, Environmental Photochemistry Part II*, ed. P. Boule, D. W. Bahnemann and P. K. J. Robertson, Springer-Verlag, Berlin, Heidelberg, 2005, vol. 2M, pp. 451–481.
178. M. Bonato, G. Allen and T. Scott, *Micro Nano Lett.*, 2008, **3**, 57.
179. S. O. Odoh, Q.-J. Pan, G. A. Shamov, F. Wang, M. Fayek and G. Schreckenbach, *Chem.-Eur. J.*, 2012, **18**, 7117.

180. Y. Li, J. Su, E. Mitchell, G.-Q. Zhang and J. Li, *Sci. China Chem.*, 2013, **56**, 1671.
181. M. I. Litter, M. E. Morgada and J. Bundschuh, *Environ. Pollut.*, 2010, **158**, 1105.
182. J. Bundschuh, M. I. Litter, V. Ciminelli, M. E. Morgada, L. Cornejo, S. Garrido Hoyos, J. Hoinkis, M. T. Alarcón-Herrera, M. A. Armienta and P. Bhattacharya, *Water Res.*, 2010, **44**, 5828.
183. M. I. Litter, R. G. Fernández, R. E. Cáceres, D. G. Cobián, D. Cicerone and A. Fernández Cirelli, *Ing. Sanit. Ambientale*, 2008, **100**, 41.
184. M. E. Morgada and M. I. Litter, *e-Terra*, 2008, 5.
185. M. T. Emett and G. H. Khoe, *Water Res.*, 2001, **35**, 649.
186. P. Mondal, S. Bhowmick, D. Chatterjee, A. Figoli and B. Van der Bruggen, *Chemosphere*, 2013, **92**, 157.
187. K. S. Santhanam and N. S. Sundaresan, in *Standard potentials in aqueous solution*, ed. J. A. Bard, R. Parsons and J. Jordan, Marcel Dekker, Inc., New York, 1985, pp. 162–171.
188. P. K. Dutta, S. O. Pehkonen, V. K. Sharma and A. K. Ray, *Environ. Sci. Technol.*, 2005, **39**, 1827.
189. J. Ryu and W. Choi, *Environ. Sci. Technol.*, 2004, **38**, 2928.
190. M. A. Ferguson, M. R. Hoffmann and J. G. Hering, *Environ. Sci. Technol.*, 2005, **39**, 1880.
191. U. K. Kläning, B. H. J. Bielski and K. Sehested, *Inorg. Chem.*, 1989, **28**, 2717.
192. X. Guan, J. Du, X. Meng, Y. Sun, B. Sun and Q. Hu, *J. Hazard. Mater.*, 2012, **215–216**, 1–16.
193. Z. Xu and X. Meng, *J. Hazard. Mater.*, 2009, **168**, 747.
194. H. Lee and W. Choi, *Environ. Sci. Technol.*, 2002, **36**, 3872.
195. E. S. Tsimas, K. Tyrovola, N. P. Xekoukoulotakis, N. P. Nikolaidis, E. Diamadopoulos and D. Mantzavinos, *J. Hazard. Mater.*, 2009, **169**, 376.
196. T. V. Nguyen, S. Vigneswaran, H. H. Ngo, J. Kandasamy and H. C. Choi, *Sep. Purif. Technol.*, 2008, **61**, 44.
197. G.-h. Moon, D.-h. Kim, H.-il Kim, A. D. Bokare and W. Choi, *Environ. Sci. Technol. Lett.*, 2014, **1**, 185.
198. L. Yu, X. Peng, F. Ni, J. Li, D. Wang and Z. Luan, *J. Hazard. Mater.*, 2013, **246–247**, 10.
199. N. Rivera-Reyna, L. Hinojosa-Reyes, J. L. Guzmán-Mar, Y. Cai, K. O'Shea and A. Hernández-Ramírez, *Photochem. Photobiol. Sci.*, 2013, **12**, 653.
200. T. Xu, Y. Cai and K. E. O'Shea, *Environ. Sci. Technol.*, 2007, **41**, 5471.
201. Z. Xu, C. Jing, F. Li and X. Meng, *Environ. Sci. Technol.*, 2008, **42**, 2349.
202. S. Zheng, Y. Cai and K. E. O'Shea, *J. Photochem. Photobiol., A*, 2010, **210**, 61.
203. D. Lu, F. Ji, W. Wang, S. Yuan, Z.-H. Hu and T. Chen, *Environ. Sci. Pollut. Res.*, 2014, **21**, 8025.
204. X. Li and W. Leng, *J. Phys. Chem. C*, 2013, **117**, 750.
205. H. Yang, W. Y. Lin and K. Rajeshwar, *J. Photochem. Photobiol., A*, 1999, **123**, 137.

206. I. K. Levy, M. Mizrahi, G. Ruano, G. Zampieri, F. G. Requejo and M. I. Litter, *Environ. Sci. Technol.*, 2012, **46**, 2299.
207. M. I. Litter, I. K. Levy, N. Quici, M. Mizrahi, G. Ruano, G. Zampieri and F. G. Requejo, in *Advanced Oxidation Technologies – Sustainable solutions for environmental treatments*, ed. M. I. Litter, R. Candal and J. M. Meichtry, CRC Press/Balkema, Taylor & Francis Group, London, 2013, pp. 23–43.
208. M. I. Litter, I. K. Levy, N. Quici, M. Mizrahi, G. Ruano, G. Zampieri and F. G. Requejo, *Proceedings 7th European Meeting on Solar Chemistry and Photocatalysis: Environmental Applications – SPEA7, 17–20 June 2012, Porto, Portugal*, Elsevier, 2013.
209. M. I. Litter, I. K. Levy, N. Quici, M. Mizrahi, G. Ruano, G. Zampieri and F. G. Requejo, *Proceedings 5th International Congress on Arsenic in the Environment, Buenos Aires*, 11–16 May 2014, CRC, Balkema, pp. 820–824.
210. I. K. Levy, M. A. Brusa, M. E. Aguirre, G. Custo, E. San Román, M. I. Litter and M. A. Grela, *Phys. Chem. Chem. Phys.*, 2013, **15**, 10335.
211. I. K. Levy, N. Quici, G. Custo, M. I. Litter, M. Brusa, M. E. Aguirre, M. A. Grela and E. San Román, *Proceedings 5th International Congress on Arsenic in the Environment, Buenos Aires*, 11-16 May 2014, CRC, Balkema, pp. 802–804.
212. N. Barrabés and J. Sá, *Appl. Catal., B*, 2011, **104**, 1.
213. A. Pintar, J. Batista and I. Muševič, *Appl. Catal., B*, 2004, **52**, 49.
214. S. Choe, K.-Y. Hwang, Y.-Y. Chang and J. Khim, *Environ. Eng. Res.*, 2001, **6**, 1.
215. H. Y. Hu, N. Goto and K. Fujie, *Water Res.*, 2001, **35**, 2789.
216. Y. M. Chen, C. W. Li and S. S. Chen, *Chemosphere*, 2005, **59**, 753.
217. K.-S. Lin, N.-B. Chang and T.-D. Chuang, *Sci. Technol. Adv. Mater.*, 2008, **9**, 1513.
218. Y.-H. Hwang, D.-G. Kim and H.-S. Shin, *J. Hazard. Mater.*, 2011, **185**, 1513.
219. B. I. Kharisov, H. V. Rasika Dias, O. V. Kharissova, V. M. Jimenez-Perez, B. Olvera Perez and B. Munoz Flores, *RSC Adv.*, 2012, **2**, 9325.
220. R. A. Crane and T. B. Scott, *J. Hazard. Mater.*, 2012, **211–212**, 112.
221. N.-U. Haque and M. Tariq, *J. Chem. Soc. Pak.*, 2010, **32**, 396.
222. S. Choe, K.-Y. Hwang, Y.-Y. Chang and J. Khim, *Environ. Eng. Res.*, 2001, **6**, 1.
223. H. Y. Hu, N. Goto and K. Fujie, *Water Res.*, 2001, **35**, 2789.
224. Y. M. Chen, C. W. Li and S. S. Chen, *Chemosphere*, 2005, **59**, 753.
225. K.-S. Lin, N.-B. Chang and T.-D. Chuang, *Sci. Technol. Adv. Mater.*, 2008, **9**, 1513.
226. Y.-H. Hwang, D.-G. Kim and H.-S. Shin, *J. Hazard. Mater.*, 2011, **185**, 1513.
227. B. I. Kharisov, H. V. Rasika Dias, O. V. Kharissova, V. M. Jimenez-Perez, B. Olvera Perez and B. Munoz Flores, *RSC Adv.*, 2012, **2**, 9325.
228. R. A. Crane and T. B. Scott, *J. Hazard. Mater.*, 2012, **211–212**, 112.
229. K. A. Karanasios, I. A. Vasiliadou, S. Pavlou and D. V. Vayenas, *J. Hazard. Mater.*, 2010, **180**, 20.

230. Y.-H. Kim, E.-D. Hwang, W. S. Shin, J.-H. Choi, T. W. Ha and S. J. Choi, *Desalination*, 2007, **202**, 286.
231. World Health Organization, *Ammonia in Drinking-water, background document for development of WHO Guidelines for Drinking-water Quality*, WHO/SDE/WSH/03.04/01, 2003.
232. J. Sawyer, *Integrated Crop Management News*, http://www.extension.iastate.edu/CropNews/2008/0502JohnSawyer.htm, accessed 6 September 2014.
233. T. Yang, K. Doudrick and P. Westerhoff, *Water Res.*, 2013, **47**, 1299.
234. P. H. Rieger, *Electrochemistry*, Chapman & Hall, New York, 1994, p. 19.
235. R. W. Fessenden and D. Meisel, *J. Am. Chem. Soc.*, 2000, **122**, 3773.
236. A. R. Cook, N. Dimitrijevic, B. W. Dreyfus, D. Meisel, L. A. Curtiss and D. M. Camaioni, *J. Phys. Chem. A*, 2001, **105**, 3658.
237. R. Gao, A. Safrany and J. Rabani, *Radiat. Phys. Chem.*, 2003, **67**, 25.
238. H. H. Mohamed, C. B. Mendive, R. Dillert and D. W. Bahnemann, *J. Phys. Chem. A*, 2011, **115**, 2139.
239. A. Kudo, K. Domen, K.-I. Maruya and T. Onishi, *Chem. Lett.*, 1987, **6**, 1019.
240. A. Kudo, K. Domen, K.-I. Maruya and T. Onishi, *J. Catal.*, 1992, **135**, 300.
241. K. T. Ranjit, T. K. Varadarajan and B. Viswanathan, *J. Photochem. Photobiol., A*, 1995, **89**, 67.
242. K. T. Ranjit and B. Viswanathan, *J. Photochem. Photobiol., A*, 1997, **108**, 73.
243. Y. Li and F. Wasgestian, *J. Photochem. Photobiol., A*, 1998, **112**, 255.
244. B. Bems, F. C. Jentoft and R. Schlogl, *Appl. Catal., B*, 1999, **20**, 155.
245. S. Tawkaew, S. Yin and T. Sato, *Int. J. Inorg. Mater.*, 2001, **3**, 855.
246. W. Gao, R. Jin, J. Chen, X. Guan, H. Zeng, F. Zhang and N. Guan, *Catal. Today*, 2004, **90**, 331.
247. J. Sá, T. Berger, K. Föttinger, A. Riss, J. Anderson and H. Vinek, *J. Catal.*, 2005, **234**, 282.
248. L. F. Liu, X. Y. Dong, F. L. Yang and J. C. Yu, *Chin. J. Inorg. Chem.*, 2008, **24**, 211.
249. N. Wehbe, M. Jaafar, C. Guillard, J.-M. Herrmann, S. Miachon, E. Puzenat and N. Guilhaume, *Appl. Catal., A*, 2009, **368**, 3.
250. J. Sá, C. A. Agüera, S. Gross and J. A. Anderson, *Appl. Catal., B*, 2009, **85**, 192.
251. J. Hirayama and Y. Kamiya, *ACS Catal.*, 2014, **4**, 2207.
252. K. Doudrick, O. Monzón, A. Mangonon, K. Hristovski and P. Westerhoff, *J. Environ. Eng.*, 2012, **138**, 852.
253. B. Ohtani, M. Kakimoto, H. Miyadzu, S. Nishimoto and T. Kagiya, *J. Phys. Chem.*, 1988, **92**, 5773.
254. K. Doudrick, T. Yang, K. Hristovski and P. Westerhoff, *Appl. Catal., B*, 2013, **136**, 40.
255. S. Rengaraj and X. Z. Li, *Chemosphere*, 2007, **66**, 930.
256. H. Kominami, H. Gekko and K. Hashimoto, *Phys. Chem. Chem. Phys.*, 2010, **12**, 15423.

257. J. Hirayama, H. Kondo, Y. Miura, R. Abe and Y. Kamiya, *Catal. Commun.*, 2012, 99.
258. M. Yamauchi, R. Abe, T. Tsukuda, K. Kato and M. Takata, *J. Am. Chem. Soc.*, 2011, **133**, 1150.
259. L. Li, Z. Xu, F. Liu, Y. Shao, J. Wang, H. Wan and S. Zheng, *J. Photochem. Photobiol., A*, 2010, **212**, 113.
260. F. Zhang, R. Jin, J. Chen, C. Shao, W. Gao, L. Li and N. Guan, *J. Catal.*, 2005, **232**, 424.
261. A. V. Lozovskii, I. V. Stolyarova, R. V. Prikhod'ko and V. V. Goncharuk, *J. Water Chem. Technol.*, 2010, **31**, 360.
262. J. A. Anderson, *Catal. Today*, 2012, **181**, 171.
263. H. Gekko, K. Hashimoto and H. Kominami, *Phys. Chem. Chem. Phys.*, 2012, **14**, 7965.
264. H. Kominami, A. Furusho, S. Murakami and H. Inoue, *Catal. Lett.*, 2001, **76**, 31.
265. R. Jin, W. Gao, J. Chen, H. Zeng, F. Zhang, Z. Liu and N. Guan, *J. Photochem. Photobiol., A*, 2004, **162**, 585.
266. D. D. B. Luiz, S. L. F. Andersen, C. Berger, H. J. José and R. D. F. P. M. Moreira, *J. Photochem. Photobiol., A*, 2012, **246**, 36.
267. H. Kominami, T. Nakaseko, Y. Shimada, A. Furusho, H. Inoue, S.-Y. Murakami, Y. Kera and B. Ohtani, *Chem. Commun.*, 2005, 2933.
268. J. Hirayama, R. Abe and Y. Kamiya, *Appl. Catal., B*, 2014, **144**, 721.
269. D. D. B. Luiz, H. J. José and R. D. F. P. M. Moreira, *J. Chem. Technol. Biotechnol.*, 2015, **90**(5), 821–829.

CHAPTER 3

Solar Photocatalytic Disinfection of Water

PILAR FERNÁNDEZ-IBÁÑEZ*[a], J. ANTHONY BYRNE[b], M. INMACULADA POLO-LÓPEZ[a], PATRICK S. M. DUNLOP[b], POPI KARAOLIA[c], AND DESPO FATTA-KASSINOS[c]

[a]Centro de Investigaciones Energéticas Medioambientales y Tecnológicas, Plataforma Solar de Almería, Ctra. Senés km 4, E04200 Tabernas, Spain; [b]Nanotechnology and Integrated BioEngineering Centre, School of Engineering, University of Ulster, Newtownabbey, Northern Ireland, BT37 0QB, UK; [c]Nireas-International Water Research Centre, University of Cyprus, P.O. Box 20537, 1678, Nicosia, Cyprus
*E-mail: pilar.fernandez@psa.es

3.1 Solar Disinfection of Water

3.1.1 Solar Spectrum and SODIS Method

For the solar spectrum at sea level, the UV (λ = 280–400 nm) is the most energetic and germicidal. It is well known that the overexposure to natural sunlight may cause damage in cells. UV-B radiation (λ = 280–320 nm) is believed to be the component of sunlight mainly responsible for most human skin cancers as it overlaps with the tail of the DNA absorption spectrum. However, the UV-A region (λ = 320–400 nm) of sunlight is also potentially carcinogenic and is certainly involved in photo-aging. Practical applications of solar disinfection have

RSC Energy and Environment Series No. 15
Photocatalysis: Applications
Edited by Dionysios D Dionysiou, Gianluca Li Puma, Jinhua Ye, Jenny Schneider, and Detlef Bahnemann

Published by the Royal Society of Chemistry, www.rsc.org

been widely demonstrated and there are a large number of scientific contributions that show the inactivation of different water pathogens under a range of environmental conditions. Nevertheless, some groups of microorganisms show significant resistance to solar disinfection, *e.g.* parasites (*Cryptosporidium* spp. and *Giardia*), spores (SRC – *Clostridium perfringens*, *Fusarium*, *Bacillus subtilis*, *etc.*), virus and phages (echovirus, norovirus, MS2, FRNA phage, *etc.*). In some cases they cannot be inactivated or require very high solar UV doses.[1]

One protocol for the application of solar disinfection for drinking water is termed the "SODIS" method. This involves the use of clear 1–2 L PET bottles which are filled with raw water and then are exposed to the sun for 6–8 h (one day) of sunshine or two consecutive days in cloudy conditions. The SODIS treated water will have a reduced load of pathogens and therefore be safer to drink; however, its consumption is recommended only the day after treatment. This method is recognised by the World Health Organization (WHO) as an appropriate Household Water Treatment intervention to disinfect water particularly in regions where lack of access to safe water is an issue, including emergency situations. It is estimated that SODIS is currently being used by more than 5.5 million people around the world, mainly in developing regions of Asia, Africa and Latin America. SODIS has been compared with other household water treatment and storage methods and it was found that SODIS was slightly less cost-effective than chlorination; however, the latter requires the distribution of sodium hypochlorite or chlorine tablets, whereas solar energy is widely and freely available.[2]

3.1.2 Oxidative Stress Caused by UV Radiation

It is accepted that one of the most important mechanisms for photobiological damage is DNA lesion resulting from the direct absorption of sunlight in the UVB photon range (280–320 nm).[3] The injury caused by UV-A and UV-B light can also be due to photonic absorption by cellular components such as chromophores. The best known intracellular chromophore is probably L-tryptophan; other potential chromophores are also considered to contain unsaturated bonds such as flavins, steroids, and quinines.[4]

In addition, it has been demonstrated that oxygen plays a critical role in the photo-inactivation of bacterial cells such as *Escherichia coli* (*E. coli*),[5] particularly at wavelengths longer than 313 nm.[6] The irradiation of intracellular chromophores with UV-A light is only toxic to cells in the presence of oxygen. Cell damage *via* chromophore light absorption is attributed to the generation of reactive oxygen species (ROS), such as hydroxyl radical ($OH^{\bullet}$), superoxide radical anion ($HO_2^{\bullet}$), hydrogen peroxide (H_2O_2), and singlet oxygen (O_2^*). Perhydroxyl radical and reactive nitrogen species (RNS) have also been proposed as reactive intermediates.[7] Kadir and Nelson[8] showed that *E. coli* inactivation was strongly dependent on dissolved oxygen concentrations and the presence of UVB wavelengths; they also found that singlet oxygen is likely an important endogenous species during solar photo-inactivation of *Enterococcus faecalis* and *E. coli*. In line with this,

Gourmelon *et al.* also observed a decrease of the toxic effect of visible light irradiation on *E. coli* under anaerobic conditions. Some ROS scavengers such as β-carotene (singlet oxygen scavenger), and superoxide dismutase (SOD) protect bacteria from UV photons *via* different mechanisms. Catalase, which eliminates hydrogen peroxide and thiourea as a hydroxyl radical scavenger, showed complete protection against light inactivation of bacteria.[9] Very recent research on *E. faecalis* inactivation under natural sunlight in filtered seawater revealed new findings about the mechanisms of photo-inactivation of these bacteria. Various methods (culturing, quantitative PCR, PMA-qPCR, LIVE/DEAD staining by PI, and cellular activity including ATP concentrations and expression of the superoxide dismutase-encoding gene, soda) were used to investigate the importance of oxygen-mediated and non-oxygen-mediated photo-inactivation mechanisms. Photo-inactivation was faster under aerobic than under anoxic conditions, suggesting that oxygen-mediated mechanisms of solar inactivation dominated. Nucleic acids were not significantly affected, and the damage to the membrane was minimal with solar photo-inactivation. The authors also suggested that the cells were actively responding to oxidative stress.[10]

Other contributions have reported on UVA oxidative stress causing damage to cell components, such as lipid peroxidation and DNA rupture generating single strand breaks (SSBs) and nucleic base modifications, which lead to lethal or mutagenic damage.[11] Some authors have reported that proteins can be the main target of ROS during solar inactivation: they have observed an increase in protein carboxylation during a very early stage of irradiation (exposure of UVA dose of 250 kJ m^{-2}), followed by massive accumulation of protein aggregates after longer periods of irradiation (exposure of about 1000 kJ m^{-2}).[12] The damaged proteins were enzymes involved in translation, transport, transcription, glycolysis, DNA-repair, protein folding, respiration, and ATP synthesis. This work suggested that the proteins which are first affected are those in the cytoplasmic membrane, and the primary reason for inactivation by solar irradiation is protein damage and not DNA damage.

Scavenging enzymes such as catalase, peroxidase and superoxide dismutase, which are found in all aerobic organisms, inhibit the action of ROS. These scavenging agents not only act against UVA photo-induced oxidative stress, but also the repair of oxidative lesions that occur during normal metabolic activity such as respiration. The main oxidative species involved are perhydroxyl radical, hydrogen peroxide, and hydroxyl radical. While intracellular absorption of UVA induces direct formation of superoxide radicals, the generation of intracellular hydroxyl radicals is attributed to Fenton and Fenton-like reactions (3.1) and (3.2):[13]

$$Fe^{2+} + H_2O_2 \rightarrow Fe^{3+} + OH^- + HO^\bullet \; (k \approx 70\ M^{-1}\ s^{-1}) \quad (3.1)$$
$$Fe(OH)^{2+} + h\nu \rightarrow Fe^{2+} + HO^\bullet \quad (3.2)$$

Intracellular Fe(II) may be either free, adsorbed to proteins, nucleic acids, and lipids,[14] or incorporated into protein moieties such as porphyrins[15] or iron–sulfur clusters.[16] Bound iron can be released from enzyme co-factors, such as iron–sulfur clusters by ROS like superoxide.[17]

Superoxide, in turn, can be produced by the adventitious reduction of molecular oxygen by reduced flavins in the electron transport chain. Moreover, superoxide and singlet oxygen can be produced by the photo-excitation of flavins, as well as of other cellular constituents such as NADH and NADPH.[18,19] Therefore, there are ample opportunities for free intracellular iron to participate in reactions inside cells exposed to light and oxygen.

In summary, the following mechanisms have been proposed to explain photo-oxidative damage over bacterial cells under sunlight:

i. intracellular light-driven Fenton reactions which damage DNA;
ii. protein damage;
iii. disruption of electron transport chain which causes increase in membrane permeability and loss of functionality.

Although the inactivation mechanisms involved in solar disinfection have been widely investigated, there is still a lack of knowledge on the role of some intracellular components.

3.2 Solar Photocatalytic Disinfection of Water

Since Matsunaga *et al.*[20] first reported the inactivation of bacteria using heterogeneous photocatalysis (also called semiconductor photocatalysis) in 1985 there have been at least 1000 research papers published in the area. The effectiveness of photocatalysis against microorganisms including bacteria (cells,[21,22] spores,[23] and biofilms[24]), viruses,[25] protozoa,[26] fungi,[27] and algae[28] has been investigated and the published work has been extensively reviewed by McCullagh *et al.*,[29] Malato *et al.*,[30] and Robertson *et al.*[31]

3.2.1 Mechanisms of TiO_2 Photocatalytic Disinfection

It is generally reported that the photocatalytic disinfection of water requires at least minutes of direct exposure to irradiation of equivalent energy to the band gap of the semiconductor (E_{bg} for TiO_2 anatase = 3.2 eV ≤ 387 nm) and it can be regarded as quite a slow process, as compared to UVC disinfection. Most papers focus on the assessment of novel materials, new reactor systems, or the effect of experimental parameters on the rate of inactivation. There are a significant number of studies which have specifically investigated the mechanism involving the generated ROS and their interaction with the biological structures on or within the microorganisms, and the resultant inactivation or loss of viability. Dalrymple *et al.* conducted a review of the mechanisms involved[32] but concluded that the exact sequence of events leading to loss of viability is not completely clear. Further insight into the effect of ROS on organisms and the mechanism of action will allow researchers to tailor materials and processes, and thereby help to improve the disinfection efficiency.

Upon excitation of the semiconductor by the absorption of irradiation equal to or greater than the band gap, electron–hole pairs are generated that can either recombine or migrate to the particle solution interface where they can take part in redox reactions. In the presence of oxygen and water, a range of ROS can be generated at the semiconductor particle–solution interface. The hydroxyl radical ($HO^•$) has been suggested to be the primary species responsible for microorganism inactivation; however, superoxide radical anion ($O_2^{•-}$), hydroperoxyl radical ($HO_2^•$), and hydrogen peroxide (H_2O_2) have been shown to contribute to the biocidal process.[33] In the discipline of biology the toxicity of "free radicals" is well known. Biological systems have enzymatic process to convert ROS into less toxic species (*e.g.* catalase and superoxide dismutase) and hydroxyl radicals play a pivotal role in the reaction of white blood cells with pathogens and apoptosis (programmed cell death). The prevalence of ROS within the environment in which microorganisms thrive and the evolution of defence systems against these active species are of particular relevance to photocatalytic disinfection. Unlike antibiotics, which target a specific biological process within the lifecycle of bacterial organisms (only), ROS attack is not specific to one site or an individual pathway. This permits the use of photocatalysis against a wide range of pathogens and the development of bacterial resistance to photocatalysis is considered to be almost impossible. However, photocatalysis must produce enough ROS to overcome the defence systems of the microorganisms to result in complete inactivation and prevent bacterial recovery and re-growth.

With photocatalysis, ROS attack on microorganisms typically occurs at the outside of the organism, and then inside, destroying the sensitive metabolic processes and genetic material. The resistance of the outer layers of the organism to ROS attack effectively dictates the ability of the organism to survive. The thick protein, carbohydrate, and lipid structures surrounding protozoa and bacterial spores gives more resistance to ROS attack than viruses, fungi, and bacteria, with resistance observed in that order respectively.[29] Due to the relative ease of culture and detection, bacteria have been the most widely studied organism, with *Escherichia coli* being the primary species used. Typically bacterial organisms are classified based upon the content of their outer cell layers, which surround an internal liquid based cytoplasmic matrix, consisting of genetic material, and the biochemical systems used for energy production, cell regulation, and reproduction. The cytoplasm is bounded by the cell membrane. This is a phospholipid bilayer, containing cross membrane proteins structures, which regulate transmission of chemicals into and out of the cytoplasm. The cell membrane maintains the integrity of the cells and is of particular importance for bacterial viability. In Gram-positive bacteria, the cell membrane is surrounded by a cell wall consisting of a thick layer (20–80 nm) of porous peptidoglycan which is responsible for structural integrity. Chains of lipoteichoic acid extend from the cell membrane through the cell wall and play roles in cell binding reactions. The cell wall in Gram-negative bacteria is much more complex. A thin layer of peptidoglycan (7–11 nm) is encapsulated by a second phospholipid bilayer, termed the outer membrane. This is populated with long extending chains of lipopolysaccharide, which elicit strong immune responses in animals, and as with the inner cell membrane contains cross membrane protein channels.

3.2.2 Microbiological Aspects of Photocatalytic Disinfection

Given the complexity of microorganisms it is perhaps understandable why the full mechanism of photocatalytic inactivation is still unknown; however, the accepted sequence of events taking place during photocatalytic inactivation of microorganisms is that prolonged ROS attack results in damage of the cell wall, followed by compromise of the cytoplasmic membrane and direct attack of intracellular components (Figure 3.1). To this end, microscopy based studies have revealed the following: formation of pores within cell wall and cell membrane structures[34] the degradation of peptidoglycan,[35] lipid peroxidation within phospholipids membranes,[34,36–38] degradation of porin proteins within cell membranes,[39] the detection of intercellular compounds and genetic material exterior to the cell confirmed,[20,40] and direct DNA damage reported *via* genetic analysis.[41] More specific modes of action discussed include membrane damage leading to inactivation of respiratory pathway chemistry[34] and/or loss of fluidity and increased ion permeability.

The general consensus relates to hydroxyl radical mediated lipid peroxidation of the outer cell wall components, and perhaps the most conclusive evidence for this was recently reported by Kubacka *et al.*[42b] They undertook a detailed and systematic investigation into the levels of a range of genetic and protein markers during TiO_2 photocatalysis and concluded that extensive radical induced cell wall modifications are the main factor responsible for the biocidal action. Clearly, the interaction between photocatalyst particles and the surface of the microorganism cell or spore will have an important role in

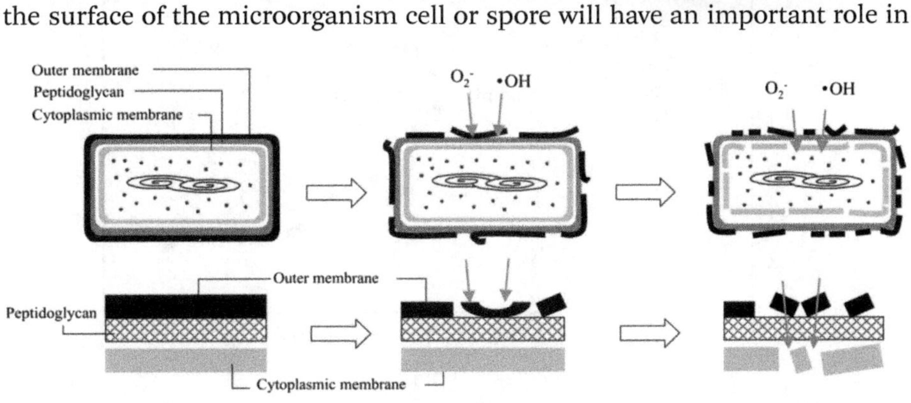

Figure 3.1 Schematic illustration of the process of *Escherichia coli* inactivation on photo-excited TiO_2. In the lower row, the part of cell envelope is magnified. Reprinted from Journal of Photochemistry and Photobiology A: Chemistry, Vol. 156, K. Sunada, T. Watanabe and K. J. Hashimoto, Studies on photokilling of bacteria on TiO_2 thin film, Pages 227–233, Copyright (2003), with permission from Elsevier. ref. 38

the disinfection process. The surface charge of TiO_2 samples has been proven to affect the inactivation kinetics of *E. coli*. TiO_2 Evonik P-25 particles in suspension are in contact with the *E. coli* cell walls and, depending on the pH, the solution will have a surface charge that can be either positive or negative. This will have implications in the surface interaction between catalyst particles and the microbial cell wall, which is intrinsically negative (Figures 3.2 and 3.3).[42a]

Operational parameters will have important effects on the observed rate of photocatalytic inactivation, depending on the type of microorganism. For instance, *E. coli* (Gram negative) is more susceptible to osmotic stress effects as compared to *E. faecalis* (Gram positive).[43,44] However, results concerning the photocatalytic inactivation of Gram positive *versus* Gram negative are sometimes conflicting. For example, some researchers report that *E. coli* is more resistant than *E. faecalis* to the TiO_2 photocatalysis process, which was

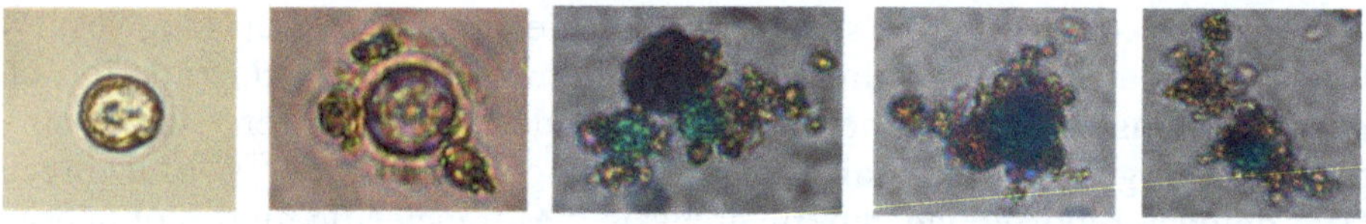

Figure 3.2 Spores of *Phytophthora capsici* in water in the presence of TiO_2 P25 Evonik slurry (100 mg L^{-1}) irradiated (UVA: 30 W m^{-2}) with sunlight. Contact time from left to right: 0, 15, 60, 120, and 300 min.

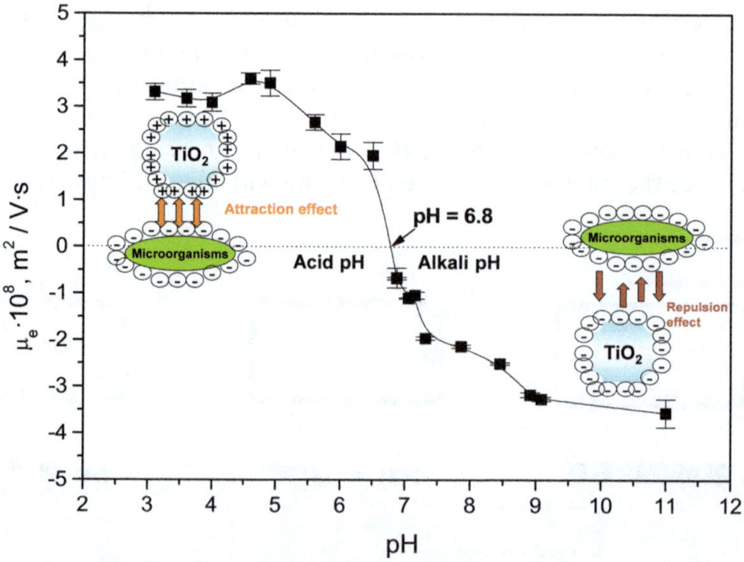

Figure 3.3 Electrophoretic mobility of P25 TiO_2 nanoparticles as a function of pH and point of zero charge (PZC = 6.8). Potential electric interactions between TiO_2 and microorganism cell. Reprinted from Water Research, Vol. 37, P. Fernández-Ibáñez, J. Blanco, S. Malato and F. J. de las Nieves, Application of the colloidal stability of TiO_2 particles for recovery and reuse in solar photocatalysis, Pages 3180–3188, Copyright (2003), with permission from Elsevier. ref. 42a

explained by the presence of the outer membrane in the Gram negative bacteria adding an extra barrier to protect against ROS attack.[45,46] In addition, the different chemical composition of the cell wall and protection mechanisms between the two types of bacteria may also be responsible for different efficiencies reported for photocatalytic inactivation.[47] Some papers report that *E. faecalis* is more resistant than *E. coli* to different solar treatments, like H_2O_2/sunlight and solar photo-Fenton at pH 3 and 5. This greater resistance was attributed to the thicker cell wall of *E. faecalis*, combined with the different internal defence mechanisms of the bacterium.[48]

Different sensitivities of *E. coli* and *Bacillus* spp. to photocatalytic treatment have been reported and explained by the morphological and biological differences between these bacterial groups. Spores are developed by some microorganisms under stress conditions; like *Bacillus* sp. The walls of these spores contain keratin with sulfide links, making the spores more resistant to more extreme environmental conditions such as high salinity, low and high pH, and oxidative conditions, the latter being generated by AOPs (advanced oxidation processes) such as photocatalysis.[49] These authors reported observed orders of photocatalytic disinfection efficiency with different faecal bacteria, with the observed rate of inactivation as below:

E. coli > *Enterococcus* sp. > all coliforms excluding *E. coli* > total Gram-negative.

Escherichia coli was found to be more vulnerable than spores of *Fusarium solani* (fungus) to photocatalytic treatment with either TiO_2 [50] or TiO_2–reduced graphene oxide composites.[51,52] This difference observed was attributed to the structure and chemical composition of these spores. Fungal spore walls are rigid structures constituted of proteins, glycoproteins, and poly-sugars, with an extra outer xylan layer.[53] This multifunctional structure confers a high resistance against different stress factors, like the photocatalytic action of TiO_2. Other authors reported differences in the UV dose required for the photocatalytic inactivation of different types of microorganism, including viral particles, spores of bacteria, and faecal bacteria.[54]

Whilst photocatalysis has great potential to be used as a biocidal technology, caution should to be exercised when conducting photocatalytic disinfection assays as organisms have the potential to recover from sub-lethal ROS exhibited stress and re-grow. Several authors have reported the need for complete inactivation of organisms and specific analysis to confirm that the treatment prevents subsequent bacterial re-growth.[21,55] Several contributions in the literature have assessed the capability of bacteria and other microorganisms to repair and recover or re-growth after a photo-oxidative treatment for periods of until 24 and 48 h post-treatment. Bacterial regrowth is considered by some to be as important as the inactivation rate or the final concentration following treatment. Some authors have defined the "EDT24" as the time necessary for total inactivation of bacteria without re-growth in a subsequent dark period referenced at 24 h (or 48 h).[50] This parameter is now being used by numerous researchers in the field and also for solar disinfection,[1] photo-Fenton,[56] and H_2O_2/sunlight.[57]

3.2.3 Inactivation of Antibiotic-Resistant Bacteria

An exciting new area for investigation is the inactivation of antibiotic-resistant bacteria in water and wastewater by photocatalytic processes. The use of antimicrobial agents such as antibiotics in large quantities was initiated in 1930s, due to their high effectiveness in the treatment of bacterial infectious diseases in humans and in veterinary practices.[58] As a result of the excessive use of these compounds and of the disposal of their unused amounts, their release in both metabolised and/or un-metabolised forms into urban wastewater systems has risen. The inability of wastewater treatment plants to completely remove them has made their entry probable into natural aquatic ecosystems such as surface waters and groundwaters through wastewater effluents, where they are a possible source of risk to environmental and human health. Wastewater effluents act as reservoirs of pathogenic antibiotic-resistant bacteria (ARB) due to the continuous exposure of a diverse bacterial community to antibiotic residues, but also represent important bioreactors where antibiotic resistance genes (ARG) are introduced and spread in pathogenic bacterial ecosystems, as environmentally relevant bacteria represent a large continuous pool of genetic material which may express as ARG, once incorporated into pathogenic bacteria.[59]

The dissemination and proliferation of ARB and of their associated genes is made possible by mechanisms of horizontal gene transfer (HGT) from one donor cell to other approximate cells which may belong to a different species, and vertical gene transfer (VGT) from one generation to the next.[60]

The proliferation and tolerance of the ARB comes about through the production and adaptation of specific mutations which provide the bacteria with physiological tools such as efflux pumps and chromosomal AmpC β-lactamases, to enable their survival in the presence of compounds which would otherwise be toxic to them, leading to the bacterial population decimation. The prevalence of various pathogenic ARB has been studied in wastewater effluents even at sub-inhibitory concentrations of antibiotics, which are the minimal selective concentrations at which there is selection of a given resistance mutation.[61] Mutations (permanent changes of the genetic material of a cell) occur in pre-existing genes of the bacterial chromosome and are positively selected by environmental forces, *i.e.* exposure to antibiotics.[62]

As a result, AOPs and more specifically photocatalytic processes, which aim at increasing the removal potential of ARB and ARGs in aquatic systems compared to the conventional biological wastewater treatments, have gained increased scientific attention.[63,64] For example, the UV TiO_2 (Aeroxide P25) photocatalytic inactivation of three types of bacteria, namely methicillin-resistant *Staphylococcus aureus* (MRSA), multi-drug resistant *Acinetobacter baumannii* (MDRAB), and vancomycin-resistant *Enterococcus faecalis* (VRE), was investigated by Tsai *et al.*[65] The results of this study have shown that the experimental reaction time of the bacteria in suspension with TiO_2 was the decisive factor; it exerted the greatest influence on the microbial survival after exposure to UVA/TiO_2 treatment and production of ROS. The shift in the proportion of ARB among the total bacterial population after treatment did not appear to correlate to an increasing resistance towards the photocatalytic treatment, although the total bacterial population was susceptible to the treatment and exhibited 1–3 log reduction. A study by Karaolia *et al.*[66] demonstrated that the proportion of the total cultivable bacteria which were resistant to a mixture of sulfamethoxazole, erythromycin, and clarithromycin showed a 1-log increase after TiO_2 photocatalysis (Aeroxide P25) under solar irradiation, despite the fact that there was a reduction in the total bacterial counts. There was no change in the proportion of antibiotic-resistant cultivable bacteria after dark treatment and photolysis after 120 min of treatment. In another study, Paul *et al.*[67] studied the residual antimicrobial activity of ciprofloxacin (100 μM) on an *E. coli* population after photocatalytic treatment under visible light ($\lambda > 400$ nm) and under UVA ($\lambda > 325$ nm) radiation, with the use of Hombikat UV100 catalyst (0.5 g L^{-1} TiO_2). Inactivation of the examined ARB was greater under UVA-TiO_2 relative to vis-TiO_2. Furthermore, the photocatalytic deactivation of the antibacterial potency of ciprofloxacin-resistant *E. coli* was stoichiometrically related to the ciprofloxacin degradation during the treatment. Xiong and Hu[68] investigated the inactivation/reactivation potential of immobilised TiO_2 of ampicillin and streptomycin-resistant *E. coli* under UVA radiation. Their study showed the impact of the illumination time as an important parameter in the inactivation of antibiotic-resistant *E. coli*, as longer illumination time left a more intense residual disinfecting effect, after 90 min of treatment. In addition, a higher radiation pulse frequency and shorter circles of illumination resulted in a higher inactivation impact of *E. coli* at the same UV dosage. Furthermore, the study demonstrated the reactivation of cells in the 4 h following treatment, with dark repair and photo-reactivation taking place after UVC illumination alone, but not after UVA/TiO_2 treatment. Dunlop *et al.*[69] reported on the efficiency of photocatalytic disinfection on an antibiotic sensitive *E. coli* strain (K12) and two antibiotic resistant strains, namely J-53R (rifampicin resistant) and HT-99 (chloramphenicol resistant), within an immobilised TiO_2 stirred tank reactor. The study showed a 2.5-log decrease in viable cell numbers (CFU mL^{-1}) within 180 min of treatment, followed by a recovery of the two strains of ARB back to their original numbers, after post-treatment incubation at 37 °C for 24 h, as the sub-lethally injured ARB employed repair mechanisms for damage reversal and regrowth. The antibiotic sensitive strain was not recovered after post-treatment incubation, suggesting the decreased susceptibility of the ARB to oxidative stress during photocatalytic treatment. The conjugated gene pair numbers after photocatalysis were four-times higher compared to the non-treated controls, indicating the efficacy of photocatalytic treatment and the importance of treatment time as an operating parameter during the treatment of antibiotic-resistant bacterial populations.

Several studies suggest the relation of bacterial ROS oxidative stress to antibiotic resistance mechanisms.[70,71] Many microorganisms adapt survival mechanisms, among which is *E. coli*, which exhibits high tolerance to oxidative stress by ROS through oxidative stress-defence mechanisms that also confer its multiple antibiotic resistance. The two major regulatory systems involved in this physiological phenomenon, OxyR and SoxRS, are targeted toward peroxides and superoxides, respectively. The *mar* region confers resistance to multiple antibiotics and some of the genes controlled by this regulon (a collection of genes under regulation by the same regulatory protein) are also members of the redox-responsive SoxRS regulon. As a result, once activated, this regulon brings about a redox-response to ROS stress and at the same time makes the bacterium resistant to multiple antibiotics.[72]

The reactivation of bacterial cells during treatment may be induced by increased sub-lethal environmental stress which instead of causing complete inactivation produces a rapid reproductive response and increase of the rate of genetic transfer among pathogenic bacteria, as they exchange ARG through transformation of extracellular or free genetic material in the environment which was released by dead or lysed bacteria.[73] This effect highlights the importance of the dose of radiation and catalyst loading needed to achieve irreversible bacterial inactivation and repression of reactivation, a consideration which when taken under account may be important for reduced energy consumption in photocatalytic applications.

Moreover, the presence of organic constituents in treated matrices such as wastewater effluents leads to decreased photocatalytic efficiency due to the ROS scavenging which protects the bacteria from the bactericidal agents. If photocatalytic treatment does not last long enough to render pathogenic bacteria completely inactivated, there is potential re-growth once sub-lethal damage is repaired, leading to increased ARG transfer and increased prevalence of ARB in treated effluents.

3.3 Novel Photocatalytic Materials for Visible Light Activity

Most research investigating photocatalytic disinfection is carried out with TiO_2, a wide band gap semiconductor that requires UVA excitation. For solar applications, visible light active materials are desirable, to increase the number of photons which can be utilised as UV is only *ca.* 4% of the solar spectrum. Narrower band gap materials that can utilise visible photons have a smaller voltage window to drive the redox reactions at the interface. Relatively few publications deal with the photocatalytic disinfection of water using novel visible light active materials. Notably, the UV activity of an undoped material may be greater than the visible light activity of a doped material and, therefore, for solar applications the efficiency should be tested under simulated solar irradiation or under real sun conditions (which includes the UV component). There are several good reviews of visible light active photocatalytic materials.[74,75]

3.3.1 Doped Materials

Matsunaga *et al.* (1985) provided one of the earliest accounts of doped TiO_2 for photoelectrochemical sterilisation of microbial cells. Platinum loaded TiO_2 powders were demonstrated to inactivate *Lactobacillus acidophilus*, *Saccharomyces cerevisiae* and *E. coli* within 2 h.[20] In addition, doping of ZnO nanostructures has been demonstrated to enhance the inactivation process either by shifting the band towards visible side or by dopant acting as co-catalyst. Several dopants including Pd,[76] Ce,[77] Co,[78] and Ag[79] have been demonstrated to enhance the bacterial inactivation. Karunakaran *et al.* have investigated the effect of a preparation method on photocatalytic inactivation activity of ZnO and Ag doped ZnO materials towards *E. coli*. The materials were synthesised by three methods, *i.e.* sol–gel,[80] combustion,[80] and microwave synthesis,[81] of which sol–gel synthesised materials have demonstrated the highest photocatalytic disinfection efficiency.

Although metal ion dopants are the most studied for visible light activity there is a lack of agreement over improvements in efficiency, with as many reports claiming enhanced activity as there are reporting a reduction in activity.[82] TiO_2 doped with various elements has been investigated for photocatalytic disinfection. Vohra *et al.* have explored Ag^+ doped P25 TiO_2 for disinfection of indoor air. *Bacillus cereus*, *Staphylococcus aureus*, *E. coli*, *Aspergillus niger*, and MS2 bacteriophage have been successfully inactivated.[83] Silver-doped TiO_2 has also been effective in water disinfection.[84] Copper and S doped TiO_2 nanoparticles have been effectively utilised for inactivation of *E. coli* and *Micrococcus lylae*, respectively.[85,86] Non-metals C, N, S, B, and the halogens as dopants for TiO_2 and ZnO have been reviewed by Rehman *et al.*[87] and also by Im *et al.*[88] Often, co-doped materials are reported to have higher rates than the single dopant regimes, as demonstrated by Li and co-workers, comparing nitrogen and carbon–nitrogen co-doping on the inactivation rate for *E. coli*.[89] Second generation photocatalysts like WO_3 [90] can also be doped to improve visible light activity, and present a further approach to developing new materials for disinfection.[91]

Rengifo-Herrera and Pulgarin reported on the photocatalytic activity of N, S co-doped, and N-doped commercial anatase (Tayca TKP 102) TiO_2 powders towards phenol oxidation and *E. coli* inactivation.[92] The doped materials did not present any enhancement as compared to undoped TiO_2 (Evonik Aeroxide P25) under simulated solar irradiation. While the N or N, S co-doped TiO_2 are coloured materials, the localised states responsible for the visible light absorption do not play an important role in the photocatalytic activity.

3.3.2 Other New Materials

Metal carbide, nitride, and sulfide materials generally have narrower band gaps than oxide materials but they tend to photo-corrode in aqueous media. Some non-metal oxide materials have been investigated, *e.g.* $ZnIn_2S_4$ was reported to show visible light photocatalytic activity for the inactivation of

E. coli (under electrochemical bias).[93] Binary metal oxide, *e.g.* bismuth vanadate, photocatalysts have also been reported for disinfection[94,95] with further enhancements in rate of disinfection reported with silver as a co-catalyst.[96] Titanium oxynitrides and niobium oxynitrides have been reported to show activity for the disinfection of water, with PdO nanoparticle modified TiON nanofibers reported for *E. coli* inactivation under visible light irradiation.[97] Perovskite materials such as $K_4Nb_6O_{17}$ and Ag/Cu modified $K_4Nb_6O_{17}$ have also been reported for the inactivation of *E. coli* under visible light irradiation.[98,99]

An interesting material is titanium dioxide–reduced graphene oxide (TiO_2–RGO) composite. This material may be easily synthesised by the photocatalytic reduction of exfoliated graphene oxide (GO) by TiO_2 under UV irradiation in the presence of a hole scavenger. Fernández-Ibáñez *et al.*[52] evaluated TiO_2–RGO composites for the disinfection of water contaminated with *E. coli* cells and *F. solani* spores under natural sunlight (Figure 3.4). Rapid water disinfection was observed with both *E. coli* and *F. solani*. An enhanced rate in the *E. coli* inactivation efficiency was observed with the TiO_2–RGO composite compared to TiO_2 (Evonik Aeroxide P25). The disinfection efficiency was also evaluated with filtered sunlight where the major part of the solar UVA was cut-off (λ > 380 nm) using a Plexiglass® screen. In this case, a much greater time was required for inactivation of *E. coli* with TiO_2 P25 but the same inactivation rate was observed for the TiO_2–RGO indicating visible light activity which could be attributed to singlet oxygen production by TiO_2–RGO composites, which would lead to *E. coli* inactivation.

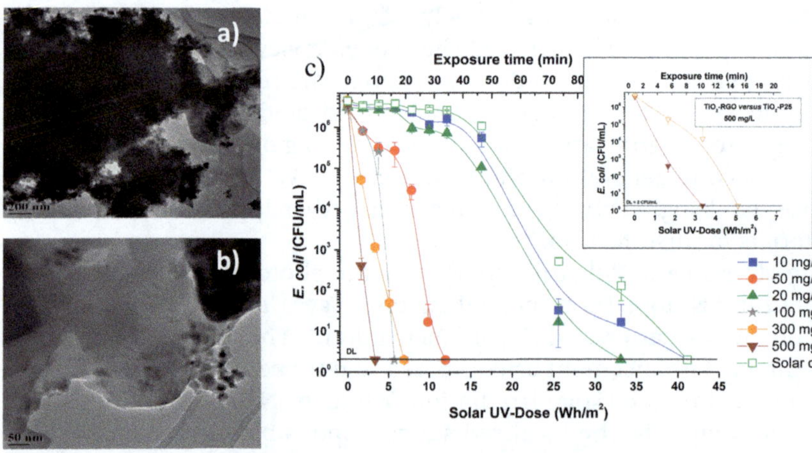

Figure 3.4 (a) TiO_2–GO aggregate before photoreduction, (b) TiO_2–RGO after UV assisted photoreduction and (c) *Escherichia coli* inactivation at several TiO_2–RGO concentrations. Insert: efficiency of TiO_2–RGO and TiO_2–P25 on the *E. coli* inactivation. Reprinted from Chemical Engineering Journal, Vol. 261, P. Fernandez-Ibanez, M. I. Polo-Lopez, S. Malato, S. Wadhwa, J. W. J.Hamilton, R. D'Sa, E. Magee, K. O'Shea, D. D. Dionysiou and J. A. Byrne, Solar photocatalytic disinfection of water using titanium dioxide graphene composites, Pages 36–44, Copyright (2015), with permission from Elsevier.[51]

3.4 Solar Photocatalytic Reactors for Water Disinfection

The solar or solar photocatalytic inactivation of microorganisms in water should be enhanced by the smart design of photo-reactors. The first photo-reactors for solar photocatalytic applications, designed at the end of the 1980s, were based on parabolic-trough collectors. One of the main advantages of such photo-reactors is that they may enhance the disinfection rate by increasing the photon flux to the water. However, to optimise photo-reactor efficiency, other factors must also be taken into account such as the total volume of treated water, reduction of the user dependence of the process and use of cheap and robust materials. A wide range of different reactor configurations have been tested for photocatalytic water disinfection at laboratory- or mini-pilot scale (10 mL to 2 L), normally to test catalyst efficiency. In some cases, research has been undertaken with pilot-scale photoreactors (>10 L). At pilot scale, one of the most promising approaches is to utilise compound parabolic collectors (CPC). CPC reactors have proven successful for both water disinfection and detoxification. They collect solar radiation under static conditions with a high collection rate of the solar diffuse-radiation. The main advantages are:

i. use of non-imaging concentration with diffuse focus;
ii. highly efficient use of the solar photon flux due to the homogeneous distribution of radiation into the absorber;
iii. utilisation of both diffuse and direct solar radiation, having high efficiency even on cloudy days;
iv. maintaining a constant concentration factor (CF = 1) for all values of sun zenith angle within the acceptance angle limit.

The photo-reactor tube should be made of UV transparent material, *e.g.* borosilicate glass has a high transmission in the UVA & B range (90%). CPC mirrors and tubes are positioned with a tilt angle equal to the local latitude and facing south in the northern hemisphere or facing north in the southern hemisphere. CPC reactors have a low environmental impact, they are easy to construct and maintain, and they have low power requirements. CPC technology is much less costly than highly concentrating systems. Demonstration photocatalytic plants have been installed for the treatment of industrial wastewater contaminated with pesticides, pharmaceuticals, *etc.*[29,100] An advantage of the CPC system is the modular system, and Polo-López and co-workers[101] reported on a semi-autonomous batch system which could be scaled up or down by the addition/removal of modules.

Some of the main parameters affecting photoreactor efficiency are discussed below:

i. *The illuminated volume/total volume ratio*: In a flow system, the solar radiation dose should be delivered un-interrupted to avoid bacterial recovery. For static batch systems, the total volume should be irradiated.

ii. *Catalyst loading in slurry/suspension reactors*: Different and conflicting results regarding the catalyst loading using suspended TiO_2 systems have been reported the literature. However, differences reported may be due to the different reactor configurations used to conduct solar photocatalytic tests and/or using different target microorganisms. This highlights that although optical phenomena inside the photoreactor play an important role in determining the optimal catalyst concentration, the type of target microorganism can also influence the final results.

iii. *Immobilised versus suspended photocatalyst*: One of the disadvantages often highlighted with photocatalytic disinfection is the need to remove suspended photocatalyst particles from the water after treatment. This post-treatment can be avoided if the catalyst is immobilised onto supporting substrates. Intense research interest has focused on the development of methodologies and materials to immobilise the catalyst onto suitable supporting substrates. Suspension systems are always more efficient due to better mass transport and better irradiated catalyst contact with the microorganisms.

iv. *Dissolved oxygen*: The photocatalytic efficiency may be limited by the amount of dissolved oxygen in the water matrix since oxygen acts as the electron acceptor. It is necessary to remove the conduction band electrons to prevent electron–hole pair recombination and thus obtain the best quantum efficiency for hole transfer to solution. In addition, the reduction of molecular oxygen yields superoxide radical anion, hydroperoxyl radical, hydrogen peroxide, and, eventually, hydroxyl radical. In addition, the oxidation of superoxide radical anion by a valence band hole can yield singlet oxygen. Therefore, the reduction of molecular oxygen plays a vital role in the production of ROS for the inactivation of microorganisms and the reactor design should consider the maintenance of at least air saturated oxygen levels in the water before and during treatment.

Finally and notably, no study has ever set out to specifically design an efficient solar photocatalytic disinfection reactor – research groups have modified existing reactors widely used for chemical treatment. In this respect there is a need for a chemical engineering approach to try and design an efficient reactor for solar photocatalytic disinfection from first principles. The existing light modelling and CPC design (see Figure 3.5) are relevant, but the key role of oxygen, mass transfer of bacteria to the catalyst surface, and the catalyst support configuration need to be specifically designed for disinfection purposes. One of the major challenges for immobilised photocatalyst systems is to provide a large irradiated catalyst surface area to reactor volume ratio.

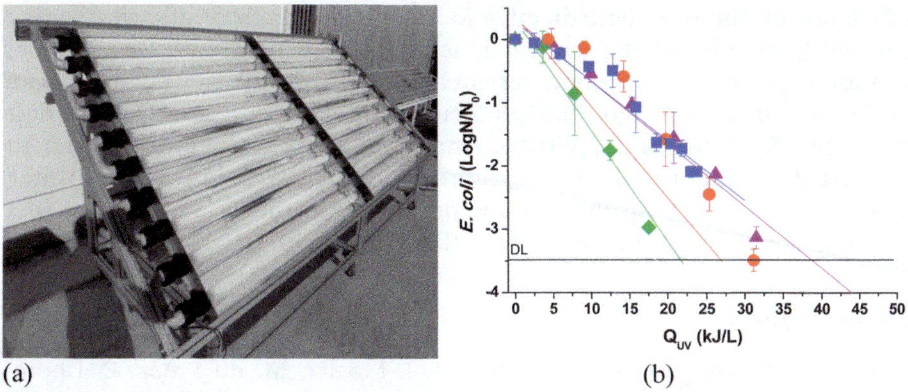

Figure 3.5 (a) Photograph of a CPC reactor for photocatalytic water disinfection at PSA facilities. (b) Photocatalytic inactivation of *Escherichia coli* under natural sunlight in real urban wastewater effluents in a CPC photoreactor at different temperatures (°C): 15 (■), 25 (▲), 35 (◆), and 45 (●). Detection limit = 2 CFU mL^{-1}. Reprinted from Catalysis Today, Vol. 240, P. Fernández-Ibáñez; J. Byrne; M. I. Polo-López; P. S. M. Dunlop, P. Karaolia, D. Fatta-Kassinos, Disinfection of urban effluents using solar TiO_2 photocatalysis: A study of significance of dissolved oxygen, temperature, type of microorganism and water matrix, Pages 30–38, Copyright (2015), with permission from Elsevier.[53]

3.5 Conclusions

The utilisation of solar energy for the disinfection of water is a very important area for research and development. According to the 2015 JMP update, in 2015, 663 million people still lack improved drinking water sources and many more rely on water that is not safe to drink due to contamination with pathogenic microorganisms. Furthermore, the re-use of wastewater is becoming more and more important due to water scarcity throughout the globe, and it is vitally important for safety reasons to ensure that water for re-use is free from pathogenic microorganisms, particularly if used for food-crop irrigation. Solar energy is free and ubiquitous on the Earth's surface. Solar irradiation of water can result in the inactivation of pathogenic microorganisms in water due to a combination of UV and thermal effects. The UV irradiation effects dominate at temperatures lower than 40 °C and the mechanisms are complex involving direct photolytic effects of UVB and indirect formation of reactive oxygen species through photochemical reactions within the microorganisms. Nevertheless, solar disinfection alone may be slow or ineffective against different microorganisms. The addition of a photocatalyst such as titanium dioxide can dramatically enhance the solar disinfection of water and the photocatalytic mechanism is non-specific through the production of reactive oxygen species, which reduces the likelihood of resistance being developed. More recent research has focused on the photocatalytic inactivation of antibiotic resistant microorganisms which are increasingly being identified as a problem of global concern. To improve upon the solar efficiency of photocatalytic disinfection, many researchers are investigating novel visible light active photocatalytic materials which can utilise more of the solar spectrum. Much more research is needed in this area and much can be learned from research that has focused on visible light active materials developed for solar water splitting. Finally, more research and development is needed to scale up photocatalytic reactors for the disinfection of water. Overall, the research involved in photocatalytic disinfection of water is very exciting, challenging, and multi-disciplinary.

References

1. K. G. McGuigan, R. N. Conroy, H. J. Mosler, M. du Preez, E. Ubomba-Jaswa and P. Fernandez-Ibañez, *J. Hazard. Mater.*, 2012, **235–236**, 29.
2. T. F. Clasen and L. Haller, *Water Quality Interventions to Prevent Diarrhoea: Cost and Cost-Effectiveness*, WHO, Geneva, 2008.
3. R. P. Sinha and D. P. Hader, *Photochem. Photobiol. Sci.*, 2002, **1**, 225.
4. R. M. Tyrrell and S. M. Keyse, *J. Photochem. Photobiol., B*, 1990, **4**, 349.
5. R. H. Reed, *Lett. Appl. Microbiol.*, 1997, **24**, 276.
6. R. B. Webb and M. S. Brown, *Photochem. Photobiol.*, 1979, **29**, 407.
7. B. Halliwell and J. M. C. Gutteridgde, *Free radicals in biology and medicine*, Oxford University Press, Oxford, UK, 3rd edn, 2007.
8. K. Kadir and K. L. Nelson, *Water Res.*, 2014, **50**, 307.
9. M. Gourmelon, J. Cillard and M. Pommepuy, *J. Appl. Bacteriol.*, 1994, **77**, 105.
10. L. M. Sassoubre, K. L. Nelson and A. B. Boehm, *Appl. Environ. Microbiol.*, 2012, **78**, 7776.
11. R. V. Miller, W. Jeffrey, D. Mitchell and M. Elasri, *ASM News*, 1999, **65**, 535.
12. F. Bosshard, C. Geiser, M. Bucheli, T. Schneider, K. Riedel and T. Egli, *Poster abstract presented at FEMS 2009, 3rd Congress of European Microbiologists*, Gothenburg, Sweden, 28 June–2 July 2009.
13. J. A. Imlay, *Annu. Rev. Microbiol.*, 2003, **57**, 395.
14. J. D. Rush, Z. Maskos and W. H. Koppenol, *FEBS Lett.*, 1990, **261**, 121.
15. R. W. Tuveson and L. J. Sammartano, *Photochem. Photobiol.*, 1986, **43**, 621.
16. I. Fridovich, *J. Exp. Biol.*, 1998, **201**, 1203.
17. J. A. Imlay, *Annu. Rev. Biochem.*, 2008, **77**, 755.
18. M. L. Cunningham, J. S. Johnson, S. M. Giovanazzi and M. J. Peak, *Photochem. Photobiol.*, 1985, **42**, 125.
19. M. L. Cunningham, N. I. Krinsky, S. M. Giovanazzi and M. J. Peak, *J. Free Radicals Biol. Med.*, 1985, **1**, 381.
20. T. Matsunaga, R. Tomoda, T. Nakajima and H. Wake, *FEMS Microbiol. Lett.*, 1985, **29**, 211.
21. P. S. M. Dunlop, J. A. Byrne, N. Manga and B. R. Eggins, *J. Photochem. Photobiol., A*, 2002, **148**, 355.
22. P. S. M. Dunlop, C. P. Sheeran, J. A. Byrne, M. A. S. McMahon, M. A. Boyle and K. G. McGuigan, *J. Photochem. Photobiol., A*, 2010, **216**, 303.
23. P. S. M. Dunlop, T. A. McMurray, J. W. J. Hamilton and J. A. Byrne, *J. Photochem. Photobiol., A*, 2008, **196**, 113.

24. N. G. Chorianopoulos, D. S. Tsoukleris, E. Z. Panagou, P. Falaras and G. J. E. Nychas, *Food Microbiol.*, 2011, **28**, 164.
25. R. J. Watts, S. Kong, M. P. Orr, G. C. Miller and B. E. Henry, *Water Res.*, 1995, **29**, 95.
26. O. Sunnotel, R. Verdoold, P. S. M. Dunlop, W. J. Snelling, C. J. Lowery, J. S. G. Dooley, J. E. Moore and J. A. Byrne, *J. Water Health*, 2010, **8**, 83.
27. C. Sichel, M. de Cara, J. Tello, J. Blanco and P. Fernández-Ibánez, *Appl. Catal., B*, 2007, **74**, 152.
28. C. A. Linkous, G. J. Carter, D. B. Locuson, A. J. Ouellette, D. K. Slattery and L. A. Smith, *Environ. Sci. Technol.*, 2000, **34**, 4754.
29. C. McCullagh, J. M. Robertson, D. W. Bahnemann and P. J. K. Robertson, *Res. Chem. Intermed.*, 2007, **33**, 359.
30. S. Malato, P. Fernandez-Ibanez, M. I. Maldonado, J. Blanco and W. Gernjak, *Catal. Today*, 2009, **147**, 1.
31. P. K. J. Robertson, J. M. C. Robertson and D. W. Bahnemann, *Hazard. Mater.*, 2012, **211**, 161.
32. O. K. Dalrymple, E. Stefanakos, M. A. Trotz and D. Y. Goswami, *Appl. Catal., B*, 2010, **98**, 27.
33. M. Cho, H. Chung, W. Choi and J. Yoon, *Water Res.*, 2004, **38**, 1069.
34. P. Wu, R. Xie, J. A. Imlay and J. K. Shang, *Appl. Catal., B*, 2009, **88**, 576.
35. J. Kiwi and V. Nadtochenko, *Langmuir*, 2005, **21**, 4631.
36. Z. Huang, P. Maness, D. M. Blake, E. J. Wolfrum, S. L. Smolinski and W. A. Jacoby, *J. Photochem. Photobiol., A*, 2000, **130**, 163.
37. M. Wainwright, *Int. J. Antimicrob. Agents*, 2000, **16**, 381.
38. K. Sunada, T. Watanabe and K. J. Hashimoto, *J. Photochem. Photobiol., A*, 2003, **156**, 227.
39. G. Carré, E. Hamon, S. Ennahar, M. Estner, M. C. Lett, P. Horvatovich, J. P. Gies, V. Keller, N. Keller and P. Andre, *Appl. Environ. Microbiol.*, 2014, **80**, 2573.
40. A. Rincon and C. Pulgarin, *Sol. Energy*, 2004, **77**, 635.
41. S. Pigeot-Rémy, F. Simonet, E. Errazuriz-Cerda, J. C. Lazzaroni, D. Atlan and C. Guillard, *Appl. Catal., B*, 2011, **104**, 390.
42. (a) P. Fernández-Ibáñez, J. Blanco, S. Malato and F. J. de las Nieves, *Water Res.*, 2003, **37**, 3180–3188; (b) A. Kubacka, M. S. Diez, D. Rojo, R. Bargiela, S. Ciordia, I. Zapico, J. P. Albar, C. Barbas, V. A. P. Martins dos Santos, M. Fernandez-Garcia and M. Ferrer, *Sci. Rep.*, 2014, **4**, 4134.
43. R. Van Grieken, J. Marugan, C. Pablos, L. Furones and A. Loópez, *Appl. Catal., B*, 2010, **100**, 212.
44. C. Sichel, J. Blanco, S. Malato and P. Fernández-Ibáñez, *J. Photochem. Photobiol., A*, 2007, **189**, 239.
45. G. Fu, P. S. Vary and C. T. Lin, *J. Phys. Chem. B*, 2005, **109**, 8889.
46. Y. Lan, C. Hu, X. Hu and J. Qu, *Appl. Catal., B*, 2007, **73**, 354.
47. T. N. Demidova and R. Hamblin, *Antimicrob. Agents Chemother.*, 2005, **49**, 2329.
48. J. Rodríguez-Chueca, M. I. Polo-López, R. Mosteo, M. P. Ormad and P. Fernández-Ibáñez, *Appl. Catal., B*, 2014, **150–151**, 619.
49. A. G. Rincón and C. Pulgarín, *Catal. Today*, 2005, **101**, 331.

50. C. Sichel, J. Tello, M. Cara and P. Fernández-Ibáñez, *Catal. Today*, 2007, **129**, 152.
51. P. Fernández-Ibáñez, M. I. Polo-López, S. Malato, S. Wadhwa, J. W. J. Hamilton, R. D'Sa, E. Magee, K. O'Shea, D. D. Dionysiou and J. A. Byrne, *Chem. Eng. J.*, 2015, **261**, 36.
52. I. García-Fernández, I. Fernández-Calderero, M. I. Polo-López and P. Fernández-Ibáñez, *Catal. Today*, 2015, **240**, 30.
53. I. García-Fernández, M. I. Polo-López, I. Oller and P. Fernández-Ibáñez, *Appl. Catal., B*, 2012, **121–122**, 20.
54. M. Agulló-Barceló, M. I. Polo-López, F. Lucena, J. Jofre and P. Fernández-Ibáñez, *Appl. Catal., B*, 2013, **136–137**, 341.
55. A. G. Rincon and C. Pulgarin, *Appl. Catal., B*, 2004, **49**, 99.
56. M. I. Polo-López, I. Oller and P. Fernández-Ibáñez, *Catal. Today*, 2013, **209**, 181.
57. F. Bichai, M. I. Polo-López and P. Fernández-Ibañez, *Water Res.*, 2012, **40**, 6040.
58. S. Rodríguez-Mozaz, S. Chamorro, E. Marti, B. Huerta, M. Gros, A. Sánchez-Melsió, C. M. Borrego, D. Barceló and J. L. Balcázar, *Water Res.*, 2015, **69**, 234.
59. F. Baquero, J. L. Martinez and R. Cantón, *Curr. Opin. Biotechnol.*, 2008, **19**, 260.
60. E. A. Auerbach, E. E. Seyfried and K. D. McMahon, *Water Res.*, 2007, **41**, 5.
61. L. Sandegren, *Upsala J. Med. Sci.*, 2014, **119**, 103.
62. A. Lupo, S. Coyne and T. U. Berendonk, *Front. Microbiol.*, 2012, **3**, 18.
63. X. Zhang, T. Zhang and H. P. Fang, *Appl. Microbiol. Biotechnol.*, 2009, **82**, 397.
64. M. Munir, K. Wong and I. Xagoraraki, *Water Res.*, 2011, **45**, 2.
65. T. Tsai, H. Chang, K. Chang, Y. Liu and C. Tseng, *J. Chem. Technol. Biotechnol.*, 2010, **85**, 1642.
66. P. Karaolia, E. Hapeshi, I. Michael, C. Drosou, N. Xekoukoulotakis and D. Fatta-Kassinos, Decontamination of antibiotics and inactivation of antibiotic-resistant bacteria in urban MBR wastewater using novel graphene-based composites, 2016, in preparation.
67. T. Paul, M. C. Dodd and T. J. Strathmann, *Water Res.*, 2010, **44**, 3121.
68. P. Xiong and J. Hu, *Water Res.*, 2013, **47**, 4547.
69. P. S. M. Dunlop, M. Ciavola, L. Rizzo, D. A. McDowell and J. A. Byrne, *Catal. Today*, 2015, **240**, 55.
70. R. R. Ariza, S. P. Cohen, N. Bachhawat, S. B. Levy and B. Demple, *J. Bacteriol.*, 1994, **176**, 1.
71. K. Poole, *Trends Microbiol.*, 2012, **20**, 5.
72. S. M. Chiang and H. E. Schellhorn, *Arch. Biochem. Biophys.*, 2012, **525**, 2.
73. M. A. S. MacMahon, J. Xu, J. E. Moore, I. S. Blair and D. A. McDowell, *Appl. Environ. Microbiol.*, 2007, **73**, 1.
74. M. Pelaez, N. T. Nolan, S. C. Pillai, M. K. Seery, P. Falaras, A. G. Kontos, P. S. M. Dunlop, J. W. J. Hamilton, J. A. Byrne, K. O'Shea, M. H. Entezari and D. D. Dionysiou, *Appl. Catal., B*, 2012, **125**, 331.

75. S. Banerjee, S. C. Pillai, P. Falaras, K. E. O'Shea, J. A. Byrne and D. D. Dionysiou, *J. Phys. Chem. Lett.*, 2014, **5**, 2543.
76. A. Khalil, M. A. Gondal and M. A. Dastageer, *Appl. Catal., A*, 2011, **402**, 162.
77. C. Karunakaran, P. Gomathisankar and G. Manikandan, *Mater. Chem. Phys.*, 2010, **123**, 585.
78. C. Xu, L. Cao, G. Su, W. Liu, X. Qu and Y. Yu, *J. Alloys Compd.*, 2010, **497**, 373.
79. C. Karunakaran, V. Rajeswari and P. Gomathisankar, *Mater. Sci. Semicond. Process.*, 2011, **14**, 133.
80. C. Karunakaran, V. Rajeswari and P. Gomathisankar, *Superlattices Microstruct.*, 2011, **50**, 234.
81. C. Karunakaran, V. Rajeswari and P. Gomathisankar, *Solid State Sci.*, 2011, **13**, 923.
82. J. W. J. Hamilton, J. A. Byrne, C. McCullagh and P. S. M. Dunlop, *Int. J. Photoenergy*, 2008, 631597.
83. A. Vohra, D. Y. Goswami, D. A. Deshpande and S. S. Block, *Appl. Catal., B*, 2006, **65**, 57.
84. M. Sökmena, F. Candana and Z. Sümer, *J. Photochem. Photobiol., A*, 2001, **143**, 241.
85. C. Karunakarana, G. Abiramasundaria, P. Gomathisankara, G. Manikandana and V. Anandib, *J. Colloid Interface Sci.*, 2010, **352**, 68.
86. J. C. Yu, W. Ho, J. Yu, H. Hoyin Yip, P. K. Wong and J. Zhao, *Environ. Sci. Technol.*, 2005, **39**, 1175.
87. S. Rehman, R. Ullah, A. M. Butt and N. D. Gohar, *J. Hazard. Mater.*, 2009, **170**, 560.
88. J. S. Im, S. Yun and Y. Lee, *J. Colloid Interface Sci.*, 2009, **336**, 183.
89. Q. Li, R. Xie, Y. W. Li, E. A. Mintz and J. K. Shang, *Environ. Sci. Technol.*, 2007, **41**, 5050.
90. A. Hameed, M. A. Gondal and Z. H. Yamani, *Catal. Commun.*, 2004, **5**, 715.
91. A. Bagabas, M. Gondal, A. Khalil, A. Dastageer, Z. Yamani and M. Ashameria, *Surf. Sci. Catal.*, 2010, **175**, 279.
92. J. A. Rengifo-Herrera and C. Pulgarin, *Sol. Energy*, 2010, **84**, 37.
93. H. Yu, X. Quan, Y. Zhang, N. Ma, S. Chen and H. Zhao, *Langmuir*, 2008, **24**, 7599.
94. B. Xie, *Chin. J. Disinfect.*, 2010, **1**, 14.
95. W. Wang, Y. Yu, T. An, G. Li, H. Y. Yip, J. C. Yu and P. K. Wong, *Environ. Sci. Technol.*, 2012, **46**, 4599.
96. A. Y. Booshehri, S. C. K. Goh, J. Hong, R. Jiang and R. Xu, *J. Mater. Chem. A*, 2014, **2**, 6209.
97. Q. Li, Y. Li, P. Wu, R. Xie and J. K. Shang, *Adv. Mater.*, 2008, **20**, 3717.
98. H. Gan, G. Zhang and H. Huang, *J. Hazard. Mater.*, 2013, **250**, 131.
99. H. Lin and H. Lin, *J. Hazard. Mater.*, 2012, **217**, 231.
100. D. Bahnemann, *Sol. Energy*, 2004, **77**, 445.
101. M. I. Polo-López, P. Fernández-Ibáñez, E. Ubomba-Jaswa, C. Navntoft, I. García-Fernández, P. S. M. Dunlop, M. Schmid, J. A. Byrne and K. G. McGuigan, *J. Hazard. Mater.*, 2011, **196**, 16.

CHAPTER 4

Solar Photocatalysis: Fundamentals, Reactors and Applications

JAVIER MARUGÁN*[a], MARÍA JOSÉ LÓPEZ-MUÑOZ[a], PILAR FERNÁNDEZ-IBÁÑEZ[b], AND SIXTO MALATO[b]

[a]Universidad Rey Juan Carlos, C/Tulipán s/n, 28933 Móstoles, Spain; [b]Plataforma Solar de Almería-CIEMAT, P.O. Box 22, 04200 Tabernas, Almería, Spain
*E-mail: javier.marugan@urjc.es

4.1 Solar Light

The Sun is an inexhaustible energy resource freely available in all locations over the Earth surface. Therefore, the use of solar light to drive photocatalytic processes has huge potential from both environmental and economic viewpoints. However, the high variability of the irradiation power and spectra as a function of the latitude, longitude, date, time, pollution and weather conditions means that the efficiency of solar processes strongly depends on the availability of the resources. Although the solar light is free, the size (and therefore the cost) of the solar collectors required for a specific application would be significantly affected by the solar average radiation flux available at a specific location. Consequently, evaluation of the technical and economic feasibility of solar photocatalytic processes requires design and engineering procedures to

RSC Energy and Environment Series No. 15
Photocatalysis: Applications
Edited by Dionysios D Dionysiou, Gianluca Li Puma, Jinhua Ye, Jenny Schneider, and Detlef Bahnemann

Published by the Royal Society of Chemistry, www.rsc.org

take into account accurately the variability of solar irradiation flux at the specific location and the dates and times of operation of the process.

The description of the solar light available for the photocatalytic process requires knowledge of the main characteristics of radiative energy:

- *Irradiance*: energy per unit time and unit of illuminated surface. Its maximum value is strongly dependent on the position of the Sun in the sky, which depends on the specific geo-coordinates (latitude and longitude), date and time. This position is characterised by the *solar vector*, whose value determines the optical path and therefore the thickness of the air layer that the beams have to cross before reaching the Earth's surface. This value is attenuated by the air depending on the atmospheric conditions, such as the presence of clouds or chemical pollution.
- *Spectrum*: relative distribution of wavelengths in the radiation beams. Atmospheric composition and conditions have differing influences depending on the wavelength, *i.e.* filtering selectively some radiation frequencies whereas other can even cross unaffected. This filtering effect is mainly produced by absorption phenomena, although scattering could also play a minor role.
- *Direction*: relative distribution of the angles between the radiation beams and the illuminated surface. When crossing the atmosphere, the solar radiation beams are subjected not only to absorption processes but also to scattering that produces changes in the direction of the photons. Consequently, not all radiation reaching the Earth's surface has the direction corresponding to the solar vector (*direct radiation*); therefore, there is an important contribution of *diffuse radiation* that can vary from nearly zero to almost total depending on the atmospheric conditions and the spectral range.

Based on these considerations, the *instant radiation energy* available for a solar photocatalytic process can be evaluated by the following steps:

(1) Estimation of the extraterrestrial irradiance based on standardised data and possible correction according to the date.
(2) Determination of the solar vector based on the geo-coordinates, date, and time.
(3) Estimation of the maximum irradiance and spectrum of the direct radiation reaching the Earth's surface.
(4) Correction of the flux, spectrum, and directional distribution of the light based on the instantaneous atmospheric conditions.

4.1.1 Extraterrestrial Irradiance and Spectrum

The total solar irradiance arriving at the top of Earth's atmosphere, corresponding to the extraterrestrial radiation, is usually referred as the *solar constant* (E_0); a value of 1366.1 W m^{-2} is the most accepted for modelling purposes.[1] This is the average solar radiation over the course of a year, with small changes during the 11 year solar cycle.[2] However, significant fluctuations occur during the year due to the Earth's varying distance from the Sun owing to the ellipticity of the orbit, taking values from 1321.0 W m^{-2} at the aphelion (4th July) to 1412.0 W m^{-2} at the perihelion (3rd January).[3]

The spectrum of the extraterrestrial solar radiation has been also standardised based on experimental data. The two most important standards are the Wehrli Standard Extraterrestrial Solar Irradiance Spectrum,[4] based on data from the top of the atmosphere, and the ASTM Standard Extraterrestrial Spectrum Reference E-490 AM0,[5] based on data from satellites; the differences between the two standards is very small. Figure 4.1 displays the ASTM E-490 standard, where the integrated spectral irradiance has been adjusted to conform to the accepted value of the solar constant of 1366.1 W m^{-2}.

A common way to simulate the Sun's spectrum is the blackbody theory. According to Planck's law, the spectral emission of a blackbody as a function of temperature can be described by the expression:

$$E_{N\lambda} = \frac{2hc^2}{\lambda^5\left(e^{\frac{hc}{\lambda k_B T}} - 1\right)} \tag{4.1}$$

where $E_{N\lambda}$ is the spectral emittance (energy per unit time, unit area of emitting surface and unit wavelength), h is the Planck constant (6.626×10^{-34} J s),

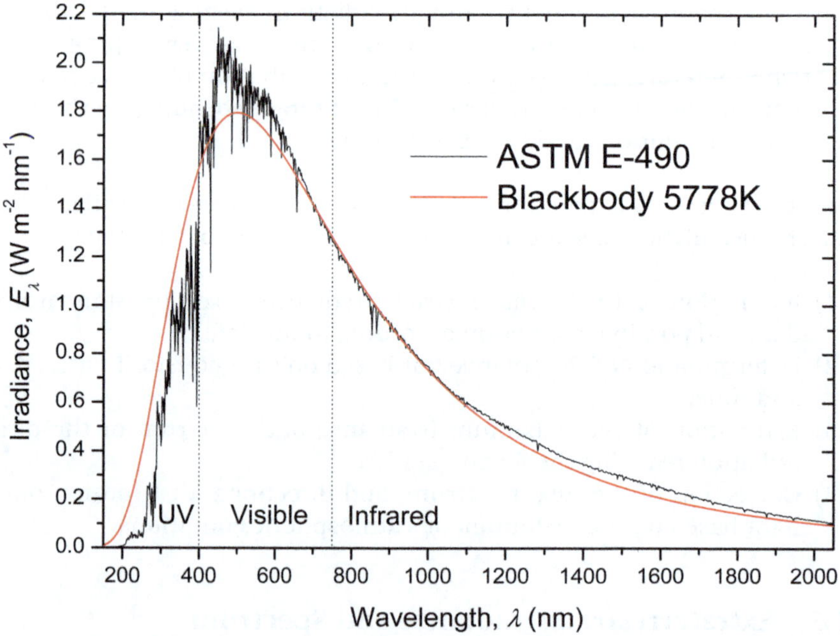

Figure 4.1 ASTM E-490 AM0 standard extraterrestrial irradiance spectrum and comparison with the theoretical emission of a blackbody at 5778 K according to Planck's law corrected with the Sun–Earth distance.

c is the speed of light in a vacuum (2.998×10^8 m s^{-1}), k_B is the Boltzmann constant (1.381×10^{-23} J K^{-1}), λ is the wavelength of the electromagnetic radiation, and T is the absolute temperature of the body. Extraterrestrial radiation reaching the top of the atmosphere can be estimated considering the radius of the Sun ($r_{Sun} = 6.96 \times 10^8$ m)[6] and the average radius of the Earth's orbit ($r_{Earth\_orbit} = 1.496 \times 10^{11}$ m)[7] according to the following expression:

$$E_{0\lambda} = E_{N\lambda} \frac{r_{Sun}^2}{r_{Earth\_orbit}^2} \tag{4.2}$$

Figure 4.1 shows the extraterrestrial irradiance simulated at 5778 K considering eqn (4.1) and (4.2). Integration of the spectral irradiance to render the total solar irradiance (solar constant) can be easily expressed by means of the Stefan–Boltzmann's law as:

$$E_0 = \int_0^{\infty} E_{0\lambda} \mathrm{d}\lambda = \frac{r_{Sun}^2}{r_{Earth\_orbit}^2} \int_0^{\infty} E_{N\lambda} \mathrm{d}\lambda = \frac{r_{Sun}^2}{r_{Earth\_orbit}^2} \sigma T^4 \tag{4.3}$$

where σ is the Stefan–Boltzmann constant (5.67×10^{-8} W m^{-2} K^{-4}). The temperature 5778 K is considered the effective temperature of the Sun,[6] as it leads to the accepted value of the solar constant and spectrum.

The comparison between the two spectra represented in Figure 4.1 shows sufficient agreement in the infrared range, with higher discrepancies in the visible, and especially in the UV region due to the absorption of the gases in the solar atmosphere (Fraunhofer lines). This is important to be consider in phenomena activated by UV-vis radiation, as in the case of photocatalytic processes.

4.1.2 Solar Vector

The solar vector that describes the Sun position regarding an observer at the Earth surface is described in spherical coordinates using two different angles defined over the celestial hemisphere: *zenith* and *azimuth*. The zenith angle (θ) defines the elevation of the Sun. It varies from 0° to 90° and it is conventionally defined as zero on the elevation perpendicular to the surface (zenith position). The value of θ is 90° at dawn, reaches a minimum value at noon and increases again to 90° at sunset. The azimuth angle (γ) varies from 0° to 360° and defines the projection of the vector on the surface with respect to the North. Figure 4.2 schematises the definition of both angles. The value of γ ranges from 90° to 270° in the Northern Hemisphere of the Earth and from 0° to 90° and 270° to 360° in the Southern Hemisphere, shifting from 0° to 360° at noon. Sometimes, the *elevation* angle is used instead of the zenith angle, this being defined with respect to the horizontal and therefore as the complementary angle of θ.

The solar vector can be determined following a fully predictive methodology, as the position of the Sun can be accurately described using astronomical calculations. Several algorithms have been described in the literature

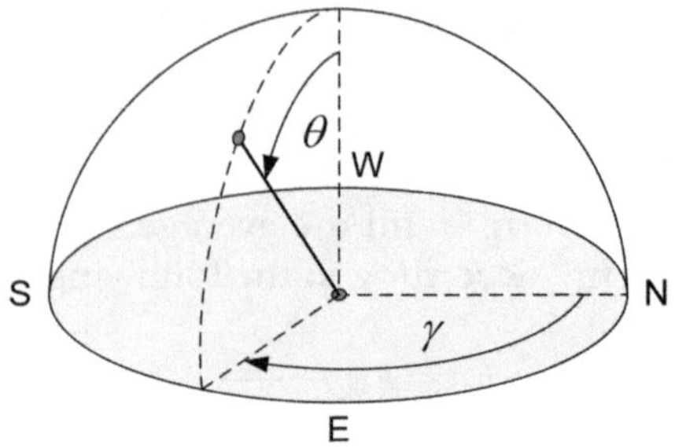

Figure 4.2 Definition of the solar vector by the zenith (θ) and azimuth (γ) angles.

with this aim, with different degrees of accuracy and complexity.[8] Applications based on a high concentration of direct solar light, such as for instance heliostat fields concentrating energy on tower receptors, could require solar tracking devices with extremely high accuracy,[9] which can be achieved using the Astronomical Almanac tables or the Multiyear Interactive Computer Almanac (MICA) available from the United States Naval Observatory.[10] Computer codes with accuracies of up to ±0.0003° are available at the websites of Plataforma Solar de Almería (PSA, Spain)[11] and the National Renewable Energy Laboratory (NREL, USA).[12]

For systems without solar tracking and non-concentrating applications, as is commonly the case for photocatalytic applications, simpler algorithms with accuracies of the order of 0.5° are enough to provide good quality predictions. To calculate declination and ascension of the Sun, geo-coordinates (latitude and longitude) and instant date and time are always required as input data. The degree of complexity (and accuracy) of the specific algorithm depends on the different astronomical factors considered (parallax, obliquity of the ecliptic, refraction, *etc.*). A detailed example of calculations can be found in the literature.[8] Final conversion from celestial into horizontal coordinates can be facilitated by direct vector analysis.[13]

4.1.3 Irradiance at the Earth Surface

Solar irradiance readily available at the Earth's surface is lower than the extraterrestrial irradiance due to the effect of the atmosphere. The intensity of radiation decreases due to absorption and scattering phenomena produced by the gases and aerosols present in the atmosphere, and this reduction strongly depends on the position of the Sun. The reason for this is that the optical path of the rays crossing the atmosphere layer (L) is longer for higher zenith angle values (Figure 4.3). The quantification of this effect is carried out through the concept of *air mass* (AM), which is the relative optical path length of the solar radiation. The value of the AM is 0 outside the atmosphere, and ranges at the Earth's surface from AM1.0 when the Sun is located at the zenith point ($\theta = 0°$) and the optical path is minimum to AM ≈ 37.5 for a zenith angle value of 90° (dawn or sunset).[14] AM1.5 and AM2.0 corresponds

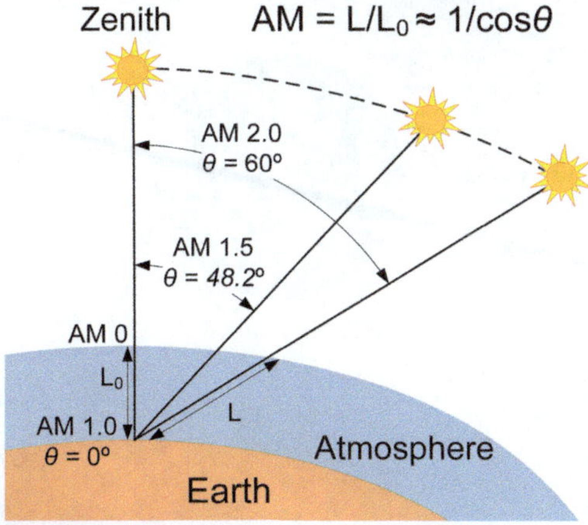

Figure 4.3 Definition of the air mass (AM) as a function of the zenith angle (θ).

to zenith angle values of 48.2° and 60°, respectively. For values of $\theta < 75°$ the AM can be practically calculated as the inverse of the cosine of θ; for higher values of the zenith angle, the error of the planar approximation becomes significant so that the curvature of the Earth's surface has to be considered. Several refinements have been proposed to calculate more accurately the AM value towards the horizon.[14]

Changes in the AM value explain the variation of the global irradiance reaching the Earth's surface not only during the day but also throughout the year. Importantly, the shortest Sun–Earth distance occurs in January during the winter in the Northern Hemisphere. Seasons are not related to the Sun–Earth distance. They are the result of the changes in AM along the year produced by the tilt of the Earth rotation axis with respect to the plane of translation around the Sun, as schematised in Figure 4.4.

The total variation in the minimum zenith angle over a one-year period is 47° (double the Earth's tilt) for any location on the Earth's surface. The minimum zenith angle at the equator goes from 23.5° to the North during the June solstice to 23.5° to the South during the December solstice, being 0° on the equinoxes. At latitude 40° North, the minimum zenith angle varies from 63.5° during the December solstice to 16.5° during the June solstice, both to the South. At latitude of 23.5° North (Tropic of Cancer) the minimum zenith angle goes from 0° during the June solstice to 47° to the South during the December solstice. In fact the tropics mark the northernmost and southernmost latitudes at which the Sun may reach the zenith. On the other hand, the Arctic and Antarctic circles are the northernmost and southernmost latitudes at which the Sun can remain continuously above or below the horizon for 24 h. In all cases, the latitude equals the minimum zenith angle at the equinoxes.

The solar industry uses AM1.5 for all standardised testing or rating of terrestrial solar cells or modules, including concentrating systems and photovoltaic applications. AM1.5 is useful to represent the overall yearly average for

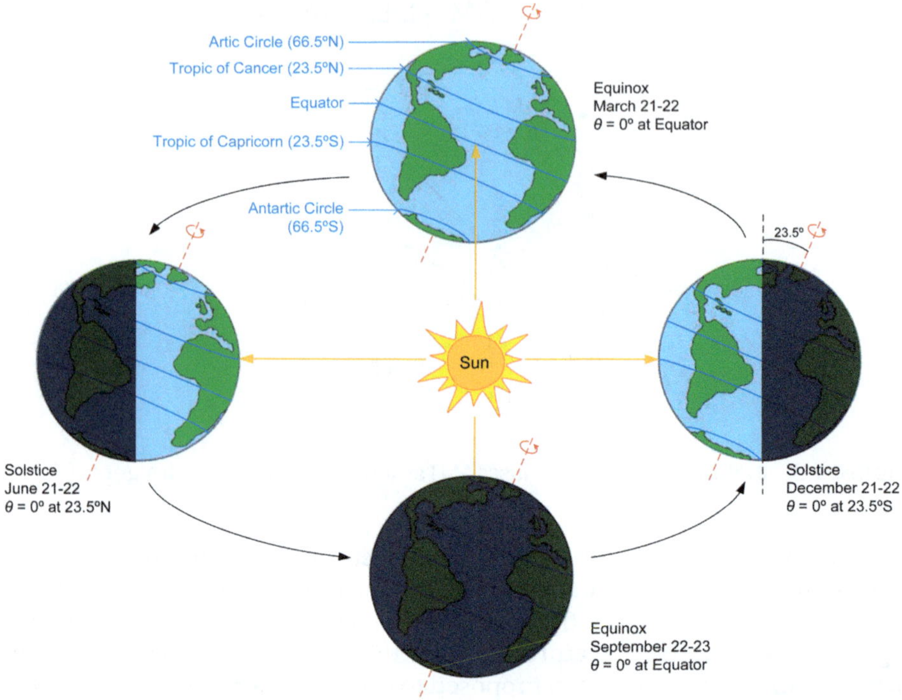

Figure 4.4 Changes in the angle of direct solar irradiation during the year.

mid-latitudes, where in summertime the AM value during the middle part of the day is less than 1.5, compensating for the higher values in the morning and evening and during other seasons of the year. Figure 4.5 represents the AM1.5 solar irradiance spectra corresponding to the ASTM G-173 standard defined for a standard atmosphere composition, temperature, pressure, and aerosol density. This spectrum corresponds to the irradiance received on an inclined plane at 37° tilt toward the equator, facing the Sun under clear day conditions with the Sun at a zenith angle of 48.2°. Notice the remarkable wavelength dependence of the atmosphere effect. In the IR range, CO_2 and H_2O lead to pronounced absorption bands that remove specific wavelengths whereas other parts of the spectrum remain unaltered. In the visible range, irradiance decreases mainly due to absorption by O_2 and O_3 and scattering produced by aerosols and Rayleigh scattering by O_2 and N_2 molecules. This latter phenomenon, consisting of the higher dispersion of shorter wavelengths is responsible for the blue colour of the sky during the middle of the day and the red colour at dawn and sunset. The highest difference between the extraterrestrial irradiance and the AM1.5 standard spectra occurs in the UV range, where the ozone layer absorbs all photons with wavelengths below 300 nm and significantly reduces the irradiance in the 300–400 nm range (Figure 4.5). Consequently, from the 102.84 W m^{-2} UV irradiance calculated by integration of the ASTM E-490 AM0 standard, only 46.10 W m^{-2} are computed from the ASTM G-173 AM1.5 standard, corresponding 30.52 W m^{-2} to direct radiation

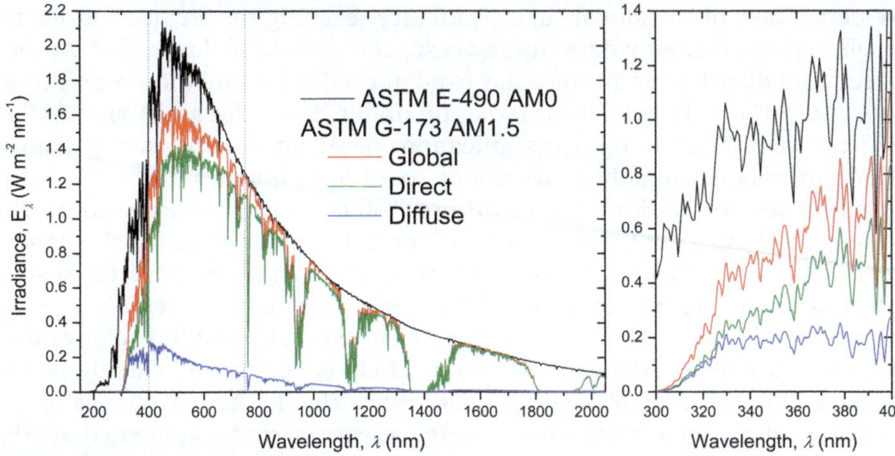

Figure 4.5 ASTM G-173 AM1.5 standard irradiance spectrum in comparison with the extraterrestrial irradiance ASTM E-490 AM0.

and 15.58 W m^{-2} to diffuse radiation. This means that 1/3 of the UV irradiance at the Earth's surface corresponds to diffuse radiation that cannot be concentrated, but on the contrary it can be collected and used, which has a strong impact on the design of solar collectors for photocatalytic applications. On the other hand, it is important to remark that although the germicidal UV-C radiation is removed by the ozone layer, a significant amount of UV-B radiation (<320 nm) still arrives to the Earth surface, playing a major role in biological processes[15] and therefore in solar disinfection applications.[16]

It also has to be considered that the effect of the AM value is not the same at all wavelengths, filtering more significantly the radiation with lower wavelengths due to Rayleigh scattering phenomena. Consequently, the higher value of the AM in the morning and evening and in winter means that not only the total solar irradiance reaching the Earth's surface is lower but also that the relative contribution of the UV radiation on the solar spectrum is dramatically reduced. As an example, the 280–400 nm UV percentage of total direct normal solar radiation has been reported to drop from 4.7% for AM1.22, to 4.1% at AM1.5, 2.5% at AM2.36, 1.6% at AM3.12, and 0.9% at AM4.5.[17] This is even more pronounced on the UV-B component; it has been reported that at latitude of 53° N the maximum value of the UV-B/UV-A ratio is above 3% in the summer solstice and decreases to a minimum of about 1% in winter.[18] This effect makes the efficiency of solar technologies based on the use of UV radiation such as photocatalytic processes especially sensitive to the date, time, and geo-coordinates.

All these considerations may be used to predict the maximum irradiance and spectrum of the solar light arriving at the Earth's surface under ideal cloudless conditions. Total solar irradiance decreases significantly under cloudy conditions. However, in this case the effect on UV is opposite to the change in AM, because water practically removes the infrared radiation but has a very high transmittance in the UV range. Consequently, the

UV percentage of the global solar irradiance reaching the Earth's surface is much higher on cloudy days. In any case, clouds have a clear effect on the directional distribution of the solar irradiance: diffuse radiation represents about 10–20% of the total solar radiation under clear skies and 100% under overcast skies. Therefore, solar applications based on absorption of UV radiation can work reasonably under cloudy conditions, but the collector should be optimised for working under diffuse non-concentrative irradiation. In any case, although the UV percentage of the global solar irradiance is higher under cloudy conditions, the absolute value of UV irradiance decreases significantly, reducing the efficiency of the photocatalytic processes.

Summarising, in addition to the predictive estimation of the solar radiation in a specific location, historical statistical data have to be analysed to consider the influence of weather conditions. This is the reason why locations at equivalent latitudes show long-term average irradiation data that differ significantly (Figure 4.6).

In any case, due to the unpredictability of weather conditions at any instant, fully reliable operation of a solar system can only be guaranteed by means of an automatic control loop with online monitoring of the instant solar irradiance and a control algorithm that regulates the flow rate and residence time in the system accordingly.

As an example, the Plataforma Solar de Almería (PSA) geo-coordinates are 37.09° N 2.36° W. Considering that PSA is located in the GMT+1 time zone, the solar vector corresponding to a specific date at a particular time, *e.g.* 30th of April 2015 15:00 h, can be estimated by a solar positioning algorithm: zenith 33.61°, azimuth 236.0°, which corresponds to an AM value of 1.2. The maximum theoretical solar radiation reaching the Earth's surface could be estimated by individual interpolating of each wavelength between the two

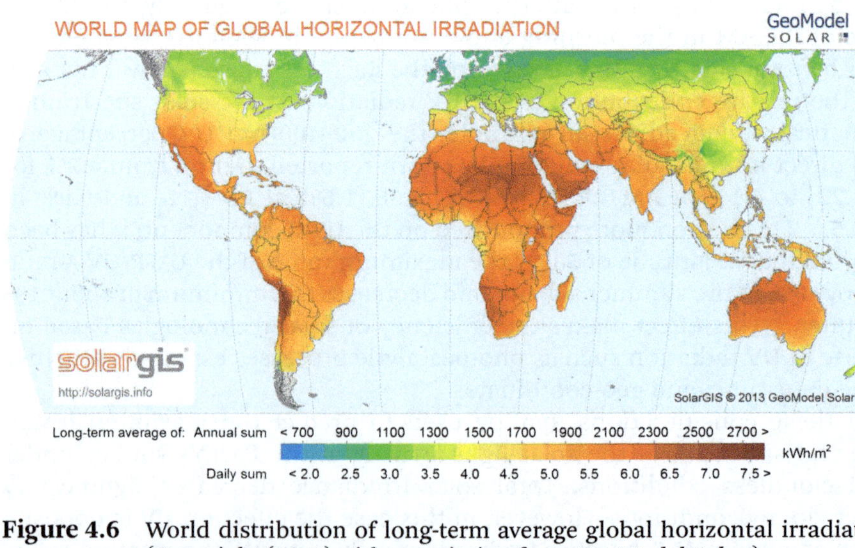

Figure 4.6 World distribution of long-term average global horizontal irradiation. (Copyright (2014) with permission from GeoModel Solar.)

solar spectra corresponding to AM0 and AM1.5. This interpolated spectrum gives a global irradiance value of approximately 1040 W m^{-2}. The correlation between the spectral intensities of the incident beams and the absorption properties of the photocatalytic material provides an estimation of the rate of photon absorption, which for instance under these conditions would be 1.48×10^{-4} Einstein L^{-1} s^{-1}. This value, corrected by the climatological atmospheric model, could be introduced into a kinetic model of the process with explicit radiation absorption effects to estimate the reaction rate of the process and therefore adjust the irradiation time required for a given conversion. Of course, this theoretical calculation is useful for the design and sizing of the solar reactor but the accurate operation of the process would require a control loop to adjust the residence and irradiation time based on the online instantaneous monitoring of the solar radiation arriving to the collector. These concepts are schematised in Figure 4.7.

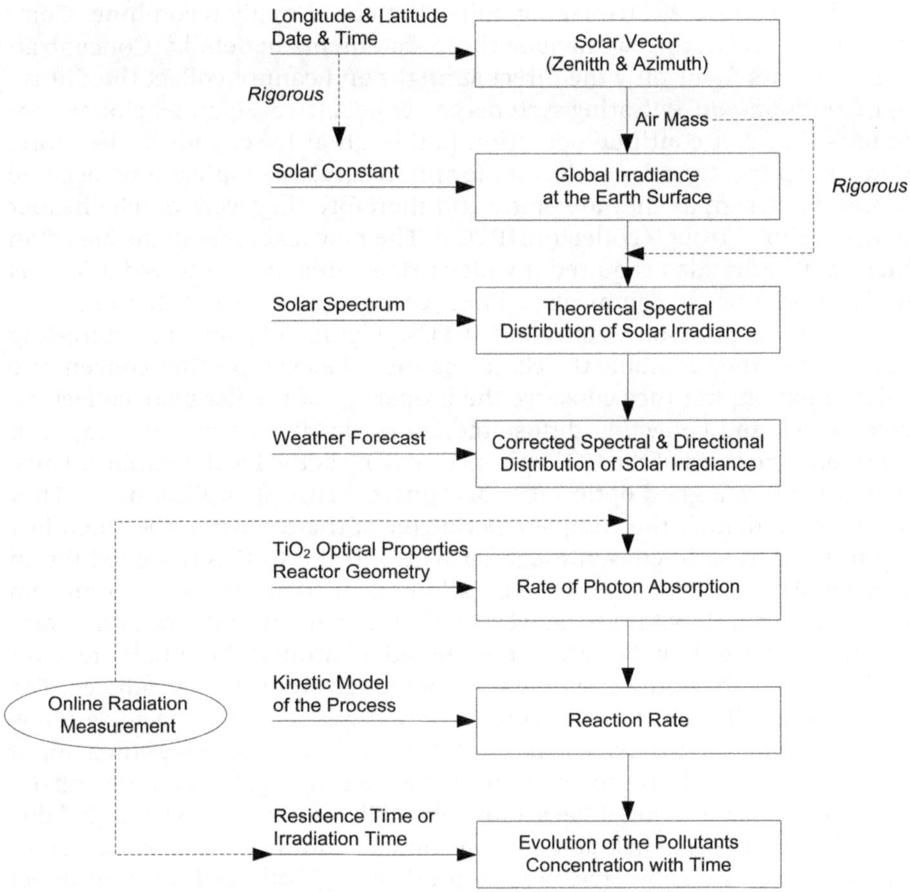

Figure 4.7 Schematic representation of the flow diagram required for estimation of the photocatalytic reaction rate based on the solar radiation theoretical calculations.

4.2 Solar Photocatalytic Reactors

4.2.1 Types of Reactors

The specific hardware needed for solar photocatalytic applications has much in common with those used for thermal applications. As a result, in the past, both reactors and photocatalytic systems have followed conventional solar thermal collector designs, such as parabolic troughs and non-concentrating collectors.[19] But at the end of the 1990s their designs begun to diverge[20] since: (i) the fluid must be exposed to ultraviolet solar radiation, and, therefore, the absorber must be UV-transparent; (ii) a temperature higher than ambient is not necessary and very often not desired for proper adsorption on TiO_2; for the photo-Fenton reactions a temperature higher than 30 °C, but lower than 50 °C, is desired to avoid rapid Fe hydroxides precipitation; and (iii) a high photon flux is not recommended, to avoid excessive electron–hole pairs that could easily recombine. Comprehensive recent reviews include these statements in detail.[21] Concentrating collectors focus only the direct sunlight and cannot collect the diffuse light, but non-concentrating systems can collect it – even under cloudy conditions they can continue operation (although at lower rates). Therefore, non-concentrating solar collectors for photocatalytic applications become a suitable design as they are static and therefore they were much cheaper than parabolic-trough collectors (PTCs). The new reactor systems based on these collectors also required a wide surface area expose to radiation, as well as resistance to chemicals, climate changes, and solar radiation.[22]

Compound parabolic collectors (CPCs, Figure 4.8) are an interesting option since they combine the characteristics of each type: they concentrate solar radiation, but they conserve the properties of the flat plate collectors, being static and collecting diffuse radiation, which even on clear days can represent one-third of the UV irradiance arriving at the Earth's surface. Thus, they constitute a good option for solar photochemical applications.[23] They do so by irradiating the complete perimeter of the receiver rather than just the front of it, as in conventional flat plates. If the CPC is designed for an acceptance angle from +90° to −90°, all incident solar diffuse radiation can be collected and $R_C = 1$ (ratio between collector aperture and tube perimeter). The light reflected by the CPC is distributed all around the tubular receiver so that almost the entire circumference of the receiver tube is irradiated. The plane of the collector should be tilted toward the equator with an angle equivalent to the local latitude, which maximises the annual-average efficiency of the collector based on the variation of the zenith angle. CPCs are considered nowadays the state of the art for solar collectors in photocatalysis[24] due to their advantages, such as turbulent flow conditions, no vaporisation of contaminants, no solar tracking, no overheating, collection of both direct and diffuse radiation, low cost, weatherproof properties, and high optical efficiency. A detailed comparison of PTC and CPC reactors for photocatalytic applications can be found in the literature.[25]

Figure 4.8 Design and geometry of compound parabolic collectors (CPC) with $R_C = 1$ (top) and photograph of one row of CPCs for water treatment (bottom).

4.2.2 Design of Solar Photocatalytic Reactors

The main technical issues associated with solar photo-reactor design for photocatalytic applications is the materials of the reflector surface, which is expected to have a high reflectance in the UV spectrum, and the photo-reactor wall composition, with desired high UV light transmission. The materials meeting these requirements are electropolished anodised aluminium and aluminium coatings for reflector surfaces; while photo-reactor tube might made be fabricated with high UV transmittance and chemical and pH resistant material, which may be fluoropolymers and low iron content glass or quartz. Although quartz has the highest UV transmission compared with all types of glasses, it is too expensive to be affordable for this application. Fluoropolymeric tubes should have high wall thickness to be resistant to pressure loses in the flow reactor, which leads to a strong decrease in UV transmittance of these materials. At present, the best choice of material for reactor tube is borosilicate glass with high transmission in the UV range (above 90%, varies depending on the thickness) and a cut-off of about 285 nm.

The reactor tube diameter also becomes a critical parameter for the design of a reactor based on processes where the catalyst has particular light absorption and scattering behaviour, which depends on the concentration of the catalyst, on the wavelength, and the colloidal properties of the catalyst suspension. The photo-reactor (tube) diameter must guarantee that light penetrates to inner layers of the reactor, so the process is not photo-limited due to limited light absorption by the catalyst suspension. Therefore larger reactor diameters will have lower optimum catalyst concentrations and *vice versa*. In particular, for the most investigated photo-catalyst, the commercial TiO_2 P25 (Evonik), the optimum TiO_2 concentration experimentally determined under sunlight is roughly 100–500 mg L^{-1} for reactor tube diameters of 25–50 mm.[21]

4.2.3 Solar Reactors for Water Disinfection

CPCs have been demonstrated to be a very promising option also for solar water disinfection with and without photocatalytic processes,[26,27] as has been proven for several water pathogens.[28,29] Nevertheless, the efficiency of these photo-reactors not only depends on solar collection system but also on other factors of design like the water volume to be treated, how the radiation is delivered into the system (wavelength and dose), and the composition (chemical and biological) of the water, which affects the behaviour of the light inside the samples. The UV-B/UV-A ratio has to be especially considered, due to its strong effect in solar disinfection applications.[16]

Other practicalities might also be considered in designing solar photocatalytic disinfection systems. Therefore, other requisites are very important when designing a solar photo-reactor for water disinfection; these are, increasing the total volume of treated water per cycle/batch, reducing the complexity of the systems functioning by making them as autonomous as possible, reducing the risks of water recontamination after treatment, reducing the end-user dependence, using local technology, and cheap and robust materials. The main principle for an effective solar photo-reactor performance is optimisation of the output of treated water in given solar exposure time.

To address this objective one must take into account the following limiting technical aspects.

4.2.3.1 *Illuminated Photo-Reactor Volume*

Large volume reactors are commonly designed as flow systems, because they permit management of large volumes of water in an easy way, preserving the quality and safety of the treated water. In recirculated water photo-reactors, the photon flux is delivered intermittently, since there are illuminated and there are dark parts like pipes, elbows, and tanks. Minimisation of these dark volumes in favour of illuminated volume is very important for photo-disinfection processes. Some research contributions have already demonstrated that the same energy dose delivered into a water sample in a continuous manner yields better disinfection results than the same energy dose applied in an intermittent way.[30,31] This effect was attributed to the dark repair mechanisms developed by some microbial agents like bacteria when the photo-oxidation process is not sufficiently damaging (such as in the absence of catalyst or very poor contact between microorganisms and immobilised catalysts). This effect depends on the nature and origin of the biological contamination, the water quality, and the kind of radiation applied, *i.e.* the wavelength range, solar flux, and total energy delivered into the system. The importance of minimising the dark volume has been already reported in the literature and CPC reactors with different ratios of illuminated to total volume have been compared. One study employed a 14L-CPC reactor with ratio of 0.3 (14 L of total volume and 4.7 L of illuminated volume)[27] and a 60L-CPC reactor with a ratio of 0.75 (Figure 4.9). Experimental data using as model of water pathogen the resistant spores of *Fusarium solani* confirmed that the 60L-CPC reactor led to enhanced inactivation of microorganisms.

4.2.3.2 *Photocatalyst Configuration*

Controversial results regarding the catalyst load using suspended TiO_2 have been observed in recent literature. As mentioned before, although optical phenomena (*i.e.* light Rayleigh scattering and absorption) inside the photo-reactor play an important role in determining the optimal catalyst concentration, the type of target microorganism influences the final results. Significantly different inactivation kinetics have been observed between different microorganisms, even if the photocatalytic conditions were optimal from the optical point of view, or even between strains of the same pathogen. Although the optical behaviour of the light traveling through the reactor and water suspensions determines the optimal photo-reactor diameter (*i.e.* best optical path length), the ideal catalyst load experimentally found to inactivate bacteria differs between publications.[32] On the other hand, the use of suspended nanoparticles requires the use of a recovery method after the treatment. Although sedimentation processes have been proved to be efficient to remove titanium dioxide nanoparticles after photocatalytic processes, even after this post-treatment there is a risk when the final application is drinking water owing to the potential presence of these catalyst particles. Therefore, for decades there has been a strong research effort in this field to develop methodologies and materials to produce new and efficient immobilised catalysts, as is explained in further sections of this chapter. Although there are many articles on this matter, the disinfection performance of supported photocatalysts is rarely more efficient than for slurry systems,[33] where the contact between TiO_2 particles and microorganism is enhanced, as has been proven in the inactivation of spores of *Phytophthora capsici* (Figure 4.10).

a)

b)

Figure 4.9 Photograph of constructed solar CPC reactor (total treatment volume 60 L) for solar TiO_2-photocatalytic disinfection of water for irrigation at Plataforma Solar de Almería facilities (South East of Spain) (a), back view of the system (b).

4.2.3.3 Flow Rate, Water Temperature, and Dissolved Oxygen

In a continuous recirculation flow reactor it is important to work in the turbulent regime to ensure that oxygen dissolves effectively in the aqueous solution. Conversely, due to the low contact between catalyst and target microorganism, it is recommended that low flow rates are used to maximise the residence time which in turn will maximise the opportunities for contact with the microorganism.[34]

On the other hand, the increase of water temperature (from 15 to 45 °C) has been shown to improve the disinfection rate for bacteria and fungi

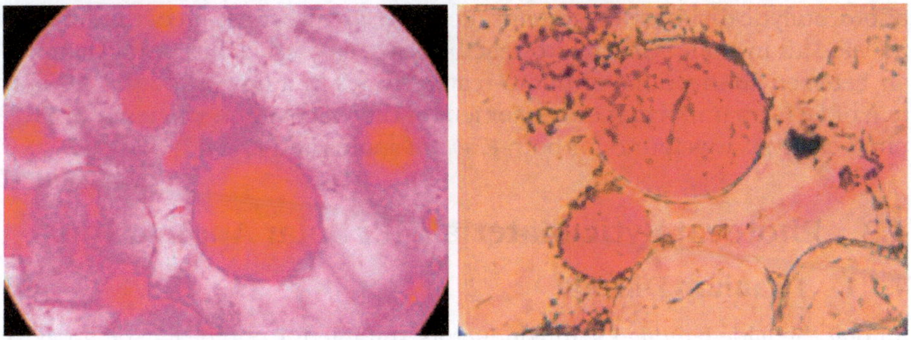

Figure 4.10 Spores of *Phytophthora capsici* (left) and in the presence of UVA irradiated TiO_2 nanoparticles (right).

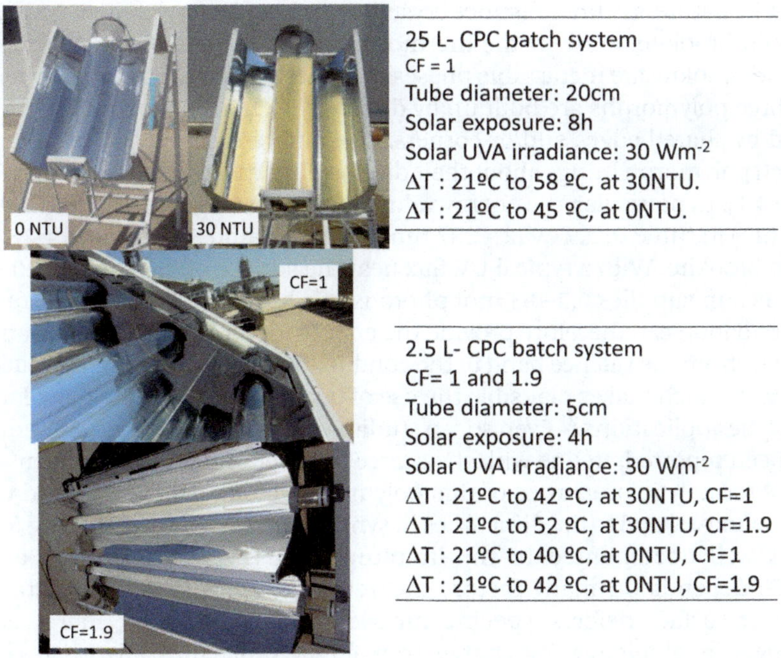

Figure 4.11 Thermal increase in different solar CPC batch reactors depending on water turbidity and CPC concentration factor (CF).

spores. The same effect of temperature has been confirmed for solar photo-Fenton with *E. faecalis*.[35] In photocatalysis the electron acceptor is normally dissolved oxygen, which in these processes can be taken from the air. In large flow reactors and static batch systems, dissolved oxygen can be rapidly consumed and must be replenished to avoid loses in photocatalytic activity. In addition, the decrease of solubility of oxygen in water with increasing temperatures is a very important effect in solar photo-reactors, where water temperature can reach up to 55 °C (Figure 4.11). Therefore, new designs of reactor must address the need for refilling of dissolved oxygen in photocatalytic disinfection systems. Recently, an investigation of the effect of injected air in a pilot CPC photo-reactor has been carried out. Upon sparging air between connections of photo-reactor tubes in a flow-through CPC system of 60 L, an important enhancement of the inactivation efficiency of TiO_2 photocatalysis was found for both *F. solani* spores and *E. coli*.[36]

4.3 Photocatalytic Materials for Solar Applications

4.3.1 Titanium Dioxide

Among the wide-bandgap semiconductors suitable for photocatalytic applications titanium dioxide (TiO_2) is the one most often used because of its ability to promote a great number of oxidation and reduction processes, high stability under different operating conditions, non-toxic nature, and low cost.[37] TiO_2 occurs in nature as three distinct crystallographic phases, namely rutile, anatase, and brookite. Rutile is the thermodynamically stable phase whereas anatase and brookite are metastable phases that transform into rutile upon heating. The three polymorphs are built up by distorted TiO_6 octahedra, which are connected by shared edges and/or corners. The different octahedral arrangement for each polymorph brings about their diverse physical and electronic properties (Table 4.1). In particular, the energy values of their band gaps vary from 3.0 eV (λ = 413 nm) in rutile to 3.2 eV (λ = 387 nm) in anatase and 3.2–3.4 eV (λ = 387–365 nm) in brookite. With a typical UV-flux near the surface of the Earth of 20–30 W m^{-2}, the Sun supplies 0.2–0.3 mol photons m^{-2} h^{-1} in the 300–400 nm range.[21] Solar radiation can therefore provide the energy for promoting the excitation of electrons from the valence band to the conduction band of the semiconductor in each case, which makes plausible the use of bare titanium dioxide for solar photocatalytic applications.[21] Even so, no studies with brookite using solar radiation have been reported, in line with the scarce number of publications focused on the photocatalytic properties of this polymorph under artificial UV and visible light, mainly related to the difficulties in synthesising brookite in its pure form.[38]

It is well documented in the literature that the photocatalytic activity of TiO_2 depends on many factors including crystalline phase, crystal size, lattice or surface defects, specific surface area, particle size, charge carrier lifetimes, or efficiency for charge transfer to molecules adsorbed on the semiconductor surface.[37,40,46–48] If pure phases are evaluated, it is generally reported that anatase is the polymorph which shows the highest activity in photocatalytic processes for water treatment. The reasons for such distinctive higher activity in comparison to pure rutile have not been fully clarified yet. For example, it has been proposed that the higher conduction band edge of anatase as compared to rutile is responsible for photoexcited electrons with a larger reductive power in anatase.[49] In addition, the surface properties of each polymorph appear to be crucial for charge trapping and separation at the surface and may affect charge transfer from the photocatalyst to adsorbed molecules.[40,50] On the other hand, synergistic effects leading to an enhanced photocatalytic activity are observed in phase mixtures of different titania polymorphs. They are generally explained by an effective separation of the photogenerated charge carriers induced by the junction of different crystalline phases.[51–55] Such is the case for the commercially available TiO_2 sample P25 supplied by Evonik (previously by Degussa), which is produced through high-temperature flame hydrolysis of $TiCl_4$ in the presence of hydrogen and oxygen. It contains anatase and rutile phases in a ratio of *ca.* 80% to 20%, with a nominal particle size of 30 nm and a surface area of 50 $m^2\ g^{-1}$. The high activity exhibited by TiO_2 P25 in comparison to other titania-based catalysts has led to extensive use as a reference in photocatalytic applications both under natural solar light and artificial radiation.

Even with the good properties as a photocatalyst mentioned above, TiO_2 has some limitations for practical applications, mainly a poor absorption of visible radiation and a rapid recombination of photogenerated electron–hole pairs.[21] Nevertheless, despite these disadvantages the great majority of photocatalytic applications performed with solar radiation report the use of bare titanium dioxide. Specifically, even though solar studies have been developed with pure anatase,[56–60] TiO_2 P25 is currently the catalyst most often used, as it has been successfully evaluated for the solar treatment of water containing a great variety of pollutants, among them herbicides,[61] pesticides and pharmaceuticals,[62] cyanides,[63] dyes,[64] sodium benzene sulfonate,[65] and arsenic.[66]

4.3.2 TiO_2 Modification for Solar Applications

The band gap value of titanium dioxide confines its ability for light absorption to the spectral UV-A region and therefore limits the extent of solar light harvesting. Ultraviolet light represents *ca.* 4–5% of total irradiance in the spectral distribution of the solar radiation reaching the Earth's surface, whereas visible light accounts for *ca.* 44% of the overall irradiance.[21] Accordingly, solar photocatalytic applications with titanium dioxide use just a small fraction of the available natural sunlight, as the energy threshold to promote electrons from the valence band to the conduction band in anatase materials is located within the UV-A region (Table 4.1). On this basis, during recent years many efforts have focused on the development of visible light active photocatalytic materials. Several approaches have been proposed, including doping of titania with metal and non-metal elements, dye sensitization, and semiconductor coupling, as recently reviewed.[67–70]

4.3.2.1 Doping of TiO_2

Doping involves the inclusion or substitution of a foreign atom into the TiO_2 lattice in such a way as to induce new electronic states and new optical transitions not seen in pure TiO_2.[46] The nature, chemical state, and load of the dopant strongly determine the effect on the final photocatalytic properties of the doped material. A great deal of research has focused on both metal and non-metal doping although TiO_2 doping with non-metal elements has attracted special attention in recent years.

Table 4.1 Properties of anatase, brookite and rutile phases.

	Anatase	Brookite	Rutile	References
Crystal structure	Tetragonal	Orthorhombic	Tetragonal	
Space group	D_{4h}^{19} $I4_1/amd$	D_{2h}^{15} $Pbca$	D_{4h}^{14} $P4_2/mnm$	A. F. Wells, 1984[39]
Lattice parameters	$a = b = 3.782$ Å $c = 9.502$ Å	$a = 5.436$ Å $b = 9.166$ Å $c = 5.135$ Å	$a = b = c = 4.584$ Å	U. Diebold, 2003[40]
Density ($g\ cm^{-3}$)	3.84	4.11	4.26	J. G. Li, 2007[41] R. C. Weast, 1984[42]
Refractive index	2.52	2.63	2.72	A. F. Wells, 1984[39]
Experimental band gap value (eV)	3.2	3.4[48]	3.0	A. Beltrán *et al.*, 2006[43] and references therein
		3.2[49]		M. Koelsch *et al.*, 2002[44] W. Wunderlicha, 2004[45]

Non-metal doping of TiO_2 was first described in 1986 by Sato[71] who reported the synthesis of N–TiO_2 with photocatalytic activity in the visible region (434 nm). However, it was not until some years later that the subject attracted increasing interest as reflected by an upward number of publications focused on N-, C-, and S-doped TiO_2 materials. Among them, the most extensively studied anion dopant has been nitrogen.

Various approaches for obtaining N-doped titania have been reported, including sol–gel synthesis,[71–78] thermal treatment of bare titanium dioxide under gaseous flow of molecular nitrogen (N_2),[79,80] and ammonia (NH_3),[81,82] reactive magnetron sputtering,[83,84] or low energy ion implantation.[85,86]

A great part of the research has focused on the elucidation of the mechanism responsible for the reduction of the optical threshold energy of N-doped TiO_2 materials, but no general agreement has been reached yet. Different proposals have been given to explain the visible light photocatalytic sensitization of N–TiO_2 materials based on both experimental results and theoretical calculations, among them: (i) the presence of impurities like NO_x;[71] (ii) a band gap narrowing originated by the mixing of N 2p with O 2p states;[79] (iii) formation of an isolated narrow band above the valence band due to substitution of TiO_2 oxygen lattice sites by nitrogen atoms;[81] (iv) localised occupied states in the energy gap induced by both substitutional-N and interstitial-N;[76] and (v) oxygen vacancies.[87–89]

Applications of N-doped materials under natural solar radiation are scarce but some promising results have been reported. For instance, Liu *et al.*[90] investigated the photocatalytic activities of nitrogen doped-TiO_2 nano-catalysts and commercial TiO_2 P25 powder for azo dyes bleaching using solar light (average light intensity of 120 W m^{-2}) and found an enhanced activity of N–TiO_2 in comparison with commercial samples.

Doping of TiO_2 with other non-metals such as sulfur,[91–94] carbon,[68,95–98] boron,[99,100] or fluoride[101–103] has also been investigated. A summary of work in this field is given by Marschall and Wang (ref. 69 and references therein). Similar arguments to those summarised for nitrogen-doping have been given to explain the origin of visible light activity of doped materials. In general terms, it is proposed that anions which are less electronegative than O, when doped into the titania lattice, will promote either the formation of discrete interstitial states or novel conduction band and/or valence band edges which in both cases result in the shift of the threshold energy of the semiconductor towards the visible range. Specifically in the case of S-dopant it has been reported that sulfur can be incorporated into the titania lattice either as an anion (S^{2-}) by replacing the lattice oxygen[91] or as a cation (S^{4+} and S^{6+}) by substituting Ti^{4+} ions.[104] Similarly the carbon dopant has also been described either as an anion that replaces oxygen substitutionally in the lattice[96,105] or as a cation that occupies an interstitial lattice site.[106–108]

In recent years co-doped titania with two[104,109–111] or even three anions[112] has attracted much attention as it has been reported that it may further expand the photo-absorption region of the TiO_2 photocatalyst and enhance its catalytic activity due to the synergetic effect of the non-metals.[113,114] The performance of co-doped materials for solar applications has been reported by Reddy *et al.*,[112] who prepared TiO_2 doped with C, N, and S by hydrolysis process using titanium isopropoxide and thiourea. The doped samples showed much higher photocatalytic activity than undoped titania for degradation of isoproturon under solar radiation. At higher loadings photocatalytic activity of co-doped samples decreased though they showed more redshift in the absorption edge. The result was ascribed to an enhanced electron–hole recombination due to the excess dopant acting as recombination centres.

Doping of TiO_2 with metals has been extensively explored for a great variety of elements including Fe, Cr, V, Ni, Nb, Mn, Cu, Al, Co, Ag, Au, Pd, *etc.* (ref. 46 and 116 and references therein). Depending on the ionic radius size of the dopant, it will be preferentially located at substitutional sites (size similar to that of Ti^{4+} ions) or at interstitial positions (ionic radius size smaller than Ti^{4+} ions). Incorporation of the metal induces the formation of new energy states within or beyond the band gap so that the electronic transitions from the valence band of titania to the intra-band gap energy levels and/or from the latter to the conduction band have been considered as responsible for the photoresponse to visible light shown by some metal doped materials.[46] Such enhanced visible light absorption, however, does not assure a good photocatalytic activity. Indeed, several authors have reported that, although metal ion doping decreases the photo-threshold energy of TiO_2, the metal ions may also serve as recombination centres for electrons and holes, thus reducing the overall activity of the photocatalyst.[76] Nevertheless, some interesting solar applications with metal-doped TiO_2 have been described. As an example, Mn/Co co-doped TiO_2 nanoparticles showed an enhanced photocatalytic activity in comparison to TiO_2 P25 under solar irradiation with *E. coli* and *Klebsiella pneumonia* reference strains.[117] In addition, Cu-and Cu/N-doped titania thin films showed a good potential for bacterial inactivation of *Escherichia coli* and *Enterococcus faecalis* and chemical degradation of methylene blue in the presence of visible light or sunlight.[118]

Noticeably, even though the main objective of doping titanium dioxide is to achieve a better profit of solar radiation, most studies performed with doped TiO_2 are focused on their photocatalytic activity under artificial visible light and reports are scarce for natural solar light as a source of irradiation; therefore, further research is needed in this field.

4.3.2.2 Semiconductor Coupling

Coupling of titanium dioxide with narrower band gap semiconductors such as CdS, Fe_2O_3, CdSe, V_2O_5, ZnO, SnO_2, or Bi_2S_3 has been also proposed as an effective approach to extend the absorption of TiO_2 to the visible region.[68,116,119–123] In this process, irradiation of the small band semiconductor (or sensitizer semiconductor) by visible light promotes electrons to its conduction band. If the sensitizer semiconductor has a more negative conduction band level than that of titania the electrons can be injected to the conduction band of TiO_2.[68] The efficiency of the process depends on both the good matching of the conduction band and valence band of the two semiconductors and the difference between their band potentials, which is the driving force for the interfacial charge transfer. Coupling of semiconductors not only has the objective of extending the photocatalyst response to visible light but also to prevent charge carrier recombination, as the energy gap between the band levels of coupled semiconductors promotes the separation between the photogenerated electron–holes pairs.[70]

One important drawback of coupled semiconductors for practical applications, which has been recently addressed, is their possible disaggregation and/or dissolution along the photocatalytic reaction. As an example, for TiO_2–CdS systems it has been reported that as the sensitizer is usually deposited on the titania surface and therefore it is directly exposed to the working medium, and so leaching of CdS and Cd^{2+} during the reaction may occur.[122,123] To address this problem, some proposals have been made, among them the use of CdSe/TiO_2 hybrid mesoporous structures[124] or CdS embedded TiO_2 hybrid nanospheres,[122] but the development of synthesis procedures that ensure the stability of the sensitizer along the photocatalytic reaction is still needed.

4.3.2.3 Dye Sensitization

This approach has been reported to sensitize TiO_2 particles to visible light through the interaction with adsorbed dyes such as 8-hydroxyquinone, methylene blue, acid red 44, reactive red dye 198 (RR 198), eosin-Y, merbromin, 2′,7′-dichlorofluorescein, Erythrosin B, Rhodamine B, Rose Bengal, or Rhodamine 6G.[21,70,125–127] If the adsorbed dye molecules are excited upon irradiation with visible light, they can inject electrons into the conduction band of the semiconductor as long the potential of the lowest unoccupied molecular orbital (LUMO) of the dye molecules is more negative than the conduction band potential of TiO_2.[70] The electrons injected into titania are subsequently scavenged by molecular oxygen to form reactive species such as superoxide (O^{2-}) and hydrogen peroxide ($^{\bullet}OOH$) radicals. In general, as the concentration of target organic species to be degraded is much higher than the adsorbed dye, the degradation of the latter is not kinetically favourable. However, in the absence of other organic molecules, the sensitized system can also be used for the oxidative degradation of the dye itself.[21,70]

As a summary, Table 4.2 shows the main features of the different procedures focused on the development of visible light active photocatalytic materials described above.

4.3.3 Mode of Utilisation

For its use in photocatalytic reactions, the catalyst can be either dispersed in the irradiated aqueous solution with a slurry arrangement,[128,129] anchored on a suitable support as a fixed or fluidised bed,[130] or used as continuous thin

Table 4.2 Summary of TiO_2 modification approaches for solar applications.

	TiO_2 doping	Semiconductor coupling	Dye sensitisation
Main elements and compounds used	Non-metal dopants and co-dopants: N, C, S, F Metal dopants: transition metals (Cu, Fe, Mn, Ni, V, Zn, Zr, *etc.*); noble metals (Ag, Au, Os, Pd, Pt, Ru, Rh, Ir, *etc.*) rare earth metals (Sc, Y, La, *etc.*)	CdS, Fe_2O_3, CdSe, ZnO, SnO_2, WO_3, or V_2O_5, Bi_2S_3	8-Hydroxyquinone, methylene blue, acid red 44, reactive red dye 198 (RR 198), eosin-Y, merbromin, 2′,7′-dichlorofluorescein, Erythrosin B, Rhodamine B, Rose Bengal, Rhodamine 6G, *etc.*
Objectives	To enhance the photocatalyst response to visible light To promote the separation between electron–hole pairs to enhance photocatalytic activity		
Drawbacks	Metal ions may also serve as recombination centres for electrons and holes, thus reducing the overall photocatalytic activity	Possible disaggregation and/or dissolution during the photocatalytic reaction	Possible oxidative degradation of the dye itself in the absence of other organic molecules Desorption of dye molecules greatly reduces practical applications

film.[131,132] It is generally reported that slurry systems exhibit greater activity than those of equivalent titania loading based on immobilised catalysts.[133–135] The slurry operation provides important advantages such as a high ratio of illuminated catalyst surface to the effective reactor volume, a rather uniform catalyst distribution, and almost no mass-transfer limitations.[136] On the other hand, the suspension made of fine semiconductor TiO_2 particles has several drawbacks for practical large-scale applications that include the slow and expensive separation of the catalyst from the suspension after the treatment of water, the tendency to aggregation of suspended particles, especially at high concentrations, and the complexity of being applied to continuous flow systems. In comparison, immobilised systems have lower operational cost as no additional catalyst separation operation is needed and also enable the contaminated water to be treated in continuous mode. However, in addition to the necessity of guaranteeing the long-term stability of the material, immobilisation of TiO_2 also offers important disadvantages, mainly a lower catalyst accessibility to photons and significant external mass transfer limitations, particularly at low fluid flow rate, because of the increasing diffusional length of the reactant from bulk solution to the catalyst surface. In addition, with increasing catalyst film thickness, the internal mass transfer may play a dominant role by limiting utilisation of the catalyst.[136,137] Table 4.3 gives a comparison of the main advantages and disadvantages of suspended and supported photocatalysts.

Nowadays, most work found in the literature on photocatalytic treatment of water, specifically those focused on solar applications, report the use of TiO_2 suspended configurations. Nevertheless, since the earlier studies by Matthews,[138] significant efforts have been made to develop catalysts based on immobilised TiO_2 using diverse supports including glass, stainless steel,

Table 4.3 Summary of main advantages and disadvantages of suspended and supported photocatalysts.

	Suspended photocatalysts	Supported photocatalysts
Advantages	High ratio illuminated catalyst surface to effective reactor volume Rather uniform catalyst distribution Few mass-transfer limitations	Catalyst separation not required, thus lowering operational costs Polluted water can be easily treated in continuous mode
Disadvantages	Slow and expensive catalyst separation operation Aggregation of suspended particles at high concentrations of catalyst Difficulties in being applied to continuous flow systems	Lower catalyst accessibility to photons Significant external mass transfer limitations at low fluid flow rate Increasing role of internal mass transfer with catalyst film thickness Possible catalyst detachment from the support

silica, zeolites, or porous lavas. As an example, Guillard *et al.* reported the immobilisation of TiO_2 on silicon wafers, soda lime glass, and on Pyrex glass plates by different sol–gel processes, using hydroxybutanedioic acid (malic acid) as a model molecule to be degraded.[139] Pyrex glass was also investigated for immobilisation of titania by a dip-coating technique for degradation of eosin,[137] while Ling *et al.*[140] reported a very good photocatalytic activity for degradation of phenol and methylene blue dye present in the water using a TiO_2 thin film immobilised on glass reactor tube. The preparation of alkoxide sol–gel derived TiO_2 films on flat stainless steel substrates and on glass beads was described by Balasubramanian *et al.*[141,142] They evaluated photocatalytic activities of TiO_2 films for the degradation of 4-chlorobenzoic acid utilising a quartz batch reactor, investigating the influence of operational parameters such as coating/heat treatment cycles of the catalysts, or mixing, recycle, aeration, and UV radiation flux of the reactor. The above-mentioned studies, as with most reported in the literature, were carried out using UV-vis artificial radiation, but some studies have also been performed with solar light. Among them, Zhao *et al.*[143] investigated the influence of pH on the photocatalytic activity of co-doped N–F–TiO_2 films under solar radiation, for degradation of the antibiotic oxytetracycline in aqueous solution. Damodar *et al.*[144] compared the photocatalytic degradation of four reactive dyes (Reactive orange-16, Turquoise blue-15, Reactive blue-4, and Reactive black B-5 dye) using TiO_2 in both suspended and immobilised systems under solar irradiation. To prepare the immobilised system, a slurry made up of TiO_2 (CDH, India) and poly(vinyl alcohol) was prepared and applied on a glass plate using a brush. The procedure was repeated until the desired thickness was obtained. The TiO_2 immobilised layer was effective for degrading the dyes and was sustained for long periods. It was observed that flow rate plays a significant role in degradation, with a lower flow rate giving better colour removal. In addition, fixed TiO_2-P25 in a sandblasted glass plate was effectively used for the degradation of [D-Leu]-microcystin-LR, a powerful toxin of widespread occurrence within cyanobacteria blooms.[145] Immobilised TiO_2 PC-500 (Millennium) over a non-woven matrix was also studied[146] for the solar photocatalytic oxidation of 2,4-dichlorophenol (DCP), and two pesticides and three phytosanitary waste products, commonly found in a French vineyard area. The photocatalytic treatment resulted in non-toxic treated effluents even though the by-products formed during photocatalysis degradation were not completely mineralised.[146] An interesting application derived from the irradiation of immobilised TiO_2 with solar light is the use of PET (poly(ethylene terephthalate)) bottles coated with titania to expose drinking water to sunlight for solar disinfection and oxidation and removal of contaminants such as arsenic.[66,147] TiO_2 immobilisation on the internal surface of the bottles was easily performed by shaking a titania suspension at pH 2.5 ($HClO_4$). After removing the excess suspension, the bottles were oven-dried for 20 min at 55 °C to promote TiO_2 adhesion and, after cooling, rinsed with distilled water to remove unfixed TiO_2. A further study on the feasibility and performance of photocatalytic TiO_2 coatings in batch-process solar disinfection (SODIS) reactors to improve the quality of drinking water in developing countries was carried out by Duffy *et al.*[148] Borosilicate glass and PET plastic SODIS reactors fitted with flexible plastic inserts coated with TiO_2 powder were shown to be 20% and 25% more effective, respectively, than standard SODIS reactors for the inactivation of *E. coli* K12.

4.4 Applications of Solar Photocatalysis

Initially, solar PTCs were designed to transform the solar radiation into thermal energy and successively transformed for photochemical applications, like solar photochemical reaction for production of fine chemicals (Figure 4.12). These systems had a concentrating power of 30 times the solar radiation; they were eventually applied for industrial processes of solar photo-oxygenation of citronellol,[149] among others. Regarding the remediation of water contaminated with hazardous and recalcitrant organic pollutants, one of the first applications using a solar TiO_2 photocatalytic plant was developed in 1993. This large water treatment plant had 158 m^2 of PTCs, located in Livermore (CA, USA); it was used to remove trichloroethylene from groundwater.[150] Although solar PTCs increased the solar irradiance for photochemical reactions compared to other photo-reactors, they could only use direct sunlight, making them quite useless in cloudy conditions, were very expensive due to the tracking system, and generally had low quantum efficiencies due to high electron–hole recombination at the catalyst.[20,151]

Figure 4.12 Photograph of an Helioman-type module consisting of four PTC of 4.5 m × 1.8, with an aperture of 70.5°, and with a geometrical concentration of 10.3.

4.4.1 Non-Concentrating Solar Reactors Applications

As explained above, in the 1990s the design of solar photocatalytic reactors for water remediation applications moved to non-concentrating systems. The most basic concept of design was an inclined plate, generally covered with immobilised catalyst film, upon which the fluid flows. In this configuration, the photons first travel through the reactant fluid and then the photocatalyst film, without any concentrated solar radiation, permitting the use of all incoming irradiance, *i.e.* direct and diffuse components of solar radiation, but also with the risk of volatile and toxic chemical loses, including the possible release of catalyst particles, due to the open design. Some of these systems had a front transparent cover to avoid this problem. Given their requirements of low cost investment and high surface requirement, these systems have been used for applications in developing countries for very localised wastewater problems, where other costly and effective wastewater treatment technologies are difficult to implement.[19] For instance, non-concentrating solar pilot plants have been applied for the photocatalytic treatment of various industrial or agricultural wastewaters containing biorefractory organic pollutants,[65,152,153] for water disinfection in farming industry,[154] and for discoloration of textile wastewater with combined biotreatment in Tunisia.[155] The interesting case of wastewater discoloration was studied with a solar thin-film fixed-bed reactor (TFFBR) of 25 m^2 based on the use of immobilised TiO_2-P25. The authors demonstrated the kinetic dependency of the removal of commercial azo dyes on flow rate, catalyst loading, and initial dye concentration. They found the best discoloration efficiency at a flow rate of 3 $m^3\ h^{-1}$, with an optimal catalyst loading of 10 g m^{-2}.[155] A similar TFFBR reactor was used for solar photocatalytic disinfection of water for fish farming industry applications.[154] This study investigated the inactivation of an aquaculture pathogen like *Aeromonas hydrophila* (Figure 4.13). Under high sunlight intensities ($\geq$600 W m^{-2}) and low flow rates (4.8 L h^{-1}), the disinfection results showed optimal performance, while lower sunlight intensities resulted in a significant reduction of the overall inactivation.

Another reported application of solar photocatalysis was the purification of wastewater from the vineyard industry.[60] The authors studied the efficacy and on-site ease-of-use for the wine grower of the photocatalytic reactor. For this purpose they used a solar TFFBR fabricated with a corrugated-steel inclined plate of 1 m^2 TiO_2-coated thin material for a total batch volume of 100 L. They evaluated different water quality parameters like TOC, toxicity (Microtox test), BOD_5, before and after purification for 8 days. The results demonstrated that the reactor was robust in operation in the field by end-users, and improved the wastewater quality, but for better performance the area to volume ratio should be three times higher.

4.4.2 CPC Solar Reactors Applications

In the mid-1990s, the use of solar CPC photo-reactors started to be increasingly applied to different water problems. This occurred because the new CPC reactors were smaller and used the whole UVA solar spectrum (including

Figure 4.13 Photograph of a thin-film fixed-bed reactor (TFFBR) for solar photocatalytic inactivation of aquaculture pathogens. (Reprinted from ref. 154. Thin-film fixed-bed reactor (TFFBR) for solar photocatalytic inactivation of aquaculture pathogen *Aeromonas hydrophila*, available at http://www.biomedcentral.com/1471-2180/12/5. Copyright (2011) Khan *et al.*; licensee BioMed Central Ltd, distributed under the terms of the Creative Commons Attribution License.)

diffuse component) with low investment and running costs compared to PTC systems. Demonstration and pilot scale CPC plants from 3 up to 150 m^2 were developed and evaluated for the removal of hazardous chemicals or water pathogens using solar photocatalysis. Examples of applications for contaminated water remediation are removal of chlorophenols, dyes, pesticides, chlorinated solvents, bacteria, fungi, and parasites. Some applications of CPCs for real wastewaters include the treatment of sanitary landfill leachate, olive mill, and urban wastewater.[21]

One of the first demonstration CPC photocatalytic plants was erected in 1999, in Madrid, by the SOLARDETOX consortium (Figure 4.14). The main objective of this plant was the application for oxidation of non-biodegradable chlorinated hydrocarbon solvents, but it was subsequently applied to other recalcitrant organic compounds like phenols, chlorophenols, pesticides, PCBs, TNT, PAHs, *etc.*[156] This demonstration facility was successfully tested for the removal of hydro-carbonated solvents using 200 mg L^{-1} of P25-TiO_2

Figure 4.14 Demonstration CPC solar plant located in Madrid.

as slurry under sunlight irradiation. The total surface area of the CPC field was 100 m^2, with batch loops of 800 L and a total treatment capacity of 2 m^3.

One of the first applications of solar CPC photo-reactors for water disinfection using TiO_2 solar photocatalysis was carried out in 1999. A solar photo-reactor of 4.5 m^2 of CPC permitted a 5-log decrease of *E. coli* and *E. faecalis* within 30 min of solar exposure.[157]

Other studies investigated different mirror geometries like compound parabolic, parabolic, and V-groove solar collectors made of anodised aluminium for water disinfection.[158] This comparative assessment proved that CPC geometry leads to the best bacterial inactivation results.

TiO_2 immobilised over glass Raschig rings placed on a tubular reactor of a CPC solar system was tested against slurry systems (Figure 4.15).[159] The best performance was found for the slurry TiO_2 system due to the favoured conditions of surface contact between bacteria and catalyst surface. However, the use of a fixed-bed reactor led to inactivation rates very similar to the slurry. Not only the titania surface area of the reactor configuration is responsible for the bacteria inactivation but also an important contribution of the mechanical stress in the system had to be considered. The main advantage of the fixed-bed TiO_2 catalyst is the outstanding stability, with stable activity after several reaction runs, which make this system highly recommended for continuous water treatment systems.

Solar TiO_2 photocatalysis could be applicable to different organic hazardous contaminants and the process efficiency can be considered linearly dependent on the energy flux, but only 5% of the whole solar spectrum is available for the TiO_2 band-gap. A realistic assumption of efficiencies of 75%

Figure 4.15 Photographs of the solar collector and details of the immobilised wall reactor and fixed-bed reactor systems.

for the solar collector efficiency and 4% for the catalyst means that 0.15% of the original solar photons are efficiently used in the process. From the standpoint of solar collecting technology, this is a rather inefficient process even considering a high added-value application. However, TiO_2 photocatalysis has the advantage over other advanced oxidation processes of using sunlight as energy source, constituting an environmentally friendly technology. TiO_2 is a cheap photostable catalyst, and the process may run at ambient temperature and pressure, using the oxygen from the air as oxidant, without necessity of further addition of chemical reagents. However, as the concentration of pollutants increases, this low photo-efficiency results in very slow kinetics that limit the practical application of the process. In such cases, solar photo-Fenton is widely applied, especially for wastewaters with medium pollutant concentrations (up to 1 g L^{-1}).[21]

4.5 Integration with Other Unit Operations

Several promising approaches have been proposed based on a real integration of the photocatalysis as part of a whole treatment train or process. We focus in this section on the combination of photocatalysis with biotreatment and membrane treatments, two well-known but also innovative options.

Photocatalytic systems combined with biological processes would undoubtedly contribute to enhance the treatment of industrial wastewater. Conventional wastewater treatment plants, which make use of activated sludge systems to reduce biodegradable chemical oxygen demand (COD), are very often not effective with industrial wastewaters due to the high COD load and the presence of recalcitrant compounds. This drawback becomes particularly critical when a significant concentration of non-biodegradable toxic compounds is also present, rendering the wastewater itself toxic to the microorganisms responsible for the biodegradation.

As with other AOPs, photocatalysis can be used as a post-treatment following conventional biological treatment with the aim to remove the remaining recalcitrant and bio-refractory contaminants at low concentrations still present in wastewater treatment plant effluents.[160] In the special case of polluted water containing a high percentage of non-biodegradable compounds in comparison with the load of biodegradable matter, photocatalysis may be used to increase the biodegradability and decrease the bio-toxicity of the original refractory contaminants prior to biological purification process. Depending on the chemical nature of the contamination, the load of the contaminants, and biodegradability parameters, any type of wastewater can be either pre-treated or post-treated integrated with economically competitive biological treatment. Any optimised integration strategy between AOPs (*i.e.* photocatalysis) and biotreatment will have to be assessed for each case study to minimise investment and treatment costs and increase the efficiency of the whole process.[161] During an ideal treatment by photocatalysis, contaminants are structurally altered, forming smaller and more biodegradable compounds until complete mineralisation is achieved. The objectives can be to (i) eliminate bio-recalcitrant and toxic compounds, (ii) increase the biodegradability of wastewater before applying conventional biological process, (iii) reduce the level of toxicity and micropollutants in the effluent, or (iv) disinfect the wastewater instead of using traditional disinfection methods such as chlorination, which is known to generate carcinogenic and mutagenic by-products such as halomethanes.[162] The percentage of mineralisation should be minimal during the treatment stage to avoid unnecessary expenditure of chemicals and energy, thereby lowering the operating cost. However, if the treatment time is too short, the reaction intermediates generated could still be structurally very similar to the original non-biodegradable and/or toxic components. Therefore, several analytical parameters must be monitored for each treatment step. Chemical parameters commonly measured are the total organic carbon, the concentration of specific pollutants and complete oxidation of heteroatoms released (Cl, N, P, *etc.*) as inorganic species (Cl^-, NO_3^-, PO_4^{3-}, *etc.*) into the media. Regarding biological assays, toxicity analyses (with organisms like *Vibrio fischeri*, *Daphnia magna*, activated sludge by respirometric assays, *etc.*) and biodegradability tests (using activated sludge) are very important to guarantee the appropriate conditions of the treated effluent to be subsequently treated by a conventional biodegradation process. These series of analytical parameters should be compatible with the engineering needs for designing the coupling strategy.

Very powerful analytical tools developed recently like gas chromatography coupled with mass spectrometry (GC-MS and GC-MS/MS), liquid chromatography coupled with mass spectrometry (LC-MS, LC-TOFF/MS, LC-MS/MS) permit us to detect, and quantify, a wide range of organic molecules (contaminants and their transformation products) in the environment (water sources, ecosystems, air, *etc.*) up to extremely low concentration levels, *i.e.* lower than nanograms per litre.[163] Actually, the so-called "contaminants of emerging concern" (CEC) are those organic molecules which have not been regulated or listed in guidelines for restricted use or discharge into the environment; they are the potential candidates to be included as listed pollutants of restricted use by the new environmental protection regulations due to their characteristics of accumulative and negative effects in ecosystems (eco-toxicity, chronic toxicity), their effects on human health & wellbeing, and on the environment in general. CEC can be very persistent in the environment, their transformation products (generated under environmental conditions in rivers, soils, *etc.*) may be even more hazardous or toxic than the original compound. They can be present in particularly low concentrations, but due to their continuous release into the natural water sources, by untreated urban wastewaters discharges and municipal wastewater treatment plant (MWTP) effluents, can accumulate in the environment causing medium- to long-term negative consequences.[164]

The efficiency of photocatalysis for treating CECs, or micro-pollutants in general, in MWTP effluents is low due to the originally low concentrations of the contaminants and pseudo-first-order kinetics ($r = k_{ap}C$). Therefore, one of the solutions for increasing process efficiency would be to increase the value of C_0. This may be achieved by combining photocatalysis with membrane processes. This strategy has attracted attention in recent years, as membrane processes will increase the concentration of contaminants in concentrates to much higher values than in raw MWTP effluents.[165] Membrane technologies, such as nanofiltration (NF) and reverse osmosis (RO) produce clean water for reuse and a concentrate. These concentrated streams usually contain organic matter, micro-pollutants and wastewater treatment residues.[166] Therefore, NF membranes could be used to treat MWTP effluents and photocatalysis to decrease the concentration of micro-pollutants in a concentrated stream.[167] The combination of NF and photocatalysis leads to important advantages in terms of improving the process efficiency. Concentration of contaminants by NF produced a practically clean effluent (permeate) and a concentrated stream with lower volume that reduces the size of the required photo-reactor and therefore the investment costs. The higher concentrations of contaminants leads to a higher reaction rate of the photocatalytic treatment (Figure 4.16), which means that a slight increase in the irradiation time is able to degrade a much higher amount of pollutants, reducing the operation costs per unit of degraded pollutant, which will have a direct impact on the economic balance of the process.

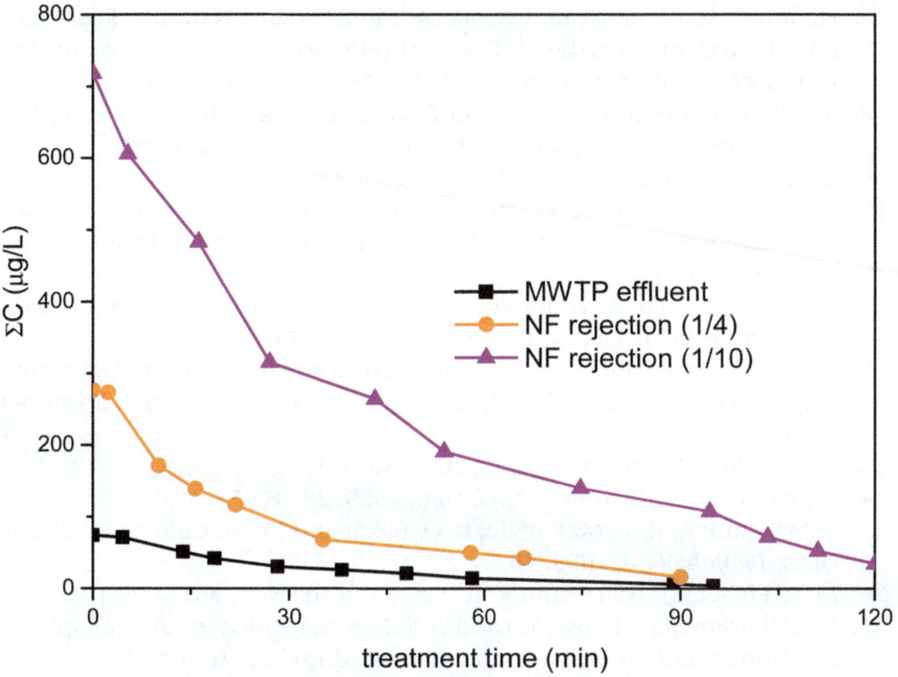

Figure 4.16 Treatment by solar photocatalysis of micropollutants in MWTP effluent or in the retentate of NF at different retentate/feed stream ratios.

Acknowledgements

The authors gratefully acknowledge the financial support of the Spanish Ministry of Economy and Competitiveness (WATER4FOOD, CTQ2014-54563-C3) and Comunidad de Madrid (REMTAVARES, S2013/MAE-2716).

References

1. C. A. Gueymard, *Sol. Energy*, 2004, **76**, 423.
2. G. Kopp and J. L. Lean, *Geophys. Res. Lett.*, 2011, **38**, L01706.
3. M. Paulescu, E. Paulescu, P. Gravila and V. Badescu, Solar radiation measurements, in *Weather Modeling and Forecasting of PV Systems Operation, Green Energy and Technology Series*, Springer, 2013, ch. 2, pp. 17–42.
4. C. Wehrli, Extraterrestrial Solar Spectrum, *Physikalisch-Meteorologisches Observatorium + World Radiation Center*, PMO/WRC, Davos Dorf, Switzerland, 1985, Publication no. 615.
5. American Society for Testing and Materials, Standard Extraterrestrial Spectrum Reference E-490-00, 2000.
6. National Aeronautics and Space Administration, National Space Science Data Center, Sun Fact Sheet, http://nssdc.gsfc.nasa.gov/planetary/factsheet/sunfact.html, accessed 19th November 2014.

7. National Aeronautics and Space Administration, National Space Science Data Center, Earth Fact Sheet, http://nssdc.gsfc.nasa.gov/planetary/factsheet/earthfact.html, accessed 19th November 2014.
8. M. Blanco-Muriel, D. C. Alarcón-Padilla, T. López-Moratalla and M. I. Lara-Coira, *Sol. Energy*, 2001, **70**, 431.
9. L. L. Vant-Hull and A. F. Hildebrandt, *Sol. Energy*, 1976, **18**, 31.
10. United States Naval Observatory, Multiyear Interactive Computer Almanac, http://aa.usno.navy.mil/software/mica/micainfo.php, accessed 19th November 2014.
11. Plataforma Solar de Almería, Solar Position Code, http://www.psa.es/sdg/sunpos.htm, accessed 19th November 2014.
12. National Renewable Energy Laboratory, Solar Position Algorithm, http://www.nrel.gov/midc/solpos/spa.html, accessed 19th November 2014.
13. A. B. Sproul, *Renewable Energy*, 2007, **32**, 1187.
14. F. Kasten and A. T. Young, *Appl. Opt.*, 1989, **28**, 4735.
15. S. Madronich, R. L. McKenzie, L. O. Björn and M. M. Caldwell, *J. Photochem. Photobiol., B*, 1998, **46**, 5.
16. M. B. Fisher, M. Iriarte and K. L. Nelson, *Water Res.*, 2012, **46**, 1745.
17. C. J. Riordan, R. L. Hulstrom and D. R. Myers, Influences of atmospheric conditions and air mass on the ratio of ultraviolet to total solar radiation, Technical Report SERI/TP-215–3895, 1990.
18. G. Seckmeyer, D. Pissulla, M. Glandorf, D. Henriques, B. Johnsen, A. Webb, A. M. Siani, A. Bais, B. Kjeldstad, C. Brogniez, J. Lenoble, B. Gardiner, P. Kirsch, T. Koskela, J. Kaurola, B. Uhlmann, H. Slaper, P. den Outer, M. Janouch, P. Werle, J. Gröbner, B. Mayer, A. de la Casiniere, S. Simic and F. Carvalho, *Photochem. Photobiol.*, 2008, **84**, 172.
19. R. Dillert, A. E. Cassano, R. Goslich and D. Bahnemann, *Catal. Today*, 1999, **54**, 267.
20. S. Malato, J. Blanco, A. Vidal and C. Richter, *Appl. Catal., B*, 2002, **37**, 1.
21. S. Malato, P. Fernández-Ibáñez, M. I. Maldonado, J. Blanco and W. Gernjak, *Catal. Today*, 2009, **147**, 1.
22. D. Bahnemann, *Sol. Energy*, 2004, **77**, 445.
23. J. A. Ajona and A. Vidal, *Sol. Energy*, 2000, **68**, 109.
24. J. Colina-Márquez, F. Machuca-Martínez and G. Li Puma, *Environ. Sci. Technol.*, 2010, **44**, 5112.
25. S. Malato, J. Blanco, D. C. Alarcón, M. I. Maldonado, P. Fernández-Ibáñez and W. Gernjak, *Catal. Today*, 2007, **122**, 137.
26. C. Navntoft, E. Ubomba-Jaswa, K. G. McGuigan and P. Fernández-Ibáñez, *J. Photochem. Photobiol., B*, 2008, **93**, 155.
27. P. Fernández-Ibáñez, C. Sichel, I. Polo-López, M. de Cara-García and J. C. Tello, *Catal. Today*, 2009, **144**, 62.
28. H. Gómez-Couso, M. Fontán-Sainz, P. Fernández-Ibáñez and E. Ares-Mazás, *Acta Trop.*, 2012, **124**, 235.
29. M. Agulló-Barceló, M. I. Polo-López, F. Lucena, J. Jofre and P. Fernandez-Ibañez, *Appl. Catal., B*, 2013, **136–137**, 341.
30. A. G. Rincón and C. Pulgarin, *Appl. Catal., B*, 2004, **49**, 99.

31. E. Ubomba-Jaswa, C. Navntoft, I. Polo-López, P. Fernández-Ibáñez and K. G. McGuigan, *Photochem. Photobiol. Sci.*, 2009, **8**, 587.
32. D. A. Keane, K. G. McGuigan, P. Fernández Ibáñez, M. I. Polo-López, J. A. Byrne, P. S. M. Dunlop, K. O'Shea, D. D. Dionysiou and S. C. Pillai, *Catal. Sci. Technol.*, 2014, **4**, 1211.
33. P. Fernández-Ibañez, J. Blanco, C. Sichel and S. Malato, *Catal. Today*, 2005, **101**, 345.
34. J. A. Byrne, P. Fernandez-Ibañez, P. S. M. Dunlop, D. M. A. Alrousan and J. W. J. Hamilton, *Int. J. Photoenergy*, 2011, 798051.
35. E. Ortega-Gómez, P. Fernández-Ibáñez, M. M. Ballesteros-Martín, M. I. Polo-López, B. Esteban-García and J. A. Sánchez-Pérez, *Water Res.*, 2012, **46**, 6154.
36. I. García-Fernández, I. Fernández-Calderero, M. I. Polo-López and P. Fernández-Ibáñez, *Catal. Today*, 2015, **240**, 30.
37. M. R. Hoffmann, S. T. Martin, W. Choi and D. W. Bahnemann, *Chem. Rev.*, 1995, **95**, 69.
38. R. Buonsanti, V. Grillo, E. Carlino, C. Giannini, T. Kipp, R. Cingolani and P. D. Cozzoli, *J. Am. Chem. Soc.*, 2008, **130**, 11223.
39. A. F. Wells, *Structural Inorganic Chemistry*, Clarendon Press, Oxford, 5th edn, 1984.
40. U. Diebold, *Surf. Sci. Rep.*, 2003, **48**, 53.
41. J. G. Li, T. Ishigaki and X. Sun, *J. Phys. Chem. C*, 2007, **111**, 4969.
42. *Handbook of Chemistry and Physics*, ed. R. C. Weast, CRC Press, Boca Raton, FL, B-154, 1984.
43. A. Beltrán, L. Gracia and J. Andrés, *J. Phys. Chem. B*, 2006, **110**, 23417.
44. M. Koelsch, S. Cassaignon, J. F. Guillemoles and J. R. Jolivet, *Thin Solid Films*, 2002, **403**, 312.
45. W. Wunderlicha, T. Oekermannb, L. Miaoc, N. T. Hued, S. Tanemurac and M. Tanemurac, *J. Ceram. Proc. Res.*, 2004, **5**, 343.
46. M. A. Henderson, *Surf. Sci. Rep.*, 2011, **66**, 185.
47. A. Fujishima, X. Zhang and D. A. Tryk, *Surf. Sci. Rep.*, 2008, **63**, 515.
48. A. L. Linsebigler, G. Q. Lu and J. T. Yates, *Chem. Rev.*, 1995, **95**, 735.
49. M. Batzill, *Energy Environ. Sci.*, 2011, **4**, 3275.
50. R. Hengerer, L. Kavan, P. Krtil and M. Grätzel, *J. Electrochem. Soc.*, 2000, **147**, 1467.
51. R. I. Bickley, T. Gonzalez-Carreño, J. S. Lees, L. Palmisano and R. J. D. Tilley, *J. Solid State Chem.*, 1991, **92**, 178.
52. R. Bacsa and J. Kiwi, *Appl. Catal., B*, 1998, **16**, 19.
53. T. Ohno, K. Sarukawa, K. Tokieda and M. Matsumura, *J. Catal.*, 2001, **203**, 82.
54. M. Yan, F. Chen, J. Zhang and M. Anpo, *J. Phys. Chem. B*, 2005, **109**, 8673.
55. G. Li and K. A. Gray, *Chem. Phys.*, 2007, **339**, 173.
56. A. M. Amat, A. Arques, F. López and M. A. Miranda, *Sol. Energy*, 2005, **79**, 393.
57. R. A. Damodar, K. Jagannathan and T. Swaminathan, *Sol. Energy*, 2007, **81**, 1.

58. D. Gumy, A. G. Rincón, R. Hajdu and C. Pulgarín, *Sol. Energy*, 2006, **80**, 1376.
59. H. Leong, P. Monash, S. Ibrahim and P. Saravanan, *Sol. Energy*, 2014, **101**, 321.
60. P. Pichat, S. Vannier, J. Dussaud and J. P. Rubis, *Sol. Energy*, 2004, **77**, 533.
61. C. Minero, E. Pelizzetti, S. Malato and J. Blanco, *Sol. Energy*, 1996, **56**, 411.
62. M. M. Ahmed, M. Brienza, V. Goetz and S. Chiron, *Chemosphere*, 2014, **117**, 256.
63. V. Augugliaro, J. Blanco Gálvez, J. Cáceres Vázquez, E. García López, V. Loddo, M. J. López Muñoz, S. Malato Rodrıguez, G. Marcì, L. Palmisano, M. Schiavello and J. Soria Ruiz, *Catal. Today*, 1999, **54**, 245.
64. S. Sakthivel, B. Neppolian, M. V. Shankar, B. Arabindoo, M. Palanichamy and V. Murugesan, *Sol. Energy Mater. Sol. Cells*, 2003, **77**, 65.
65. T. Zhang, T. Oyama, S. Horikoshi, J. Zhao, H. Hidaka and N. Serpone, *Sol. Energy*, 2001, **71**, 305.
66. A. H. Fostier, M. S. Silva Pereira, S. Rath and J. R. Guimaraẽs, *Chemosphere*, 2008, **72**, 319.
67. M. Peláez, N. T. Nolan, S. C. Pillai, M. K. Seery, P. Falaras, A. G. Kontosd, P. S. M. Dunlop, J. W. J. Hamilton, J. A. Byrne, K. O'Shea, M. H. Entezari and D. D. Dionysiou, *Appl. Catal., B*, 2012, **125**, 331.
68. R. Daghrir, P. Drogui and D. Robert, *Ind. Eng. Chem. Res.*, 2013, **52**, 3581.
69. R. Marschall and L. Wang, *Catal. Today*, 2014, **225**, 111.
70. S. Rehman, R. Ullah, A. M. Butt and N. D. Gohar, *J. Hazard. Mater.*, 2009, **170**, 560.
71. S. Sato, *Chem. Phys. Lett.*, 1986, **123**, 126.
72. S. Sakthivel, M. Janczarek and H. Kisch, *J. Phys. Chem. B*, 2004, **108**, 19384.
73. J. L. Gole, J. D. Stout, C. Burda, Y. Lou and X. Chen, *J. Phys. Chem. B*, 2004, **108**, 1230.
74. Z. Lin, A. Orlov, R. M. Lambert and M. C. Payne, *J. Phys. Chem. B*, 2005, **109**, 20948.
75. S. Sato, R. Nakamura and S. Abe, *Appl. Catal., B*, 2005, **284**, 131.
76. C. di Valentin, E. Finazzi, G. Pacchioni, A. Selloni, S. Livraghi, M. C. Paganini and E. Giamello, *Chem. Phys.*, 2007, **339**, 44.
77. D. Mitoraj and H. Kisch, *Chem.-Eur. J.*, 2010, **16**, 261.
78. V. Etacheri, M. K. Seery, S. J. Hinder and S. C. Pillai, *Chem. Mater.*, 2010, **22**, 3843.
79. R. Asahi, T. Morikawa, T. Ohwaki, K. Aoki and Y. Taga, *Science*, 2001, **293**, 269.
80. D. Li, H. Haneda, S. Hishita and N. Ohash, *Mater. Sci. Eng., B*, 2005, **117**, 67.
81. H. Irie, Y. Watanabe and K. Hashimoto, *J. Phys. Chem. B*, 2003, **107**, 5483.
82. J. Zhang, Y. Wang, Z. Jin, Z. Wu and Z. Zhang, *Appl. Surf. Sci.*, 2008, **254**, 4462.
83. J. M. Mwabora, T. Lindgren, E. Avendano, T. F. Jaramillo, J. Lu, S. E. Lindquist and C. G. Granqvist, *J. Phys. Chem. B*, 2004, **108**, 20193.

84. G. Abadias, F. Paumier, D. Eyidi, P. Guerin and T. Girardeau, *Surf. Interface Anal.*, 2010, **42**, 970.
85. H. Shen, L. Mi, P. Xu, W. Shen and P. N. Wang, *Appl. Surf. Sci.*, 2007, **253**, 7024.
86. J. Li, M. Xinxin, S. Mingren, X. Li and S. Zhenlun, *Thin Solid Films*, 2010, **519**, 101.
87. T. Ihara, M. Ando and S. Sugihara, *Photocatalysis*, 2001, **5**, 19.
88. I. Nakamura, N. Negishi, S. Kutsuna, T. Ihara, S. Sugihara and K. Takeuch, *J. Mol. Catal. A: Chem.*, 2000, **161**, 205.
89. Y. Guo, X. W. Zhang, W. H. Weng and G. R. Han, *Thin Solid Films*, 2007, **515**, 7117.
90. Y. Liu, X. Chen, J. Li and C. Burda, *Chemosphere*, 2005, **61**, 11.
91. T. Umebayashi, T. Yamaki, S. Yamamoto, A. Miyashita, S. Tanaka, T. Sumita and K. Asai, *J. Appl. Phys.*, 2003, **93**, 5156.
92. T. Ohno, T. Mitsui and M. Matsumura, *Chem. Lett.*, 2003, **32**, 364.
93. R. Bacsa, J. Kiwi, T. Ohno, P. Albers and V. Nadtochenko, *J. Phys. Chem. B*, 2005, **109**, 5994.
94. C. Yu, W. K. Ho, J. G. Yu, H. Yip, P. K. Wong and J. C. Zhao, *Environ. Sci. Technol.*, 2005, **39**, 1175.
95. D. E. Gu, Y. Lu, B. C. Yang and Y. D. Hu, *Chem. Commun.*, 2008, 2453.
96. S. U. M. Khan, M. Al-Shahry and W. B. Ingler Jr., *Science*, 2002, **297**, 2243.
97. H. Irie, Y. Watanabe and K. Hashimoto, *Chem. Lett.*, 2003, **32**, 772.
98. Y. Park, W. Kim, H. Park, T. Tachikawa, T. Majima and W. Choi, *Appl. Catal., B*, 2009, **91**, 355.
99. A. Zaleska, J. W. Sobczak, E. Grabowska and J. Hupka, *Appl. Catal., B*, 2008, **78**, 92.
100. D. Chen, D. Yang, Q. Wang and Z. Jiang, *Ind. Eng. Chem. Res.*, 2006, **45**, 4110.
101. J. C. Yu, J. Yu, W. Ho, Z. Jiang and L. Zhang, *Chem. Mater.*, 2002, **14**, 3808.
102. A. Hattori, K. Shimota, H. Tada and S. Ito, *Langmuir*, 1999, **15**, 5422.
103. D. Li, N. Ohashi, S. Hishita, T. Kolodiazhnyi and H. Haneda, *J. Solid State Chem.*, 2005, **178**, 3293.
104. T. Ohno, M. Akiyoshi, T. Umebayashi, K. Asai, T. Mitsui and M. Matsumura, *Appl. Catal., A*, 2004, **265**, 115.
105. Y. Choi, T. Umebayashi and M. Yoshikawa, *J. Mater. Sci.*, 2004, **39**, 1837.
106. T. Ohno, T. Tsubota, K. Nishijima and Z. Miyamoto, *Chem. Lett.*, 2004, **33**, 750.
107. S. Sakthivel and H. Kisch, *Angew. Chem., Int. Ed.*, 2003, **42**, 4908.
108. Y. Li, D. S. Hwang, N. H. Lee and S. J. Kim, *Chem. Phys. Lett.*, 2005, **404**, 25.
109. F. Wei, L. Ni and P. Cui, *J. Hazard. Mater.*, 2008, **156**, 135.
110. D. Li, N. Ohashi, S. Hishita, T. Kolodiazhnyi and H. Haneda, *J. Solid State Chem.*, 2005, **178**, 3293.
111. Y. Xie, Y. Li and X. Zhao, *J. Mol. Catal. A: Chem.*, 2007, **277**, 119.
112. P. A. K. Reddy, P. V. L. Reddy, V. M. Sharma, B. Srinivas, V. D. Kumari and M. Subrahmanyam, *J. Water Resour. Prot.*, 2010, **2**, 235.
113. H. M. Lu, T. Takata and Y. Lee, *Chem. Mater.*, 2004, **16**, 846.
114. H. Y. Liu and L. Gao, *Chem. Lett.*, 2004, **33**, 730.

115. M. Pelaez, A. A. de la Cruz, E. Stathatos, P. Falaras and D. D. Dionysiou, *Catal. Today*, 2009, **144**, 19.
116. H. Zhang, G. Chen and D. W. Bahnemann, *J. Mater. Chem.*, 2009, **19**, 5089.
117. D. Venieri, A. Fraggedaki, M. Kostadima, E. Chatzisymeona, V. Binas, A. Zachopoulos, G. Kiriakidis and D. Mantzavinos, *Appl. Catal., B*, 2014, **154–155**, 93.
118. M. B. Fisher, D. A. Keane, P. Fernández-Ibáñez, J. Colreavy, S. J. Hinder, K. G. McGuigan and S. C. Pillai, *Appl. Catal., B*, 2013, **130–131**, 8.
119. J. W. Shi, X. Yan, H. J. Cui, X. Zong, M. L. Fu, S. H. Chen and L. Z. Wang, *J. Mol. Catal. A: Chem.*, 2012, **356**, 53.
120. L. Jianhua, Y. Rong and L. Songmei, *Rare Met.*, 2006, **25**, 636.
121. L. Wu, J. C. Yu and X. Fu, *J. Mol. Catal. A: Chem.*, 2006, **244**, 25.
122. X. Guo, C. Chen, W. Song, X. Wang, W. Dia and W. Qin, *J. Mol. Catal. A: Chem.*, 2014, **387**, 1.
123. A. H. Zyoud, N. Zaatar, I. Saadeddin, C. Ali, D. Park, G. Campet and H. S. Hilal, *J. Hazard. Mater.*, 2010, **173**, 318.
124. L. Liu, J. Hensel, R. C. Fitzmorris, Y. Li and J. Z. Zhang, *J. Phys. Chem. Lett.*, 2009, **1**, 155.
125. A. K. Jana, *J. Photochem. Photobiol., A*, 2000, **132**, 1.
126. D. Chatterjee and A. Mahata, *J. Photochem. Photobiol., A*, 2002, **153**, 199.
127. J. Moon, C. Y. Yun, K. W. Chung, M. S. Kang and J. Yi, *Catal. Today*, 2003, **87**, 77.
128. W. Gernjak, M. I. Maldonado, S. Malato, J. Cáceres, T. Krutzler, A. Glaser and R. Bauer, *Sol. Energy*, 2004, **77**, 567.
129. D. Robert, A. Piscopo and J. V. Weber, *Sol. Energy*, 2004, **77**, 553.
130. P. K. Song, Y. Irie and Y. Shigesato, *Thin Solid Films*, 2006, **496**, 121.
131. S. A. Tomás, A. Luna-Resendis, L. C. Cortés-Cuautili and D. Jacinto, *Thin Solid Films*, 2009, **518**, 1337.
132. L. Andronic and A. Duta, *Thin Solid Films*, 2007, **515**, 6294.
133. R. L. Pozzo, M. A. Baltanás and A. E. Cassano, *Catal. Today*, 1997, **39**, 219.
134. A. E. Cassano and O. M. Alfano, *Catal. Today*, 2000, **58**, 167.
135. R. L. Pozzo, M. A. Baltanás, J. L. Giombi and A. E. Cassano, *Catal. Today*, 2000, **62**, 175.
136. D. F. Chen, F. Li and A. K. Ray, *AIChE J.*, 2000, **46**, 1034.
137. S. Zhou and A. K. Ray, *Ind. Eng. Chem. Res.*, 2003, **42**, 6020.
138. R. W. Matthews, *J. Phys. Chem.*, 1987, **91**, 3328.
139. C. Guillard, B. Beaugiraud, C. Dutriez, J. M. Herrmann, H. Jaffrezic, N. Jaffrezic-Renault and M. Lacroix, *Appl. Catal., B*, 2002, **39**, 331.
140. C. M. Ling, A. R. Mohamed and S. Bhatia, *Chemosphere*, 2004, **57**, 547.
141. G. Balasubramanian, D. D. Dionysiou, M. T. Suidan, V. Subramanian, I. Baudin and J.-M. Laîné, *J. Mater. Sci.*, 2003, **38**, 823.
142. G. Balasubramanian, D. D. Dionysiou, M. T. Suidan, I. Baudin and J.-M. Laîné, *Appl. Catal., B*, 2004, **47**, 73.
143. C. Zhao, M. Pelaez, X. Duan, H. Deng, K. O'Shea, D. Fatta-Kassinos and D. D. Dionysiou, *Appl. Catal., B*, 2013, **134–135**, 83.

144. R. A. Damodar, K. Jagannathan and T. Swaminathan, *Sol. Energy*, 2007, **81**, 1.
145. W. F. Domingues Vilela, A. Minillo, O. Rocha, E. M. Vieira and E. Bessa Azevedo, *Sol. Energy*, 2012, **86**, 2746.
146. T. Janin, V. Goetz, S. Brosillon and G. Plantard, *Sol. Energy*, 2013, **87**, 127.
147. S. J. Hug, L. Canonica, M. Wegelin, D. Gechter and U. Von Guten, *Environ. Sci. Technol.*, 2001, **35**, 2114.
148. E. F. Duffy, F. Al Touati, S. C. Kehoe, O. A. McLoughlin, L. W. Gill, W. Gernjak, I. Oller, M. I. Maldonado, S. Malato, J. Cassidy, R. H. Reed and K. G. McGuigan, *Sol. Energy*, 2004, **77**, 649.
149. M. Oelgemoller, N. Healy, L. de Oliveira, C. Jung and J. Mattay, *Green Chem.*, 2006, **8**, 831.
150. M. S. Mehos and C. S. Turchi, *Environ. Prog.*, 1993, **12**, 194.
151. D. J. Goswami, *Sol. Energy Eng.*, 1997, **119**, 101.
152. C. Guillard, J. Disdier, C. Monnet, J. Dussaud, S. Malato, J. Blanco, M. I. Maldonado and J. M. Herrmann, *Appl. Catal., B*, 2003, **46**, 319.
153. G. Zayani, L. Bousselmi, F. Mhenni and A. Ghrabi, *Desalination*, 2009, **246**, 344.
154. S. J. Khan, R. H. Reed and M. G. Rasul, *BMC Microbiol.*, 2012, **12**, 1.
155. L. Bousseimi, S. U. Geissen and H. Schroeder, *Water Sci. Technol.*, 2004, **49**, 331.
156. J. Blanco, S. Malato, P. Fernández, A. Vidal, A. Morales, P. Trincado, J. C. Oliveira, C. Minero, M. Musci, C. Casalle, M. Brunotte, S. Tratzky, N. Dischinger, K. H. Funken, C. Sattler, M. Vincent, M. Collares-Pereira, J. F. Mendes and C. M. Rangel, *Sol. Energy*, 1999, **67**, 317.
157. A. Vidal, A. I. Diaz, A. El Hraiki, M. Romero, I. Muguruza, F. Senhaji and J. Gonzalez, *Catal. Today*, 1999, **54**, 283.
158. O. A. McLoughlin, S. C. Kehoe, K. G. McGuigan, E. F. Duffy, F. Al Touati, W. Gernjak, I. Oller Alberola, S. Malato Rodríguez and L. W. Gill, *Sol. Energy*, 2004, **77**, 657.
159. C. Sordo, R. Van Grieken, J. Marugan and P. Fernandez-Ibañez, *Water Sci. Technol.*, 2010, **61**, 507.
160. I. Oller, S. Malato and J. A. Sánchez-Pérez, *Sci. Total Environ.*, 2011, **409**, 4141.
161. V. Sarria, S. Parra, N. Adler, P. Péringer, N. Benitez and C. Pulgarin, *Catal. Today*, 2002, **76**, 301.
162. L. Rizzo, *Water Res.*, 2011, **45**, 4311.
163. S. D. Richardson and T. A. Ternes, *Anal. Chem.*, 2014, **86**, 2813.
164. B. Du, A. E. Price, W. C. Scott, L. A. Kristofco, A. J. Ramirez, C. K. Chambliss, J. C. Yelderman and B. W. Brooks, *Sci. Total Environ.*, 2014, **466–467**, 976.
165. A. Pérez-Gonzalez, A. M. Urtiaga, R. Ibáñez and I. Ortiz, *Water Res.*, 2012, **46**, 267.
166. P. Westerhoff, H. Moon, D. Minikata and J. Crittenden, *Water Res.*, 2009, **43**, 3992.
167. S. Miralles-Cuevas, I. Oller, A. Ruiz Aguirre, J. A. Sánchez Pérez and S. Malato Rodríguez, *Chem. Eng. J.*, 2014, **239**, 68.

CHAPTER 5

Combined Photocatalysis–Separation Processes for Water Treatment Using Hybrid Photocatalytic Membrane Reactors

TEIK-THYE LIM*[a] AND RONN GOEI[b]

[a]School of Civil and Environmental Engineering, Nanyang Technological University, 50 Nanyang Avenue, Singapore 639798, Republic of Singapore; [b]Nanyang Environment and Water Research Institute (NEWRI), Nanyang Technological University, 1 Cleantech Loop, CleanTech One, Singapore 637141, Republic of Singapore
*E-mail: cttlim@ntu.edu.sg

5.1 Introduction

Water shortage is a major global issue facing both developed and developing countries due to expanding world population, rapid urbanization and industrialization. Climate change, which affects water precipitation pattern, is another factor that aggravates the water scarcity problem. To overcome the adversity of water scarcity, water reclamation and reuse has been considered

RSC Energy and Environment Series No. 15
Photocatalysis: Applications
Edited by Dionysios D Dionysiou, Gianluca Li Puma, Jinhua Ye, Jenny Schneider, and Detlef Bahnemann

Published by the Royal Society of Chemistry, www.rsc.org

as a sustainable solution. To fully realize the potential of water reclamation and to prevent secondary pollution to aquatic systems and food chain, the presence of different types of recalcitrant pollutants in domestic and industrial wastewater needs to be addressed. Various chemicals have been detected in the treated effluents from municipal wastewater treatment plants. They include pesticides, dyes, pharmaceuticals, hormones, drugs, antibiotics, personal care products, disinfection byproducts, surfactants, chelating agents, frame retardants, perfluorinated chemicals, *etc.* These compounds are found bioaccumulated in living organisms and crops, and some of them are endocrine disruptors.

The conventional treatment technologies based on biological processes are not effective in removing these recalcitrant and biologically toxic pollutants. Thus, a novel and robust wastewater treatment is required to overcome the problem caused by the different emerging contaminants. To remove the biorefractory micropollutants as well as other dissolved organic compounds, carbon adsorption, NF/RO membrane separation and advanced oxidation processes (AOPs) are the widely adopted advanced water treatment processes. Carbon adsorption using adsorbents such as activated carbon simply transfers the pollutants from one phase to another without removing then from the environment. Membrane separation process produces a concentrate stream which contains the constituents originally present in the feedwater as well as chemicals used for membrane cleaning. AOPs are among the very few viable technologies that can degrade the recalcitrant pollutants or transform them into environmentally benign products which can be degraded in the natural environmental systems. Photocatalysis using TiO_2 as photocatalyst is an AOP with the following advantages compared to other chemical oxidation processes:

- The system avoids the use of hazardous chemicals (*e.g.* chlorine, ozone, hydrogen peroxide).
- The oxidant generated is $HO^{\bullet}$ which is a green oxidant and its reaction products with organic compounds usually are less toxic and more hydrophilic (*i.e.* have a lower tendency of biosorption by living organisms).
- The system can be more energy efficient than photolysis and sonolysis.

$HO^{\bullet}$ is the second most powerful oxidant after fluorine gas (Table 5.1). It is non-selective and therefore it, along with its other co-generated reactive oxygen species (ROS), can degrade a wide range of pollutants. Klavarioti *et al.*[1] compared the performances of various AOPs toward degradation of pharmaceuticals, and the removal efficiencies are as follows: heterogeneous photocatalysis (32%), ozonation (30%), Fenton/photo-Fenton (13%), UV/H_2O_2 (12%), electrolysis (8%), sonolysis (4%) and wet air oxidation (1%). The various pollutants that have been reported as susceptible to photocatalytic degradation in water with UV-irradiated TiO_2 include pesticides, dyes, pharmaceuticals, hormones, drugs, antibiotics, personal care products,

Table 5.1 Standard reduction potentials in aqueous medium for the commonly used oxidants in water treatment.[4,5]

Oxidant	Reduction reaction	E^o (V_{NHE})
Fluorine (F_2)	$F_2(g) + 2H^+ + 2e^- \rightarrow 2HF$	3.05 [5]
Hydroxyl radical ($HO^•$)	$HO^• + H^+ + e^- \rightarrow H_2O$	2.73 [4]
Ozone (O_3)	$O_3(g) + 2H^+ + 2e^- \rightarrow O_2 + H_2O$	2.07 [5]
Hydrogen peroxide (H_2O_2)	$H_2O_2 + 2H^+ + 2e^- \rightarrow 2H_2O$	1.78 [5]
Perhydroxyl radical ($HO_2^•$)	$HO_2^• + 3H^+ + 3e^- \rightarrow 2H_2O$	1.65 [5]
Permanganate (MnO_4^-)	$MnO_4^- + 8H^+ + 5e^- \rightarrow Mn^{2+} + 4H_2O$	1.51 [5]
Hypochlorous acid (HOCl)	$HOCl + H^+ + 2e^- \rightarrow Cl^- + H_2O$	1.49 [5]
Chlorine (Cl_2)	$Cl_2(g) + 2e^- \rightarrow 2Cl^-$	1.36 [5]
Diatomic oxygen	$O_2(g) + 4H^+ + 4e^- \rightarrow 2H_2O$	1.23 [5]
Chlorine dioxide (ClO_2)	$ClO_2(g) + H^+ + e^- \rightarrow HClO_2$	1.19 [5]

phthalates, plasticizers, disinfection byproducts, surfactants, chelating agents, frame retardants, algal toxins, taste and odor compounds, and natural organic matters.[2,3] Despite the proven efficiency in laboratory studies, the large-scale application of photocatalyst in water and wastewater treatment is limited. One of the hindrances to the large-scale application is the inherent mass transfer limitation phenomenon that controls the overall efficiency of the heterogeneous photocatalysis process, especially in the system with the immobilized TiO_2.

The mass transfer limitation associated with the heterogeneous photocatalysis process can be overcome by coupling photocatalysis and enhanced mass transfer process enabled by the membrane separation process, in a hybrid photocatalytic membrane reactor (PMR) system. In this system, photocatalyst is deposited on the membrane surface. Mass transfer of pollutants to the photocatalyst is promoted through forced permeation driven by the transmembrane pressure across the membrane. This chapter presents an overview of the respective applications of TiO_2 and membrane separation in water treatment, the fabrication of the photocatalytic TiO_2-coated ceramic membrane and the potential synergism of the coupled photocatalysis–membrane separation processes, and discussions of the recent studies, prospects, challenges and possible solutions.

5.2 TiO_2 and Application for Water Treatment

Titanium dioxide (TiO_2) exists either in its natural polymorphs (*i.e.* anatase, brookite or rutile) or in its artificial polymorphs (*i.e.* TiO_2(B), TiO_2(H), TiO_2(R), *etc.*).[6] Anatase and rutile polymorphs of TiO_2 exist in the tetragonal structure, while brookite and TiO_2(B) exist in the orthorhombic structure. The Ti–Ti distance in anatase is wider than those in the octahedral arrangement of rutile. The number of neighboring octahedrons that are in contact with the octahedral structure of anatase and rutile are eight (four sharing a corner and four sharing an edge) and ten (eight sharing a corner and two sharing an edge) respectively. This would lead to a denser rutile structure.

Although anatase is thermodynamically less stable than rutile, the formation of anatase is much favored at lower temperatures (<600 °C).[6] The formation of TiO_2 crystals at a lower temperature would yield TiO_2 with a higher surface area and, hence, a higher surface density of the active sites for effective adsorption and photocatalytic degradation reaction. Therefore, anatase was found to be the most photocatalytically active polymorph of TiO_2.[7–9] The high photocatalytic activity of anatase could also be ascribed to its triangular atomic arrangement that allows effective adsorption of pollutants.[10] Both anatase and brookite polymorph are transformed into the rutile structure when subjected to a higher temperature of heat treatment.[11] The differences in the electronic and optical responses of different TiO_2 polymorphs are discussed in detail elsewhere.[12] TiO_2 shows an amphoteric characteristic, *i.e.* its surface charge is pH dependent. The iso-electric point (pH_{pzc}) of TiO_2 is between pH 5 and 7. At pH < pH_{pzc}, the surface charge of TiO_2 will be positive ($\equiv Ti{-}OH_2^+$), while at pH > pH_{pzc} the surface charge will turn negative ($\equiv Ti{-}OH^-$).[13]

TiO_2 has appeared as the leading photocatalyst for uses in water due to its good photocatalytic activity, high chemical and biological stability, relatively low cost and low toxicity. In addition, it has well-located conduction and valence band positions for production of ROS in water. Its bandgap of 3.0–3.2 V allows it to be excited using many commercial UV lamps including low-pressure and medium-pressure mercury arc lamps, xenon arc lamps, fluorescent UV lamps, metal halogen lamps, UV-LEDs, and excimer lamps. Figure 5.1 shows the mechanism of photocatalysis while Figure 5.2 shows the band positions and redox potentials for various aqueous-phase photocatalytic reactions at pH 0–14.

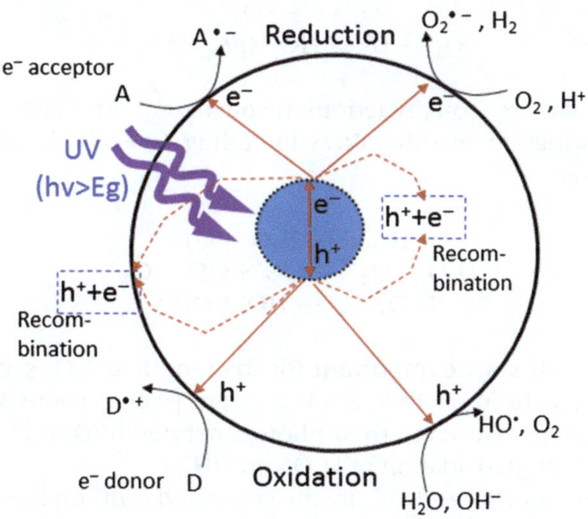

Figure 5.1 Schematic diagram of photocatalytic redox processes on the photoexcited semiconductor.

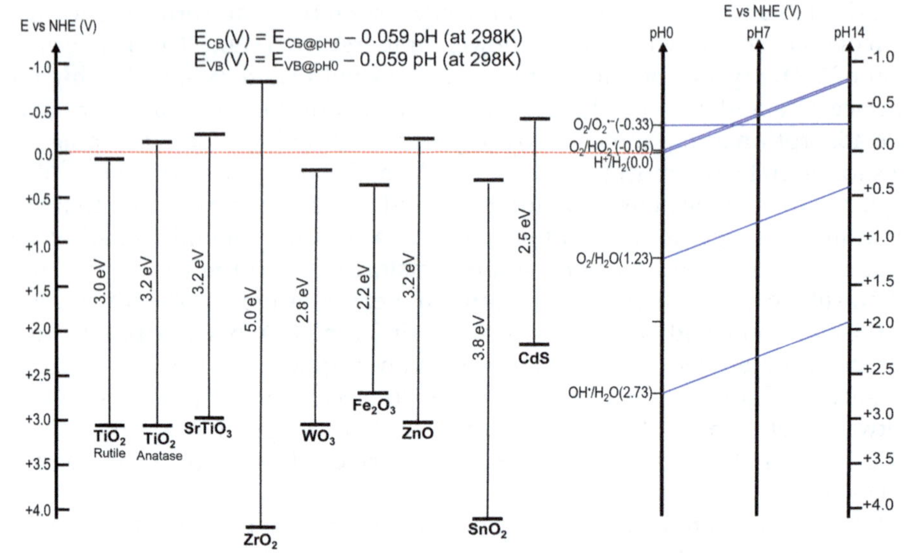

Figure 5.2 Band positions of various semiconductors and redox potentials of the various processes associated with photocatalysis in water at various pH.

The valence band of TiO_2 has oxidative potentials which are high enough to oxidize water molecule to $HO^{\bullet}$ *via* photocatalytic oxidation route. Meanwhile, the conduction band of TiO_2 is also sufficiently negative to reduce the ground-state diatomic oxygen molecule to superoxide radical ($O_2^{\bullet-}$) which is likely to exist as hydroperoxyl radical ($HO_2^{\bullet}$) at circumneutral pH:

$$O_2 + e^- \rightarrow O_2^{\bullet-} \quad (5.1)$$

$$O_2^{\bullet-} + H^+ \rightarrow HO_2^{\bullet}\ (pK_a = 4.8) \quad (5.2)$$

Subsequent spontaneous reactions involving $O_2^{\bullet-}$ and $HO_2^{\bullet}$ in water will lead to the formation of other ROS including $HO^{\bullet}$ *via* the photocatalytic reduction route:

$$2HO_2^{\bullet} \rightarrow H_2O_2 + O_2 \quad (5.3)$$

$$H_2O_2 + O_2^{\bullet-} \rightarrow HO^{\bullet} + OH^- + O_2 \quad (5.4)$$

$$H_2O_2 + e^- \rightarrow HO^{\bullet} + OH^- \quad (5.5)$$

The above reactions are important for the modified TiO_2 semiconductors whose bandgaps are lower than 3.0 V, and are photoresponsive under visible-light irradiation. However, their photogenerated h^+ could be too weak to produce $HO^{\bullet}$ through oxidation of H_2O and OH^-.

The formation of $HO^{\bullet}$ and $O_2^{\bullet-}$ by the separated e^-/h^+ on the photoexcited TiO_2 is the important route for triggering a series of photocatalytic degradation or transformation of the pollutants in water. Direct photoreduction and photo-oxidation of the pollutants on the surface of photo-excited TiO_2 are not likely to be the dominant processes due to the limited mass transfer rate of the electron acceptors and electron donors to the TiO_2 surface.

TiO_2 can be prepared in different shapes and morphologies such as nanocrystals/nanodots, nanowires/nanorods (when TiO_2 was synthesized with templating agent), nanosheets or nanofilms. Nanosized TiO_2 demonstrates unique optical, electronic, mechanical, thermal and structural properties which depend on its size, shape and microstructure. Different methods used to prepare and deposit various types of TiO_2 nanostructures include sol–gel, micelles/inverse-micelles, hydrothermal, solvothermal, direct oxidation, chemical vapor deposition, physical vapor deposition, electro-deposition, electrochemical anodization, sonochemical and microwave irradiation methods.[14,15]

The performance of TiO_2 photocatalyst for photocatalytic degradation of pollutants is strongly influenced by the crystal phase, crystallinity, lattice defect, particle size, porosity, overall surface area and the density of hydroxyl groups on the surface of the photocatalysts.[16,17] Particle size is an important parameter for photocatalytic efficiency. A change in particle size will change the surface/bulk ratio, thus modifying the significance of volume and surface e^-/h^+ recombination. According to some literature data, the optimum particle size of TiO_2 lies around 10 nm,[18] although other studies report an improved TiO_2 activity for particles below 10 nm diameter.[19,20] Around this particle size, there exists the best compromise between the e^-/h^+ recombination rate and interfacial charge transfer rate from TiO_2 to electron donors and acceptors. TiO_2 with smaller crystallite size would induce blue-shift in the light absorption spectrum (due to quantum confinement effect) and favors surface recombination of photogenerated e^-/h^+.

In view of the above findings, for the fabrication of TiO_2-coated membranes, the crystallite size of the anchored TiO_2 photocatalyst must be optimized to achieve a high photocatalytic activity while not incurring a high membrane resistance due to small pore sizes. In addition, there is a need to produce a thin and uniform layer of nanosized photocatalyst, with a thickness of several tenths of nanometers, over the underlying highly porous membrane layer in order to achieve high water permeability of the TiO_2-coated membrane. However, if the coated TiO_2 layer is too thin, it will reduce its UV absorbance and photocatalytic activity.[21]

5.3 Separation Process with Ceramic Membrane

Membrane separation process has been used widely in water filtration, medical separation process, gas separation, chemical purification and many other applications. The use of membrane separation for water reclamation and purification has gained increasing interest in the recent decades due to the increasingly stringent environmental protection regulations, and increasing need for wastewater reclamation for various reuses including indirect and direct potable reuse. Membrane filtration is a reliable technology for effective removal of colloids, microorganisms and organic pollutants in water and wastewater treatment. The advantages of membrane technology are high system compacity (small footprint), low energy consumption, easy operational control, easy up-scaling and combination with other processes, good permeate quality and reliable performance. The main inherent drawback associated with the membrane separation is membrane fouling which leads to a higher operational and overall energy consumption in the membrane treatment process.[22–24] In addition, during membrane separation, the pollutants in the feedwater are merely separated from the water (without degradation) by the membrane. This results in generation of the secondary waste stream associated with the membrane retentate. Therefore, one of the recent areas of interest in membrane development is to impart reactivity to the membrane such that it would degrade the pollutants in the feedwater and meanwhile has the potential of self-cleaning by degrading the deposited foulants.

Polymeric membranes have been used extensively in water and wastewater treatment process. However, the disadvantages of the polymeric membrane are its poor chemical and thermal durability for use as the reactive membrane, and it is also photodegradable under UV irradiation. This limits their use for the PMR process. The need to fabricate a new type of membrane that is chemical- and UV-resistant has led to interest in the development of inorganic membranes for applications in PMR.

In recent years, ceramic membranes have been used increasingly in the water industry including in municipal water and wastewater treatment plants. This is because a ceramic membrane has the following features: high permeability (due to good surface wettability and relatively uniform pore size), good physical stability under high temperature and high pressure system, good chemical and biological stability, excellent UV resistance, robust mechanical strength, high abrasion resistance, long working life, easy cleaning and reusability even after high-temperature treatment.[25–30] Thus, it could be an excellent substrate for fabricating the photocatalytic membrane.

Permeability and separation factor are two most important performance indicators of a membrane. For a porous ceramic membrane, the two functional characteristics are controlled by the surface porosity, pore size distribution and thickness of the membrane. Typically, a porous ceramic membrane consists of the following asymmetric layers:

- macroporous support (bottom layer) – to provide mechanical support;
- mesoporous intermediate layers – to bridge the pore size differences between the macroporous support layer and the microporous top layer;
- microporous (or dense) top layer – where actual separation takes place.

The commonly used materials for making ceramic membrane are Al_2O_3, TiO_2, ZrO_2 and SiO_2, and a mixture of these materials. Among the different types of ceramic porous support materials, α-Al_2O_3 is the most commonly chosen due to its high chemical stability under extreme pH conditions.[31]

Such ceramic membranes are usually fabricated through multiple steps, starting from the support layer, followed by multiple coating of intermediate layers and finishing with coating of a final separation layer. Each step involves high-temperature sintering followed by cooling to room temperature. Often, one or more different ceramic materials are used to make the multiple layers in order to obtain the desired pore sizes for each layer. Different types of ceramic membrane elements can be fabricated, such as flat discs, plates, tubes, capillaries, corrugated systems, hollow fibers or multi-channel monolithic elements. They have different specific surface/volume ratios (typically above 500 m^2 m^{-3}).[32]

In a PMR process, the membrane is used as an active participant in a chemical reaction in the reactor chamber. The membrane not only plays the role of a separator but also as part of the reactor, such as catalyst support or membrane contactor where the desired reactions take place. Therefore, it is important to develop a photocatalytic TiO_2-coated membrane that allows effective removal of environmental pollutants through simultaneous degradation and permselective separation process based on molecular sizes. The membrane system should be operated under high water flux conditions which can be maintained or restored after appropriate cleaning.

5.4 Fabrication of TiO_2-Coated Ceramic Membrane

Coating of TiO_2 photocatalytic layers on porous ceramic substrate is an important step for the fabrication of the photocatalytic TiO_2 membrane. Different coating techniques have been used to prepare and coat the TiO_2 thin-film layer on the ceramic membrane support, on both internal and external surface as well as in the pores. These include dip coating,[33,34] spin coating,[35–37] slip casting,[38,39] *in situ* hydrothermal/solvothermal synthesis,[40,41] aerosol spraying,[42–44] chemical vapor deposition,[45,46] physical vapor deposition,[47,48] radio-frequency magnetron sputtering,[49] spray pyrolysis,[44,50] reactive sputtering,[51] electrochemical deposition,[52,53] plasma spraying,[54,55] *etc.* Different coating techniques would result in TiO_2 thin films of various thicknesses and morphologies. The final microstructure and the performance of the TiO_2 layer as photocatalytic and permselective layer strongly depend on the preparation condition. The goal of coating is to produce a thin TiO_2 layer with strong adherence, good uniformity in thickness and pore size, defect-free surface and minimal pore blockage of the underlying layer.

Among the various coating techniques, the dip-coating technique based on coating of TiO_2 sol is the most commonly adopted. The sol is a stable dispersion of nanometer-sized colloidal particles in a liquid solution. Different types of titanium precursors such as titanium alkoxides or non-alkoxide titanium source can be used to prepare the TiO_2 sol.[14,56,57] Drying of the sol at the ambient pressure (thermal evaporation of the liquid) will cause shrinkage in the dispersion producing an intermediate xerogel phase. A final gel is produced by complete removal of the water or solvent at a higher temperature.

Ordered mesoporous TiO_2 nanostructure can be prepared using an evaporation-induced self-assembly method. In this process, a structure directing agent (SDA) is added during preparation of the TiO_2 sol. SDAs are molecules that possess both hydrophobic and hydrophilic segments and exhibit partial affinity toward both polar and non-polar surfaces. The SDA assembles to form a micellar structure that can be removed during calcination steps, leaving behind an ordered mesoporous TiO_2 structure. The addition of SDA in the sol allows fine-tuning the porosity and porous structure of the resulting TiO_2 layer.

Table 5.2 summarizes the common types of SDAs that have been used in the preparation of porous TiO_2 layer. They include block copolymers and cationic, anionic and non-ionic surfactants. A non-ionic SDA has generally been preferred due to the weaker interactions between the SDA and the TiO_2 scaffold allowing easy removal of the SDA during calcination step. In addition, the use of a non-ionic SDA increases the structural stability of the mesoporous structure.[58–60] Alginate fibers were also found to effectively disperse and stabilize TiO_2 nanoparticles in their matrix.[61] During the preparation of the sol, the hydrated head group of the SDA (*e.g.* poly(ethylene oxide) of P-123 copolymer) interacts directly with the Ti alkoxide precursor. This interaction promotes the formation of a micellar structure even at a concentration lower than the critical micelle concentration.[62] The resulting pore diameter is relatively uniform. The wall thickness of the resulting microstructure depends on the overall concentration of SDA in the sol

Table 5.2 Common types of structure directing agents (SDAs) used in the preparation of mesoporous TiO_2.

Class	Examples (scientific/trade names)	References
Polymer	Polyethylene glycol (PEG)	65–67
Di-block copolymer	Poly(ethylene-*co*-butylene)-*block*-poly(ethylene oxide)/KLE	68–70
Tri-block copolymer	Poly(ethylene oxide)-*block*-polybutadiene-*block*-poly(ethylene oxide)/PEO-PB-PEO	71
	Polystyrene-*block*-polybutadiene-*block*-polystyrene/SBS	72
	Poly(ethylene oxide)-*block*-poly(propylene oxide)-*block*-poly(ethylene oxide)/Pluronic® P-123	62,73–76
	Poly(ethylene oxide)-*block*-poly(propylene oxide)-*block*-poly(ethylene oxide)/Pluronic® F-127	77–79
Cationic surfactant	Cetyl(trimethyl)ammonium bromide/CTAB	80–82
	Poly(ethyleneimine)/PEI	83
Anionic surfactant	Sodium dodecyl sulfate/SDS	81
Non-ionic surfactant	Polyethylene glycol *p*-(1,1,3,3-tetramethylbutyl)-phenyl ether/Triton X-100	84–86
	Polyoxyethylene (20) sorbitan monolaurate/Tween 20	87 and 88
	Polyoxyethylene (20) sorbitan monooleate/Tween 80	33,89,90
	Poly(ethylene glycol) hexadecyl ether/Brij58	78 and 91

(a thicker wall with a lower concentration of SDA). Generally, the increase in the SDA chain length would result in a structure with a larger porosity. The increase in SDA concentration has a similar effect as the increase in SDA chain length.[63,64]

Quite recently, binders such as hydroxypropyl cellulose (HPC) and poly(vinyl alcohol) (PVA) were added into the TiO_2 sol.[38] The binders could help to adjust the viscosity of the sol, lower the surface tension and increase the mechanical strength of the resulting membrane.[92,93] In addition, PVA can also be used as the pore fillers during membrane fabrication to minimize infiltration of TiO_2 sol into the pores of the underlying membrane support.[73]

Figure 5.3 presents a flow diagram of the typical preparation steps of micro/mesoporous membranes *via* the sol–gel technique. In a typical preparation, a thickened sol is coated onto the porous support. The asymmetrical structure is obtained through coating of two or more layers with different porosities. Usually, more than one layer of coating is required for each type of the sol to ensure the smoothness and uniformity of the coating. A mesoporous intermediate layer is coated prior to the coating of the top layer to reduce the surface roughness. This will result in a multilayered ceramic membrane with a hierarchically porous architecture. The presence of porosity gradient is important to prevent early clogging of the composite membrane during the filtration process. This is due to the better mixing and improved contact between pollutants and internal porous structure of the photocatalytic membrane as induced by the asymmetric structure.[94]

Zhao and coworkers[95] investigated different preparation conditions that would affect the microstructure of the resulting TiO_2 thin film such as amounts of the SDA and hydrolysis retardants added, aging time, pH and calcination temperature. The possibility of high-temperature calcination while

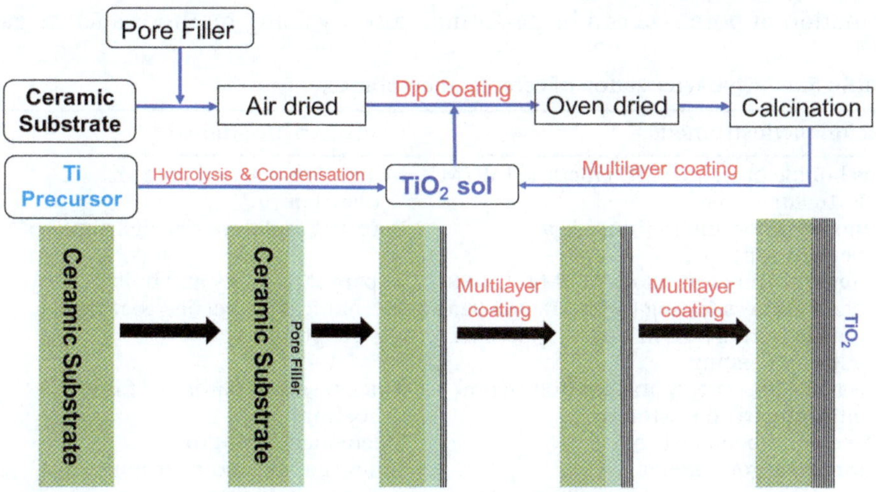

Figure 5.3 Flow diagram illustrating preparation of microporous/mesoporous ceramic membranes *via* the sol–gel technique.

maintaining a high-porosity TiO_2 was demonstrated by Lee *et al.*[96] using a mixture of soft and hard templates as the SDA. The materials were able to withstand a calcination temperature of 1000 °C while maintaining a high porosity.

A recent study by Goei and Lim[34] demonstrated the preparation of a TiO_2 photocatalytic membrane in which each of the photocatalytic layers were prepared using different types of SDAs. Pluronic P-123 block copolymer and non-ionic SDAs such as Triton X-100, Tween 20 and Tween 80 were used in the preparation of the different types of TiO_2 sols. These SDAs exhibited a similar micellar assembly with high hydrophilic characteristic (as shown by their high hydrophile–lipophile numbers). The TiO_2 sols prepared using different SDAs exhibited good compatibility among them, allowing the flexibility of constructing the TiO_2 membrane that exhibited favorable characteristics of membrane separation process, *i.e.* high permeability, good membrane rejection, or both.

Heat treatment plays an important role in the fabrication of a crack-free membrane. A lower calcination temperature, while preventing transformation of anatase into rutile, would yield a membrane with a smaller pore and crystallite size. By controlling the heating rate during the calcination step, one can avoid the development of thermally induced stress arising from the unequal thermal expansion of the composite membrane or layer shrinkage due to thermal decomposition of organic additives. For the multilayer coating process, the repeated thermal treatments performed after each coating could cause a partial collapse of the porous structure thus reducing its photocatalytic degradation (PCD) efficiency.[91] It was suggested that the calcination temperature is the ultimate control of the porous structure of the resulting membrane.[97]

Table 5.3 summarizes different characterization techniques that can be used to investigate the various characteristics of ceramic membrane. Determination of pore size can be performed directly using methods such as gas

Table 5.3 Characterization of ceramic membranes.

Technique/instrument	Characteristic studied
Gas bubble pressure (GBP) method (ASTM F316-80)	Mean pore size and pore size distribution[98]
Nano-permporometry (humid gas permeation)	Pore size and pore size distribution[99]
Water absorption method (ASTM C373-88)	Apparent porosity and bulk density[38]
Pencil hardness testing (ASTM D3363-92a)/ tensile testing (ASTM C633-01)/scratch adhesion testing	Mechanical properties/coating strength
Spectral ellipsometry and environmental ellipsometric porosimetry	Thickness and porosity of the coating[100]
Three-point bend test	Mechanical strength
Static angle goniometer	Contact angle and hydrophilicity/ hydrophobicity
Zeta potential analyzer	Electrical potential difference (streaming potential)

bubble pressure method, permporometry or nano-permporometry. The pore size can also be quantified indirectly by measuring the pure water permeability, or by estimation using the water absorption method or the molecular weight cut-off (MWCO) value of the membrane. Evaluating the mechanical properties of the ceramic composite membrane is important to ensure the integrity of the membrane. Properties such as hardness, roughness, tensile strength and adhesion strength of TiO_2 skin layer to the underlying support layer need to be characterized. The photocatalytic TiO_2-coated ceramic membrane can exhibit a photo-induced superhydrophilicity phenomenon.[101–103] This is caused by hydroxylation of the surface of TiO_2 layer when subjected to the UV irradiation. This phenomenon can be observed through contact angle measurement whereby the typical contact angle of the UV-excited TiO_2 layer is <12°. Lastly, the measurement of zeta potential is essential to ensure electrical charge compatibility between the target pollutants and the membrane.

5.5 Performance of Photocatalytic Ceramic Membrane

The photonic efficacy of the TiO_2 coated on a nonporous substrate is limited due to the mass-transfer limitation of the pollutants to the surface of the immobilized photocatalysts. However, when the TiO_2 photocatalyst is coated on the porous substrate that functions as membrane, the passage of water driven by the applied pressure gradient through the TiO_2-coated membrane can overcome the mass transfer limitation by the forced permeation process which delivers the pollutants to the TiO_2 surface. The overall improvement of the direct contact between the target pollutants and the photocatalysts will result in an enhanced photocatalytic efficiency of the hybrid treatment system.

The hypothetical synergism arising from coupling of photocatalysis and membrane filtration is as illustrated in Figure 5.4. Accumulation of bulk chemicals (macromolecules) and biofoulant (due to microbial growth and biofilm formation) on the membrane is a major operating problem of the membrane separation process for water and wastewater treatment. With photocatalytic TiO_2 coating, the membrane becomes less susceptible to fouling because the UV-excited TiO_2 can degrade the foulants or dislodge them from the membrane surface. For the micropollutants, they can be adsorbed on the membrane and simultaneously degraded *via* the direct oxidation by the photogenerated h^+ or indirect photocatalytic degradation by the photogenerated $HO^\bullet$. Moreover, the photo-induced superhydrophilicity of the TiO_2 surface can have a potential to increase the water permeability of the TiO_2-coated photocatalytic membrane.

Synergistic coupling of photocatalysis and membrane filtration was first investigated by Anderson *et al.*[25] in which they examined different sol–gel synthesis techniques to obtain a membrane with controllable pore size. Since then, several research efforts have aimed to improve the structure

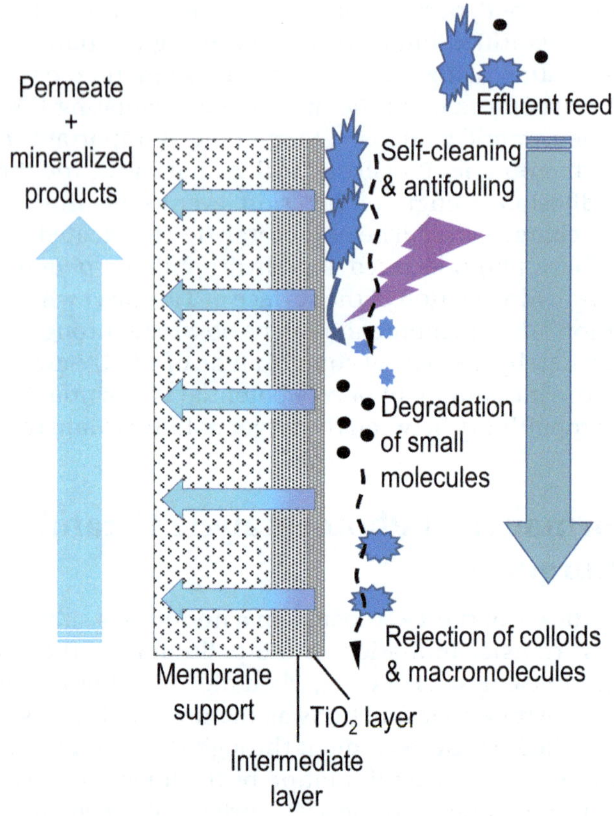

Figure 5.4 Schematic illustration of coupled photocatalytic degradation and separation processes in a photocatalytic membrane reactor with cross-flow mode.

and performance of the photocatalytic membrane. Table 5.4 summarizes selected works on the TiO_2-based photocatalytic ceramic membranes as well as other TiO_2-coated substrates.

Goei and Lim[34,73,115] prepared several TiO_2-coated photocatalytic membranes and evaluated the coupled photocatalysis–membrane separation process using a PMR as illustrated in Figure 5.5. The PMR can be operated in either a dead-end filtration mode or a cross-flow filtration mode. The residence time of the target pollutants inside the membrane cell was determined by controlling the feed flow rate. The water sampling points were at the inlet of the cell (sampling point 1), the dead-end outlet (sampling point 2) and the cross-flow outlet (sampling point 3). The trans-membrane pressure was monitored on-line using pressure transducers. Aeration or auxiliary oxidants could be introduced in the feed tank to improve the overall photocatalytic activity of the membrane.

The synergistic effect of combining photocatalysis and membrane separation process was evident from the observed increased water permeability

Table 5.4 Selected works on photocatalytic TiO_2-coated ceramic membrane and other substrates: preparation, properties and performances.

References	Fabrication	Physicochemical properties	Performances
Anderson *et al.*[25]	• Two types of membrane were prepared from particulate sols and polymeric sols, respectively • TiO_2 sols (both particulate and polymeric sols) were coated on γ-Al_2O_3 substrate	• The size of the particles in the particulate sols strongly depended on pH • For the polymeric sols, a transparent gel could only be obtained when $[H_2O]/[Ti] \leq 4$; acid does not significantly affect the properties of the polymeric gels • The membrane pores whose diameter are in the 1.8–5.0 nm range	• No performance evaluation was carried out
Zhang *et al.*[104]	• TiO_2/Al_2O_3 composite membranes were prepared from colloidal titania sols	• Uniform pore sizes were around 50–100 nm • The dissolution of TiO_2 microstructure was observed after six times of usage resulting in membrane with larger pore size • Membrane consisted of TiO_2 tubules with average thickness of 50 μm	• Different retention rates were observed for model pollutant Direct Black 168 when treated at different pH • Immediate decline in flux was observed followed by a stable sustainable flux thereafter
Choi *et al.*[64]	• TiO_2 sol was prepared using acetic-acid based sol–gel route with non-ionic surfactant as SDA • TiO_2 sol was dip-coated on borosilicate glass • Among the non-ionic SDAs, Tween 80 showed the best control of pore structure and film thickness	• TiO_2 crystallite size range of 8.3–15.6 nm • Slow hydrolysis reaction enhanced the structure of the resulting TiO_2 • Capping effect from the SDA reduced the hydrolysis and condensation reaction rate	• Higher removal of methylene blue for higher loading of SDA up to 2.0 molar ratios (with respect to TiO_2) was achieved, above which photocatalytic degradation efficiency was reversed
Choi *et al.*[33]	• TiO_2 sol was prepared *via* acetic-acid based sol–gel route with Tween 80 as SDA • TiO_2 sol was dip-coated on alumina substrate	• The composite membrane exhibited a hierarchical change in pore diameter from 2 to 11 nm (change in porosity was from 46% to 69%) • The TiO_2 layer was 0.9 μm thick	• Hierarchical layer improved water permeability of the composite membrane without sacrificing its separation and photocatalytic activity • The composite membrane showed some antibacterial action against *E. coli*

(*continued*)

Table 5.4 (*continued*)

References	Fabrication	Physicochemical properties	Performances
Ding *et al.*[105]	• TiO_2 sol was prepared *via* triethanolamine based sol–gel route • TiO_2 sol was dip-coated on tubular α-alumina substrate	• Membrane thickness increased linearly with the square root of the dipping time • A minimum of 30 s dipping time was required to form a defect-free composite membrane • The thickness of TiO_2 layer was 5.9 μm with and average pore size 60 nm	• Water permeability of 860 L m^2 h^{-1} bar^{-1} was observed • Bovine serum albumin (MW 67 000 da) rejection was >90%
Bosc *et al.*[106]	• TiO_2 sol was prepared *via* acid-catalyzed sol–gel route with two types of SDAs (*i.e.* polystyrene latex spheres and P-123 and F-127 triblock copolymers)	• Two types of isolated spherical macropores and mesopores were generated • Interconnected micropores from aggregation of anatase crystals were formed (1.5 nm pore size)	• No test was made on the performance of the composite membrane
Wang *et al.*[107]	• Two types of hydrolyzed anatase powder with poly(acrylic acid) or poly(ethylene glycol) as SDA were coated on tubular rutile support • Solid loading was 10–20 wt%	• The obtained average pore size was 0.10–0.12 μm • Poly(acrylic acid) was superior to SDA	• Water permeability of 740 L m^2 h^{-1} Bar^{-1} was observed • 600 °C was the optimum sintering temperature for maximum photocatalytic activity
Syafei *et al.*[108]	• Commercial Degussa P25 was suspended in acetylacetone solution with Triton X-100 as SDA • P-25 suspension was coated on TAMI ATZ (alumina, titania, zirconia) ceramic membrane	• Commercial membranes with MWCO 1, 15 and 30 kDa were used	• Higher flux decline for the TiO_2-coated membrane which could be attributed to the adsorption of organic pollutant • Higher humic acid removal by the TiO_2 coated membrane was attributed to the membrane rejection

Alem *et al.*[38]	• TiO_2/Al_2O_3 composite membrane was prepared from TiO_2 colloidal sol with HPC and PVA added as binder • TiO_2 layer was coated using slip casting method	• Optimum thickness of the membrane with good integrity was 1 μm • Composite membrane with large surface area (75 $m^2\ g^{-1}$) and small pore size (4.7 nm) was fabricated	• Membrane showed permeability of 30 mL $min^{-1}\ bar^{-1}\ cm^{-2}$ with Knudsen diffusion observed at low trans-membrane pressure • Effective removal of methyl orange in the batch photoreactor was observed
Alem *et al.*[109]	• Multilayer TiO_2/Al_2O_3 composite membrane was prepared using TiO_2 polymeric and TiO_2 colloidal sol as the top and intermediate layer, respectively • Aging to increase the viscosity of the sol and thus minimizing sol infiltration to the underlying layer	• Composite membrane with improved surface area (83 $m^2\ g^{-1}$) and smaller pore size (4 nm) was fabricated	• Addition of top TiO_2 layer had a minimum impact on the membrane permeability • Improved PCD of methyl orange in the batch photoreactor due to the presence of TiO_2 top layer prepared from polymeric sol
Ma *et al.*[110]	• Si–TiO_2 sol was prepared by sol–gel method with TEOS as the silica source • Si–TiO_2 sol was coated on α-Al_2O_3 tubular support with an unsymmetrical layered structure	• Thin separation layer (thickness 800 nm) was achieved • TiO_2 crystallite size range of 8–10 nm • MWCO was approximately 20 kDa	• Composite membrane was tested using a PMR operated in a cross-flow mode • Water permeability of 23 L $m^2\ h^{-1}\ bar^{-1}$ was observed. Higher flux was achieved when operated under UV irradiation • Synergistic effect of photocatalysis and membrane separation was demonstrated by improved permeability and increased dye removal
Ma *et al.*[111]	• Ag–TiO_2 sol was prepared by sol–gel method • Ag–TiO_2/hydroxyapatite ceramic membrane was fabricated. Hydroxyapatite layer was coated first followed by Ag–TiO_2 layer	• Average thickness of the composite membrane was 10 μm, wherein the thickness of Ag–TiO_2 layer was 10–30 nm • Mean pore size of the composite membrane was ~800 nm	• Antibacterial activity was observed from the composite membrane

(continued)

Table 5.4 (*continued*)

References	Fabrication	Physicochemical properties	Performances
Papageorgiou *et al.*[61]	• Ca-alginate/TiO_2 fiber was prepared by homogenization of alginate hollow fibers and P25 powder • Ca-alginate/TiO_2 fiber was coated on γ-Al_2O_3 tubular ceramic membrane using a chemical vapor deposition technique	• TiO_2 can be dispersed effectively using Ca-alginate hollow fibers • TiO_2 nanoparticles of 10–15 nm size were coated on the outer surface of the tubular membrane. Elongated TiO_2 particles (25 nm wide and 60 nm long) were found deposited on the internal membrane surface	• Higher photocatalytic activity was observed for TiO_2 powders dispersed in Ca-alginate fibers • Threefold increase in photocatalytic activity (against methyl orange) of TiO_2/Ca-alginate pretreated hybrid UF membrane was observed. Higher flux was also observed for membrane pretreated with the fibers
Romanos *et al.*[112–114]	• TiO_2 was coated on α-Al_2O_3 tubular membrane using a chemical vapor deposition technique • Two types of membrane were prepared: *via* pyrolytic nanoparticle deposition and *via* physico-sorption and surface reaction on the ceramic membrane	• The pore sizes of the resulting membranes were 3.9 and 5.1 nm respectively for two approaches • TiO_2 was deposited on both sides of the tubular membrane	• Irradiation of both sides of the tubular membrane was demonstrated resulting in a cleaner effluent • Double-side active TiO_2 UF membrane exhibited similar performance as of the NF membrane in terms of flux and removal of methyl orange
Goei *et al.*[73]	• TiO_2 sol was prepared by sol–gel method using P-123 as the SDA • Hierarchically porous TiO_2 membrane was prepared by multilayers coating of different TiO_2 sols	• P-123 could control the resulting porous structure of TiO_2 layer • The range of TiO_2 crystallite sizes was 6–14 nm • BET surface areas of the TiO_2 layers were 12–200 $m^2\ g^{-1}$ • TiO_2 membranes with different configurations were prepared	• High water permeability was observed (up to 116 $L\ m^2\ h^{-1}\ bar^{-1}$) • High removal of Rhodamine B (RhB) (up to 2700 $mg\ m^{-2}$) was observed • By altering the configurations of the membrane, a membrane with desired performance parameter could be obtained

Goei and Lim[34]	• TiO_2 sol was prepared by sol–gel method using various types of SDAs (P-123, Triton X-100, Tween 20 and Tween 80) • Six types of hierarchically porous TiO_2 membranes were prepared, four of which were fabricated by coating of TiO_2 layers prepared with different SDAs	• Different type of SDAs could control the resulting porous structure of TiO_2 layer • The range of TiO_2 crystallite sizes was 6–14 nm • BET surface areas of the TiO_2 layers were 12–155 $m^2\ g^{-1}$	• High water permeability was observed (up to 155 L $m^2\ h^{-1}\ bar^{-1}$) • High removal of RhB (up to 3050 mg m^{-2}) was observed • Good compatibility between TiO_2 layers prepared by different SDAs allowed flexibility of constructing TiO_2 membrane with certain desired performance (high flux, high photocatalytic activity, or both)
Goei and Lim[115]	• Different types of Ag-decorated TiO_2 sols were prepared by sol–gel method • Six types of hierarchically porous Ag-decorated TiO_2 membranes were prepared	• P-123 and Ag could control the resulting porous structure of TiO_2 layer • TiO_2 CS range of 3–15 nm • BET surface area of the TiO_2 layer range of 2–146 $m^2\ g^{-1}$	• Ag could be used to retard the growth of TiO_2 crystals and to impart an additional antibacterial property to the membrane • Ag could improve the photocatalytic activity of TiO_2 • The best performing membrane could remove RhB with an initial rate of 1007 mg $m^{-2}\ h^{-1}$

of the PMR when UV irradiation was introduced.[73] Figure 5.6 shows a common phenomenon of declining membrane permeability with operating time during membrane filtration process. However, the decline can be minimized if the membrane is coated with photocatalytic TiO_2 and irradiated with UV. This observation could be attributed to the good membrane wettability induced by the superhydrophilicity property of the UV-irradiated membrane (as discussed in Section 5.4) and the self-cleaning phenomenon on the surface of the photocatalytic membrane (Figure 5.4). The photo-induced superhydrophilicity and membrane self-cleaning phenomena could increase the water flux while operating at a relatively low trans-membrane pressure, thus reducing overall energy cost of the treatment process.

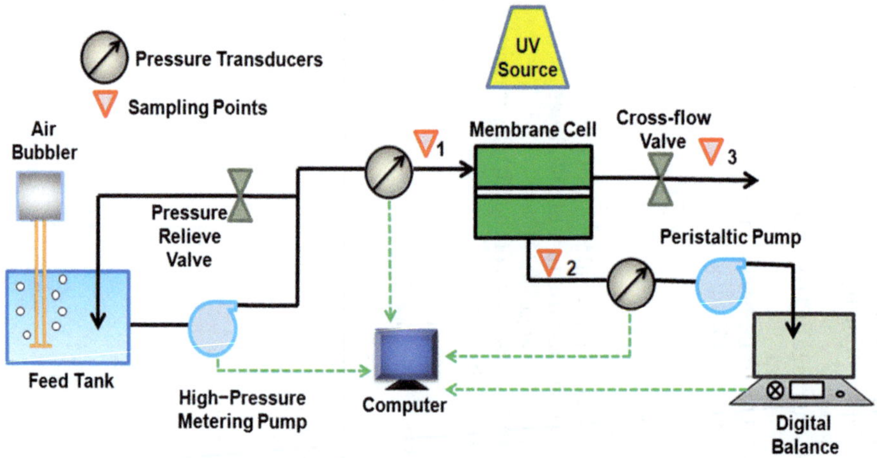

Figure 5.5 Schematic diagram of photocatalytic membrane reactor (PMR). (Adapted from Goei and Lim.[115])

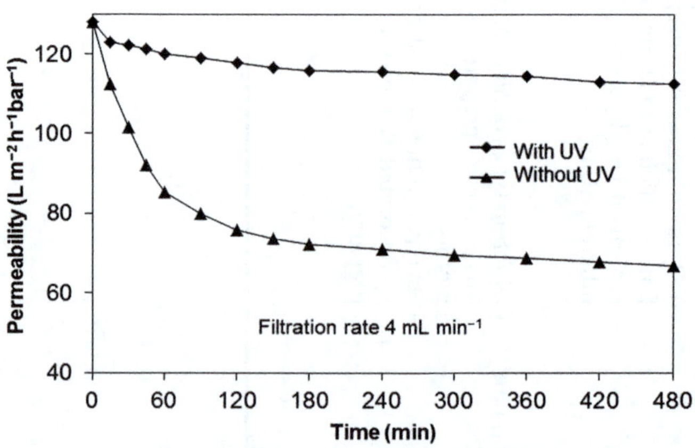

Figure 5.6 Comparison of permeability of photocatalytic membrane with and without UV irradiation.

The water flux through the membrane could be further improved by increasing the applied trans-membrane pressure, which would make for a more energy intensive operation and reduce the residence time of the target pollutant in the PMR (*i.e.* degradation efficiency). In recent work by Mozia *et al.*,[94] the effects of different process parameters (*i.e.* TiO_2 loading, feed cross-flow velocity and trans-membrane pressure) on fouling and stability of MF/UF TiO_2 membranes were investigated. A recent review by Zangeneh *et al.* summarizes the effects of different operating parameters and different modified TiO_2 structures and configurations on the performance of the photocatalytic reactor.[116]

The addition of other semiconductors during the preparation of TiO_2 photocatalyst could enhance the overall photocatalytic performance of TiO_2. Binary oxides such as TiO_2–SiO_2,[117,118] TiO_2–MoO_2,[119] TiO_2–RuO_2,[120,121] TiO_2–WO_3[122,123] and TiO_2–ZrO_2[118,124] were reported to have better performance as photocatalyst than pure TiO_2, resulting in a more efficient membrane process with less fouling. A recent study of Khan *et al.*[124] reported that the use of TiO_2–ZrO_2 binary oxide in a submerged photocatalytic membrane reactor resulted in a reduction in membrane fouling caused by humic acid as compared to the case whereby pure TiO_2 particle was coated. The introduction of secondary or even tertiary metal oxide into TiO_2 lattice could also alter and control the physicochemical characteristics of the resulting composite. The ability to control the characteristics of the photocatalytic layer is important in tailor-making composite membrane to treat a specific compound or to suit a specific operating condition.

The commercial ceramic membrane currently in use does not inactivate the pathogens that come into contact with the membrane. Bacteria cells deposited on the membrane surface can grow into a biofilm which is more difficult to remove than individual bacteria cells in suspension.[125] UV-irradiated TiO_2 can display photocatalytic disinfection functionality.[126] The photocatalytic disinfection capability of TiO_2 has been tested against different types of pathogens such as bacteria, fungi, algae, protozoa and virus.[127,128] The mechanisms of photocatalytic inactivation of the pathogenic cells are direct oxidation of coenzymes in cells, destruction of cell structure and damage to the cell DNA.[129,130] Thus, coating of photocatalytic TiO_2–ceramic membrane offers the added function of anti-biofouling.

The inclusion of Ag nanoparticles (AgNPs) into a TiO_2 photocatalytic ceramic membrane could further enhance the antibacterial property and inhibit the growth of biofilm. In comparison with other bactericidal nanoparticles, AgNPs exhibit bactericidal activity with minimal biocompatibility issues.[131–133] Furthermore, AgNPs can act as an electron traps to prevent e^-/h^+ recombination and thus enhance the overall photocatalytic activity of the AgNP–TiO_2 composite membrane.[134–137] A recent study of Goei and Lim[115] reported that the synergistic actions of Ag-decorated TiO_2 photocatalytic membrane and UV irradiation resulted in a higher overall *E. coli* removal (up to 3-log increment). The study also found that an increase of the Ag loading in the membrane improved the overall photocatalytic activity of the membrane, suggesting the role of Ag in enhancing the life-time of the photogenerated e^-/h^+ pairs.

5.6 Future Outlook and Challenges

Membrane separation and photocatalysis are two technologies that can remove almost all kinds of organic pollutants from the treated water. The synergistic coupling of the two processes in a hybrid photocatalytic membrane reactor has been proven in laboratory pilot scale studies. Although a ceramic membrane is costly to manufacture, the technology for mass-scale production of ceramic membranes has improved and the cost of production has been decreasing over the years. Functionalizing the ceramic membrane with catalysts such as TiO_2 will impart an added functionality to ceramic, whose value-add could overcome the additional membrane manufacturing cost.

It is anticipated that during operation of PMR, membrane fouling may develop faster than the rate of membrane self-cleaning through photocatalytic degradation, especially when treating water that contains a high concentration of pollutants. In this case, the feedwater to PMR has to be pretreated to remove the particulate solids and macromolecules, and to reduce turbidity of the water. Operating the PMR in cross-flow mode will reduce membrane fouling propensity.

The main challenge associated with the field scale application of PMR essentially lies in the system design and installation, particularly the incorporation of UV lamps in the membrane module. The hybrid ultrafiltration–photocatalytic membrane reactor developed by Romanos and co-workers, in which the TiO_2 was irradiated by UV fluorescent lamps and UV light emission diodes, has advanced the design of PMR for pilot scale application.[112] Nevertheless, to accommodate the light sources and their illumination design, the system compacity (*i.e.* the inherent advantage of a membrane process) has to be compromised.

In order not to severely reduce the membrane compacity or increase the system footprint, an innovative photonic technology such as fiber-optic bundle can be used to transmit low-energy UV into the membrane module and irradiate the membrane surface. Optical fiber has been applied to transmit the sunlight and UV into solar photoreactors.[138,139] In the context of a photocatalytic ceramic membrane of tubular module or multichannel monolith, the interior surface can be coated with photocatalyst while the side-glowing optical fiber can be slotted through the membrane lumen. The side-glowing optical fiber emits UV onto the surface of photocatalyst-coated membrane and that can activate photocatalytic reaction inside the membrane module. Such an optical fiber-incorporated membrane module does not reduce membrane compacity in the PMR system. The presence of optical fiber in the membrane lumen can also increase flow hydrodynamics along the membrane surface and therefore minimize foulant deposition on the membrane surface.

In summary, PMR has the potential for field-scale applications in water and wastewater treatment. It provides the potential benefits of energy-saving operation and smaller facility footing. The proper treatment process design, advance in the photonic technology and innovative PMR module development are the keys to its successful applications.

References

1. M. Klavarioti, D. Mantzavinos and D. Kassinos, *Environ. Int.*, 2009, **35**, 402–417.
2. T.-T. Lim, P.-S. Yap, M. Srinivasan and A. G. Fane, *Crit. Rev. Environ. Sci. Technol.*, 2011, **41**, 1173–1230.
3. T. Fotiou, T. M. Triantis, T. Kaloudis, L. M. Pastrana-Martínez, V. Likodimos, P. Falaras, A. M. T. Silva and A. Hiskia, *Ind. Eng. Chem. Res.*, 2013, **52**, 13991–14000.
4. W. Choi, *Catal. Surv. Asia*, 2006, **10**, 16–28.
5. A. J. Bard, R. Parsons and J. Jordan, *Standard Potentials in Aqueous Solutions*, Marcel Dekker Inc., New York, 1985.
6. J. M. Herrmann, *Catal. Today*, 1999, **53**, 115–129.
7. U. I. Gaya and A. H. Abdullah, *J. Photochem. Photobiol., C*, 2008, **9**, 1–12.
8. K. Tanaka, T. Hisanaga, A. P. Riviera, D. F. Ollis and H. Al-Ekabi, *Photocatalytic Purification and Treatment of Water and Air*, Elsevier, Amsterdam, Netherlands, 1993.
9. K. M. Schindler and M. Kunst, *J. Phys. Chem.*, 1990, **94**, 8222–8226.
10. C. Di Valentin, E. Finazzi, G. Pacchioni, A. Selloni, S. Livraghi, M. C. Paganini and E. Giamello, *Chem. Phys.*, 2007, **339**, 44–56.
11. A. L. Linsebigler, G. Lu and J. T. Yates Jr, *Chem. Rev.*, 1995, **95**, 735–758.
12. M. Landmann, E. Rauls and W. G. Schmidt, *J. Phys.: Condens. Matter*, 2012, **24**, 1–6.
13. W. Choi, *Catal. Surv. Asia*, 2006, **10**, 16–28.
14. X. Chen and S. S. Mao, *Chem. Rev.*, 2007, **107**, 2891–2959.
15. N. Liu, I. Paramasivam, M. Yang and P. Schmuki, *J. Solid State Electrochem.*, 2012, **16**, 3499–3504.
16. M. Răileanu, M. Crişan, N. Drăgan, D. Crişan, A. Galtayries, A. Brăileanu, A. Ianculescu, V. S. Teodorescu, I. Niţoi and M. Anastasescu, *J. Sol-Gel Sci. Technol.*, 2009, **51**, 315–329.
17. M. V. Shankar, K. K. Cheralathan, B. Arabindoo, M. Palanichamy and V. Murugesan, *J. Mol. Catal. A: Chem.*, 2004, **223**, 195–200.
18. Z. Zhang, C.-C. Wang, R. Zakaria and J. Y. Ying, *J. Phys. Chem. B*, 1998, **102**, 10871–10878.
19. R. van Grieken, J. Aguado, M. J. López-Muñoz and J. Marugán, *J. Photochem. Photobiol., A*, 2002, **148**, 315–322.
20. M. Anpo, T. Shima, S. Kodama and Y. Kubokawa, *J. Phys. Chem.*, 1987, **91**, 4305–4310.
21. H. Choi, E. Stathatos and D. D. Dionysiou, *Appl. Catal., B*, 2006, **63**, 60–67.

22. P. Le-Clech, V. Chen and T. A. G. Fane, *J. Membr. Sci.*, 2006, **284**, 17–53.
23. M. Gander, B. Jefferson and S. Judd, *Sep. Purif. Technol.*, 2000, **18**, 119–130.
24. A. Seidel and M. Elimelech, *J. Membr. Sci.*, 2002, **203**, 245–255.
25. M. A. Anderson, M. J. Gieselmann and Q. Xu, *J. Membr. Sci.*, 1988, **39**, 243–258.
26. C. H. Chang, R. Gopalan and Y. S. Lin, *J. Membr. Sci.*, 1994, **91**, 27–45.
27. C. Guizard, A. Julbe, A. Larbot and L. Cot, *J. Alloys Compd.*, 1992, **188**, 8–13.
28. P. R. Bhave, *Inorganic Membrane Synthesis, Characteristics, and Applications*, Van Nostrand Reinhold, New York, 1991.
29. *Fundamental of Inorganic Membrane Science and Technology*, ed. A. J. Burggraaf and L. Cot, Elsevier Science, Amsterdam, 1996.
30. A. Buekenhoudt, in *Inorganic Membranes: Synthesis, Characterisation and Applications*, ed. R. Mallada and M. Menendez, Elsevier B.V., Jaca, Spain, 2008, pp. 1–31.
31. T. Tsuru, D. Hironaka, T. Yoshioka and M. Asaeda, *Sep. Purif. Technol.*, 2001, **25**, 307–314.
32. J. Caro, K. J. Caspary, C. Hamel, B. Hoting, P. Kölsch, B. Langanke, K. Nassauer, T. Schiestel, A. Schmidt, R. Schomäcker, A. Seidel-Morgenstern, E. Tsotsas, I. Voigt, H. Wang, R. Warsitz, S. Werth and A. Wolf, *Ind. Eng. Chem. Res.*, 2007, **46**, 2286–2294.
33. H. Choi, A. C. Sofranko and D. D. Dionysiou, *Adv. Funct. Mater.*, 2006, **16**, 1067–1074.
34. R. Goei and T.-T. Lim, *Ceram. Int.*, 2014, **40**, 6747–6757.
35. R. Fretwell and P. Douglas, *J. Photochem. Photobiol., A*, 2001, **143**, 229–240.
36. J. H. Pan and W. I. Lee, *Chem. Mater.*, 2006, **18**, 847–853.
37. Z. Wang, U. Helmersson and P. O. Käll, *Thin Solid Films*, 2002, **405**, 50–54.
38. A. Alem, H. Sarpoolaky and M. Keshmiri, *J. Eur. Ceram. Soc.*, 2009, **29**, 629–635.
39. V. T. Zaspalis, W. Van Praag, K. Keizer, J. R. H. Ross and A. J. Burggraaf, *J. Mater. Sci.*, 1992, **27**, 1023–1035.
40. P. S. Shen, Y. C. Tai, P. Chen and Y. C. Wu, *J. Power Sources*, 2014, **247**, 444–451.
41. W. T. Zhan, H. W. Ni, R. S. Chen, Z. Y. Wang, Y. W. Li and J. H. Li, *Thin Solid Films*, 2013, **548**, 299–305.
42. S. H. Cho and Y. J. Yoon, *Thin Solid Films*, 2013, **547**, 91–94.
43. M. Kubo, Y. Ishihara, Y. Mantani and M. Shimada, *Chem. Eng. J.*, 2013, **232**, 221–227.
44. L. Kavan and M. Grätzel, *Electrochim. Acta*, 1995, **40**, 643–652.
45. A. Mills, N. Elliott, I. P. Parkin, S. A. O'Neill and R. J. Clark, *J. Photochem. Photobiol., A*, 2002, **151**, 171–179.
46. V. Diesen, M. Jonsson and I. P. Parkin, *Chem. Vap. Deposition*, 2013, **19**, 355–362.
47. E. M. Moser, S. Chappuis and J. Olleros, *Surf. Coat. Technol.*, 2013, **227**, 2–9.

48. K. W. Weng and Y. P. Huang, *Surf. Coat. Technol.*, 2013, **231**, 201–204.
49. R. Dholam, N. Patel, M. Adami and A. Miotello, *Int. J. Hydrogen Energy*, 2008, **33**, 6896–6903.
50. A. Ennaoui, B. R. Sankapal, V. Skryshevsky and M. C. Lux-Steiner, *Sol. Energy Mater. Sol. Cells*, 2006, **90**, 1533–1541.
51. K. O. Awitor, A. Rivaton, J. L. Gardette, A. J. Down and M. B. Johnson, *Thin Solid Films*, 2008, **516**, 2286–2291.
52. M. Hepel and I. D. Kumarihamy, *Int. J. Hydrogen Energy*, 2007, **32**, 2693–2702.
53. S. Palmas, A. M. Polcaro, J. R. Ruiz, A. Da Pozzo, M. Mascia and A. Vacca, *Int. J. Hydrogen Energy*, 2010, **35**, 6561–6570.
54. Z. Znamirowski, K. Nitsch and L. Pawłowski, *Surf. Coat. Technol.*, 2013, **220**, 271–275.
55. J. J. Zou, H. He, L. Cui and H. Y. Du, *Int. J. Hydrogen Energy*, 2007, **32**, 1762–1770.
56. A. Guillén-Santiago, S. A. Mayén, G. Torres-Delgado, R. Castanedo-Pérez, A. Maldonado and M. D. L. L. Olvera, *Mater. Sci. Eng. B: Solid State Mater. Adv. Technol.*, 2010, **174**, 84–87.
57. N. C. Raut, T. Mathews, S. T. Sundari, T. N. Sairam, S. Dash and A. K. Tyagi, *J. Nanosci. Nanotechnol.*, 2009, **9**, 5298–5302.
58. M. H. Habibi, M. N. Esfahani and T. A. Egerton, *Int. J. Photoenergy*, 2007, **2007**, 1–8.
59. M. H. Habibi and M. Nasr-Esfahani, *Dyes Pigm.*, 2007, **75**, 714–722.
60. M. H. Habibi, M. Nasr-Esfahani and T. A. Egerton, *J. Mater. Sci.*, 2007, **42**, 6027–6035.
61. S. K. Papageorgiou, F. K. Katsaros, E. P. Favvas, G. E. Romanos, C. P. Athanasekou, K. G. Beltsios, O. I. Tzialla and P. Falaras, *Water Res.*, 2012, **46**, 1858–1872.
62. Q. L. Wu and S. E. Rankin, *J. Sol-Gel Sci. Technol.*, 2011, **60**, 81–90.
63. J. Zhang, Z.-L. Wang, J. Liu, S. Chen and G.-Y. Liu, *Self-Assembled nanostructure*, Kluwer Academic/Plenum Publishers, New York, 2003.
64. H. Choi, E. Stathatos and D. D. Dionysiou, *Thin Solid Films*, 2006, **510**, 107–114.
65. N. Arconada, A. Durán, S. Suárez, R. Portela, J. M. Coronado, B. Sánchez and Y. Castro, *Appl. Catal., B*, 2009, **86**, 1–7.
66. Y. Chen and D. D. Dionysiou, *Appl. Catal., A*, 2007, **317**, 129–137.
67. H. S. Jahromi, H. Taghdisian, S. Afshar and S. Tasharrofi, *Surf. Coat. Technol.*, 2009, **203**, 1991–1996.
68. T. Coquil, C. Reitz, T. Brezesinski, E. J. Nemanick, S. H. Tolbert and L. Pilon, *J. Phys. Chem. C*, 2010, **114**, 12451–12458.
69. D. Fattakhova-Rohlfing, M. Wark, T. Brezesinski, B. M. Smarsly and J. Rathouský, *Adv. Funct. Mater.*, 2007, **17**, 123–132.
70. E. Ortel, S. Sokolov, C. Zielke, I. Lauermann, S. Selve, K. Weh, B. Paul, J. Polte and R. Kraehnert, *Chem. Mater.*, 2012, **24**, 3828–3838.
71. E. Ortel, A. Fischer, L. Chuenchom, J. Polte, F. Emmerling, B. Smarsly and R. Kraehnert, *Small*, 2012, **8**, 298–309.

72. M. B. Zakaria, N. Suzuki, N. L. Torad, M. Matsuura, K. Maekawa, H. Tanabe and Y. Yamauchi, *Eur. J. Inorg. Chem.*, 2013, 2330–2335.
73. R. Goei, Z. Dong and T. T. Lim, *Chem. Eng. J.*, 2013, **228**, 1030–1039.
74. J. Liu, T. An, G. Li, N. Bao, G. Sheng and J. Fu, *Microporous Mesoporous Mater.*, 2009, **124**, 197–203.
75. J. Procházka, L. Kavan, V. Shklover, M. Zukalová, O. Frank, M. Kalbáč, A. Zukal, H. Pelouchová, P. Janda, K. Mocek, M. Klementová and D. Carbone, *Chem. Mater.*, 2008, **20**, 2985–2993.
76. L. Sun, T. An, S. Wan, G. Li, N. Bao, X. Hu, J. Fu and G. Sheng, *Sep. Purif. Technol.*, 2009, **68**, 83–89.
77. J. Wang, H. Li, H. Li, H. Wang and Q. Cai, *J. Nanosci. Nanotechnol.*, 2013, **13**, 1493–1497.
78. Y. Castro and A. Durán, *J. Sol-Gel Sci. Technol.*, 2013, 1–9.
79. R. Rahimi, M. M. Moghaddas, S. Zargari and R. Rahimi, 2013, pp. 56–61.
80. H. U. Lee, S. C. Lee, J. H. Seo, W. G. Hong, H. Kim, H. J. Yun, H. J. Kim and J. Lee, *Chem. Eng. J.*, 2013, **223**, 209–215.
81. M. M. Rashad and A. E. Shalan, *J. Mater. Sci.: Mater. Electron.*, 2013, **24**, 3189–3194.
82. C. Shao, G. Zhou, Z. Li, Y. Wu, D. Xu and B. Sun, *Chem. Eng. J.*, 2013, **230**, 227–235.
83. Y. L. Chang and H. I. Hsiang, *Ceram. Trans.*, 2010, **223**, 9–22.
84. H. Xu, F. Wu, M. Li and Z. Liang, *J. Sol-Gel Sci. Technol.*, 2013, **67**, 394–405.
85. S. Sanguanruang, R. Leotphayakkarat, N. Fangern, N. Koonsaeng and C. Chawengkijwanich, 2011, pp. 2863–2870.
86. H. Li, H. Xu, S. Wang, Y. Pan, Q. Wu and C. Liu, 2011, pp. 1731–1734.
87. Y. Chen and D. D. Dionysiou, *Appl. Catal., B*, 2008, **80**, 147–155.
88. M. L. García-Benjume, M. I. Espitia-Cabrera and M. E. Contreras-García, *Mater. Charact.*, 2009, **60**, 1482–1488.
89. H. Choi, Y. J. Kim, R. S. Varma and D. D. Dionysiou, *Chem. Mater.*, 2006, **18**, 5377–5384.
90. C. Han, M. Pelaez, V. Likodimos, A. G. Kontos, P. Falaras, K. O'Shea and D. D. Dionysiou, *Appl. Catal., B*, 2011, **107**, 77–87.
91. N. Arconada, Y. Castro and A. Durán, *Appl. Catal., A*, 2010, **385**, 101–107.
92. A. Larbot, J. P. Fabre, C. Guizard and L. Cot, *J. Membr. Sci.*, 1988, **39**, 203–212.
93. J. Sekulic-Kuzmanovic, PhD thesis, University of Twente, The Netherlands, 2004.
94. S. Mozia, K. Szymański, B. Michalkiewicz, B. Tryba, M. Toyoda and A. W. Morawski, *Sep. Purif. Technol.*, 2015, **142**, 137–148.
95. L. Zhao, Y. Yu, L. Song, M. Ruan, H. Hu and A. Larbot, *Appl. Catal., A*, 2004, **263**, 171–177.
96. J. Lee, M. Christopher Orilall, S. C. Warren, M. Kamperman, F. J. Disalvo and U. Wiesner, *Nat. Mater.*, 2008, **7**, 222–228.
97. P. Puhlfürß, A. Voigt, R. Weber and M. Morbeé, *J. Membr. Sci.*, 2000, **174**, 123–133.

98. H. Y. Ha, S. W. Nam, T. H. Lim, I. H. Oh and S. A. Hong, *J. Membr. Sci.*, 1996, **111**, 81–92.
99. T. Tsuru, T. Hino, T. Yoshioka and M. Asaeda, *J. Membr. Sci.*, 2001, **186**, 257–265.
100. C. Boissiere, D. Grosso, S. Lepoutre, L. Nicole, A. B. Bruneau and C. Sanchez, *Langmuir*, 2005, **21**, 12362–12371.
101. T. Watanabe, A. Nakajima, R. Wang, M. Minabe, S. Koizumi, A. Fujishima and K. Hashimoto, *Thin Solid Films*, 1999, **351**, 260–263.
102. H. Irie, S. Washizuka, N. Yoshino and K. Hashimoto, *Chem. Commun.*, 2003, 1298–1299.
103. M. Kazemi and M. R. Mohammadizadeh, *Thin Solid Films*, 2011, **519**, 6432–6437.
104. H. Zhang, X. Quan, S. Chen, H. Zhao and Y. Zhao, *Sep. Purif. Technol.*, 2006, **50**, 147–155.
105. X. Ding, Y. Fan and N. Xu, *J. Membr. Sci.*, 2006, **270**, 179–186.
106. F. Bosc, P. Lacroix-Desmazes and A. Ayral, *J. Colloid Interface Sci.*, 2006, **304**, 545–548.
107. Y. H. Wang, X. Q. Liu and G. Y. Meng, *Mater. Res. Bull.*, 2008, **43**, 1480–1491.
108. A. D. Syafei, C. F. Lin and C. H. Wu, *J. Colloid Interface Sci.*, 2008, **323**, 112–119.
109. A. Alem, H. Sarpoolaky and M. Keshmiri, *Ceram. Int.*, 2009, **35**, 1837–1843.
110. N. Ma, X. Quan, Y. Zhang, S. Chen and H. Zhao, *J. Membr. Sci.*, 2009, **335**, 58–67.
111. N. Ma, X. Fan, X. Quan and Y. Zhang, *J. Membr. Sci.*, 2009, **336**, 109–117.
112. G. E. Romanos, C. P. Athanasekou, V. Likodimos, P. Aloupogiannis and P. Falaras, *Ind. Eng. Chem. Res.*, 2013, **52**, 13938–13947.
113. G. E. Romanos, C. P. Athanasekou, F. K. Katsaros, N. K. Kanellopoulos, D. D. Dionysiou, V. Likodimos and P. Falaras, *J. Hazard. Mater.*, 2012, **211–212**, 304–316.
114. P. Falaras, G. Romanos and P. Aloupogiannis, European Patent EP2409954, 2012.
115. R. Goei and T. T. Lim, *Water Res.*, 2014, **59**, 207–218.
116. H. Zangeneh, A. A. L. Zinatizadeh, M. Habibi, M. Akia and M. Isa, *J. Ind. Eng. Chem.*, 2015, **26**, 1–36.
117. R. M. Mohamed and I. A. Mkhalid, *J. Alloys Compd.*, 2010, **501**, 143–147.
118. F. U. Xianzhi, L. A. Clark, Q. Yang and M. A. Anderson, *Environ. Sci. Technol.*, 1996, **30**, 647–653.
119. A. Katrib, A. Benadda, J. W. Sobczak and G. Maire, *Appl. Catal., A*, 2003, **242**, 31–40.
120. D. P. Debecker, B. Farin, E. M. Gaigneaux, C. Sanchez and C. Sassoye, *Appl. Catal., A*, 2014, **481**, 11–18.
121. E. Kusmierek and E. Chrzescijanska, *J. Photochem. Photobiol., A*, 2015, **302**, 59–68.

122. J. He, Q. Luo, Q. Z. Cai, X. W. Li and D. Q. Zhang, *Mater. Chem. Phys.*, 2011, **129**, 242–248.
123. B. Lu, X. Li, T. Wang, E. Xie and Z. Xu, *J. Mater. Chem. A*, 2013, **1**, 3900–3906.
124. S. Khan, J. Kim, A. Sotto and B. Van der Bruggen, *J. Ind. Eng. Chem.*, 2015, **21**, 779–786.
125. J. W. Costerton, P. S. Stewart and E. P. Greenberg, *Science*, 1999, **284**, 1318–1322.
126. A. Fujishima, T. N. Rao and D. A. Tryk, *J. Photochem. Photobiol., C*, 2000, **1**, 1–21.
127. H. A. Foster, I. B. Ditta, S. Varghese and A. Steele, *Appl. Microbiol. Biotechnol.*, 2011, **90**, 1847–1868.
128. N. Yao and K. Lun Yeung, *Chem. Eng. J.*, 2011, **167**, 13–21.
129. T. Matsunaga, R. Tomoda, T. Nakajima and H. Wake, *FEMS Microbiol. Lett.*, 1985, **29**, 211–214.
130. C. Hu, J. Guo, J. Qu and X. Hu, *Langmuir*, 2007, **23**, 4982–4987.
131. O. Akhavan and E. Ghaderi, *Surf. Coat. Technol.*, 2010, **204**, 3676–3683.
132. D. Gumy, C. Morais, P. Bowen, C. Pulgarin, S. Giraldo, R. Hajdu and J. Kiwi, *Appl. Catal., B*, 2006, **63**, 76–84.
133. Q. Li, S. Mahendra, D. Y. Lyon, L. Brunet, M. V. Liga, D. Li and P. J. J. Alvarez, *Water Res.*, 2008, **42**, 4591–4602.
134. A. Kubacka, M. Ferrer, A. Martínez-Arias and M. Fernández-García, *Appl. Catal., B*, 2008, **84**, 87–93.
135. M. K. Seery, R. George, P. Floris and S. C. Pillai, *J. Photochem. Photobiol., A*, 2007, **189**, 258–263.
136. K. V. Subba, B. Lavédrine and P. Boule, *J. Photochem. Photobiol., A*, 2003, **154**, 189–193.
137. J. Yu, J. Xiong, B. Cheng and S. Liu, *Appl. Catal., B*, 2005, **60**, 211–221.
138. N. J. Peill and M. R. Hoffmann, *Environ. Sci. Technol.*, 1995, **29**, 2974–2981.
139. H. Lin and K. T. Valsaraj, *J. Appl. Electrochem.*, 2005, **35**, 699–708.

CHAPTER 6

Process Integration. Concepts of Integration and Coupling of Photocatalysis with Other Processes

SIXTO MALATO*[a], PILAR FERNÁNDEZ-IBÁÑEZ[a], ISABEL OLLER[a], INMACULADA POLO[a], MANUEL I. MALDONADO[a], SARA MIRALLES-CUEVAS[a], AND RICARDO SÁNCHEZ[a]

[a]Plataforma Solar de Almería-CIEMAT, P.O. Box 22, 04200 Tabernas, Almería, Spain
*E-mail: sixto.malato@psa.es

6.1 Introduction

Industrial wastewaters have very different characteristics from domestic sewage, such as high salinity, substantial toxicity, great variability in organic and inorganic content, different pH and temperature *etc.* Therefore, it is very difficult to treat many industrial wastewaters in domestic (municipal) wastewater treatment plants. Proper characterization of the key pollutant sources is needs to be focused to develop effective treatment solutions. When treating different and complicated wastewaters (*e.g.* landfill leachates, pharmaceutical, textile industries or olive mill effluents), the combination of processes (mainly physical, biological and chemical oxidation treatment) and the study

RSC Energy and Environment Series No. 15
Photocatalysis: Applications
Edited by Dionysios D Dionysiou, Gianluca Li Puma, Jinhua Ye, Jenny Schneider, and Detlef Bahnemann

Published by the Royal Society of Chemistry, www.rsc.org

of them as a whole results in increased treatment efficiency and costs savings. A detailed feasibility study should increase the efficiency of the process and reduce costs.

Complete mineralization is generally expensive because intermediates formed during treatment are frequently more resistant compounds and all transformation reactions involved consume energy and chemical reagents, and hence increase operating costs.[1] One attractive and well-studied alternative is to apply these chemical oxidation processes in a pretreatment to convert the initially persistent organic compounds into more biodegradable intermediates, which would then be treated in a biological oxidation process with a considerably lower cost. Therefore, the main role of the chemical pretreatment is partial oxidation of the biologically resistant part to produce biodegradable reaction intermediates.[2–4] The percentage of mineralization should be minimized during the pretreatment stage to avoid unnecessary expenditure of chemicals and energy, thereby lowering the operating cost. Since pretreatment does not always improve the biodegradability or reduce toxicity, the other option, namely using chemical oxidation as a polishing step, should also be explored, as stated above. The best option for remediation of a specific wastewater is not an easy choice and the combination of chemical oxidation and biotreatment is not always the more convenient approach. The water quality standards to be met affect the choice of the processes to be combined, jointly with overall costs. The main factors to be taken into account are: (i) the detailed composition of wastewater including natural organic matter (NOM), (ii) conventional treatment options available on-site, (iii) overall treatment efficiency concerning organics and toxicity removal and (iv) the overall costs.

Usually, conventional treatment possibilities and capabilities are well known. However, the efficiency of new technologies (photocatalysis) combined with conventional ones is very often unknown. Therefore, up-scaled studies are very useful when new wastewater treatment technologies are applied to industrial wastewater treatment. When photocatalysis is applied, sometimes its effect is not significant or is even detrimental. There are various reasons for this: (i) formation of stable by-products, (ii) lack of selectivity for the target contaminants, (iii) poor selection of treatment conditions with excessive pre-oxidation (effluent with too little metabolic value for the biomass), (iv) oxidant and/or catalyst remained in the pretreated wastewater (harmful for the microorganisms), (v) increase in toxicity and (vi) low concentration of target contaminants that leads to slow kinetics (very usual when treating micro-contaminants). Therefore, a step-by-step research methodology is needed, applying advanced analytical tools and bioassays to infer the reaction steps, kinetics, heteroatom release (as cations and anions) in the wastewater matrix, biodegradability, toxicity, *etc.*[5–8]

This chapter highlights how to apply photocatalysis in real cases as part of integrated processes for treating different wastewaters. It also highlights the utilization of solar irradiation in photocatalytic reactors optimizing chemical and energy consumption without reducing effectiveness. Integration of processes involves not only photoreactors and bioreactors but also other more conventional processes such as coagulation, sedimentation or filtration to separate solids or organics as sludge from industrial wastewater. Industrial wastewater streams can be a good example for integrating photocatalysis with other processes. Membrane processes coupled with photo-processes could also be applied for the treatment of various types of wastewater streams such as effluents from secondary treatment from municipal wastewater treatment plants. Developing and testing innovative combined treatment solutions could allow comparisons of old and new technologies and it would likely result in improved performance of wastewater treatment systems (better effluent quality, more operational stability, higher energy efficiency and in particular reduction in their capital and operating costs) and thus to an overall beneficial impact on the environment.

6.2 Treatment of Biorecalcitrant Wastewater by Integrating Solar Photocatalysis and Other Processes

Studies with model effluents cannot be applied directly to real wastewater as it must be considered that the real wastewater contains other constituents that compete with pollutants for the oxidizing agent ($HO^\bullet$), usually other organics, inorganic salts and NOM. When treating target pollutants, oxidation may lead to the formation of by-products that could seriously decrease process efficiency, since these species could be less reactive with $HO^\bullet$. In their pioneer work, Scott and Ollis[2] have identified several types of wastewater where a combined strategy based on AOPs (advanced oxidation processes)/biological degradation has and important potential:

(i) biorecalcitrant wastewater containing large macromolecules (as soluble polymers) so persistent because their size and absence of active centres, type I;
(ii) biodegradable wastewater containing small concentrations of biorecalcitrant substances, type II;
(iii) toxic wastewater containing inhibiting compounds to biological cultures, type III; and
(iv) wastewater containing specific substances and their metabolites that would accumulate and inhibit growth of the microorganisms, type IV.

Sections 6.2 and 6.3 will focus on wastewaters relevant to the aforementioned types summarized above, but here we will focus mainly on types I and III wastewaters. A combination of AOPs and bioprocesses presents an additional difficulty, as the chemical oxidant and bioculture must not come into contact with each other. For example, hydrogen peroxide causes an adverse effect on the microorganisms at concentrations >50 mg L^{-1}. The AOP reaction time must also be taken into account to avoid an over-degraded effluent, because the metabolic value of the highly oxidized products for the microorganisms is minimal.[4,6,9–11] It would be necessary to avoid large doses of oxidant in the degradation of easily biodegradable intermediates.[12–14]

Agriculture is the central water consuming activity, and intense agricultural activities use pesticides as a common practice that has significantly increased pollution in water and soils.[15,16] Either in their parent form or their metabolites resulting from diverse biotic or abiotic processes, pesticides are normally still toxic in the long term, biorecalcitrant in water treatments, and consequently represent a serious danger both to the environment and human health (wastewater type III). Pesticides are usually disposed of by throwing away solutions used for fumigating or for dosing pesticides with irrigation water. These chemical solutions are prepared just before use and the equipment must be cleaned afterwards, very often draining of unused pesticide. It is therefore necessary to eliminate this source of contamination before it enters the water cycle, where diffuse pollution would be very often impossible to address. Pesticide solutions are prepared in water at very high concentration, very often in emulsions with foams where the pesticides are suspended and not dissolved. Photocatalysis, and other AOPs in general, have been described as suitable methods for oxidizing these compounds[17–20] because hydroxyl radicals can mineralize any organic pollutant. Several combined processes have been suggested for the treatment of different industrial wastewater, but they are still not in routine industrial use. One of the reasons is that most investigations were limited to evaluation of overall parameters such as biological oxygen demand (BOD), chemical oxygen demand (COD) and dissolved organic carbon (DOC), working with model compounds and model wastewaters or demineralized water. The concepts and examples presented in this chapter intend to fill this gap by suggesting analytical tools, bioassays methodology and strategies for a combined AOP/aerobic biological treatment of real wastewaters, summarized in Figure 6.1. Notice that the active ingredients (considered as a whole) were all eliminated after a certain amount of mineralization of the initial DOC. Experiments were focused on the impact of the effluent (from the combined system photocatalysis/biotreatment) in any sewage system. Therefore, the studies focused on the by-products, and to investigate if they would cause any substantial change in toxicity and biodegradability behaviour. Toxicity was evaluated using two different bioassays: *Vibrio fischeri* and respirometry tests.[7] It was revealed that toxicity decreased from 96% to 70% after the elimination of the parent compounds (Figure 6.1(A)). Dilutions (1 : 10) showed the same tendency, dropping from 80% to 20%, confirming that the treatment has no detrimental effect on this site, as photocatalysis was able to gradually reduce the toxicity. Toxicity was also studied with a respirometry assay using municipal activated sludge (AS) as the presence of different microorganism makes AS more resistant to toxic effluents than *Vibrio fischeri*. AS did not show any toxicity after the parent compounds were eliminated. However, the inhibition of 56% at the end of the treatment correlates well with the results obtained in the *Vibrio fischeri* assay.

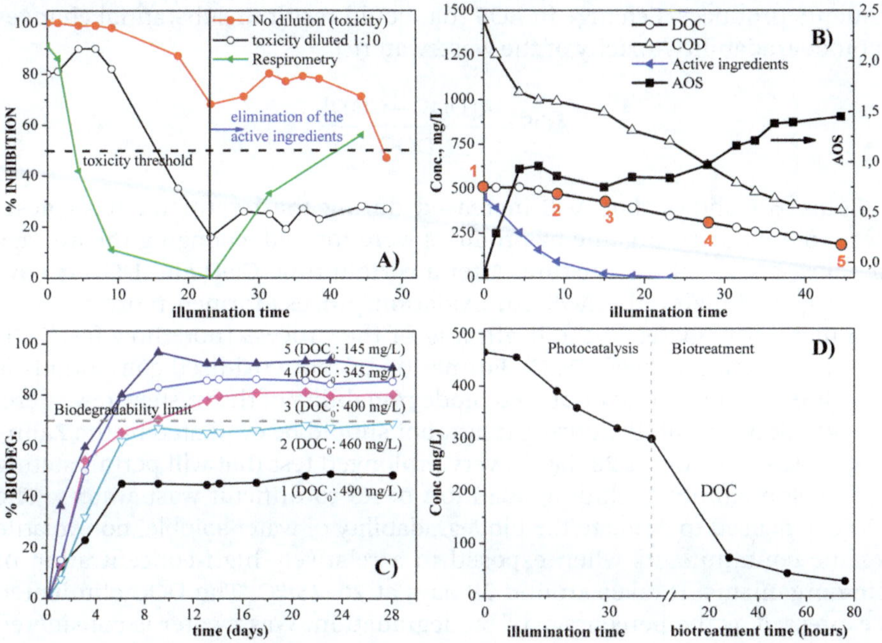

Figure 6.1 Experimental steps for designing a wastewater (containing toxic and biorecalcitrant pesticides) treatment by integrating photocatalysis and biotreatment. (A) Toxicity evaluation, (B) evaluation of degradation of contaminants, mineralisation (COD and DOC) and average oxidation state (AOS), (C) biodegradability assessment, (D) combination of photocatalysis and biotreatment.

This means that the compounds present at around 25 min of treatment did not have an important toxic impact. Toxicity is related to the intermediates formed in advanced stages of the treatment, therefore the total elimination of the active ingredients might be a key point of the treatment and it should not be continued after this stage. In general, when water will be disposed to a natural water reservoir (lake, river) it is advisable to conduct battery tests (several bioassays) with organisms selected from different trophic levels, including some fresh/seawater algae, or *Daphnia magna*, since their sensitivity will differ considerably from each other. But heterotrophic biomass is, in general, more robust to inhibitory pollutants and therefore when water would be sent to sewage or specific biotreatment, activated sludge inhibition tests/respirometric activity experiments are recommended since they more or less simulate the situation in conventional biological treatment systems.

The COD was also measured (Figure 6.1(B)) during treatment to determine the average oxidation state (AOS) of the wastewater according to eqn (6.1), where COD and DOC are expressed in molar concentrations. AOS is a useful parameter in AOP of a complex wastewater consisting of the parent compounds and their oxidation by-products. It also provides indirect information on biodegradability.[10] Any variation in the composition of the effluents produces a change in AOS that could result in substantial changes in biodegradability/toxicity of the wastewater:

$$AOS = \frac{4(DOC - COD)}{DOC} \quad (6.1)$$

Figure 6.1 shows that AOS increased during the treatment. It suggests that more oxidized organic by-products were formed, changing the average chemical nature of the mixture. After a certain time COD and DOC behave similarly, stabilizing the AOS. An oxidation process formed more oxidized organic intermediates at the beginning of the process (note how fast COD decreases compared with DOC). Formation of more oxidized compounds is an indirect indication of improved biodegradability of the wastewater. Therefore, AS biodegradability during treatment should be evaluated by the Zahn–Wellens (Z–W) test over 28 days, a very prolonged test that will permit stating of biodegradability, including adaption of AS to difficult wastewaters. The Z–W test is used to evaluate the biodegradability of water-soluble, non-volatile organic contaminants when exposed to a relatively high concentration of microorganisms. It takes around 28 days at 20–25 °C. The DOC eliminated is expressed as the percentage of biodegradation. Wastewater is considered biodegradable when the biodegradation percentage is over 70%.[7] The results are summarized in Figure 6.1(C). Among the different samples taken during the treatment, three of them (1–3) contained parent compounds (pesticides). Sample 1 (highest concentration of the active ingredients) and sample 2 were non-biodegradable, whereas samples 3–5 were biodegradable according to the Z–W test. But sample 3 was biodegradable after a long period of more than 8 days. These results demonstrate that the most suitable point for combining AOP with the biological treatment was after the total elimination of the parent compounds, somewhere between 15 and 30 min of treatment. Regarding the former results, different considerations were taken into account: (i) biodegradability was enhanced when active ingredients of the pesticides were removed and (ii) the composition of the real wastewater is usually variable and therefore it is very difficult to adjust exactly the treatment time until good biodegradability is attained. Consequently, it is logical to adopt a compromise solution and extend AOP treatment to a larger mineralization degree to ensure biodegradability of the remaining DOC. Therefore, the proposed strategy for the combined system would be: extending the AOP process up to 40–50% of mineralization before discharging the partially oxidized solution in the biotreatment. Figure 6.1(D) shows the combined AOP/aerobic biological treatment applied to biorecalcitrant wastewater. A detailed study using different analytical tools and bioassays is therefore recommend for adequate design of any treatment, as the treatment time by AOP should be decided based on several parameters, and not only AOS.

Landfilling is the most widespread method for municipal solid waste (MSW) disposal, receiving up to 95% solid residues worldwide.[21] As a result of percolation of rainfall combined with degradation of the organic fraction and other compounds transfer, leachates are generated. These are considered to be highly threatening to the environment and human health due to the large amounts of organic compounds they contain (biodegradable and refractory to biodegradation). In addition to leachates' inherent complexity, their quality is greatly dependent on diverse factors such as age, precipitation, seasonal weather variation, waste type and surrounding population.[22] Conventional biological treatments and classical physicochemical methods are regarded as the most appropriate technologies for landfill leachate management. However, biological treatment is hampered by the presence of biorefractory compounds and the limited amount of biodegradable organics (wastewater type I, especially in the case of stabilized landfills). Advanced oxidation processes (AOPs), ozonation and electro-oxidation, have been proposed as an effective alternative for the mineralization of recalcitrant organics in landfill leachate.[23–27] However, these treatments are not economically acceptable for application to large-scale effluents treatment. Hence, strategies based on integrated chemical–physical–biological processes must be studied as they can ameliorate the drawbacks of individual processes, improving the overall treatment efficiency. The remediation strategy should include a preliminary physicochemical stage due to the solids presence, inherent to any landfill leachate.

Application of a photocatalytic process requires the lowest quantities of suspended solids as possible so that radiation penetration is favoured, and efficiency and operating costs are highly reduced. Accordingly, a preliminary physicochemical treatment of the received landfill leachates based on different coagulants should be tested; Table 6.1 shows typical results of this treatment.[28,29] Therefore, the intention is to remove as much organic matter, along with suspended solids. After this physicochemical pretreatment, the initial COD was reduced but toxicity and biodegradability enhancements were not observed after the conditioning step. A progressive discoloration also took place.

A final AOP treatment was performed to study the evolution of acute toxicity and biodegradability. The aim of these tests was to verify whether it was able to improve LL biodegradability. During the assays, a substantial discoloration was observed. Figure 6.2(A) shows the evolution of AOS, inhibition percentage and short-term biodegradability of the samples taken during the photo-treatment. Biodegradability is measured by the CODb (biodegradable

Table 6.1 Typical landfill leachates before and after pretreatment (coagulation).

	Landfill leachate (LL)	LL after pretreatment
pH	7.4	3
Conductivity (mS cm^{-1})	77.3	80
COD (g O_2 L^{-1})	42.6	29.2
DOC (g L^{-1})	15.6	13.3
Cl^- (g L^{-1})	40.2	41
SO_4^{2-} (g L^{-1})	15.7	16
Toxicity (%I)	50	50
Biodegradability (CODb/COD)	0	0

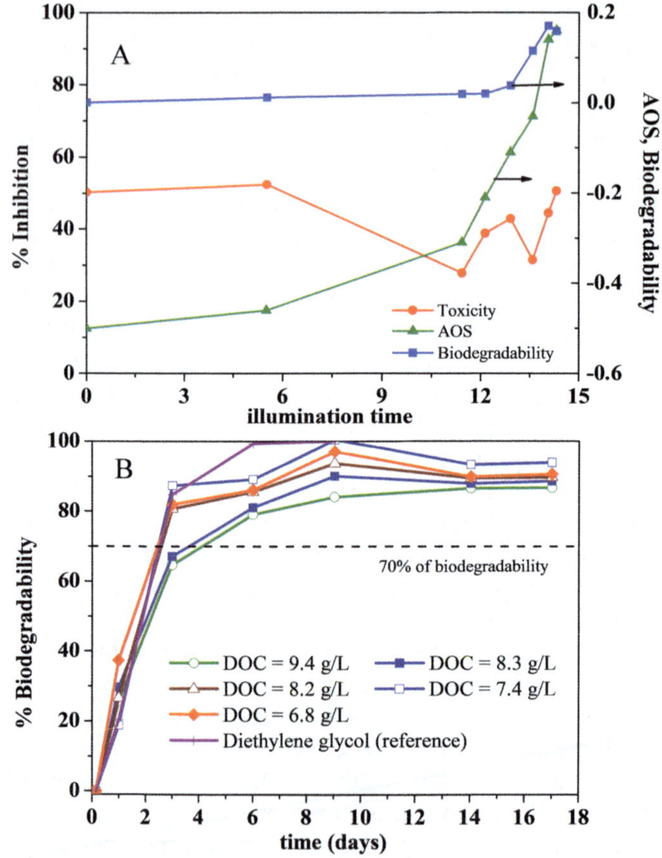

Figure 6.2 AOS, toxicity and short-term biodegradability (A) assessment during photocatalytic treatment of LL and Zahn–Wellens test results (B).

COD fraction)/COD ratio by respirometry using municipal AS.[6] When it is higher than 0.1, a sample can be considered to be slowly biodegradable. These values indicated no significant toxicity changes and made it interesting to study the long-term biodegradability of each sample to find the best point for a possible combination with a subsequent biological treatment. Regarding AOS profile, it can be said that more oxidized intermediates were formed to the end of the photo-treatment (AOS continuously increased). Toxicity (by respirometry using municipal AS) evolution did not present important changes but the inhibition was always lower than that of the initial sample. These biodegradability results provide information about the immediate response of the activated sludge since each test has approximately a duration of 30 min. Nevertheless, and with the objective of having more information regarding the long-term behaviour of conventional biomass from a municipal wastewater treatment plant against this partially treated effluent, a Zahn–Wellens test was carried out using some of the samples which had been already evaluated by respirometry (Figure 6.2(B)). With this biological assay, more specific conclusions can be extracted regarding biomass adaptation, adsorption phenomena, sludge age and hydraulic retention time for the correct operation of a possible bioreactor. As can be observed, all the partially treated samples attained 70% of biodegradability after a quite low residence time (not higher than five days). This means that even less oxidized samples could be considered biodegradable after a very short adaptation period of the conventional AS. In addition, and as happened for the respirometric assays, leachate biodegradability increased progressively as the photo-treatment took place, reaching higher biodegradability values in shorter times, at lower DOC concentrations. Although lowest oxidized samples showed high biodegradability percentages after four days of treatment, scaling-up to conventional biological reactors could require a previous biomass adaptation stage to reduce residence time and sludge age. Taking this consideration into account, photo-treatment is recommended until achieving a mineralization percentage of 27% (final DOC of 8.2 g L^{-1}). Afterwards, the partially oxidized effluent could be completely treated in a conventional biological treatment plant with the corresponding reduction in overall operating costs. Consequently, an appropriate treatment strategy for this type of water remediation would consist of a physicochemical pretreatment, followed by an AOP (performed until an extent of initial DOC removal of between 20% and 30%), and the last step would consist of a conventional biological process.

6.3 Partially Biorecalcitrant Wastewater Treatment by Integrating Solar Photocatalysis and Other Processes

A biological pretreatment followed by an AOP may be more useful when the original wastewater contains a considerable amount of biodegradable compounds. Under these circumstances a pre-oxidation step will not lead to any significant improvement of biodegradability and it will only cause needless consumption of chemicals. Such integrated systems are particularly favourable for effluents that initially contain some biodegradable fractions which could easily be removed first and, therefore, not compete for the chemical oxidant (wastewater type II).

A typical case is shown in Figure 6.3. An initial biodegradability test (Zahn–Wellens standard methodology) of the original wastewater showed significant biodegradability but one contaminant remains almost constant, demonstrating it is biorecalcitrant and cannot be degraded by conventional biotreatment (Figure 6.3(A)). The Zahn–Wellens test showed that the specific biorecalcitrant contaminant persisted. Otherwise, most of the organic content of the wastewater could be successfully biodegraded. It highlights the need to remove the non-biodegradable compound afterward by an AOP. Working in a bioreactor using AS, after a few days, >96% of the initial DOC was removed (Figure 6.3(B)). On the first day, interestingly, the

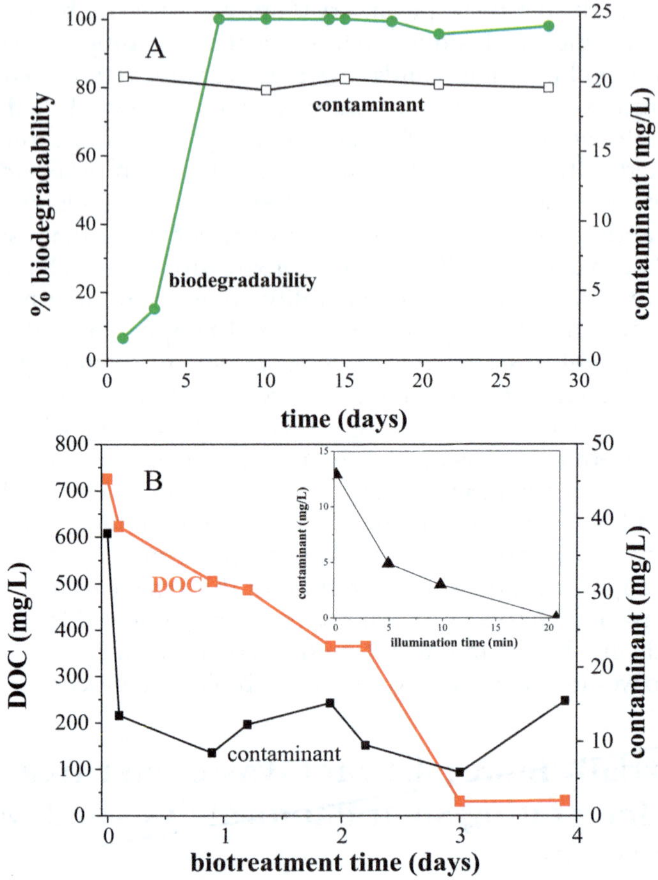

Figure 6.3 Zahn–Wellens test results (A) and biotreatment (B) followed by AOP (inset of B) of partially biorecalcitrant wastewater.

contaminant concentration was reduced. This decrease could only be related to the adsorption of the compound on the activated sludge, as afterwards it remained constant until the end of the process – very typical behaviour of these contaminants in AS. It also demonstrated that the contaminant was less toxic for biomass as biotreatment of the DOC was successful. However, it was biorecalcitrant. The biotreated wastewater was subsequently treated by AOP (inset Figure 6.3(B)). The non-biodegradable compound was degraded after 20 min of illumination time, reaching a DOC concentration low enough for disposal. The overall efficiency of the combined biotreatment/AOP in elimination of dissolved organic carbon (DOC) was over 95%. It could be concluded that biorefractory compounds are very often not toxic to AS, and biotreatment prior to AOP may be successful. It would avoid long treatment times and high chemical consumption in the AOP step, the main drawbacks of these advanced treatments. In such cases, the AOP step is more suitable as a final polishing step.

6.4 Removal of Micropollutants in Water and Wastewater by Integrating Solar Photocatalysis and Membrane Nanofiltration

Recent years have seen an increased application of nanofiltration and reverse osmosis units (NF/RO) for the treatment of effluents from municipal waste water treatment plants (MWTP) or other water matrix (groundwater, surface and drinking water). One of the main objectives was to remove micropollutants (priority and emerging contaminants) from these streams.[30] Membranes have been demonstrated to be the most promising treatments for the removal of these compounds from wastewater.[31–33] The concentrate and the membrane cleaning systems together with the operation conditions (pressure, temperature and pH) and feed type are key parameters that affect membrane life time, process and costs. Concentrates contain NOM, micro-contaminants and MWTP residuals (*i.e.*, microbial by-products, biodegraded organics that did not reached mineralization and other chemicals used for membrane operation) and they cannot be discharged into the environment (wastewater type IV).[34,35] AOPs have been proposed as a suitable treatment of NF/RO retentate[36–38] but it would be important to take into account different aspects that could optimize process integration.

In general, photocatalysis rates are slow with low loads of the contaminants ($C_{\mu P}$, priority and emerging contaminants). Degradation of contaminants at low concentrations follows pseudo-first-order kinetics ($r = kC$). One way to increase efficiency would be to increase C_0 ($C_{\mu P}$). The combination of photocatalysis and other AOPs with membrane technologies under these conditions (Figure 6.4) has attracted attention during the last few years, as the concentrations in retentate are much higher than in raw effluents.[39]

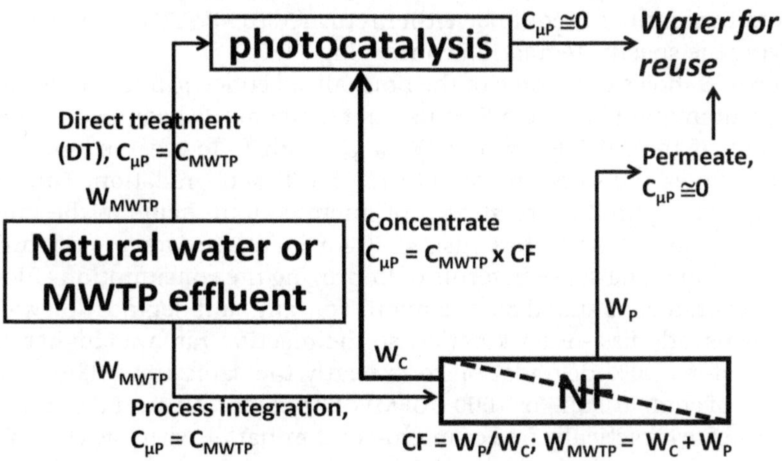

Figure 6.4 General diagram of combination membrane–AOP systems for micropollutant treatment; W = mass of water; CF = concentration factor.

Nanofiltration and photocatalytic treatment of membrane retentate, combined in this sequence, is a more efficient tertiary treatment than photocatalysis directly applied to the MWTP effluent. Micro-contaminants would be concentrated in the retentate of NF ($C_{\mu P} = C_{MWTP}$), discharging a permeate free of contaminants. The concentrate (W_C) is then treated by photocatalysis and results compared to direct photocatalytic treatment (DT) of the NF feed-in (W_{MWTP}).

Experiments were conducted with real MWTP effluent, evaluated using LC-QTRAP-MS. MWTP effluent was pretreated before NF by conventional filtration (through a 75 μm sand filter and two 25 and 5 μm microfilters). The main characteristics of effluent (W_{MWTP}) used in this study, after filtration, were (concentration in mg L^{-1}): Na^+ (257), Mg^{2+} (61), Ca^{2+} (72), SO_4^{2-} (149), Cl^- (419), DOC (30), conductivity 2.2 mS cm^{-1}. The main characteristics of the concentrate (W_C) used in this study, after filtration, were (mg L^{-1}): Na^+ (727), Mg^{2+} (209), Ca^{2+} (213), SO_4^{2-} (481), Cl^- (1200) and DOC (61), conductivity 6.5 mS cm^{-1}. The initial mass of water was concentrated fourfold ($3W_C = W_P$; $W_{MWTP} = 4W_C$, with CF = 4). Thirty-five micro-contaminants were detected in the MWTP effluent. Table 6.2 shows their starting concentrations.

The starting concentration of the sum of the 35 micro-contaminants detected in the MWTP effluent was 46 800 ng L^{-1} and increased to 154 760 ng L^{-1} in the concentrate. As the CF was 4, it can be inferred that some of the contaminants were adsorbed on the NF membrane, as $C_{\mu P}$ in W_P was negligible. Degradation in W_C of caffeine, carbamacepine, naproxen, 4-AAA, norfloxacine, 4-FAA, antipirine, ofloxacin and sulfametoxazole can be seen in Figure 6.5. In the first minutes of treatment degradation was very fast. These results show that micro-contaminants in concentrates from NF effluents can be easily degraded with a low detrimental impact on degradation of concentration of salts and DOC present in the concentrate. DOC mineralization was negligible but, in any case, at the end of the treatment the concentrate free of micro-contaminants could be either returned to MWTP or ready for reuse satisfying the specific regulation concerning DOC.

Figure 6.6 shows the profile of the normalized concentration of the sum of micro-contaminants in the NF concentrate stream (W_C) compared to direct treatment of the MWTP effluent (W_{MWTP}). Both followed the same trend, needing almost the same treatment time for 90% degradation. Thus, similar energy is required for treatment of micro-contaminants in the range of hundreds of μg L^{-1}. Therefore, using NF for increasing micro-contaminant concentration made sense in terms of improving the consumption efficiency of $HO^•$ generated. As stated before, micro-contaminant degradation was governed by pseudo-first-order kinetics, so the reaction rate was higher in W_C, and, therefore, 90% degradation took nearly the same time. Thus, for the purposes of comparison, for 1000 L of MWTP effluent, NF would produce 750 L of permeate (practically free of micro-contaminants, with low DOC, free of microorganisms, and with low salinity) and 250 L of concentrate that would have to be treated by photocatalysis, while the direct treatment would require photocatalytic application to the whole 1000 L volume. The nanofiltration pretreatment made it possible to run photocatalysis at lower flow rates

Table 6.2 Initial concentration (ng L^{-1}) of each micro-contaminant in W_{MWTP} and W_C determined by LC-QTRAP-MS; 4-AAA (4-acetylaminoantipyrine), 4-FAA (4-formylaminoantipyrine), 4-AA (4-aminoantipyrine).

	Compound	$C_{\mu P}$ in W_{MWTP}	$C_{\mu P}$ in W_C
1	Caffeine	20 200	54 600
2	Carbamacepine	6600	28 600
3	Naproxen	4500	11 400
4	4-AAA	4300	16 900
5	Norfloxacine	2300	12 600
6	4-FAA	2300	6000
7	Antipirine	1300	4200
8	Ofloxacin	860	4300
9	Sulfametoxazole	870	3300
10	Atenolol	560	2400
11	Ciprofloxacin	450	1500
12	4-AA	180	420
13	Eritromicin	380	1200
14	Ranitidin	260	1100
15	Sulfapiridin	260	840
16	Fenofibric acid	270	790
17	Trimethoprim	210	720
18	Keteprophen	160	740
19	Venlafaxin	160	780
20	Azithromycin	270	650
21	Nadolol	100	380
22	Citalopram	60	260
23	Metronidazol	40	40
24	Lincomicin	30	140
25	Pirimidone	40	140
26	Propranolol	30	130
27	Indomethacin	40	120
28	Sotalol	30	110
29	Clorfenvinphos	30	80
30	Famotidin	30	100
31	Metoprolol	20	90
32	Terbutaline	10	70
33	Mepivacaine	10	20
34	Salbutamol	10	10
35	Sulphatiazol	10	30
	All	$\Sigma C_{\mu P}$ in $W_{MWTP} = 47$ μg L^{-1}	$\Sigma C_{\mu P}$ in $W_C = 155$ μg L^{-1}

and with higher starting micropollutant concentrations, and substantially reduced the surface area of solar collectors and reagents needed per cubic meter of treated effluent. *Vibrio fischeri* toxicity analyses were performed before and after the application of the treatment and it was always observed that toxicity did not substantially change during oxidation. In W_{MWTP}, inhibition was less than 15–20% throughout the process and in W_C remained between 28% and 35%.[40] These results reinforce the idea that treatment of extremely low concentrations of contaminants, such as those found in effluents from MWTPs, requires different operating concepts from the application of photocatalysis to high-organic-load industrial wastewaters.

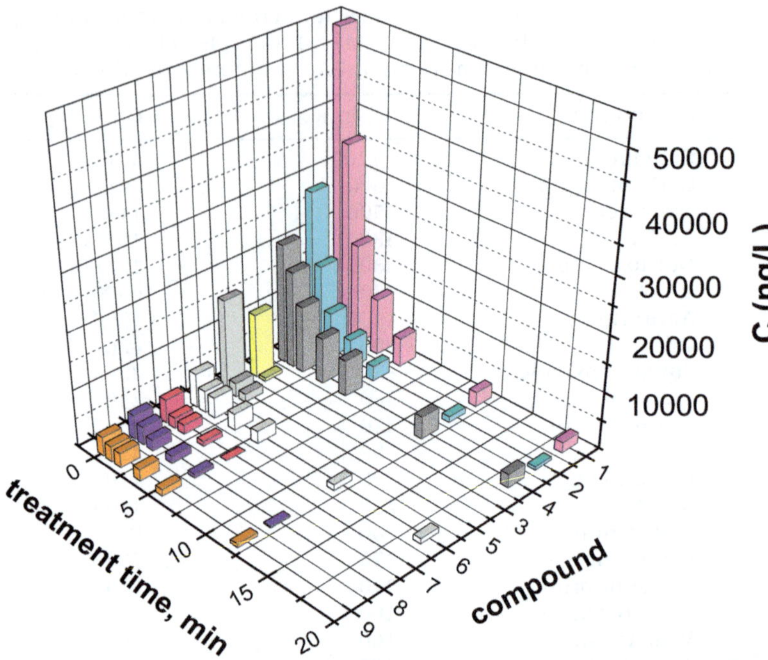

Figure 6.5 Degradation profile for the nine micro-contaminants at higher concentration present in W_C. Notation (1–9) as in Table 6.2.

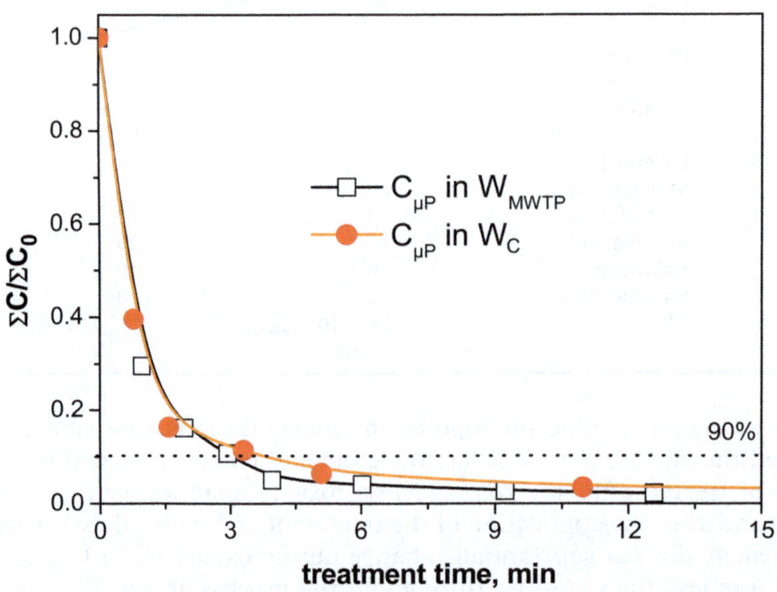

Figure 6.6 C/C_0 over treatment time for photocatalytic treatment of MWTP effluent ($\Sigma C_{\mu P}$ in $W_{MWTP} = 47\ \mu g\ L^{-1}$) and NF concentrate ($\Sigma C_{\mu P}$ in $W_C = 155\ \mu g\ L^{-1}$); contaminants as in Table 6.2.

6.5 Conclusions and Recommendations

Concepts of integration and coupling of photocatalysis with biotreatment, coagulation, sedimentation and nanofiltration have been presented and evaluated taking into account not only the disappearance of the parent compounds but also mineralization of contaminants. The basic parameters (necessary for relevant information on process kinetics and for assuring consistent results) to be evaluated are mineralization (COD and DOC), concentration of the original pollutant (LC-UV and/or LC-MS) and concentration of inorganic ions. LC-MS is progressively gaining acceptance, mainly for the determination of micro-contaminants. But these techniques are not enough to optimize photo-treatment time and chemical consumption on photocatalysis, the major concerns for the cost and energy efficiency benefits. The evolution of biocompatibility of the wastewater during the treatment is the other key factor. Toxicity and biodegradability could be considered critical indicators in order to make decisions. Accordingly, recent research has focused on the development of representative, cost-effective and quantitative test bioassays, which can detect different effects. Both Z–W and respirometric tests using AS are very useful for evaluating photocatalysis tests in order to properly integrate with biotreatment. Very often, toxicity changes continuously during the treatment and, therefore, it is not a suitable way to determine the moment when biodegradability is most enhanced, but toxicity can predict substantial changes in wastewater composition, identifying when biodegradability assessment methods should be applied. Last, but not least, is the combination of photocatalysis with membrane technologies as a way to increase C_0 of contaminants when treating rejection streams from membrane processes. When increasing C_0, the photocatalytic rate ($r = kC$) also increases.

It is necessary to push new concepts in environmental engineering, and more precisely in water issues innovation, to achieve long-term water quality and availability and viability of water resources. The continuous increase in demand forces water agencies to look for alternative sources. Some form of intermediate wastewater treatment must be introduced between the source and urban treatment centres, to ensure that recalcitrant pollutants are removed. Therefore, an important increase is expected in applications of such combined processes described in this chapter in the future, considering also the expected increases in water reuse applications.

Acknowledgements

The authors wish to thank the Spanish Ministry of Economy and Competitiveness for funding under the TRICERATOPS Project (Reference: CTQ2015-69832-C4-1-R).

References

1. I. Muñoz, S. Malato, A. Rodríguez and X. Doménech, *J. Adv. Oxid. Technol.*, 2008, **11**, 270.
2. J. P. Scott and D. F. Ollis, *J. Adv. Oxid. Technol.*, 1997, **2**(3), 374.
3. D. F. Ollis, *Water Sci. Technol.*, 2001, **44**(5), 117.
4. D. Mantzavinos and E. Psillakis, *J. Chem. Technol. Biotechnol.*, 2004, **79**, 431.
5. V. Augugliaro, M. Litter, L. Palmisano and J. Soria, *J. Photochem. Photobiol., C*, 2007, **7**, 123.
6. L. Rizzo, *Water Res.*, 2011, **45**, 4311.
7. I. Oller, S. Malato and J. A. Sánchez-Pérez, *Sci. Total Environ.*, 2011, **409**, 4141.
8. *Innovative and Integrated Technologies for the Treatment of Industrial Wastewater*, ed. A. López, C. Di Iaconi, G. Mascolo and A. Police, IWA Publishing, London, 2012.
9. J. Bandara, C. Pulgarin, P. Péringer and J. Kiwi, *J. Photochem. Photobiol., A*, 1997, **111**, 253.
10. C. Pulgarín, M. Invernizzi, S. Parra, V. Sarria, R. Polaina and P. Péringer, *Catal. Today*, 1999, **54**, 341.
11. A. M. Amat, A. Arqués, A. García-Ripoll, L. Santos-Juanes, R. Vicente, I. Oller, M. I. Maldonado and S. Malato, *Water Res.*, 2009, **43**, 784.
12. S. Esplugas, S. Contreras and D. Ollis, *J. Environ. Eng.*, 2004, **130**, 967.
13. A. Kyriacou, K. E. Lasaridi, M. Kotsou, C. Balis and G. Pilidis, *Proc. Biochem.*, 2005, **40**, 1401.
14. C. Sirtori, A. Zapata, I. Oller, W. Gernjak, A. Agüera and S. Malato, *Water Res.*, 2009, **43**, 661.
15. *The pesticide manual*, ed. C. D. S. Tomlin, British Crop Protection Council, 15th edn, Surrey, UK, 2009.
16. CFDI, Confederation of the Food and Drink Industries of the EU. Report on Managing Environmental Sustainability in the European Food & Drink Industries, 2014.
17. P. R. Gogate and A. B. Pandit, *Adv. Environ. Res.*, 2004, **8**, 501.
18. M. Pera-Titus, V. García-Molina, M. A. Baños, J. Giménez and S. Esplugas, *Appl. Catal., B*, 2004, **47**, 219.
19. J. J. Pignatello, E. Oliveros and A. MacKay, *Crit. Rev. Environ. Sci. Technol.*, 2006, **36**, 1.
20. M. A. Shannon, P. W. Bohn, M. Elimelech, J. G. Georgiadis, B. J. Mariñas and A. M. Mayes, *Nature*, 2008, **452**, 301.
21. P. R. O'Leary, G. Tchobanoglous and F. Kreith, *Handbook of Solid Waste Management*, McGraw-Hill, 2002.
22. J. Wiszniowski, D. Robert, J. Surmacz-Gorska, K. Miksch and J. V. Weber, *Environ. Chem. Lett.*, 2006, **4**, 51.
23. J. P. V. Vilar, S. M. S. Capelo, T. F. C. V. Silva and R. A. R. Boaventura, *Catal. Today*, 2011, **161**, 228.
24. A. Z. Gotvajn, T. Tisler and J. Zagorc-Koncan, *J. Hazard. Mater.*, 2009, **162**, 1446.

25. M. Umar, H. A. Aziz and M. S. Yusoff, *Waste Manage.*, 2010, **30**, 2113.
26. T. Poznyak, G. L. Bautista, I. Chaírez, R. I. Córdova and L. E. Ríos, *J. Hazard. Mater.*, 2008, **152**, 1108.
27. P. B. Moraes and R. Bertazzoli, *Chemosphere*, 2005, **58**, 41.
28. C. Amor, E. De Torres-Socías, J. A. Peres, M. I. Maldonado, I. Oller, S. Malato and M. S. Lucas, *J. Hazard. Mater.*, 2015, **286**, 261.
29. E. De Torres-Socías, L. Prieto-Rodríguez, A. Zapata, I. Fernández-Calderero, I. Oller and S. Malato, *Chem. Eng. J.*, 2015, **261**, 60.
30. I. Michael, L. Rizzo, C. S. McArdell, C. M. Manaia, C. Merlin, T. Schwartz, C. Dagot and D. Fatta-Kassinos, *Water Res.*, 2012, **47**, 957.
31. B. V. Bruggen and C. Vandecasteeleet, *Environ. Pollut.*, 2003, **122**, 435.
32. P. Westerhoff, H. Moon and D. Minikata, *J. Crittenden Water Res.*, 2009, **43**, 3992.
33. L. Nghiem, A. I. Schäfer and M. Elimelech, *Environ. Sci. Technol.*, 2005, **39**, 7698.
34. A. Pérez-Gonzalez, A. M. Urtiaga, R. Ibáñez and I. Ortiz, *Water Res.*, 2012, **46**, 267.
35. L. Rizzo, C. M. Manaia, C. Merlin, T. Schwartz, C. Dagot, M. C. Ploy, I. Michael and D. Fatta-Kassinos, *Sci. Total Environ.*, 2013, **447**, 345.
36. E. Dialynas, D. Mantzavinos and E. Diamadopoulos, *Water Res.*, 2008, **42**, 4603.
37. S. B. Abdelmelek, J. Greaves, K. P. Ishida, W. J. Cooper and W. Song, *Environ. Sci. Technol.*, 2011, **45**, 3665.
38. A. Y. Bagastyo, D. J. Batstone, I. Kristiana, W. Gernjak, C. Joll and J. Radjenovic, *Water Res.*, 2012, **46**, 6104.
39. S. Miralles-Cuevas, I. Oller, J. A. Sánchez Pérez and S. Malato, *Environ. Sci. Pollut. Res.*, 2015, **22**, 846.
40. S. Miralles-Cuevas, F. Audino, I. Oller, R. Sánchez-Moreno, J. A. Sánchez Pérez and S. Malato, *Sep. Purif. Technol.*, 2014, **122**, 515.

CHAPTER 7

Photocatalytic Purification and Disinfection of Air

ALEXANDER VORONTSOV*[a]

[a]Boreskov Institute of Catalysis, Novosibirsk 630090, Russia
*E-mail: a-voronts@yandex.ru, voronts@catalysis.ru

7.1 Introduction

Air quality is one of the key factors that determine human health and longevity. The air that we breathe can be contaminated with various impurities. According to the typical size of airborne particles, the air contaminants are divided into molecular impurities consisting of diverse inorganic and organic compounds in the vapor state with a molecular size below 10 nm and mechanical impurities that have typical particles sizes ranging from 100 nm to 0.1 mm and consist of solid and liquid aerosols. Such aerosols can contain non-living inorganic and organic materials as well as biological matter including bacteria, viruses, pollen, fungi and even small insects and mites. The major air pollutants and their typical concentrations are listed in Table 7.1.

More complete lists of air pollutants acknowledged to represent an air pollution problem can be found in the literature.[1] Among molecular air pollutants, volatile organic compounds (VOCs) represent the largest set of compounds, with over 150 entities. These compounds in low concentration (below few ppm) are the typical target of photocatalytic air purification devices.

RSC Energy and Environment Series No. 15
Photocatalysis: Applications
Edited by Dionysios D Dionysiou, Gianluca Li Puma, Jinhua Ye, Jenny Schneider, and Detlef Bahnemann

Published by the Royal Society of Chemistry, www.rsc.org

Table 7.1 Major air pollutants according to the US EPA (http://www.epa.gov).

Pollutant type	Pollutant name	Typical concentrations
Solid aerosol	Asbestos and mineral fibers	0.01–0.1 fibers per liter
	Solid particles $PM_{2.5}$, PM_{10}	Level of mg m^{-3}
Liquid aerosol	Smoke	Up to tens of mg m^{-3}
Biological aerosol	Bacteria	Level of μg m^{-3}
	Mold and mildew	Level of μg m^{-3}
	Viruses	Level of μg m^{-3}
	Mites	Level of μg m^{-3}
	Pollen	Up to 10^4 grains per m^3
Molecular pollutants	Carbon monoxide	0.2 ppm to hundreds ppm
	Formaldehyde	Few ppb to tens ppm
	Nitrogen oxides	ppb to ppm level
	Sulfur dioxide	Up to few ppm
	Pesticides	ppb level
	Radon and other radionuclides	ppt to ppb level
	Volatile organic compounds (VOCs)	Few ppb to hundreds ppm
	Ozone	ppb level

Photocatalytic processes are the result of a natural combination of photochemistry and catalysis. Photocatalytic reactions proceed as a result of absorption of ultraviolet (UV), visible or infrared light quanta by photocatalyst species designated here as Pc. An excited species of photocatalyst (Pc*) initiates chemical transformations of reactants R into products P and return into their ground state after performing each cycle of chemical transformations.[2] The following sequence of chemical reactions reflects the essence of photocatalytic transformations:

$$\mathrm{Pc} + h\nu \rightarrow \mathrm{Pc}^* \tag{7.1}$$

$$\mathrm{Pc}^* \rightarrow \mathrm{Pc} \tag{7.2}$$

$$\mathrm{Pc}^* + \mathrm{R} \rightarrow \mathrm{Pc} + \mathrm{P} \tag{7.3}$$

The difference between usual photochemical and photocatalytic reactions lies in the species that absorb light. In photochemical reactions, the reactants themselves absorb light.

Nature has employed photocatalytic reactions for billions of years to carry out the process of photosynthesis. First reports on photocatalytic reactions carried out artificially by humans date back to mid-nineteenth and start of the twentieth century. Edmond Becquerel reported the production of current between electrodes with deposited photocatalyst in 1839.[3] Carl Renz observed the reduction of oxides of titanium, cerium, niobium and zinc upon solar irradiation in organic liquids.[4] However, true chemical transformations taking place in contact with irradiated photocatalyst were reported possibly for the first time for uranyl salts catalyzed decomposition of oxalic acid.[5]

Since the energy of light quanta for UV light is above 3 eV, species of photocatalyst excited with UV quanta (Pc*) can drive almost any reduction and oxidation reactions and under suitable conditions also initiate reactions *via* energy transfer. It is possible to oxidize molecules of pollutants directly and initiate radical processes as the following reaction scheme shows:

$$RH + Pc^* \rightarrow R^\bullet + H^+ + Pc^- \tag{7.4}$$

$$Pc^- + O_2 + H_2O \rightarrow Pc + HO_2^\bullet + OH^- \tag{7.5}$$

$$R^\bullet + O_2 \rightarrow RO_2^\bullet \tag{7.6}$$

$$R^\bullet + HO_2^\bullet \rightarrow ROOH \tag{7.7}$$

$$RO_2^\bullet + ROOH + O_2 \rightarrow \ldots \rightarrow CO_2 + H_2O \tag{7.8}$$

Peroxyl radicals $RO_2^\bullet$ are quickly formed upon reaction of alkyl radical $R^\bullet$ with oxygen of air as eqn (7.6) demonstrates. Alternatively, hydroperoxide ROOH is formed, in Reaction (7.7), with participation of the superoxide radical ($HO_2^\bullet$). Further transformations of peroxyl radicals and hydroperoxides proceed in the dark but can include participation of excited photocatalyst. The final products of photocatalytic destruction of any organic compound are inorganic compounds, which are carbon dioxide and water for organics containing C, H and O atoms (see eqn (7.8)). The whole process of transformation of organic air impurities into inorganic products is called mineralization.

Besides the direct oxidation shown in eqn (7.4) an alternative route for destruction of air pollutants exists that is of immense importance for disinfection. Water molecules are always present in atmospheric air and upon adsorption on the photocatalyst can be oxidized with the formation of hydroxyl radicals ($^\bullet OH$):

$$H_2O + Pc^* \rightarrow {}^\bullet OH + H^+ + Pc \tag{7.9}$$

Hydroxyl radicals ($OH^\bullet$) and superoxide radicals ($HO_2^\bullet$) can initiate destruction of organic materials including those in the microorganisms according to eqn (7.10):

$$^\bullet OH + RH \rightarrow R^\bullet + H_2O \tag{7.10}$$

The radicals and other molecular entities containing reactive oxygen are called reactive oxygen species (ROS). Such species are produced over the surface of UV light irradiated photocatalysts and can travel significant distances before they meet organic matter and oxidize it. This causes strong inhibition of microbiological objects in the vicinity of photocatalytic air purification devices, an effect known from real life examples of application of photocatalytic devices for air treatment.

Note that the process of photocatalytic oxidation described above is more suitable for air treatment than for water purification.[6] This is due to an order of magnitude larger oxidation rate in the gas phase compared to aqueous solutions.

In this chapter we consider the basics of photocatalytic processes of air purification and disinfection and perspectives for implementation of photocatalytic devices in indoor air treatment. Solid state photocatalysts are operational in the processes of photocatalytic air purification.

7.2 Photocatalytic Reactions for Air Purification

Molecular and mechanical contaminants can be removed from air by means of destructive and non-destructive methods. Non-destructive approaches towards air purification employ filters that capture contaminants. Vapor phase molecules are adsorbed on the surface of adsorbents or absorbed with liquids. Mechanical air impurities are filtered out using various filters including HEPA filters and electrofilters. The non-destructive approach is simple and the contaminants captured can be recuperated when they have significant value. For indoor air purification, the non-destructive approach is widely used but has serious disadvantages which necessitate its combination with destructive air cleaning. Problems with non-destructive air purification include bacterial and fungal growth in the filters and re-emission of captured molecular contaminants at increased temperature and air humidity. Non-destructive processes of air cleaning have poor capability in removing molecules of very low boiling point pollutants such as carbon monoxide and formaldehyde.

Diverse methods of destructive air purification are being developed. There are methods intended for destruction of pollutants in large concentrations based on catalytic combustion or thermal plasma. The methods for indoor airborne pollutants abatement should be, however, economic and thus need and cause minimal air heating. Room temperature destructive methods of air purification are based on non-thermal plasma (NTP) processes including ozone generation, room temperature catalytic reactions, photocatalytic reactions, and neutralization reactions proceeding after pollutants adsorption or absorption.

All the mentioned destructive methods of indoor air purification have specific advantages that delimit their applications. Neutralization reactions are used to treat airstreams containing chemical agents that can be hydrolyzed. The neutralization filters need to be replaced or regenerated after saturation. Room temperature catalytic reactions can be applied to treat specific contaminants such as carbon monoxide oxidation over Au/TiO_2 catalysts. Photocatalytic reactions and destruction with NTP can be applied to diverse diluted molecular and biological air pollutants that can be oxidized or decomposed.

As mentioned in the Introduction, photocatalytic transformations of organic air pollutants are dominated by the oxidation and destruction processes. Complete mineralization of pollutants without formation of any intermediates in the gas phase has been reported for some cases. Ethylene is oxidized quantitatively into CO_2 and H_2O as the following reaction scheme shows:[7]

$$C_2H_4 + O_2 \xrightarrow{h\nu, TiO_2} CO_2 + H_2O \tag{7.11}$$

Photocatalytic oxidation of alcohols usually produces significant concentrations of gaseous byproducts the main component of which is the corresponding aldehyde. Oxidation of ethanol vapors present at an initial concentration above 20 ppm over TiO_2 Degussa P25 in dry air produced diverse byproducts as Reaction (7.12) shows:[8]

$$C_2H_5OH + O_2 \xrightarrow{h\nu, TiO_2} CH_3CHO + CH_3COOH + HCOOH + CO_2 + CH_3COOC_2H_5 + HCOOCH_3 + HCOOC_2H_5 \quad (7.12)$$

Photocatalytic oxidation of acetaldehyde at concentrations above 300 ppm produces acetic acid as well as CO_2.[9] A higher concentration of acetaldehyde results in a higher yield of acetic acid.

Photocatalytic oxidation of benzene vapors at a concentration of 514 ppm over TiO_2 P25 in a humidified airstream results in formation of CO_2 and CO products, with selectivity to CO being 7–10%.[10] Photocatalytic oxidation of ethylene also produces toxic CO byproduct but only if ethylene is present at a very large concentration.[11] Photocatalytic oxidation of toluene vapors (concentration of 13 000 ppm) over anatase TiO_2 gave rise to mainly benzaldehyde with benzene, CO_2, benzyl alcohol and benzoic acid being the minor gaseous products.[12]

The key conditions for the release of byproducts of photocatalytic oxidation are the high concentration of the contaminant and insufficient residence time inside photocatalytic reactor. Alberici and Jardim reported that many vapor phase organic compounds, including trichloroethylene (TCE), isooctane, acetone and methanol, were completely oxidized into CO_2 without formation of any intermediate in an annular flow reactor with a flow rate 200 mL min^{-1}, which corresponded to a residence time slightly above 2 min.[13] Also, byproducts are usually not observed if initial concentration of air pollutant is below 1 ppm.

Photocatalytic oxidation of heteroatomic and halogenated compounds is often accompanied by formation of diverse byproducts. Photocatalytic oxidation of sulfur-containing organic compounds trimethylene sulfide (C_3H_6S), propylene sulfide, thiophene (C_4H_4S) and methyl disulfide (MeSSMe) in an annular photoreactor with a residence time of around 1 min resulted in the release of CO_2 and SO_2.[14] Photocatalytic oxidation of diethyl sulfide in a reactor with a low residence time produced various gaseous products shown in the reaction scheme below:[15]

$$C_2H_5SC_2H_5 + O_2 \xrightarrow[TiO_2]{h\nu} Et_2S_2 + CH_3CHO + C_2H_5OH + C_2H_4 + CO_2 \quad (7.13)$$

Reaction (7.13) can be envisaged as an attack of ROS at carbon and sulfur atoms and breakage of a C–S bond. Diverse heavier products such as H_2SO_4, Et_2S_3, $EtSO_2H$, $EtSO_3H$, Et_2SO and Et_2SO_2 are accumulated on the photocatalyst surface.

Of special interest are oxidation reactions of chlorine-containing organic compounds, which demonstrated extremely high quantum yields due to chain reactions. The apparent quantum yield of TCE oxidation was reported to be 0.97.[16] Photocatalytic oxidation of perchloroethylene was reported to produce trichloroethylene and HCl as major products and dichloroacetic acid, monochloroacetic acid, CCl_4, chloroform and phosgene as minor products.[17]

Chlorinated organic compounds such as trichloroethylene can be employed to enhance photocatalytic degradation of hydrocarbons in binary mixtures. It has been demonstrated that addition of TCE into airborne mixture with toluene, ethylbenzene, *m*-xylene, methyl ethyl ketone, acetaldehyde, butyraldehyde, methyl *t*-butyl ether, methyl acrylate, 1,4-dioxane and hexane at a concentration of 50 mg m^{-3} increased their oxidation rate over UV irradiated TiO_2.[18] The effect was very pronounced with ethylbenzene, acetone and methyl acrylate. However, several compounds such as methanol strongly inhibit photocatalytic oxidation of TCE and CH_2Cl_2.[19]

Inorganic compounds such as NH_3, NO and CO are converted during photocatalytic oxidation into deeper oxidation products – eventually HNO_3, H_2O and CO_2, with CO being directly oxidized into CO_2. Nitrogen oxides designated as NO_x, which is a mixture of NO and NO_2, are removed from air in a process of photocatalytic oxidation over TiO_2 film.[20] The reaction proceeds through consecutive steps with formation of NO_2 and HNO_3 as the following reaction scheme demonstrates:

$$NO \xrightarrow[O_2+h\nu]{TiO_2} NO_2 \xrightarrow[H_2O+O_2+h\nu]{TiO_2} HNO_3 \quad (7.14)$$

Besides oxidation, photocatalysts can drive other types of reactions leading to removal of airborne pollutants. Yamashita *et al.* reported photocatalytic decomposition of NO into N_2, O_2 and N_2O over TiO_2/Zeolite-Y catalyst.[21] Anpo *et al.* communicated that NO was transformed into N_2, N_2O and NO_2 over UV irradiated Ag^+/ZSM-5 photocatalyst.[22,23] Cho *et al.* reported photocatalytic decomposition of ozone over Pt/TiO_2 and TiO_2 photocatalysts:[24]

$$2O_3 \xrightarrow[h\nu]{Pt/TiO_2} 3O_2 \quad (7.15)$$

In contrast to bare TiO_2, 1% Pt loaded titania demonstrated a higher ozone decomposition rate even at high air humidity. Addition of ethanol to the Pt/TiO_2 photocatalyst increased ozone conversion significantly.

Subbotina *et al.* reported that NO is reduced with CO in NO–CO gas mixtures to produce N_2 and CO_2 over UV illuminated 2.5% MoO_3/SiO_2 photocatalyst.[25] The studies leaved unanswered the question of whether this reaction can proceed when NO and CO are present in air.

The transformation of N_2O into N_2, CO_2 and CO over Ag/TiO_2 photocatalyst in the presence of methanol vapors is also a reductive process.[26] However, the practical question of activity in the presence of air was not studied for this case as well.

To sum up, there are various photocatalytic reactions that can remove almost any pollutant from air. However, intermediate products are often formed and special measures must be undertaken to avoid their release into purified air.

7.3 Photocatalysts and Their Supports for Air Purification

Titanium dioxide is by far the most widely used photocatalyst for transformations of air pollutants. It has been applied at least since the 1960s. TiO_2 photocatalyst can be prepared in the form of powder, granular or thin film. Loading on supports is usually required to avoid removal of photocatalyst with the airflow or as a result of mechanical effects. Diverse organic and inorganic materials have been tested as supports for TiO_2 based photocatalysts.

TiO_2 can be loaded onto woven or nonwoven fiber textiles by impregnation with solution containing 25 g L^{-1} of TiO_2 and 0.05 g L^{-1} dioctyl sulfosuccinate used as dispersing agent and subsequent drying at ambient temperature and at 105 °C.[27] Felted polyester fabric was suggested as a material for photocatalytic filters that could simultaneously remove VOCs and dust.[28] The procedure for coating included obtaining TiO_2 sol and dip coating followed by drying at 20–30 °C. Raillard *et al.*[29] loaded TiO_2/SiO_2 photocatalyst onto a nonwoven support composed of cellulose and synthetic fibers bound together with an organic mixture. Adjimi *et al.*[30] supported TiO_2/SiO_2 photocatalyst on cellulose paper and found it active in ethanol oxidation. The stability of the photocatalyst–support composite is an important issue for organic supports which is, however, seldom investigated. Thevenet *et al.* reported that utilization of colloidal silica binder for attaching TiO_2 powder to organic and inorganic fibers resulted in a significant decrease of photocatalytic activity in acetylene oxidation.[31] Therefore, application of binders for fixing TiO_2 should be carried out with care and investigations are always needed to elucidate if catalyst activity is retained. Nonwoven perforated filters loaded with TiO_2 have been tried for production by Ahlström for photocatalytic applications.[32] Dong *et al.*[33] loaded woven cotton fabric with TiO_2 photocatalyst using aminosilicone binder that was added to a TiO_2 dipping suspension at a concentration of 10 g L^{-1}.

Activated carbon fiber (ACF) matt is a prospective photocatalyst support material that was tested in several works.[34,35] It has the advantages of a porous structure that is penetrable by the air flow and it has adsorbent capabilities for the removal and accumulation of peak pollutant concentrations and byproducts from photocatalytic oxidation. TiO_2 can be loaded onto ACF by dipping into its aqueous suspension. However, SEM images demonstrate[35] that such loading procedure is far from ideal since the deposition is strongly non-uniform.

Woven or nonwoven fiberglass materials, being inorganic materials, possess the advantage of high stability under the action of ROS generated during photocatalytic oxidation. Spitskiy and Parshina[36] synthesized ZnO directly onto a woven fiberglass support by loading it with zinc nitrate and heating at 300 °C. This resulted in strongly attached ZnO that was not removed by washing with water. Mineralization of acetone was obtained in an annular reactor over the developed photocatalytic composite.

Slimen *et al.*[37] used very thin anodized titanium mesh as support for TiO_2 that was loaded by dip coating and calcination at 550 °C for 3 h. It can be concluded that despite the use of a vast variety of photocatalyst supports the issues of mechanical and chemical stability as well as highest activity are still ongoing.

All photocatalysts can be classified into those working only with ultraviolet (UV) light and those which are active in visible or both ultraviolet and visible light. Table 7.2 lists major UV light active photocatalysts and their performance in the destruction of air pollutants. Most studies deal with TiO_2 in anatase polymorph, but mixtures with rutile and brookite are also active. Very recently TiO_2 anatase with dominant exposed face (001) was prepared and showed much increased activity in photocatalytic oxidation compared to anatase with naturally exposed (001), (101) and (100) faces.

To improve the performance of TiO_2 photocatalyst in oxidation reactions it can be loaded or mixed with other materials. Table 7.2 shows that TiO_2 can loaded onto adsorbents such as activated carbon, zeolites and silica, prepared in mixed form with other oxides like ZrO_2, WO_3 and La_2O_3, loaded with noble metals or their mixture as well as surface modified with inorganic acids. These approaches often result in activity up to one order of magnitude greater than that of pure TiO_2.

Table 7.2 Photocatalysts for air purification working under ultraviolet light.

Photoactive component	Support	Air contaminants	Best activity[a]	References
TiO_2 powder (anatase or anatase + rutile)	None, glass, quartz	Acetone, benzene, 2-propanol, toluene, ethanol, NO_x, formaldehyde, acetaldehyde, methanol, hexane, acetylene, NH_3	QE up to 180% reported	20,31,33, 38–47
F–TiO_2 71% (001)	None	Acetone, acetaldehyde	7× that of TiO_2	48 and 49
Mg^{2+}/TiO_2 (001)	None	Acetaldehyde	QE 8.5%	50
TiO_2 film	None	Ethylene, TCE, acetaldehyde, acetic acid, toluene	QE 3–124%	9,51–54
TiO_2	SiO_2	TCE	QE 13%	55
TiO_2/Y-Zeolite	None	Toluene, benzene	2.57× that of TiO_2	56
TiO_2 75%	Mordenite	Pyridine	QE 49–67%	57
TiO_2	Carbon, activated carbon	Ethanol, acetaldehyde, NO, BTEX, SO_2, methanol, 1-propanol, 1-butanol	—	34,35,58,59
TiO_2	$SiMgO_x$	H_2S	—	60
TiO_2	SnO_2	Acetaldehyde	4× of TiO_2	61
SiO_2/TiO_2	None	Toluene, acetaldehyde, benzene	7× that of TiO_2	62–64
ZrO_2/TiO_2	None	Acetone, formaldehyde	2.5× that of TiO_2	62,65,66

(continued)

Table 7.2 (*continued*)

Photoactive component	Support	Air contaminants	Best activity[a]	References
PO_4^{3-}/TiO_2	None	Acetaldehyde	2× that of TiO_2	67
SO_4^{2-}/TiO_2	Quartz, glass	Ethanol, acetaldehyde, toluene, ethylene, acetone	4× that of TiO_2	68–71
$H_3PW_{12}O_{40}/TiO_2$	Quartz	2-Propanol	1.5× that of TiO_2	72
Ag/TiO_2	None	N_2O	—	26
$Ag/Pt/TiO_2$		Toluene	2× that of TiO_2	73
Au/TiO_2	None	NO, CO, dimethyl ether	2.5× that of TiO_2	74–76
Pd/TiO_2	None	Toluene	2.5× that of TiO_2	77
Pt/TiO_2	None	Benzene, ethylene, methanol, acetaldehyde, acetone, CO, TCE	3× that of TiO_2	7,39,78–82
$Pt/SO_4^{2-}/TiO_2$	Glass	Acetone	7× that of TiO_2[b]	83
Rh/TiO_2	None	Benzene	2.5× that of TiO_2	84
Bi_2O_3/TiO_2	None	Toluene	2.2× that of TiO_2	85
CeO_2/TiO_2	None	Toluene	3.5× that of TiO_2	86
SnO_2/TiO_2	None	Benzene[c]	8× that of TiO_2	87
VO_x/TiO_2	$MgSiO_3$	Toluene, xylene	Thermal + photo	88
WO_3/TiO_2	None	2-Propanol, butyl acetate, toluene	3.5× that of TiO_2	89–92
La_2O_3/TiO_2	None	Acetone	1.5× that of TiO_2	93
Y_2O_3/TiO_2	None	Acetone	2.5× that of TiO_2	93
$Ca_2Nb_2O_7$	None	Benzene	= TiO_2	94
ZnO	None	Acetone	—	95
SnO_2	None	1-Butene	3× that of TiO_2 P25	96
$CdSnO_3{\cdot}3H_2O$	None	Benzene, cyclohexane, acetone	10× that of TiO_2 P25	97
$AgCl/Al_2O_3$	None	NO	—	98

[a]QE – quantum efficiency, photonic efficiency.
[b]At [acetone] < 20 ppm.
[c]Incomplete mineralization.

Besides TiO_2-based materials, several other compounds have been found that demonstrated good photocatalytic activity in mineralization of air pollutants. SnO_2 with primary particles of 5 nm demonstrated very high photocatalytic activity under UV irradiation while SnO_2 with 22 nm particles was inactive.[96] ZnO has long been investigated as an oxidation photocatalyst but usually demonstrates activity inferior to that of TiO_2. Some novel materials like $CdSnO_3$ have been tested recently that showed very promising activity in oxidation of organic air contaminants.

The second class of air purification photocatalysts includes materials possessing activity under visible light. Table 7.3 reveals major types of visible light active Pcs. Doping was suggested for extending the action spectrum of TiO_2 into the visible light range. Nitrogen, carbon and sulfur doped TiO_2 has been given much attention during the last decade due to the ease of preparation of these materials. Nitrogen doped TiO_2 has absorption in the whole range of visible light. In *et al.*[99] found that the highest activity is observed at 0.5% N content and they observed about 6% conversion of MTBE with a concentration of 68 ppm in an air flow of 44 mL min^{-1} under 50 mW cm^{-2} irradiance in the range 420–770 nm. The photocatalytic activity of undoped pristine TiO_2 under these conditions was at least one order of magnitude lower. However, in terms of quanta utilization, their activity is limited to a quantum yield below 2%. This is in contrasts to quantum yields close to 100% for UV light active photocatalysts.

Therefore, different classes of visible light active Pc are being developed. Many of the materials exhibit activity just a few times higher than that of

Table 7.3 Photocatalysts for air purification working under visible light or combined visible and ultraviolet light.

Photocatalyst	Air contaminants	Spectral range (λ, nm) (temperature[a] (°C))	Best activity	References
N-doped TiO_2	MTBE, acetaldehyde	420–550 (43–55)	Over 10× that of TiO_2	99 and 100
Fe_2O_3/S-doped TiO_2	Acetic acid, formic acid, acetaldehyde	>430	4× that of TiO_2	101
UO_2^{2+}/TiO_2	Ethanol	>420	—	102
Ag_3VO_4/TiO_2	Benzene	$900 > \lambda > 450$	4–7× that of TiO_2	103
Fe^{3+}/TiO_2	Toluene	455	QE ≈ 1%	104
Mn^{4+}/TiO_2 (001)	Acetaldehyde		QE 4.5%	105
$PO_4^{3-}/\alpha\text{-}Fe_2O_3$	Acetaldehyde	>420	—	106
$(Ag_{0.75}Sr_{0.25})(Nb_{0.75}Ti_{0.25})O_3$	Acetaldehyde	>400	QE 1.5% at $\lambda = 440$ nm	107
$Pt/V_2O_5/SmVO_4$	Acetone, methanol, 2-propanol, benzene	Xe lamp light transmitted through glass, 130–140 °C	7× that of TiO_2	108
$SO_4^{2-}/MoO_x/MgF_2$	Acetone	Same as above	6× that of TiO_2	109
$V_2O_5/GdVO_4$	Acetone	Same as above	Same as TiO_2	110
$Pt/NaNbO_{3-x}N_x$	2-Propanol	419	QE 8.6%	111
N-doped $(BiO)_2CO_3$	NO	>420	—	112
$WO_3/CaFe_2O_4$	Acetaldehyde	Y-43 filter (Asahi)	QY 1.5%	113

[a]Temperature is given if it is different from ambient.

TiO_2 in visible light, which is in terms of quantum yield just below 1%. Several photocatalysts such as Mn^{4+} and UO_2^{2+} doped TiO_2 demonstrated very high activity. We can conclude that the development of photocatalysts with useful characteristics such as high quantum yield in solar visible light, high stability in oxidation of diverse classes of air pollutants and absence of production of oxidation byproducts is still far from completion.

7.4 Kinetics of Photocatalytic Oxidation

Knowledge of the kinetics of photocatalytic oxidation is very important for development of insights into the oxidation process as well as for practical applications in air cleaning. Studies on kinetics include investigations on the time course of pollutant concentration during a reaction run in a static reactor as well as the effects of various conditions on the reaction rate.

Piera *et al.*[114] reported that kinetic curves of ethanol and acetaldehyde concentrations during photocatalytic oxidation of these individual compounds in a recirculation static reactor follows half-order kinetics according to the following equation:

$$\sqrt{C_0} - \sqrt{C} = kt \tag{7.16}$$

where C_0 is initial concentration, C is concentration at time t, and k is the reaction rate coefficient. Half-order kinetics can correspond to the case of dissociative adsorption of the molecule to be oxidized over two surface sites. Most probably, such kinetic curves result from the combined effects of reaction and adsorption.

Several studies deal with the effects of diverse factors on the rate of photocatalytic oxidation of gaseous air impurities. These factors include reaction temperature, pollutant initial concentration, initial air humidity, air flow rate or contact time, oxygen concentration and spectral irradiance at the photocatalyst position. Below we consider the influence of all these factors.

The temperature is not usually varied widely in research on photocatalytic oxidation. This is associated with the dominant role of thermal oxidation processes hindering the contribution of photo-initiated reactions at temperature above 150–200 °C. Most studies report an increase of the rate of photocatalytic oxidation with an increase in temperature. Over the temperature range 2–48 °C, the rate of photocatalytic oxidation of ethylene on TiO_2 increases.[115] The apparent activation energy was only 3.4 kcal mol^{-1}. The rate of oxidation is usually maximal at 80–120 °C over pure TiO_2. A further increase in temperature results in a decrease of the rate of photocatalytic oxidation of gaseous organic compounds.[79] However, such a decrease can be absent over platinized TiO_2 as was observed for acetone oxidation in the temperature range 40–140 °C.[116] The rate of oxidation of acetaldehyde increased but the contribution of photo-initiated processes towards the total oxidation rate can decrease and the predominant route of oxidation over Pt/TiO_2 is dark catalysis at high temperatures. For oxidation of ethanol vapors, the rate of oxidation over Pt/TiO_2 decreases when the temperature reaches about 100 °C but then increases again due to the growing contribution of thermal catalysis.[117]

The effect of pollutant concentration on the rate of its oxidation is usually well described by the Langmuir–Hinshelwood equation:[118]

$$W = \frac{kKC}{1+KC} \tag{7.17}$$

where W is the rate of oxidation, k is the surface reaction rate coefficient, K is the pollutant adsorption constant and C is the gas phase pollutant concentration. Sometimes, more complex multisite kinetic models are required to adequately describe the dependence of rate on concentration. This was the case for acetone oxidation over TiO_2 and Pt/TiO_2 when the concentration was varied over a very wide range. The following two-site Langmuir–Hinshelwood equation described the dependence very well:[116,119]

$$W = \frac{k_1 K_1 C}{1+K_1 C} + \frac{k_2 K_2 C}{1+K_2 C} \tag{7.18}$$

where k_1 and k_2 are the rate coefficients, and K_1 and K_2 are the adsorption constants for the sites of the first and the second type, respectively.

Demeestere *et al.*[120] suggested taking into account competition for the adsorption sites on TiO_2 surface of intermediate oxidation products and even CO_2 and describe TCE oxidation rate with the following equation:

$$W = \frac{kK_{\mathrm{TCE}}[\mathrm{TCE}]}{1+K_{\mathrm{TCE}}[\mathrm{TCE}]+K_{\mathrm{P}}[P]+K_{\mathrm{CO_2}}[\mathrm{CO_2}]} \tag{7.19}$$

where k is the rate coefficient, K_{TCE}, K_{P} and $K_{\mathrm{CO_2}}$ are adsorption constants of TCE, intermediate products and CO_2, respectively; [TCE], [P] and [CO_2] are the concentration of TCE, intermediate products and CO_2, correspondingly.

Photocatalytic oxidation of ethanol vapors over TiO_2 Hombifine N was studied in a flow-circulation reactor which provides excellent mixing of reaction media.[121] The reaction proceeded in two stages according to the following scheme:

$$\mathrm{CH_3CH_2OH} \xrightarrow[\mathrm{TiO_2}]{\mathrm{O_2}+h\nu} \mathrm{CH_3CHO} \xrightarrow[\mathrm{TiO_2}]{\mathrm{O_2}+h\nu} 2\mathrm{CO_2} \tag{7.20}$$

The rate of the first and the second stages was measured at different concentrations of ethanol and acetaldehyde. A three site kinetic model was required to describe the reaction rate of ethanol and acetaldehyde oxidation at their varied concentrations as the next equations demonstrate:

$$W_{\mathrm{E}} = \left(W_{\mathrm{E}}^0 + \frac{k_{\mathrm{E}}^0 K_{\mathrm{E}} C_{\mathrm{E}}}{1+K_{\mathrm{E}}C_{\mathrm{E}}+K_{\mathrm{A}}C_{\mathrm{A}}} \right) I^{\alpha} \tag{7.21}$$

$$W_{\mathrm{A}} = \left(\frac{k_{\mathrm{A1}}^0 K_{\mathrm{A1}} C_{\mathrm{A}}}{1+K_{\mathrm{A1}}C_{\mathrm{A}}} + \frac{k_{\mathrm{A}}^0 K_{\mathrm{A}} C_{\mathrm{A}}}{1+K_{\mathrm{A}}C_{\mathrm{A}}+K_{\mathrm{E}}C_{\mathrm{E}}} \right) I^{\beta} \tag{7.22}$$

where W_E is the ethanol oxidation rate, W_A is the acetaldehyde oxidation rate; W_E^0 is the rate of ethanol oxidation over sites accessible to ethanol molecules only, k_E^0 is the rate coefficient for ethanol oxidation on sites accessible to both ethanol and acetaldehyde molecules; K_E and K_A are adsorption constants for ethanol and acetaldehyde, respectively; C_E and C_A are ethanol and acetaldehyde concentration; k_{A1}^0 and K_{A1} are, respectively, rate coefficient and adsorption constant for acetaldehyde adsorption on sites accessible to acetaldehyde only. The exponents α and β equal to 0.64 and 0.46, respectively, for the dependence of ethanol and acetaldehyde oxidation rate on irradiance I (E s^{-1} cm^{-2}).

The influence of the oxygen concentration in the gas phase on the oxidation rate also follows the Langmuir–Hinshelwood model. Benzene conversion over TiO_2 increased when O_2 concentration increases and the conversion reaches a steady value at $[O_2] \geq 40\%$.[10] Shang *et al.*[122] reported a linear dependence of inverse reaction rate on inverse oxygen concentration. A very small Langmurian oxygen adsorption constant of 0.0024 m^3 g^{-1} was obtained for heptane photocatalytic oxidation.

Interestingly, photocatalytic decomposition of gaseous organic compounds can take place even in the complete absence of oxygen. Photocatalytic transformation of acetic acid over TiO_2 produces CO_2, CH_4, C_2H_6 and H_2O.[123] Apparently, two processes, photo-Kolbe decarboxylation and oxidation with extraction of surface oxygen from TiO_2 lattice, proceed in inert atmospheres. Undoubtedly, the oxidation process cannot be sustainable in anaerobic conditions.

The rate of oxidation grows and depends linearly on the photon flux striking the photocatalyst surface at low light intensity. At higher levels of irradiance, nonlinearity arises due to the increasing contribution of recombination of charge carriers. The effect of irradiance level, I, on the reaction rate coefficient, k, is described by the following equation:[122]

$$k = k_0 I^\alpha \tag{7.23}$$

where k_0 is rate coefficient independent of irradiance, α is the order coefficient, taking values from 0.5 to 1.

Ohko *et al.* reported that the rate of oxidation of gaseous 2-propanol levels off if mass transport limiting conditions are created, namely, pollutant concentration < 1 ppm and photon flux above 10^{15} cm^{-2} s^{-1}.[124] At an isopropanol concentration of 100 ppm, a photon flux above 2×10^{16} cm^{-2} s^{-1} was required to reach mass transport limiting conditions. However, if some kind of air mixing with linear speed of 0.5 m s^{-1} is applied the rate of mass transport increases so that even at an initial concentration 0.1 ppm a light flux up to 10^{16} cm^{-2} s^{-1} leads to a linear increase in the rate increasing photon flux.

A very interesting observation has been reported by Ohko *et al.* on acetaldehyde autooxidation.[44] They found that at very low irradiance levels acetaldehyde photocatalytic oxidation over TiO_2 can have a quantum yield significantly above 100%. It is suggested that the observed QY of about 180% at photon flux 10^{12} cm^{-2} s^{-1} originates from the following chain mechanism of acetaldehyde oxidation into acetic acid:

$$H_2O + h\nu \xrightarrow{TiO_2} H^+ + {}^{\bullet}OH \tag{7.24}$$

$$CH_3CHO + {}^{\bullet}OH \rightarrow CH_3CO^{\bullet} + H_2O \tag{7.25}$$

$$CH_3CO^{\bullet} + O_2 \rightarrow CH_3COOO^{\bullet} \tag{7.26}$$

$$CH_3COOO^{\bullet} + CH_3CHO \rightarrow CH_3COOOH + CH_3CO^{\bullet} \tag{7.27}$$

$$CH_3COOOH + CH_3CHO \rightarrow 2CH_3COOH \tag{7.28}$$

This mechanism can explain the formation of acetic acid from acetaldehyde with a QY up to 200%.

Relative humidity (RH) of ambient air can vary from a few per cent in dry winter time to about 100% in humid summer weather. Several of studies elucidate the effect of air relative humidity on the rate of photocatalytic oxidation. Obee and Hay report that an increase in water concentration from 1000 to 25 000 ppm steadily decreases the rate of ethylene photocatalytic oxidation.[115] The negative effect of water vapor on the rate of photocatalytic oxidation can be taken into account in the Langmuir–Hinshelwood model by assuming that water molecules are in competition for reactive adsorption centers. Accordingly, the following simplified equation describes the effect:

$$W = \frac{kKC}{1 + KC + K_w C_w} \tag{7.29}$$

where K_w is the water adsorption constant and C_w is water vapor concentration.[96] However, at very low air humidity with $[H_2O] < 1000$ ppm, the rate of photocatalytic oxidation decreases significantly.[53] Therefore, the above equation is valid for a limited range of $[H_2O]$. For photocatalytic oxidation of trichloroethylene, cyclohexene, benzene and toluene, increasing RH up to about 50% resulted in a steady increase of reaction rate.[125,126] Obviously, water increases the rate of oxidation if it is required for the formation of deep oxidation products as for trichloroethylene oxidation. The next equation demonstrates the stoichiometry of this reaction:

$$CCl_2CHCl + 1.5O_2 + H_2O = 2CO_2 + 3HCl \tag{7.30}$$

Photocatalyst deactivation is a widely observed phenomenon. It is usually seen for near room temperature photocatalytic oxidation of aromatic compounds but actually occurs for any air pollutants. At elevated temperature, thermal processes contribute and can cause strong catalyst deactivation even for compounds that do not deactivate TiO_2 at room temperature. Several studies report photocatalyst deactivation at ambient temperature. Lichtin and Sadeghi reported that benzene was always oxidized incompletely over TiO_2.[127] Toluene deactivated TiO_2 and SiO_2/TiO_2 photocatalysts during several hours of oxidation.[63] The process of deactivation is described by the exponential function shown below:

$$W = W_0 e^{-\alpha t} + W_F \tag{7.31}$$

where W is the rate of photocatalytic oxidation at time t, W_0 is the initial rate of photocatalytic oxidation, W_F is the rate of oxidation after complete deactivation, and α is the deactivation rate parameter. The parameter α decreases with an increase in air RH and increases with increasing pollutant concentration.[128]

Cao *et al.* reported partial deactivation of TiO_2 photocatalysts during photocatalytic oxidation of toluene[129] and 1-butene.[130] The rate of oxidation decreased about twice during the period of 1–4 h in dry air. In humidified air, the rate decreased for some of the TiO_2 photocatalysts. It was reported that toluene causes stronger TiO_2 deactivation than benzene and the deactivation is due to the deposition of carbon-containing compounds in amounts of up to similar to 0.9 carbon atoms per Ti site.[125] Ethanol caused a gradual decrease of photocatalytic activity of TiO_2 when several runs of its oxidation are carried out consecutively over the same photocatalyst.[114]

An interesting observation was reported for photocatalytic oxidation of a 250 ppm benzene airstream on titanate nanotubes mixed with carbon nanotubes.[131] Stable benzene conversion close to 5% and CO_2 production at 50 ppm was observed, which contrasts with the continuous decline in activity over TiO_2 Degussa P25.

Using infrared spectroscopy of the surface and temperature-programmed desorption it has been demonstrated that the deactivation is caused by the accumulation of the products of partial oxidation over the photocatalyst surface. An increase in air relative humidity alleviates the catalyst deactivation since water molecules displace partial oxidation products and decrease their formation rate. The photocatalyst activity can be recovered by irradiation of photocatalyst in pure air for time that is long enough for complete oxidation of all adsorbed organic compounds. Alternatively, irradiation in an ozone enriched humid air stream can be used and this method was reported to provide higher recovery of photocatalytic activity than UV irradiation alone.[27]

Photocatalytic oxidation of compounds containing N, P, S, Si and other elements usually exhibit catalyst deactivation. For example, deactivation was observed during triethylamine oxidation over TiO_2 P25 film.[53] Deactivation was more pronounced and proceeded faster at higher concentration of triethylamine and lower concentration of water vapor in air. Diethyl sulfide deactivated TiO_2 photocatalyst when 5–14 μmol of this organic compound, depending on its initial concentration, reacted over 1 m^2 of photocatalyst surface.[132] Octamethylcyclotetrasiloxane photocatalytic oxidation at an initial concentration of 500 ppb over TiO_2 loaded paper (Ahlstrom, France) resulted in a conversion decrease from 98% to 30% during three days of reaction.[133] Silicon-containing products were found on the surface of deactivated photocatalyst.

However, deactivation that was caused by accumulation of non-volatile inorganic products produced during photocatalytic oxidation of heteroatomic compounds is removed by prolonged irradiation of deactivated photocatalyst in pure air only partially. Inorganic and organic products are removed from deactivated TiO_2 by washing with water so that initial activity is restored almost completely.[132] Some loss in activity arises from etching of photocatalyst with accumulated inorganic acids like H_2SO_4, which are capable of dissolving TiO_2.

As pointed out above, increasing reaction temperature usually results in a multifold increase of the rate of photocatalytic oxidation. Excessive increase of reaction temperature can result in photocatalyst deactivation as several studies report. Acetone photocatalytic oxidation is accompanied by TiO_2 deactivation at elevated temperature. Increasing the water content shifts deactivation onset from 80 to 120 °C, while a rise of acetone concentration from 50 to 2000 ppm shifts the deactivation onset from 120 to 100 °C.[134] The deactivation was accompanied by dark discoloration of TiO_2. It was demonstrated that keeping the photocatalyst at high temperature without acetone does not cause deactivation.

Luo and Falconer discovered that acetone and acetaldehyde undergo oligomerization over TiO_2 P25 surface at temperatures close to 400 °C and higher.[135] Aldol condensation and cyclization processes catalyzed by surface acid centers give rise to diverse products of lower volatility which accumulate over TiO_2 surface and contribute to both an increase of photocatalytic activity at moderate temperature and photocatalyst deactivation at $T > 100$ °C. El-Maazawi *et al.* discovered that aldol condensation of acetone over TiO_2 P25 proceeds even at room temperature in the absence of moisture.[136]

We can see that the kinetics of photocatalytic oxidation have been investigated relatively thoroughly. Clearly, the best photocatalysts are to have a very high pollutant adsorption constant together with a high enough reaction rate coefficient.

7.5 Photocatalytic Destruction of Microbiological Objects

Reactive oxygen species as well as holes and electrons produced in irradiated photocatalysts are toxic for such microbiological organisms as bacteria, viruses and fungi. Therefore, photocatalytic processes are applicable for disinfection of water, air and surfaces. The subject of photocatalytic disinfection of air has already been reviewed, for example, by Chen *et al.*[137] In this section, brief but experimentally detailed, summaries of recent examples are given for photocatalysis disinfection capabilities and air purification from biological objects.

The vast majority of studies on photocatalytic disinfection deals with processes taking place in aqueous suspensions. For example, Sousa *et al.*[138] demonstrated inactivation of antibiotic resistant strains of *Escherichia coli*

(*E. coli*) under UVA fluorescent irradiation in suspensions of TiO_2 P25 (Evonik, Germany) and over paint coatings. The paint was formulated from TiO_2 P25, $CaCO_3$ and silicates extenders, dispersing agents, a binder, thickeners and a polymer extender. The large fraction of TiO_2 ensured a porous structure of the film formed from this paint. Prior activation of the film with UVA irradiation for 24 h was needed to develop photocatalytic activity. Irradiation of bacteria suspensions for 40 min both loaded over paint film and in TiO_2 suspension resulted in 98.7–99.9% viability loss, with the photocatalyst suspension being more active. The dark loss was only 3–5% while UVA irradiation without photocatalyst addition resulted in 43.7–56.8% inactivation.

However, to successfully demonstrate photocatalytic disinfection of air, the reaction has to be carried out at the photocatalyst–air interface. Several works deal with inactivation and mineralization of bacteria loaded in dried state over photocatalytic coatings. Such studies are of special importance for demonstration of inactivation and mineralization of microorganisms in air streams. Eight hours of UVA irradiation resulted in about 99% inactivation of *Bacillus subtilis* spores loaded over TiO_2 Degussa P25 film.[139] From 0.19×10^{14} to 2.27×10^{14} photons were needed to inactivate each CFU. UVA irradiation for 30 min of slides coated with TiO_2 and Pt/TiO_2 resulted in inactivation of *Mycobacterium smegmatis* up to 98% and *Vaccinia* virus up to 96%, with platinized TiO_2 being consistently more active than pure TiO_2.[140]

Complete mineralization of bacterial mass into CO_2 over TiO_2 and Pt/TiO_2 coated slides has been demonstrated by Kozlova *et al.* for *Mycobacterium smegmatis*.[141] Inactivation of *Bacillus thuringiensis*, vaccinia and influenza A H3N2 viruses was noted to proceed up to 99.8% after 30 min UVA irradiation. Platinization was beneficial for both mineralization and inactivation. Very long irradiation for over 24 h was required to achieve complete mineralization of bacteria into CO_2. The above studies clearly testify to the ability of UVA irradiated TiO_2 and modified TiO_2 to kill all kinds of microorganisms including highly resistant bacterial spores.

Even objects as large as pollen can be destroyed using photocatalytic oxidation. *Platanus* and *Cedrus* pollen with typical sizes 20 and 80 μm, respectively, underwent inactivation and almost complete destruction under UVA irradiation for 24 h in the presence of TiO_2 photocatalyst film.[142] This result makes photocatalytic treatment a prospective means for removal of allergenic particles from air and their complete destruction. The disadvantage of this first research on pollen was that conversion into complete oxidation products such as CO_2 was not demonstrated and remains a subject of future research.

7.6 Reactors for Photocatalytic Air Treatment

This section considers reactors that can be used for investigations on photocatalytic processes as well as for solving practical tasks of photocatalytic air purification in small rooms. Diverse types of reactors are used for research purposes. Table 7.4 exemplifies reactor types and construction details for gas phase photocatalytic reactions. Two major classes of reactors are static (batch) and flow reactors. The literature shows that annular flow reactors are most frequently used in research. The annular design is so prominent due to the high degree of utilization of the light of a lamp situated at the center of the reactor. It is important to keep the gap between the annulus equal to or below 5 mm because larger hydraulic diameters caused a significant decrease of pollutant conversion in the annular reactor due to mass transport resistances.[143]

Figure 7.1(A) shows a scheme of cross section of a flow annular reactor. Because of the low internal volume and required residence time of tens of seconds a typical flow rate in such a reactor should be below 1 L min^{-1} to obtain a conversion above 50%. For this reason, annular flow reactors are used for research purposes only. However, if one loads photocatalyst over a proper support and places this loaded photocatalyst between the annuli it is possible to attain flow rate up to tens of m^3 per hour.[152] Figure 7.1(B) reveals the design of such a reactor where photocatalyst can be loaded onto quartz tubes or granules or any other support transparent to lamp irradiation. Owing to the good contact between the photocatalyst and the passing air high flow rates and powerful light sources can be used.

For reaction kinetics studies, a flow-circulation reactor, where the reaction mixture is recirculated tens of times faster than it is fed, is usually considered to be very convenient because it simulates very closely the ideal stirred-tank

Table 7.4 Types of research photocatalytic reactors employed for air treatment.

Type	Construction and geometry	References
Static fixed bed	Glass plate coated with Pc film	52 and 57
	Cylindrical vessel coated with Pc film inside	17
Static recirculation fixed bed	Monolith coated with Pc film	38
	Pc powder spread on flat glass frit	114
	Pc coated glass rings in glass tube	144
Static fluidized bed	Annular with lamp on the axis, recirculation mode	145
Flow fixed bed	Annular, 1–5 mm gap between cylindrical lamp and Pc layer	14,39,40,77,143,146
	Tubular loaded with granules surrounded by UV lamps	10,62,147
	Flat plate covered with Pc film, single pass	16,115,148
	Optical fibers coated with 1.5 μm TiO_2 film, inserted into a tube	149
	Zigzag shaped rectangular channel with Pc powder on the bottom	120
Flow fluidized bed	Flattened tube	55 and 150
Vibrofluidized bed	Cylindrical tube, flow mode	151 and 152
Flow-circulation reactor	Flat glass loaded with Pc film reactor included in a flow-circulation loop	121

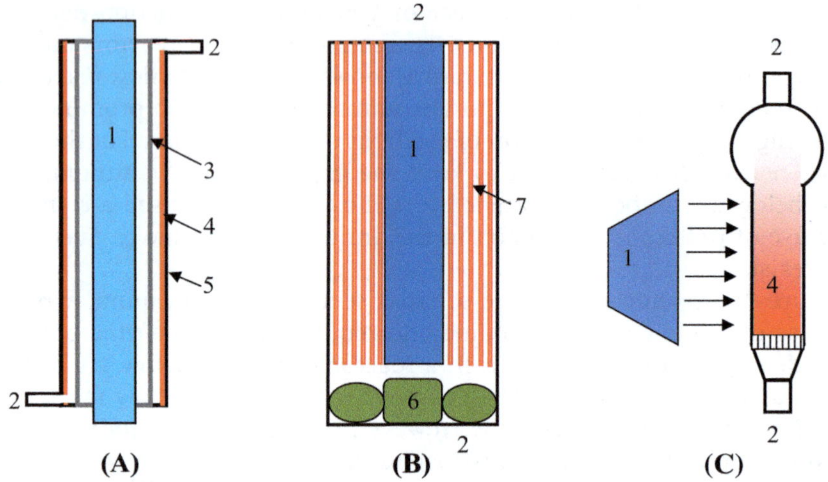

Figure 7.1 Cross-sectional view of typical research photocatalytic reactors: (A) flow annular reactor; (B) flow annular reactor with filled gap; (C) fluidized bed reactor. (1) UV lamp, (2) air inlet and outlet, (3) inner annulus typically made of quartz, (4) photocatalyst layer, (5) outer annulus, (6) fan, (7) UV translucent carriers with photocatalyst.

reactor. Fluidized and vibrofluidized bed reactors also have this advantage due to the intensive movements of photocatalyst particles in the flowing gas. Figure 7.1(C) demonstrates a fluidized photocatalytic reactor that is irradiated externally with a flat beam of light. Granules of photocatalyst with a typical size of tens of micrometers are loaded over a glass frit and fluidized by passing an appropriate flux of air. Typically, to obtain high conversion in a fluidized bed reactor one needs to use a very powerful light source (~100 to 1000 W) because the fluidization process requires high flow rates. In a vibrofluidized photocatalytic reactor,[152] one can obtain very high conversions with a light source of a few watts because no high flow rate is required for fluidization.[152] As a result of granule rotation, fluidized and vibrofluidized bed reactors realize the periodic photocatalyst irradiation regime that possesses advantages of more economic utilization of photogenerated charges which can be consumed during the dark periods of the photocatalyst.

Special requirements for photocatalytic reactions must be met for air purification photocatalytic reactors. A good contact of the light quanta, photocatalyst and molecules of pollutants is a pre-requisite for the good performance of a photocatalytic reactor for air purification. The choice of irradiation geometry, contact time and photocatalyst–support composite determines to a great extent the reaction rate and its stability. Inorganic porous photocatalyst support media are preferable to organic fibers due to the inherent high stability of inorganic supports towards oxidation. For example, glass fiber filter loaded by dip coating with TiO_2 and irradiated with fluorescent 6 W UVA lamps was used for removal of NO and BTEX at ppb level.[153] A residence time of several minutes was needed to obtain conversion levels above 50%. Park and Kim reported a photocatalytic reactor consisting of four felted fabric sheets coated with TiO_2 and three UVA or UVC lamps between the sheets.[28] They obtained about 10% toluene conversion with a toluene initial concentration of 7 ppm and flow rate of 0.5 m^3 h^{-1}. Yang *et al.* reported a photocatalytic reactor with TiO_2 photocatalyst loaded onto foam nickel tubes with one end of each tube being sealed.[154] UV lamps (254 nm) were located at the center of each tube.

Slimen *et al.* commercialized their Ti mesh based photocatalyst support with loaded TiO_2 photocatalyst in the prototype air purifier Titan Power developed with the U-Vix corporation.[37] This air purifier has six UVC lamps each surrounded by TiO_2 loaded Ti mesh.

Figure 7.2 shows a photocatalytic air cleaning system certified for medical use. This unit has dimensions of 470 × 170 × 160 mm^3 and can be attached to a wall. Two ventilators situated at the air exits of the purifier provide a flow rate through the cylindrical photocatalytic filter from 40 to 55 m^3 h^{-1} in night time and daytime regimes, respectively. A UVA fluorescent lamp situated at the axis of the felt photocatalytic filter with loaded TiO_2 photocatalyst has a power of 24 W. A dust filter surrounds the photocatalytic filter and removes major mechanical air contaminants before air enters the photocatalytic layer. Two more commercially produced air purifiers with analogous design and U-shaped and annular fluorescent lamps are described in the literature.[152] These units are commercially produced by enterprises in Moscow, Lipetsk and Novosibirsk (Russian Federation).

Modesto *et al.*[155] constructed a flow photocatalytic reactor that was tested for air purification from nebulized bacteria. The reactor is a rectangular tube with cross-section 10 × 4.5 cm^2 and length 42 cm. The photocatalyst was coated over six glass plates which were attached to the walls of the rectangular tube. Four UVA fluorescent lamps of 4 W each were distributed uniformly along the length of the tube. The air residence time inside the tube during the air purification experiments was 14 s. Pure TiO_2 as well as TiO_2 loaded with Ag, Pd and Fe was used for the glass plate coating. Suspensions of *Escherichia coli*, *Staphylococcus aureus* or *Bacillus subtilis* with a concentration of

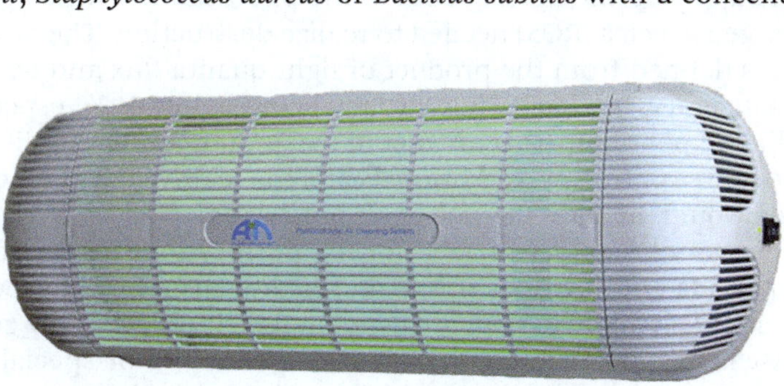

Figure 7.2 Photocatalytic air purifier L-5524M manufactured by Ambilife Ltd. (Russia).

10^5 CFU mL^{-1} were nebulized at the rate of 0.25 mL min^{-1} into the air flow entering the reactor at the rate of 8 L min^{-1}. Due to the design of the reactor, relatively poor contact between the air flow and the photocatalyst was realized. However, inactivation of 35–50% of bacteria was obtained for pure TiO_2 photocatalyst and as high as 47–94% for metal loaded photocatalysts.

To improve the microorganisms' contact with the irradiated photocatalyst and their inactivation efficiency it was suggested to use combination of a photocatalytic filter with negative air ionization.[156] A special negative ion generator produced anions in air with a concentration of about 5×10^5 cm^{-3}. The negative ions were injected from the thin wires into the contaminated air stream entering into the photocatalytic filter block. Photocatalyst (TiO_2 Degussa P25) was loaded over the mechanical particle fiber filter. Irradiation was provided by an 8 W UVA lamp located above the photocatalytic filter. Aerosols of *Escherichia coli*, *Candida famata* and λ virus phage were created and treated with and without negative ion addition. The efficiency of microorganism removal always increased as a result of turning on the negative ion generator. The highest inactivation of about 90% was observed for λ virus phage followed by *C. famata* and *E. coli*.

Paschoalino and Jardim constructed a photocatalytic reactor for indoor air disinfection based on an annular space around a cylindrical UV lamp.[157] Photocatalytic TiO_2 was coated onto the internal wall of PVC tube surrounding the 30 W germicidal lamp. Air flow of m^3 h^{-1} was circulated through the reactor with a fan. The reactor was tested in a meeting and coffee room of 67 m^3 volume. The reactor demonstrated higher inactivation with photocatalyst and the germicidal lamp on than with UV lamp alone. However, the efficiency for fungi removal decreased when the reactor was operated for the second time. It is reported that the activity was restored after washing the photocatalyst in 30% H_2O_2.

7.7 Combined Methods of Air Purification

The rate of photocatalytic destruction of air pollutants relies entirely on the rate of supplying the light quanta to the photocatalyst which generates reactive oxygen species (ROS) needed to realize destruction. The destruction rate, thus, is defined from the product of light quanta flux and destruction quantum efficiency, the later value usually being below 50%, especially for strongly diluted pollutants that are met in real indoor air. The addition of an alternative source of ROS and additional catalysts to drive the reactions of air purification would increase the rate of air purification.[158]

The choice of precursors of ROS for air purification is limited to gaseous compounds and includes ozone and hydrogen peroxide vapors. Very active species like atomic oxygen are generated inside electrically generated plasma discharges.[158] To destroy diluted air pollutants of special interest is the non-thermal plasma (NTP) which produces oxidizing species such as atomic oxygen and ozone with a high energetic efficiency. NTP usually employs two different geometries called dielectric barrier discharge (DBD) and corona discharge (CD). DBD is generated by applying alternating high voltage between electrodes, one or two of which are covered with insulator. CD is produced at the sharp places of a coronating electrode, which can be fed with direct or alternating as well as pulsed high voltage.

It is possible to make several different geometric combinations of photocatalytic oxidation, non-thermal plasma and thermal (dark) catalyst. Notably, the presence of the dark catalyst or destroying ozone photocatalyst is vital in order to destroy the ozone produced by NTP before air is released into the building or atmosphere. Below are listed the combinations which were applied in the literature:

1. Photocatalyst is located inside the zone of DBD or CD and should be irradiated with a lamp. While some quantity of light is always emitted by NTP, experiments demonstrated that this light is not enough to drive a photocatalytic reaction at a rate needed for air treatment. Due to the low contact time, ozone is usually not destroyed efficiently over the photocatalyst and additional ozone destruction catalyst is required for ozone elimination.
2. Photocatalyst with a lamp that is not in the discharge zone and placed upstream of the discharge. In this case additional dark ozone destruction catalyst is required at the place downstream NTP to avoid ozone release.
3. Photocatalyst with a lamp that is not in the discharge and placed downstream of the NTP discharge zone. In this case the photocatalytic section can efficiently utilize the ozone produced in NTP for oxidation of pollutants and simultaneously decompose O_3 into O_2.
4. Dark ozone destruction catalyst can be placed upstream the discharge zone while the photocatalytic filter can be located before or after the discharge. This geometry can be used for research setups but in practice is inefficient for air purification.
5. The ozone destruction catalyst is placed downstream the discharge before the photocatalytic filter.
6. The ozone destruction catalyst is placed downstream the discharge after the photocatalytic filter.

The synergistic effect of photocatalytic oxidation and DBD was studied for the case of acetone vapor photocatalytic oxidation.[159] When DBD served as a source of UV excitation for TiO_2 photocatalyst no additional reaction was observed since atmospheric DBD is a too weak a source of UV light. Combining photocatalytic oxidation and DBD resulted in 1.5-times increased rate compared to the sum of these processes applied separately.

Besides electrical discharges, ozone is produced as a result of short-wavelength irradiation of air stream. Ozone enriched air produced by such an ozone generator was used for downstream located photocatalytic air treatment to remove toluene vapor.[160] MnO_2 loaded on TiO_2–SiO_2 support served as ozone destruction catalyst. A synergistic effect was observed when ozone destruction catalyst was mixed with TiO_2 photocatalyst.

Additions of ozone into the flow reactor feed was also practiced for photocatalytic oxidation of hexane.[161] Addition of ozone led to only a ~10% increase in the conversion of hexane compared to photocatalytic oxidation. An increase of ozone concentration from 0 to 14 ppm resulted in a 1.5-fold increase of toluene oxidation rate in a flow reactor.[162]

In the work of Yuan *et al.*,[163] however, the major role in photocatalytic destruction of acetaldehyde vapors was played by ozone. The best catalyst was TiO_2/Mn-ZSM-5, which produced a conversion 55% without UV light and 79% in the presence of both 254 nm UV light and 131 ppm of ozone.

Jeong *et al.* used an UVC lamp with 254 and 185 nm radiation for excitation of a TiO_2 coated annular reactor in the process of toluene and benzene photodegradation.[164] It is reported that irradiation with 254 + 185 nm lamp resulted in 99.9% conversion of toluene and 94% conversion of benzene while 365 or 254 nm irradiation of the same electric power (4 W) resulted in 72–73% and 68–71% conversion of toluene and benzene, correspondingly. Moreover, photocatalyst deactivation was not observed for the 254 + 185 nm irradiation even for a toluene concentration 10 ppm, which led to strong TiO_2 deactivation under 365 or 254 nm illumination. CO_2, CO and water soluble organic compounds were registered as products of toluene degradation under 185 + 254 nm irradiation with CO_2 accounting for 49–79%. Ozone produced under 185 nm irradiation can be destroyed to give oxygen with an ozone destruction MnO_2 catalyst. Yang *et al.*[154] discovered a synergistic effect of vacuum UV lamp (6% 185 + 80% 254 nm output) and photocatalytic oxidation with 254 nm light source for the case of air purification from formaldehyde. Photocatalyst TiO_2 was loaded onto foam nickel tubes with a 254 nm lamp in the axis. Conversion of 190 ppb formaldehyde was increased up to three-fold with the vacuum UV lamp turned on. The effect is attributed to ozone produced (100–300 ppb) by the vacuum UV lamp.

Kataoka *et al.* examined the combination of photocatalytic reaction with microwave irradiation.[165] A positive effect of microwave heating on the rate of ethylene oxidation was reported only for a RH level of 15% but not for RH 5% and 0%. The effect is caused by removal of excess surface water under the action of microwave irradiation.

Hydrogen peroxide is a powerful oxidant that can facilitate photocatalytic oxidation. The effect of gas phase H_2O_2 in concentrations up to 1000 ppm has been evaluated for photocatalytic oxidation of acetone and benzene vapors over TiO_2 Degussa P25.[166] It turned out that photocatalytic oxidation of acetone is suppressed by additions of gaseous H_2O_2 into the flow reactor feed. This was due to the gas phase reaction between acetone and H_2O_2 with the formation of nonvolatile peroxide. Photocatalytic oxidation of benzene in concentrations close to 130 ppm was greatly accelerated in the steady state upon addition of 1000 ppm gaseous H_2O_2. The rate increased by a factor of 6.5 and photocatalyst deactivation as well as catalyst dark color did not develop in the presence of gaseous hydrogen peroxide.

7.8 Conclusions

The process of gas phase photocatalytic oxidation of air contaminants has been developed for over 40 years. A great deal of insight has already been obtained for the field to be applicable for air purification. It is hoped that the development of novel materials to be used as photocatalysts and to build nanoscale architectures to obtain the desired course of photocatalytic transformations will be the way to solve diverse problems such as fast photocatalyst deactivation, insignificant selectivity towards deep oxidation products, low sensitivity and quantum yields in visible and solar light and very low reaction rates.

List of Abbreviations and Designations

ACF	Activated carbon fibers
CD	Corona discharge
CFU	Colony forming unit
BTEX	Benzene, toluene, ethylbenzene, *o*-xylene
DBD	Dielectric barrier discharge
HEPA	High efficiency particles arresting
MTBE	Methyl tertiary butyl ether
NTP	Non-thermal plasma
Pc	Photocatalyst
Pc*	Excited photocatalyst
QE	Quantum efficiency
QY	Quantum yield
ROS	Reactive oxygen species
SEM	Scanning electron microscopy
TCE	Trichloroethylene
UV	Ultraviolet
UVA	Ultraviolet light in the A range (330–400 nm)
[X]	Concentration of species X in the gas phase
λ	Wavelength

Acknowledgements

Work on this chapter has been supported in part *via* base budget project V.45.3.2, grant of Presidium RAS 24.49 and the Russian Foundation for Basic Research projects 13-08-01286 and 15-08-01936.

References

1. M. Lim, Y. Zhou, L. Wang, V. Rudolph and G. Q. (Max) Lu, *Asia-Pac. J. Chem. Eng.*, 2009, **4**, 387–402.
2. V. N. Parmon, *Catal. Today*, 1997, **39**, 137–144.
3. E. Becquerel, *Comptes Rendus*, 1839, **9**, 561–567.

4. C. Renz, *Helv. Chim. Acta*, 1921, **4**, 961–968.
5. A. Berthoud, *Trans. Faraday Soc.*, 1926, **21**, 554.
6. Y. Paz, *Appl. Catal., B*, 2010, **99**, 448–460.
7. X. Fu, L. A. Clark, W. A. Zeltner and M. A. Anderson, *J. Photochem. Photobiol., A*, 1996, **97**, 181–186.
8. M. R. Nimlos, E. J. Wolfrum, M. L. Brewer, J. A. Fennell and G. Bintner, *Environ. Sci. Technol.*, 1996, **30**, 3102–3110.
9. I. Sopyan, M. Watanabe, S. Murasawa, K. Hashimoto and A. Fujishima, *J. Photochem. Photobiol., A*, 1996, **98**, 79–86.
10. H. Einaga, S. Futamura and T. Ibusuki, *Phys. Chem. Chem. Phys.*, 1999, **1**, 4903–4908.
11. D. Park, *J. Catal.*, 1999, **185**, 114–119.
12. G. Martra, S. Coluccia, L. Marchese, V. Augugliaro, V. Loddo, L. Palmisano and M. Schiavello, *Catal. Today*, 1999, **53**, 695–702.
13. R. M. Alberici and W. F. Jardim, *Appl. Catal., B*, 1997, **14**, 55–68.
14. M. C. Canela, R. M. Alberici, R. C. R. Sofia, M. N. Eberlin and W. F. Jardim, *Environ. Sci. Technol.*, 1999, **33**, 2788–2792.
15. A. V. Vorontsov, E. V. Savinov, L. Davydov and P. G. Smirniotis, *Appl. Catal., B*, 2001, **32**, 11–24.
16. I. Salvadó-Estivill, D. M. Hargreaves and G. Li Puma, *Environ. Sci. Technol.*, 2007, **41**, 2028–2035.
17. N. Fukami, M. Yosida, B.-D. Lee, K. Taku and M. Hosomi, *Chemosphere*, 2001, **42**, 345–350.
18. O. Dhennezel and D. Ollis, *J. Catal.*, 1997, **167**, 118–126.
19. N. N. Lichtin, M. Avudaithai, E. Berman and A. Grayfer, *Sol. Energy*, 1996, **56**, 377–385.
20. N. Negishi, K. Takeuchi and T. Ibusuki, *J. Mater. Sci.*, 1998, **33**, 5789–5794.
21. H. Yamashita, Y. Ichihashi, M. Anpo, M. Hashimoto, C. Louis and M. Che, *J. Phys. Chem.*, 1996, **100**, 16041–16044.
22. M. Anpo, M. Matsuoka and H. Yamashita, *Catal. Today*, 1997, **35**, 177–181.
23. M. Anpo, S. G. Zhang, H. Mishima, M. Matsuoka and H. Yamashita, *Catal. Today*, 1997, **39**, 159–168.
24. K.-C. Cho, K.-C. Hwang, T. Sano, K. Takeuchi and S. Matsuzawa, *J. Photochem. Photobiol., A*, 2004, **161**, 155–161.
25. I. R. Subbotina, B. N. Shelimov, V. B. Kazansky, A. A. Lisachenko, M. Che and S. Coluccia, *J. Catal.*, 1999, **184**, 390–395.
26. T. Sano, N. Negishi, D. Mas and K. Takeuchi, *J. Catal.*, 2000, **194**, 71–79.
27. W. Wang, L.-W. Chiang and Y. Ku, *J. Hazard. Mater.*, 2003, **101**, 133–146.
28. O.-H. Park and C.-S. Kim, *J. Appl. Polym. Sci.*, 2004, **91**, 3174–3179.
29. C. Raillard, V. Héquet, P. Le Cloirec and J. Legrand, *J. Photochem. Photobiol., A*, 2004, **163**, 425–431.
30. S. Adjimi, N. Sergent, J.-C. Roux, F. Delpech, M. Pera-Titus, K. Chhor, A. Kanaev and P.-X. Thivel, *Appl. Catal., B*, 2014, **154–155**, 123–133.
31. F. Thevenet, O. Guaïtella, J. M. Herrmann, A. Rousseau and C. Guillard, *Appl. Catal., B*, 2005, **61**, 58–68.

32. P.-F. Biard, A. Bouzaza and D. Wolbert, *Appl. Catal., B*, 2007, **74**, 187–196.
33. Y. Dong, Z. Bai, R. Liu and T. Zhu, *Atmos. Environ.*, 2007, **41**, 3182–3192.
34. T. Reztsova, C.-H. Chang, J. Koresh and H. Idriss, *J. Catal.*, 1999, **185**, 223–235.
35. T. Guo, Z. Bai, C. Wu and T. Zhu, *Appl. Catal., B*, 2008, **79**, 171–178.
36. S. V. Spitskiy and A. D. Parshina, *Russ. J. Gen. Chem.*, 2010, **80**, 2762–2767.
37. H. Slimen, T. Ochiai, K. Nakata, T. Murakami, A. Houas, Y. Morito and A. Fujishima, *Ind. Eng. Chem. Res.*, 2012, **51**, 587–590.
38. M. Sauer and D. F. Ollis, *J. Catal.*, 1994, **149**, 81–91.
39. X. Fu, W. A. Zeltner and M. A. Anderson, *Appl. Catal., B*, 1995, **6**, 209–224.
40. S. A. Larson, J. A. Widegren and J. L. Falconer, *J. Catal.*, 1995, **157**, 611–625.
41. O. d' Hennezel, P. Pichat and D. F. Ollis, *J. Photochem. Photobiol., A*, 1998, **118**, 197–204.
42. D. S. Muggli, J. T. McCue and J. L. Falconer, *J. Catal.*, 1998, **173**, 470–483.
43. T. Noguchi, A. Fujishima, P. Sawunyama and K. Hashimoto, *Environ. Sci. Technol.*, 1998, **32**, 3831–3833.
44. Y. Ohko, D. A. Tryk, K. Hashimoto and A. Fujishima, *J. Phys. Chem. B*, 1998, **102**, 2699–2704.
45. C. Wu, Y. Yue, X. Deng, W. Hua and Z. Gao, *Catal. Today*, 2004, **93–95**, 863–869.
46. X. Gao, J. Li and W. Gao, *Colloid J.*, 2008, **70**, 392–395.
47. F. Thevenet, C. Guillard and A. Rousseau, *Chem. Eng. J.*, 2014, **244**, 50–58.
48. Q. Xiang, K. Lv and J. Yu, *Appl. Catal., B*, 2010, **96**, 557–564.
49. Y. Luan, L. Jing, Y. Xie, X. Sun, Y. Feng and H. Fu, *ACS Catal.*, 2013, **3**, 1378–1385.
50. M.-V. Sofianou, M. Tassi, N. Boukos, S. Thanos, T. Vaimakis, J. Yu and C. Trapalis, *Catal. Today*, 2014, **230**, 125–130.
51. B. R. Weinberger and R. B. Garber, *Appl. Phys. Lett.*, 1995, **66**, 2409.
52. A. Hattori, K. Shimoda, H. Tada and S. Ito, *Langmuir*, 1999, **15**, 5422–5425.
53. A. Huang, L. Cao, J. Chen, F.-J. Spiess, S. L. Suib, T. N. Obee, S. O. Hay and J. D. Freihaut, *J. Catal.*, 1999, **188**, 40–47.
54. M. C. Blount, D. H. Kim and J. L. Falconer, *Environ. Sci. Technol.*, 2001, **35**, 2988–2994.
55. L. A. Dibble and G. B. Raupp, *Environ. Sci. Technol.*, 1992, **26**, 492–495.
56. M. Takeuchi, M. Hidaka and M. Anpo, *J. Hazard. Mater.*, 2012, **237–238**, 133–139.
57. S. Sampath, H. Uchida and H. Yoneyama, *J. Catal.*, 1994, **149**, 189–194.
58. C. H. Ao and S. C. Lee, *J. Photochem. Photobiol., A*, 2004, **161**, 131–140.
59. J. Araña, J. M. Doña-Rodríguez, C. G. i Cabo, O. González-Díaz, J. A. Herrera-Melián and J. Pérez-Peña, *Appl. Catal., B*, 2004, **53**, 221–232.
60. R. Portela, R. F. Tessinari, S. Suárez, S. B. Rasmussen, M. D. Hernández-Alonso, M. C. Canela, P. Ávila and B. Sánchez, *Environ. Sci. Technol.*, 2012, **46**, 5040–5048.

61. A. Hattori, Y. Tokihisa, H. Tada, N. Tohge, S. Ito, K. Hongo, R. Shiratsuchi and G. Nogami, *J. Sol-Gel Sci. Technol.*, 2001, **22**, 53–61.
62. X. Fu, L. A. Clark, Q. Yang and M. A. Anderson, *Environ. Sci. Technol.*, 1996, **30**, 647–653.
63. R. Méndez-Román and N. Cardona-Martínez, *Catal. Today*, 1998, **40**, 353–365.
64. Z. Liu, F. Chen, P. Fang, S. Wang, Y. Gao, F. Zheng, Y. Liu and Y. Dai, *Appl. Catal., A*, 2013, **451**, 120–126.
65. J. C. Yu, J. Lin and R. W. Kwok, *J. Phys. Chem. B*, 1998, **102**, 5094–5098.
66. Q. Huang, W. Ma, X. Yan, Y. Chen, S. Zhu and S. Shen, *J. Mol. Catal. Chem.*, 2013, **366**, 261–265.
67. L. Jing, Y. Cao, H. Cui, J. R. Durrant, J. Tang, D. Liu and H. Fu, *Chem. Commun.*, 2012, 10775.
68. D. S. Muggli and L. Ding, *Appl. Catal., B*, 2001, **32**, 181–194.
69. S. Yamazaki, N. Fujinaga and K. Araki, *Appl. Catal., A*, 2001, **210**, 97–102.
70. J. C. Yu, J. Yu and J. Zhao, *Appl. Catal., B*, 2002, **36**, 31–43.
71. E. Barraud, F. Bosc, D. Edwards, N. Keller and V. Keller, *J. Catal.*, 2005, **235**, 318–326.
72. K. Pruethiarenun, T. Isobe, S. Matsushita and A. Nakajima, *Appl. Catal., A*, 2012, **445–446**, 274–279.
73. A. Zielińska-Jurek and A. Zaleska, *Catal. Today*, 2014, **230**, 104–111.
74. J. Hernández-Fernández, A. Aguilar-Elguezabal, S. Castillo, B. Ceron-Ceron, R. D. Arizabalo and M. Moran-Pineda, *Catal. Today*, 2009, **148**, 115–118.
75. J. H. Byeon and Y.-W. Kim, *ACS Appl. Mater. Interfaces*, 2014, **6**, 763–767.
76. G. Schubert, A. Gazsi and F. Solymosi, *J. Catal.*, 2014, **313**, 127–134.
77. C. Belver, M. J. López-Muñoz, J. M. Coronado and J. Soria, *Appl. Catal., B*, 2003, **46**, 497–509.
78. M. Sadeghi, W. Liu, T. G. Zhang, P. Stavropoulos and B. Levy, *J. Phys. Chem.*, 1996, **100**, 19466–19474.
79. J. L. Falconer and K. A. Magrini-Bair, *J. Catal.*, 1998, **179**, 171–178.
80. A. V. Vorontsov, E. N. Savinov and Z. Jin, *J. Photochem. Photobiol., A*, 1999, **125**, 113–117.
81. J. Lee and W. Choi, *J. Phys. Chem. B*, 2005, **109**, 7399–7406.
82. E. A. Kozlova and A. V. Vorontsov, *High Energy Chem.*, 2008, **42**, 583–585.
83. D. Kozlov and A. Vorontsov, *J. Catal.*, 2008, **258**, 87–94.
84. H. Einaga, T. Ibusuki and S. Futamura, *Environ. Sci. Technol.*, 2004, **38**, 285–289.
85. M. N. Gómez-Cerezo, M. J. Muñoz-Batista, D. Tudela, M. Fernández-García and A. Kubacka, *Appl. Catal., B*, 2014, **156–157**, 307–313.
86. M. J. Muñoz-Batista, M. N. Gómez-Cerezo, A. Kubacka, D. Tudela and M. Fernández-García, *ACS Catal.*, 2014, **4**, 63–72.
87. G. Zuo, Z. Cheng, H. Chen, G. Li and T. Miao, *J. Hazard. Mater.*, 2006, **128**, 158–163.
88. J. Blanco, P. Avila, A. Bahamonde, E. Alvarez, B. Sánchez and M. Romero, *Catal. Today*, 1996, **29**, 437–442.

89. K. Y. Song, M. K. Park, Y. T. Kwon, H. W. Lee, W. J. Chung and W. I. Lee, *Chem. Mater.*, 2001, **13**, 2349–2355.
90. V. Keller, P. Bernhardt and F. Garin, *J. Catal.*, 2003, **215**, 129–138.
91. A. Kubacka, G. Colón and M. Fernández-García, *Catal. Today*, 2009, **143**, 286–292.
92. J. Yang, X. Zhang, H. Liu, C. Wang, S. Liu, P. Sun, L. Wang and Y. Liu, *Catal. Today*, 2013, **201**, 195–202.
93. J. Lin and J. C. Yu, *J. Photochem. Photobiol., A*, 1998, **116**, 63–67.
94. S. Liang, L. Wen, G. Liu, S. Zhu, R. Yuan and L. Wu, *Catal. Today*, 2013, **201**, 175–181.
95. W. Xie, Y. Li, W. Shi, L. Zhao, X. Zhao, P. Fang, F. Zheng and S. Wang, *Chem. Eng. J.*, 2012, **213**, 218–224.
96. L. Cao, F.-J. Spiess, A. Huang, S. L. Suib, T. N. Obee, S. O. Hay and J. D. Freihaut, *J. Phys. Chem. B*, 1999, **103**, 2912–2917.
97. Y. Chen, D. Li, J. Chen, J. Wang, S. Meng, J. Xian, X. Fu and Y. Shao, *Appl. Catal., B*, 2013, **129**, 403–408.
98. Y. Yamashita, N. Aoyama, N. Takezawa and K. Yoshida, *Environ. Sci. Technol.*, 2000, **34**, 5211–5214.
99. S. In, A. Orlov, F. Garcia, M. Tikhov, D. S. Wright and R. M. Lambert, *Chem. Commun.*, 2006, 4236–4238.
100. H. Li, S. Yin, Y. Wang and T. Sato, *Environ. Sci. Technol.*, 2012, **46**, 7741–7745.
101. K. Nishijima, B. Ohtani, X. Yan, T. Kamai, T. Chiyoya, T. Tsubota, N. Murakami and T. Ohno, *Chem. Phys.*, 2007, **339**, 64–72.
102. P. A. Kolinko, T. N. Filippov, D. V. Kozlov and V. N. Parmon, *J. Photochem. Photobiol., A*, 2012, **250**, 72–77.
103. J. Wang, H. Ruan, W. Li, D. Li, Y. Hu, J. Chen, Y. Shao and Y. Zheng, *J. Phys. Chem. C*, 2012, **116**, 13935–13943.
104. M. Nakamura, A. Ono, E. Bae, N. Murakami and T. Ohno, *Appl. Catal., B*, 2013, **130–131**, 264–269.
105. M.-V. Sofianou, M. Tassi, V. Psycharis, N. Boukos, S. Thanos, T. Vaimakis, J. Yu and C. Trapalis, *Appl. Catal., B*, 2015, **162**, 27–33.
106. W. Sun, Q. Meng, L. Jing, D. Liu and Y. Cao, *J. Phys. Chem. C*, 2013, **117**, 1358–1365.
107. D. Wang, T. Kako and J. Ye, *J. Am. Chem. Soc.*, 2008, **130**, 2724–2725.
108. Y. He, T. Sheng, J. Chen, R. Fu, S. Hu and X. Wu, *Catal. Commun.*, 2009, **10**, 1354–1357.
109. Y. He, T. Sheng, Y. Wu, J. Chen, R. Fu, S. Hu and X. Wu, *J. Hazard. Mater.*, 2009, **168**, 551–554.
110. Y. He, Y. Wu, T. Sheng and X. Wu, *J. Hazard. Mater.*, 2010, **180**, 675–682.
111. H. Shi, G. Chen and Z. Zou, *Appl. Catal., B*, 2014, **156–157**, 378–384.
112. F. Dong, H. Liu, W.-K. Ho, M. Fu and Z. Wu, *Chem. Eng. J.*, 2013, **214**, 198–207.
113. M. Miyauchi, Y. Nukui, D. Atarashi and E. Sakai, *ACS Appl. Mater. Interfaces*, 2013, **5**, 9770–9776.

114. E. Piera, J. A. Ayllón, X. Doménech and J. Peral, *Catal. Today*, 2002, **76**, 259–270.
115. T. N. Obee and S. O. Hay, *Environ. Sci. Technol.*, 1997, **31**, 2034–2038.
116. A. Vorontsov, I. V. Stoyanova, D. V. Kozlov, V. I. Simagina and E. N. Savinov, *J. Catal.*, 2000, **189**, 360–369.
117. J. C. Kennedy III and A. K. Datye, *J. Catal.*, 1998, **179**, 375–389.
118. A. Sirisuk, C. G. Hill and M. A. Anderson, *Catal. Today*, 1999, **54**, 159–164.
119. S. Yamazaki, S. Tanaka and H. Tsukamoto, *J. Photochem. Photobiol. Chem.*, 1999, **121**, 55–61.
120. K. Demeestere, A. D. Visscher, J. Dewulf, M. V. Leeuwen and H. V. Langenhove, *Appl. Catal., B*, 2004, **54**, 261–274.
121. A. V. Vorontsov and V. P. Dubovitskaya, *J. Catal.*, 2004, **221**, 102–109.
122. J. Shang, Y. Du and Z. Xu, *Chemosphere*, 2002, **46**, 93–99.
123. D. S. Muggli and J. L. Falconer, *J. Catal.*, 1999, **187**, 230–237.
124. Y. Ohko, A. Fujishima and K. Hashimoto, *J. Phys. Chem. B*, 1998, **102**, 1724–1729.
125. H. Einaga, S. Futamura and T. Ibusuki, *Appl. Catal., B*, 2002, **38**, 215–225.
126. S. B. Kim and S. C. Hong, *Appl. Catal., B*, 2002, **35**, 305–315.
127. N. N. Lichtin and M. Sadeghi, *J. Photochem. Photobiol., A*, 1998, **113**, 81–88.
128. M. M. Ameen and G. B. Raupp, *J. Catal.*, 1999, **184**, 112–122.
129. L. Cao, Z. Gao, S. L. Suib, T. N. Obee, S. O. Hay and J. D. Freihaut, *J. Catal.*, 2000, **196**, 253–261.
130. L. Cao, A. Huang, F.-J. Spiess and S. L. Suib, *J. Catal.*, 1999, **188**, 48–57.
131. Z.-R. Tang, F. Li, Y. Zhang, X. Fu and Y.-J. Xu, *J. Phys. Chem. C*, 2011, **115**, 7880–7886.
132. A. V. Vorontsov, E. N. Savinov, C. Lion and P. G. Smirniotis, *Appl. Catal., B*, 2003, **44**, 25–40.
133. L. Lamaa, C. Ferronato, S. Prakash, L. Fine, F. Jaber and J. M. Chovelon, *Appl. Catal., B*, 2014, **156–157**, 438–446.
134. A. V. Vorontsov, E. N. Kurkin and E. N. Savinov, *J. Catal.*, 1999, **186**, 318–324.
135. S. Luo and J. L. Falconer, *J. Catal.*, 1999, **185**, 393–407.
136. M. El-Maazawi, A. N. Finken, A. B. Nair and V. H. Grassian, *J. Catal.*, 2000, **191**, 138–146.
137. F. Chen, X. Yang, H. K. C. Mak and D. W. T. Chan, *Build. Environ.*, 2010, **45**, 1747–1754.
138. V. M. Sousa, C. M. Manaia, A. Mendes and O. C. Nunes, *J. Photochem. Photobiol., A*, 2013, **251**, 148–153.
139. S. M. Zacarías, M. L. Satuf, M. C. Vaccari and O. M. Alfano, *Ind. Eng. Chem. Res.*, 2012, **51**, 13599–13608.
140. A. S. Safatov, S. A. Kiselev, V. Y. Marchenko, A. A. Sergeev, M. O. Skarnovich, E. K. Emelyanova, M. A. Smetannikova, G. A. Buryak and A. V. Vorontsov, *Chem. Eng.*, 2008, **16**, 275.
141. E. A. Kozlova, A. S. Safatov, S. A. Kiselev, V. Y. Marchenko, A. A. Sergeev, M. O. Skarnovich, E. K. Emelyanova, M. A. Smetannikova, G. A. Buryak and A. V. Vorontsov, *Environ. Sci. Technol.*, 2010, **44**, 5121–5126.

142. M. Sapiña, E. Jimenez-Relinque and M. Castellote, *Environ. Sci. Technol.*, 2013, **47**, 11711–11716.
143. W.-K. Jo, J.-H. Park and H.-D. Chun, *J. Photochem. Photobiol., A*, 2002, **148**, 109–119.
144. J. M. Coronado, M. E. Zorn, I. Tejedor-Tejedor and M. A. Anderson, *Appl. Catal., B*, 2003, **43**, 329–344.
145. G. Qijin, W. Qingming and Z. Bin, *Ind. Eng. Chem. Res.*, 2012, **51**, 15360–15373.
146. M. Addamo, V. Augugliaro, S. Coluccia, M. Faga, E. Garcialopez, V. Loddo, G. Marci, G. Martra and L. Palmisano, *J. Catal.*, 2005, **235**, 209–220.
147. V. Augugliaro, S. Coluccia, V. Loddo, L. Marchese, G. Martra, L. Palmisano and M. Schiavello, *Appl. Catal., B*, 1999, **20**, 15–27.
148. C. T. Brigden, S. Poulston, M. V. Twigg, A. P. Walker and A. J. Wilkins, *Appl. Catal., B*, 2001, **32**, 63–71.
149. W. Choi, J. Y. Ko, H. Park and J. S. Chung, *Appl. Catal., B*, 2001, **31**, 209–220.
150. M. Hajaghazadeh, V. Vaiano, D. Sannino, H. Kakooei, R. Sotudeh-Gharebagh and P. Ciambelli, *Catal. Today*, 2014, **230**, 79–84.
151. A. V. Vorontsov, E. N Savinov and P. G. Smirniotis, *Chem. Eng. Sci.*, 2000, **55**, 5089–5098.
152. A. V. Vorontsov, D. V. Kozlov, P. G. Smirniotis and V. N. Parmon, *Kinet. Catal.*, 2005, **46**, 437–444.
153. C. H. Ao, S. C. Lee and J. C. Yu, *J. Photochem. Photobiol., A*, 2003, **156**, 171–177.
154. L. Yang, Z. Liu, J. Shi, Y. Zhang, H. Hu and W. Shangguan, *Sep. Purif. Technol.*, 2007, **54**, 204–211.
155. O. Modesto, P. Hammer and R. F. P. Nogueira, *J. Photochem. Photobiol., A*, 2013, **253**, 38–44.
156. K.-P. Yu, G. W.-M. Lee, S.-Y. Lin and C. P. Huang, *J. Aerosol Sci.*, 2008, **39**, 377–392.
157. M. P. Paschoalino and W. F. Jardim, *Indoor Air*, 2008, **18**, 473–479.
158. A. V. Vorontsov, M. N. Lyulyukin and A. S. Besov, *Global J. Environ. Sci. Technol.*, 2012, **2**, 1–10.
159. A. S. Besov and A. V. Vorontsov, *Plasma Chem. Plasma Process.*, 2007, **27**, 624–634.
160. K. Sekiguchi, A. Sanada and K. Sakamoto, *Catal. Commun.*, 2003, **4**, 247–252.
161. P. Zhang and J. Liu, *J. Photochem. Photobiol., A*, 2004, **167**, 87–94.
162. K. Yu and G. Lee, *Appl. Catal., B*, 2007, **75**, 29–38.
163. J. Yuan, X. Huang, M. Chen, J. Shi and W. Shangguan, *Catal. Today*, 2013, **201**, 182–188.
164. J. Jeong, K. Sekiguchi, W. Lee and K. Sakamoto, *J. Photochem. Photobiol., A*, 2005, **169**, 279–287.
165. S. Kataoka, D. T. Tompkins, W. A. Zeltner and M. A. Anderson, *J. Photochem. Photobiol., A*, 2002, **148**, 323–330.
166. A. V. Vorontsov, *Catal. Commun.*, 2007, **8**, 2100–2104.

CHAPTER 8

Self-Cleaning Photocatalytic Activity: Materials and Applications

NIALL B. McGUINNESS[a,b], HONEY JOHN[c], MAHESWARI K. KAVITHA[c], SWAGATA BANERJEE[d], DIONYSIOS D. DIONYSIOU[e], AND SURESH C. PILLAI*[a,b]

[a]Nanotechnology Research Group, Department of Environmental Sciences, Institute of Technology Sligo, Sligo, Ireland; [b]Centre for Precision Engineering, Materials and Manufacturing Research, Institute of Technology Sligo, Sligo, Ireland; [c]Department of Chemistry, Indian Institute of Space Science and Technology, Thiruvananthapuram, Kerala, India; [d]Department of Biochemistry, Adamas University, Adamas Knowledge City, Barasat – Barrackpore Road, Kolkata, India; [e]Department of Biomedical, Chemical and Environmental Engineering, University of Cincinnati, Cincinnati, Ohio, USA
*E-mail: pillai.suresh@itsligo.ie

8.1 Introduction to Self-Cleaning Materials

There has been growing interest focused on self-cleaning materials due to their ever increasing role in a range of applications,[1,2] especially within the manufacture of textiles,[3–6] tiles[7,8] and glass,[9–12] along with plastic[5,13,14] and metal[15–18] based components and products. The attractiveness of these types

RSC Energy and Environment Series No. 15
Photocatalysis: Applications
Edited by Dionysios D Dionysiou, Gianluca Li Puma, Jinhua Ye, Jenny Schneider, and Detlef Bahnemann

Published by the Royal Society of Chemistry, www.rsc.org

of materials stems from their utilisation in many applications and their ease of maintenance. These materials can be categorised as possessing either hydrophilic or hydrophobic surfaces, where the superhydrophilic surface causes water droplets to spread, producing a film on the surface. In contrast, the superhydrophobic surface causes the water droplets to repel from the material, while transporting contaminants away in tow.[19]

8.2 Mechanism of Self-Cleaning Activity

Certain superhydrophilic materials possess photocatalytic properties, which permits the exploitation of light for the decomposition of contaminants on the surface. Titanium dioxide (titania) is one such material, which when exposed to light of a specific wavelength (<390 nm) excites an electron from the valence band to the conduction band (e^-_{CB}), thus leaving a hole present in the valence band (h^+_{VB}) (Figure 8.1).

After this step has been accomplished (see eqn (8.1)), various pathways can be taken by the charge carriers. The electron produced *via* photoexcitation, which is present within the conduction band, can recombine with the hole present in the valance band *via* the charge-recombination reaction to produce radiation (eqn (8.2)):

$$TiO_2 + h\nu \rightarrow h^+_{VB} + e^-_{CB} \tag{8.1}$$

$$h^+_{VB} + e^-_{CB} \rightarrow \text{radiation} \tag{8.2}$$

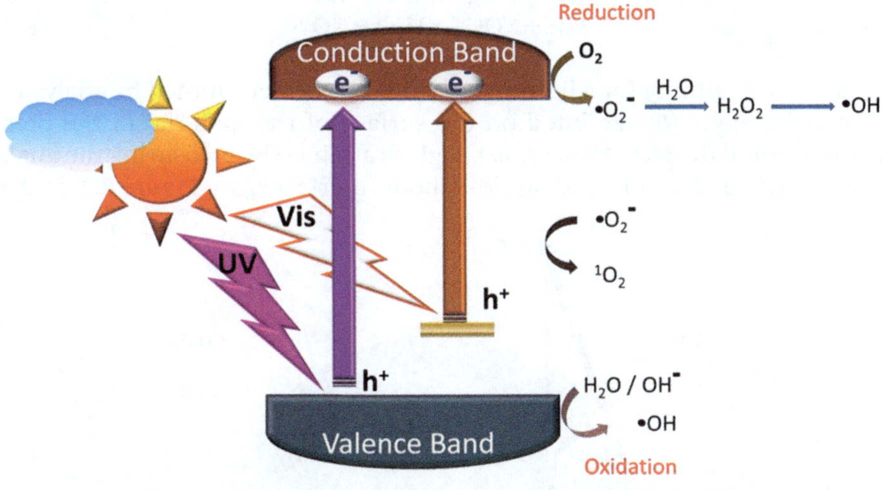

Figure 8.1 Representation of various reactions occurring during the photoexcitation of titanium dioxide using UV/vis light.[72] (Reprinted with permission from S. Banerjee, S. C. Pillai, P. Falaras, K. E. O'Shea, J. A. Byrne and D. D. Dionysiou, *J. Phys. Chem. Lett.*, 2014, 5, 2543. Copyright (2014) American Chemical Society.)

When this does not occur, the charge carriers are capable of migrating to the material's surface. Here the photoexcited electron reduces oxygen present in the atmosphere, producing ROS (reactive oxygen species), a superoxide radical anion (eqn (8.3)) or a hydroperoxide radical (eqn (8.4)):

$$O_2 + e^-_{CB} \rightarrow O_2^{-\bullet} \tag{8.3}$$
$$O_2^{-\bullet} + H^+ \rightarrow {}^{\bullet}OOH \tag{8.4}$$

The superoxide radical anion and the hydroperoxide radical can react with volatile organic compounds (VOCs) to produce water and carbon dioxide (eqn (8.5) and (8.6)):

$$O_2^{-\bullet} + VOC \rightarrow H_2O + CO_2 \tag{8.5}$$
$${}^{\bullet}OOH + VOC \rightarrow H_2O + CO_2 \tag{8.6}$$

In addition, a hydroperoxide radical can react with another hydroperoxide radical to yield hydrogen peroxide (eqn (8.7)), which can then form a hydroxyl radical or promote charge separation by accepting a photoexcited electron:

$${}^{\bullet}OOH + {}^{\bullet}OOH \rightarrow H_2O_2 + O_2 \tag{8.7}$$

Furthermore, the valence band hole is capable of oxidising water or a hydroxide anion to yield a hydroxyl radical (eqn (8.8)), which can oxidise VOCs to yield water and carbon dioxide (eqn (8.9)):

$$H_2O + h^+_{VB} \rightarrow {}^{\bullet}OH + H^+ \tag{8.8}$$
$${}^{\bullet}OH + VOC \rightarrow H_2O + CO_2 \tag{8.9}$$

A material's affinity for a liquid (water/oil) can be determined by analysing the contact angle (θ) exhibited on the surface of the material in the presence of a liquid droplet. The contact angle is attained by measuring the angle between the solid–liquid and liquid–vapour interfaces, see Figure 8.2, and is

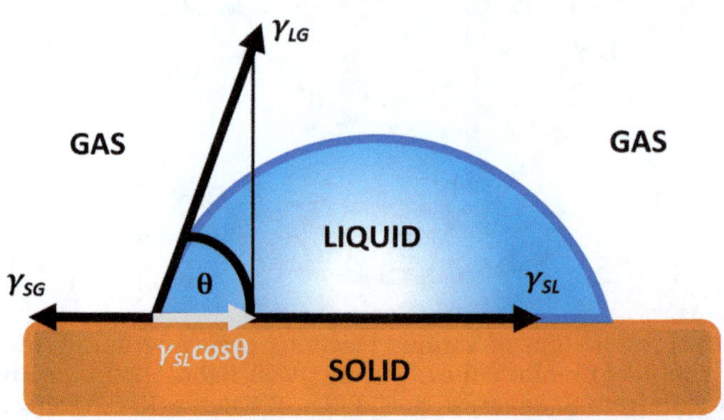

Figure 8.2 Diagram illustrating the contact angle measurement of a surface.

determined by the balance between the adhesive forces, occurring between the surface and liquid, and the cohesive forces, occurring within the liquid droplet.[20,21]

Assuming the surface is ideal (perfectly flat, rigid and defect free) Young's equation (eqn (8.10)) can be employed to predict the surface contact angle if the surface energy, γ, between each phase, solid (S), liquid (L) and gas (G), is known:

$$\gamma_{SG} = \gamma_{SL} + \gamma_{LG} \cos\theta \tag{8.10}$$

Once the contact angle of a material has been determined, a degree of wettability can then be assigned to it, which corresponds to the liquid droplet's ability to spread across the surface. The surface can also be deemed to demonstrate either hydrophilic/phobic or oleophilic/phobic tendencies, upon the employment of a water or oil droplet, respectively, during analyses (Table 8.1).[20]

Notably, eqn (8.10) is valid only for ideal, inert, rigid and smooth surfaces. The wettability behaviour of rough surfaces is generally described by the models proposed by Wenzel[22] and Cassie–Baxter.[23] For a rough surface, the Wenzel model assumes that a liquid droplet can come into contact with the entire surface through complete penetration into the surface cavities (Figure 8.3a). Under such circumstances, the Wenzel model correlates the "apparent" contact angle, θ_{app}, for a liquid droplet to the "true" contact angle, θ_s, multiplied by the roughness factor, r, of the surface (eqn (8.11)):

$$\cos\theta_{app} = r\cos\theta_s \tag{8.11}$$

where r can be defined as the ratio of geometric surface area to the actual surface area (eqn (8.12)):

$$r = \frac{\text{Geometric surface area}}{\text{Actual surface area}} \tag{8.12}$$

Table 8.1 Relationship between contact angle and wettability behaviour.

Contact angle measurement	Material designation	Degree of wetting	Strength of forces: Solid–liquid interactions	Strength of forces: Liquid–liquid interactions
$\theta < 10°$	Superhydrophilic/ superoleophilic	Perfect wetting	Strong	Weak
$0° < \theta < 90°$	Hydrophilic/ oleophilic	High wettability	Strong Weak	Strong Weak
$\theta > 90°$	Hydrophobic/ oleophobic	Low wettability	Weak	Strong
$\theta > 150°$	Superhydrophobic/ superoleophobic	Perfect non-wetting	Weak	Strong

According to the Wenzel model, the hydrophilicity ($\theta < 90°$) or hydrophobicity ($\theta > 90°$) of a specific surface is governed by the nature of that surface and either property will increase due to an increase in surface roughness ($r > 1$). Although these trends are observed in most cases, the Wenzel model cannot satisfactorily deal with all heterogeneous surfaces. The Cassie–Baxter model was then proposed in 1944.[23] This model postulates that the liquid droplet applied to the surface of the material does not penetrate into the small cavities present, but spreads over the surface, producing solid–liquid and liquid–vapour interfaces (Figure 8.3b). The θ_{app} of the liquid droplet on a rough surface can be related to θ_s (eqn (8.13)):

$$\cos\theta_{app} = f_s \cos\theta_s + f_v \cos\theta_v \tag{8.13}$$

In this equation, θ_v is the contact angle of the liquid in air, while f_s and f_v are representative of the liquid droplets area fraction, which is in contact with the surface and also in contact with the vapour within the surface cavities, respectively. If $\theta_v = 180°$ and $f_s + f_v = 1$ then eqn (8.13) can be rephrased (eqn (8.14)):

$$\cos\theta_{app} = -1 + f_s(\cos\theta_s + 1) \tag{8.14}$$

Miwa *et al.* analysed the impact of surface roughness on the water contact angle of superhydrophobic surfaces.[24] Miwa *et al.'s* model combines the previously discussed Wenzel and Cassie–Baxter states and is generally known as the Miwa–Hashimoto model[25,26] (Figure 8.3c). Roughness parameters of various superhydrophobic materials were investigated and compared and it was observed that the sliding angles decrease with increasing contact angle depending on the surface roughness. This model can be represented as:

$$\cos\theta' = r_{MH} f_{MH} \cos\theta + f_{MH} - 1 \tag{8.15}$$

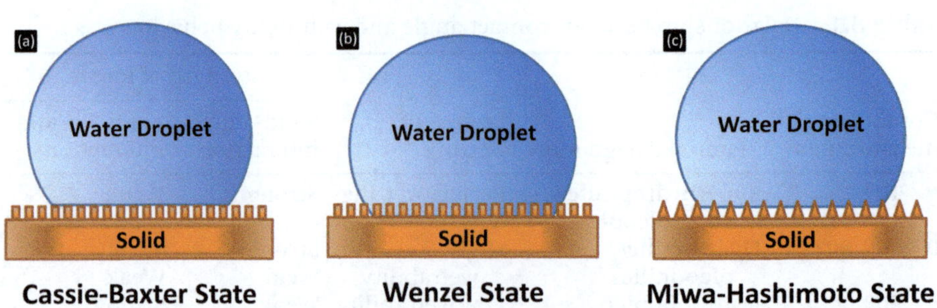

Figure 8.3 Diagrams displaying the three states that occur due to the effect of surface structure on the wettability of a solid surface: (a) Wenzel, (b) Cassie–Baxter and (c) Miwa–Hashimoto.

where θ' denotes the equilibrium contact angles on a rough surface and θ represents the same on a flat surface; r_{MH} is the ratio of the side area to the bottom area and f_{MH} is the fraction of surface area of the substrate in contact with the liquid.[25] Detailed analyses of surface models of various self-cleaning materials are given in a recent review article by Banerjee *et al.*[26]

The superhydrophilicity experienced by titanium dioxide can be explained by a combination of two processes. Firstly *via* the generation of light-induced hydroxylation[9] on the titanium dioxide surface and secondly *via* the photoinduced removal of the hydrocarbon layer[27] on the surface, present due to oxidation by the atmosphere.

8.2.1 Light-Induced Hydroxylation of the Surface

The first mechanism occurs when UV light produces photoexcited electrons in the material. At the surface these conduction band electrons reduce titanium(IV) to titanium(III), while the corresponding valence band holes oxidise the O^{2-} anions, causing the oxygen atoms to be ejected, leaving vacancies. It is at these oxygen vacancy sites that hydroxide anions and water molecules can then be absorbed, giving the surface its hydrophilic tendencies[9,21] (Scheme 8.1).

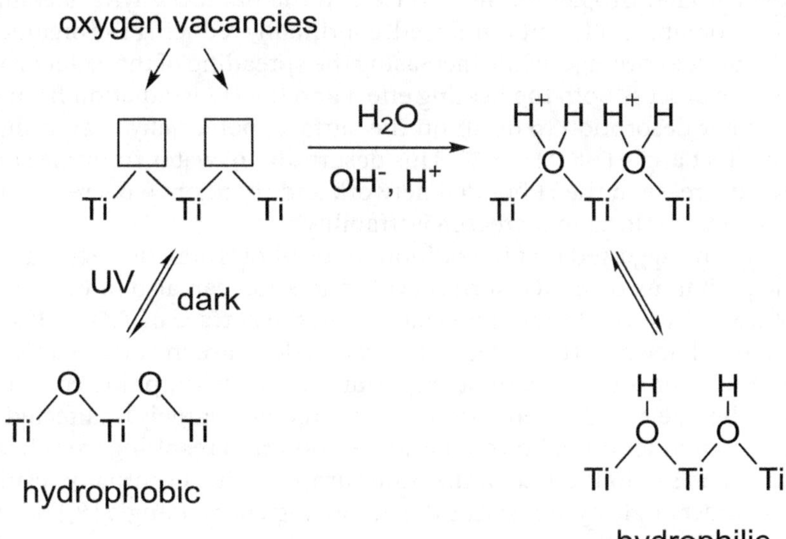

Scheme 8.1 Mechanism of photoinduced hydroxylation on a titanium dioxide material creating a hydrophilic surface. During photocatalysis, the electrons ejected will reduce Ti(IV) to Ti(III). The corresponding holes oxidise O^{2-} anions, which are removed and "oxygen vacancies" are created. These oxygen vacancies facilitate the surface's affinity towards water molecules and effectively transform the surface from hydrophobic into hydrophilic.

Under UV irradiation, the rate of hydrophilic conversion of a TiO_2 film electrode increased when a high positive electrode potential was applied and decreased upon the addition of hole scavenging agents.[28] These observations suggested that the diffusion of photogenerated holes is crucial for the hydrophilic conversion. Subsequently, Sakai *et al.* proposed that the positive holes created under UV illumination of TiO_2 surface diffuse to the surface, where they can be trapped at the lattice oxygen sites.[29] This can in turn weaken the Ti–O bond and leads to formation of new hydroxyl bonds in the presence of water molecules. When the TiO_2 surface is stored in the dark, it gradually reverts to a less hydrophilic state, due to the slow desorption of surface hydroxyl groups.

The importance of photoinduced electron–hole generation was further emphasised by Yan *et al.* It has been demonstrated that the hydrophilic conversion occurs in two stages and the decrease in contact angle of a water droplet can only occur when the illuminating light possesses energy that is greater than the band gap energy of TiO_2.[30]

8.2.2 Photo-Oxidation of Adsorbed Hydrocarbons on the Surface

The second mechanism of self-cleaning is proposed to occur when UV light irradiates a water droplet on the surface and the photocatalytic decomposition of the organic pollutant is induced, continually reducing the organic pollutant's surface coverage, while increasing the spreading of the water droplet. During this process both the heating effect and the UV irradiation from light causes water desorption to occur on the surface, potentially decreasing the water contact angle (WCA) to <5°. This desorption of water from the surface causes a decrease in the H-bonded network and therefore a decrease in surface tension, resulting in increased wettability.[27,31]

It has been suggested that in addition to the photocatalytic degradation of organic pollutants on a TiO_2 surface, UV irradiation can also induce desorption of weakly attached water molecules from the surface of TiO_2.[32] The subsequent breakdown in H-bonded network and decrease in surface tension of the water cluster is believed to be important for the hydrophilic conversion. Notably, the free energy of cohesion of water remains largely unaffected with variation in temperature; however, water evaporation resulting from heating causes a decrease in the size of the water droplet. This eventually results in a wide contact angle hysteresis, as the advancing contact angle (θ_A) becomes larger than the receding contact angle (θ_R).

As discussed above, different mechanisms have been suggested to explain the photoinduced hydrophilicity; however, no consensus mechanism has been proposed to elucidate the phenomenon. Water contact angles for several metal oxides such as tungsten oxide and vanadium oxide have been found to decrease under UV-illumination, but no photocatalytic activity was reported.[33] This suggests that the photocatalytic degradation of pollutants is not the exclusive mechanism for hydrophilic conversion. Additionally, alkaline wash, vacuum UV treatments *etc.* also improve surface wettability by removing contaminants from the surface; however, these treatments cannot produce superhydrophilic surface.[33] These observations indicate that a combination of various mechanisms is required to explain the photoinduced hydrophilic conversion.

In a recent report, Emeline *et al.* demonstrated that the intensity and the wavelength of the light govern the efficiency of the photoinduced superhydrophilic conversion and point out the involvement of photogenerated electron–hole pairs in the initial step of the phenomenon.[34] This study also reports that the surface acidity and external hydrate layers play crucial roles in trapping charge carriers and dictate the efficiency of hydrophilic conversion.[34]

8.3 Photocatalytic Materials

Photocatalytic materials can be synthesised using various methods such as sol–gel,[35–37] chemical vapour deposition,[38–40] electrochemical,[41–44] pyrolysis[45–47] and solvothermal/hydrothermal synthesis.[48–50] Sol–gel synthesis involves the hydrolysis and condensation of metal alkoxides or metal chlorides to produce metal oxides.[35–37] Chemical vapour deposition creates a ceramic material by depositing the solid material onto a substrate from a gaseous mixture of the desired metal halide and carbon, nitrogen, oxygen or many other materials.[38–40] Electrochemical synthesis requires electrodes to be submerged in an electrolytic medium followed by the application of a specific potential or current to yield the material as a recoverable sediment[44] or as coatings/particles deposited on the substrate.[41–43,51] During pyrolysis, solutions containing metal chlorides are atomised using a nebuliser, which are passed through a furnace, in the presence of an inert atmosphere, to produce the ceramic material.[45–47] Solvothermal synthesis requires the use of metal precursors in a closed system, which experiences moderate to high pressure and temperature, to produce metal oxides.[48–51]

8.3.1 Titanium Dioxide

The most commonly discussed and utilised photocatalytic semiconductor is titanium dioxide. It has been employed during the fabrication of self-cleaning materials,[52–56] anti-abrasive materials,[57,58] materials for air/groundwater remediation,[53,59,60] antifogging glass,[51,54,55] solar cells[61–63] and antibacterial surfaces.[64–68] Titania is the most popular material for photoinduced self-cleaning; not only does it possess excellent photocatalytic activity and low toxicity, it also demonstrates a high rate of physical and chemical stability. Furthermore, it is very accessible and inexpensive.[69–72]

In 1997, Wang *et al.* reported the photogeneration of a highly hydrophilic and oleophilic surface on anatase TiO_2 coated glass, yielding a material possessing both self-cleaning and antifogging properties, when exposed to UV irradiation.[9] Since this work, there has been much development with regard to the enhancement of these initial properties *via* the incorporation of organic compounds and metals. In addition, there has been a great increase in the number of possible applications. More recently in 2013, Sawai *et al.* demonstrated the UV irradiation induced underwater superoleophobicity (approx. 167° oil contact angle) of an amphiphilic surface prepared from TiO_2 sol–gel.[73] The same year, Afzal *et al.* fabricated self-cleaning cotton fabrics coated with monolayer films of porphyrin/anatase TiO_2 catalyst, which possessed superior self-cleaning properties, when compared to the bare TiO_2 film.[74] This material demonstrated photocatalytic degradation of methylene blue (MB), red wine and coffee stains. Furthermore, a copper(II) porphyrin/anatase TiO_2 catalyst[66] exhibited significant photostability under visible light, when compared to porphyrin/anatase TiO_2 catalyst cotton fabrics.[74] Also of note, thin films of anatase TiO_2 nanowire arrays with an exposed highly reactive (001) facet have been synthesised *via* a fluorine free hydrothermal technique.[75] These oriented anatase TiO_2 thin films exhibited high transparency, photocatalytic activity and superhydrophilicity upon UV irradiation, properties which are advantageous for applications in self-cleaning coatings.

Two polymorphs of TiO_2 have been of great interest for some time, these being the metastable anatase and stable rutile crystal structures. The anatase and rutile polymorphs of TiO_2 possess band gaps (E_g) of 3.20 and 3.03 eV, respectively.[20,76] It is known that, as pure phases, anatase is more photocatalytically active than rutile,[77,78] but when the phases are mixed the material demonstrates an even greater photo-activity, due to the synergistic effects of the two mixed crystal structures.[76] One significant issue, prominent during the fabrication of TiO_2 coated tiles (requiring approx. 1000 °C), is the anatase into rutile transformation (ART) which occurs at approx. 600 °C.[79]

8.3.2 Rapid Testing of Self-Cleaning Photocatalytic Activity

Mills *et al.* have recently described a rapid and inexpensive method for determining the photocatalytic activity of commercially available self-cleaning materials using indicator ink and a digital scanner (Figure 8.4), which can be performed in the laboratory or in the field.[80] This was developed further in 2014 by utilising a smart phone camera possessing a colour monitoring app (RGB), in conjunction with a resazurin (Rz) based photocatalyst activity indicator ink (*paii*) to assess the activity of commercial self-cleaning films. During this study, the photocatalysed reduction of the red dye was monitored by measuring the RGB colour components of the sample during irradiation, which produced results correlating well with conventional laboratory performed methods.[81] The redox dye basic blue 66 (BB66) was also employed and was found to respond >6-times faster than the Rz *paii* on low activity self-cleaning tiles.[82] The redox dyes Rz, BB66 and acid violet 7 (AV7) were also used to determine the photocatalytic activity of both self-cleaning glass and paint in a very short test time (<10 min). These methods are inexpensive and when used

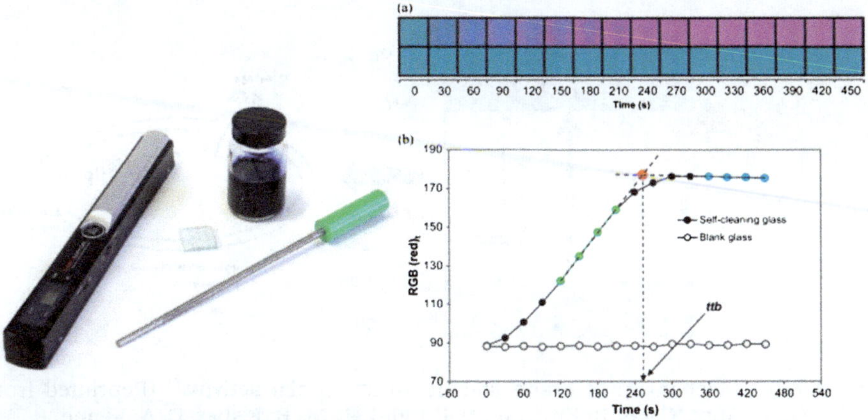

Figure 8.4 (Left) Handheld document scanner, Rz ink, wire wound rod (K-bar #3) and self-cleaning glass employed to perform self-cleaning photocatalytic testing. (Right (a)) Images recorded at 30 s irradiation intervals for (top) self-cleaning and (bottom) plain glass. (Right (b)) Plot of RGB (red) variation against irradiation time, extracted from images in (a).[80] (J. Photochem. Photobiol., A, Vol. 272, A. Mills, J. Hepburn, D. Hazafy, C. O'Rourke, J. Krysa, M. Baudys, M. Zlamal, H. Bartkova, C. E. Hill, K. R. Winn, M. E. Simonsen, E. G. Sogaard, S. C. Pillai, N. S. Leyland, R. Fagan, F. Neumann, C. Lampe and T. Graumann, A simple, inexpensive method for rapid testing of the photocatalytic activity of self-cleaning surfaces, Page 18, Copyright (2013), with permission from Elsevier.)

together it is possible to probe a wide range of activities not previously covered by current ISO tests, in the absence of advanced analytical instrumentation.[83]

8.3.3 Photocatalytic Antibacterial Activity

Titania coated ceramics, while possessing superhydrophilic properties, are also capable of demonstrating antibacterial activity. This makes them extremely useful materials for applications in public facilities, which accommodate practices in the fields of medicine and education, and in areas of the food service/preparation industry, where hygiene is a main concern. Titania (TiO_2) can also be employed during the fabrication of construction materials, such as glass and tiles, due to its antibacterial properties, which provides an inhospitable environment for the growth of microorganisms such as *Escherichia coli*,[45,65] *Lactobacillus acidophilus*,[68] *Candida albicans*,[68] *Pseudomonas aeruginosa*[66] and *Staphylococcus aureus*[64] (Figure 8.5). Many other semiconductors possess desirable properties that can be employed for the previously stated applications, either alone or in conjunction with another material. Such semiconductors can be composed of metal oxides or metal sulfides such as zinc oxide (E_g = 3.37 eV)[84] or zinc sulfide (E_g = 3.60 eV).[85] These can also be used for their antimicrobial properties against pathogens such as *E. coli*,[42,86] *S. aureus*,[86] *Bacillus subtilis*[42] and *Salmonella typhii*.[42]

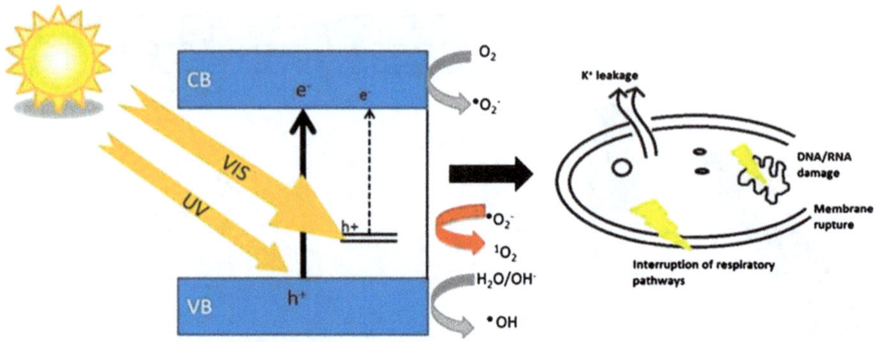

Figure 8.5 Mechanism of photocatalytic antimicrobial activity.[71] (Reprinted from Appl. Catal. B: Environ., Vol. 130–131, M. B. Fisher, D. A. Keane, P. Fernandez-Ibanez, J. Colreavy, S. J. Hinder, K. G. McGuigan and S. C. Pillai, Nitrogen and copper doped solar light active TiO_2 photocatalysts for water decontamination, Page 8, Copyright (2013), with permission from Elsevier.)

8.4 Semiconductor Doping and TiO_2/SiO_2 Composites

Important research and development has been focusing on enhancing the potential of TiO_2 in commercial applications. This has been performed by attempting to reduce the band gap of the material, which would permit both UV (<390 nm) and visible light absorption, upon irradiation of the material. This is hugely beneficial as the solar spectrum consists of approx. 40% visible light as opposed to 4% UV light.[87] Successful methods currently employed to reduce the band gap are the use of non-metal dopants within the material, particularly nitrogen, which can yield the doped material photocatalytic activity at just less than 500 nm.[88,89] Carbon[90,91a] and also fluorine, while in combination with nitrogen,[82] can be utilised to dope the TiO_2 material with beneficial results, including antibacterial self-cleaning properties (Figure 8.6). Li *et al.* have also demonstrated that the combination of nitrogen and fluorine as dopants can enable the material to exhibit photocatalytic activity, irrespective of the type of irradiation employed.[92] The utilisation of metal dopants, such as silver,[93,94] chromium,[95,96] iron[96,97] and various other transition metals,[96] can have a similar effect on photocatalytic activity. Importantly, the photoinduced superhydrophilic character of pure TiO_2 generally reverts back gradually to its hydrophobic character in the absence of light illumination. However, for practical applications, it is desirable that the hydrophilic character is maintained for a long period of time, even in the absence of light.

8.4.1 Self-Cleaning Activity

TiO_2/SiO_2 composite films, synthesised by using the sol–gel route,[98,99] radiofrequency magnetron sputtering[100] or rapid flame deposition,[101] were found to exhibit enhanced superhydrophilicity and improved maintenance of

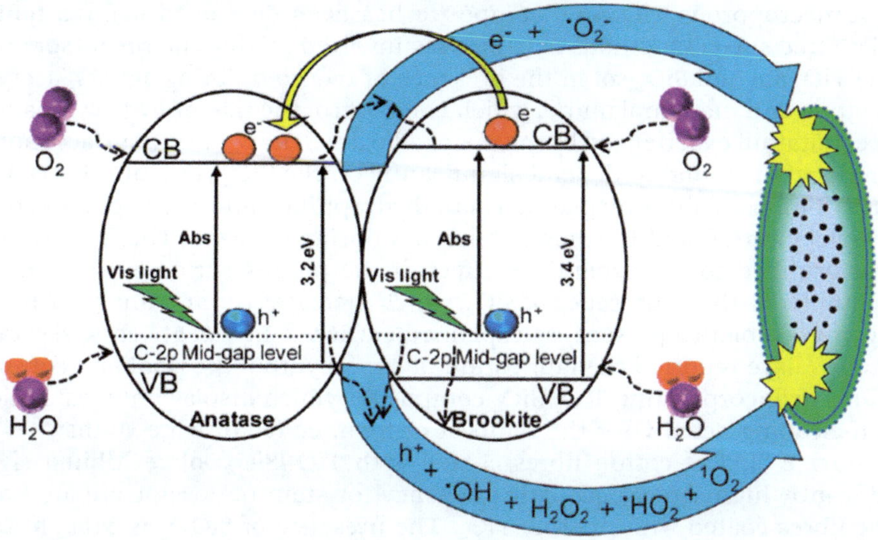

Figure 8.6 Schematic representation of visible light-induced photocatalytic bacterial killing using carbon-doped anatase–brookite heterojunctions.[91b] (Reprinted with permission from V. Etacheri, G. Michlits, M. K. Seery, S. J. Hinder and S. C. Pillai, *ACS Appl. Mater. Interfaces*, 2013, **5**, 1663. Copyright (2013) American Chemical Society).

hydrophilicity in the dark. The TiO_2/SiO_2 composite films exhibited increased Lewis acidity, resulting from the excess positive charge generated due to the doping of silicon atoms into the TiO_2 lattice.[102] Increased acidity of the composite film also accounts for the increase in the amount of hydroxyl groups at the surface, inducing enhanced hydrophilicity. Additionally, the presence of SiO_2 can decrease the refractive index of the TiO_2/SiO_2 composite film, while increasing the extent of light absorption by the film, yielding an increase in the photocatalytic activity.[103] The photocatalytic activities of TiO_2/SiO_2 composite films were found to be closely related to their hydrophilic properties.[104] Addition of 30–40 mol% SiO_2 was found to be optimum for increasing the amount of surface hydroxyl groups and decreasing the competitive adsorption of contaminants on the surface.[104] Formation of a highly porous structure has appeared as an alternative approach to achieve an increase in surface roughness and superhydrophilicity. A superhydrophilic TiO_2/SiO_2 composite film has been obtained *via* spin coating using a TiO_2 sol containing 0.5 wt% PEG (poly(ethylene glycol)). This film demonstrated a WCA that changed from 15° to 3° within 0.16 s.[105] In the presence of an increasing amount of PEG 2000 in the TiO_2 precursor sol, the surface roughness of a TiO_2/SiO_2 bilayer film was found to increase and maximum surface roughness occurred at 0.5 wt% PEG 2000. Glass slides coated with the resultant bilayer films demonstrated excellent antifogging properties compared to the uncoated slides and, furthermore, retained hydrophilicity even after storage in the dark for a long period of time.[95]

A macroporous TiO_2/SiO_2 composite has been developed using a template-free sol–gel method. This process involved mixing the precursors of the TiO_2 sol and SiO_2 sol in the presence of two complexing agents, acetyl acetone and diethanolamine, which can control the rate of hydrolysis and precipitation experienced by the metal alkoxide. The resulting surface morphology was found to depend significantly on the SiO_2 content. The TiO_2/SiO_2 composite films displayed a superhydrophilic nature as applied water droplets were found to spread within 1 s during analyses. This behaviour was ascribed to the synergistic effect of the porous surface's roughness combined with the presence of SiO_2, which increases the amount of surface hydroxyl moieties present, as explained earlier.[106] In recent years, several groups have reported the fabrication of superhydrophilic wool and cotton fibres by incorporating TiO_2/SiO_2 composites which display photocatalytic self-cleaning activity for the removal of food, coffee or wine stains[3,107,108] (Figure 8.7). The cotton fibres coated with TiO_2/SiO_2 sols exhibited significantly higher photocatalytic efficiency for stain removal, compared to the fibres coated with pristine TiO_2. The presence of SiO_2 was thought to provide a higher surface area for stain adsorption along with possessing a greater surface acidity. Moreover, the hydrophobic surface of cotton, functionalised with pristine TiO_2, was found to prevent the spreading of stain droplets and resulted in a smaller but darker spot. Conversely, in the presence of a silica coating, the surface became more hydrophilic. This was due to an increase in the amount of surface hydroxyl groups and increase in adsorbed water, facilitating the spreading of stain droplets on the surface, thus causing the stain to appear lighter.[108] Cellulosic fibres functionalised with $Au/TiO_2/SiO_2$ nanocomposites have shown significant self-cleaning properties under visible light irradiation. This composite demonstrated a durable self-cleaning property, after 20 washing cycles under simulated visible light irradiation for 20 h.[109]

8.4.2 Antireflective Properties

Self-cleaning coatings with low surface reflection/antireflection properties and high light transmission ability are greatly beneficial due to their possible applications in flat panel displays, solar energy collectors and green houses.[110–112] However, self-cleaning coatings based on TiO_2, when applied to transparent substrates, usually increase the surfaces reflection, due to the high reflective index of TiO_2 (2.52 for anatase; 2.76 for rutile). Therefore, to lower the refractive index of the coating, the amount of TiO_2 employed during synthesis should be kept low, which is not favourable for self-cleaning action.[113] Antireflective and self-cleaning coatings have been developed using a layer-by-layer assembly of raspberry like SiO_2 (core)–TiO_2 (shell) nanoparticles (NPs) on a glass support[114] (Figure 8.8).

These coatings displayed superhydrophilicity both in the presence and in the absence of UV illumination. The hydrophilic nature of the coatings, in the absence of UV irradiation, primarily arises from SiO_2 due to the increased number of surface hydroxyl groups, the surface roughness and adsorbed

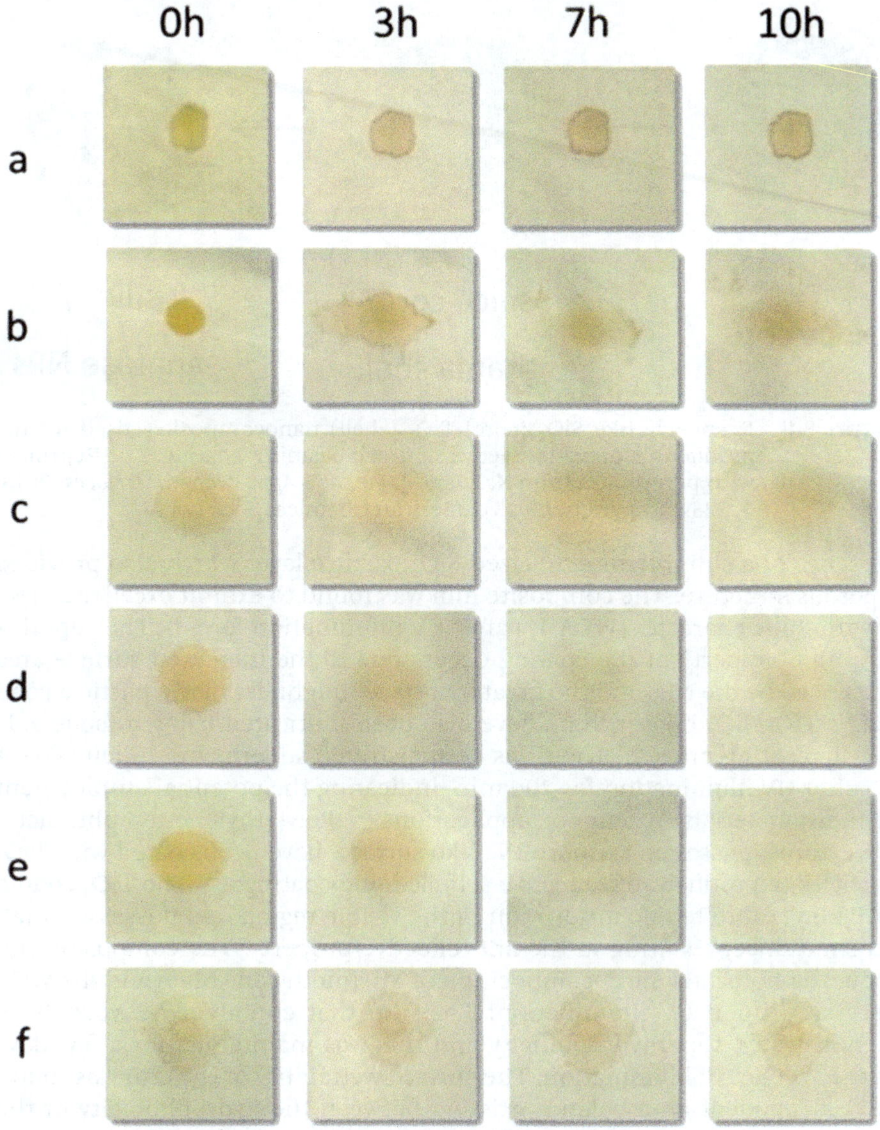

Figure 8.7 Removal of stains from coffee spilt on wool samples: (a) pristine wool; fabric coated with (b) TiO_2, (c) TiO_2/SiO_2 (70 : 30), (d) TiO_2/SiO_2 (50 : 50), (e) TiO_2/SiO_2 (30 : 70) and (f) SiO_2.[107] (Reprinted from Appl. Surf. Sci., Vol. 275, E. Pakdel, W. A. Daoud and X. Wang, Self-cleaning and superhydrophobic wool by TiO_2/SiO_2 nanocomposite, Page 397, Copyright (2013), with permission from Elsevier.)

water molecules. Under UV illumination, the photocatalytic activity of TiO_2 further enhances the hydrophilic property of the coatings.[114] Self-cleaning coatings with antireflection properties were also developed using core–shell SiO_2/TiO_2 particles *via* stepwise electrostatic deposition on a polyelectrolyte modified glass substrate.[113] The antireflection property was attributed to the

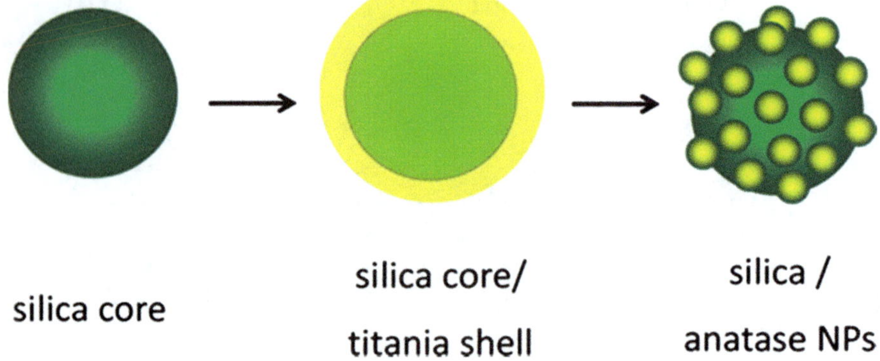

Figure 8.8 Raspberry like SiO_2 (core)–TiO_2 (shell) nanocomposites used for the production of antireflective and self-cleaning coatings.[114] (Reprinted with permission from X. Li and J. He, *ACS Appl. Mater. Interfaces*, 2013, 5, 5282. Copyright (2013) American Chemical Society.)

presence of a sub-micrometre sized SiO_2 particle layer which also provided a porous structure. The composite film was found to exhibit excellent superhydrophilic character (WCA 1°) after UV illumination for 1 h. The superhydrophilic property of the coating occurs due to the increased surface area possessed by the NPs and the greater surface roughness of the particle coating.[113] TiO_2/SiO_2 bilayer films have also been fabricated using a facile sol–gel dip coating process. The films demonstrated superhydrophilicity (WCA 2°) after UV illumination for 200 min, indicating the potential employment of these materials in outdoor applications.[115] Poly(ethylene terephthalate) (PET) films, possessing a moth-eye-like surface, have been coated with TiO_2 particles to produce antireflective self-cleaning coatings.[116] The TiO_2 coated PET films exhibited no absorption in the visible region and possessed high transmittance, resulting in the antireflective property. The composite film induced photocatalytic decomposition of MB (methylene blue) and the WCA decreased upon UV illumination. The film that contained >2 vol% TiO_2 demonstrated superhydrophilicity and this was maintained for a few days in the absence of illumination. The surface wettability of the film was found to be dependent on a balance existing between the hydrophobicity of the PET film and the superhydrophilicity of the TiO_2 particles. The antireflection and self-cleaning properties of these coatings have the potential to be used in the application such as solar cells and flat panel displays.[116] Low cost antireflection coatings that possess self-cleaning properties have been produced by employing a self-assembled block copolymer, in combination with silica sol–gel and anatase TiO_2 nanocrystals[117] (Figure 8.9). In solution phase, the block copolymer undergoes spontaneous self-assembly, followed by condensation with silica which results in an inverse opal-type structure, loaded with TiO_2. The synthetic strategy permits tuning of the coatings porosity, thus varying the refractive index and antireflective properties by changing the organic to inorganic volume fractions. Due to the presence of

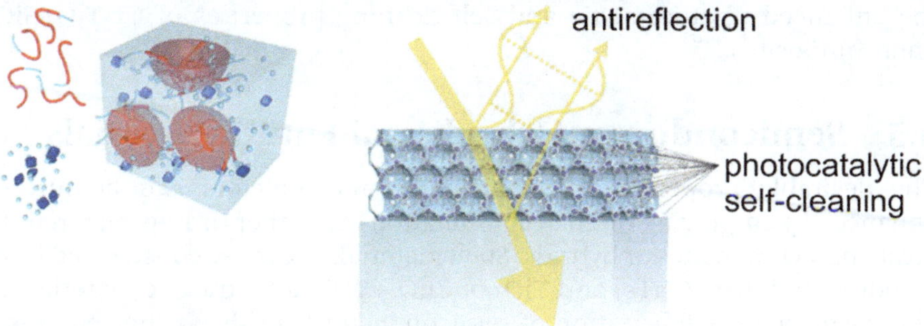

Figure 8.9 Antireflection coating prepared using self-assembled block copolymer with sol–gel silica and anatase titania nanocrystals.[117] (Reprinted with permission from S. Guldin, P. Kohn, M. Stefik, J. Song, G. Divitini, F. Ecarla, C. Ducati, U. Wiesner and U. Steiner, *Nano Lett.*, 2013, **13**, 5329. Copyright (2013) American Chemical Society.)

the hydroxyl groups on the surface, higher surface roughness SiO_2 coatings usually display hydrophilicity in the absence of light irradiation. However, increasing surface coverage by the pollutants tends to decrease the natural hydrophilicity.[117]

8.4.3 Metal Doped Coatings

Various methods have been utilised to fabricate superhydrophilic TiO_2/SiO_2 composite films that exhibit high transmittance, antireflective and antifogging properties, attained from the photocatalytic activity of TiO_2 and superhydrophilicity of SiO_2. Multifunctional coatings of TiO_2/SiO_2 that demonstrate self-cleaning, antifogging, antireflection with high mechanical and chemical endurance have been prepared by employing a TiO_2 bilayer and a methyl functionalised SiO_2 matrix.[118] Chromium-doped TiO_2/SiO_2 composite films have been prepared using a sol–gel dip coating technique.[119] The Cr-doped composite film exhibited redshifted absorption when compared to the TiO_2/SiO_2 composite. This has been attributed to the lower band gap energy of the doped material, which is due to the electronic transition from the d orbital of the Cr^{3+} ion to the TiO_2 conduction band. The Cr-doped TiO_2/SiO_2 film exhibited superhydrophilicity after 120 min of irradiation using a xenon lamp (300–500 nm) and this was also maintained in the dark. Surface hydroxyl groups in the Cr-doped TiO_2/SiO_2 film play an important role in maintaining the superhydrophilic nature of the TiO_2/SiO_2 composite films, especially in the absence of light irradiation. The photoinduced self-cleaning effect observed in TiO_2/SiO_2 composites is attributed to the combined hydrophilic nature of the SiO_2 surface, along with the photocatalytic activity of TiO_2, which together cause the oxidation of organic contaminants. Doped Cr^{3+} ions can potentially act as hole trapping sites, enhancing charge carrier separation. The photogenerated holes can then produce hydroxyl radicals from the surface adsorbed hydroxyl ions, which therefore accounts for the enhanced photocatalytic and self-cleaning properties of Cr–TiO_2/SiO_2 nanocomposites.[119]

8.5 Semiconductor Hybrids and Future Materials

The desirable properties of individual semiconductors can be further enhanced by employing them in combination with other various materials to create new composites or hybrids. Such materials are those constructed from carbon nanotubes (CNTs) and TiO_2 or ZnO. CNTs are attractive materials to employ during the fabrication of semiconductor hybrids, as they possess a larger work function which enhances the electron flow from the semiconductor oxide to the CNT. The flow of electrons from the metal oxide produces negatively charged species on the CNT surface and positively charged species in the semiconductor material. This excess negative and positive charge formed at the interface results in an area vacant of mobile charge carriers, termed the depletion layer, which maintains charge separation in the hybrid materials.[120,121] This charge transfer capability, along with the large surface area of CNTs, increases the electron–hole pair recombination time by trapping the electrons generated during photocatalysis. Kongkanand and Kamat reported that every 32 carbons can store one electron and therefore nanohybrids may store many such electrons, leading to increase in recombination time. The recombination time of electron–hole pairs in CNT–metal oxide hybrids has been measured using laser flash photolysis as approx. 10^{-9} s, while the time required for the degradation of pollutants is in the range of 10^{-8}–10^{-3} s.[122] Many published papers have reported the photocatalytic activity of such hybrids.[87,123–125] When composites or hybrids of CNTs and metal oxides are prepared, the CNTs are in close contact with the metal oxide, producing a barrier junction. This is a very effective way of reducing the recombination of electron–hole pairs, as the CNTs act as electron acceptors, while the semiconductor material acts as the electron donor.[126] In addition, due to the larger surface area of CNTs, more electrons are capable of reducing oxygen molecules to superoxide radicals, yielding a higher rate of organic contaminant decomposition. In conjunction with this, the corresponding holes generate hydroxyl radicals which are also involved in pollutant decomposition. Together, the two processes induce the formation of oxygen vacancies on the metal oxide, and since these vacancies are occupied by water molecules this induces the generation of more hydroxyl groups on the surface of the hybrid, thereby increasing the photocatalytic activity (Figure 8.10). The presence of more hydroxyl groups on the surface makes the hybrid more hydrophilic, which increases the self-cleaning capacity of the hybrid.

8.5.1 Carbon Nanotube Hybrids of TiO_2 or ZnO

When a hybrid is prepared with TiO_2 and CNT, the composition of TiO_2 and CNT plays a significant role in the photocatalytic activity. Many research groups have studied the effect of multi-walled nanotube (MWNT)/TiO_2 hybrids

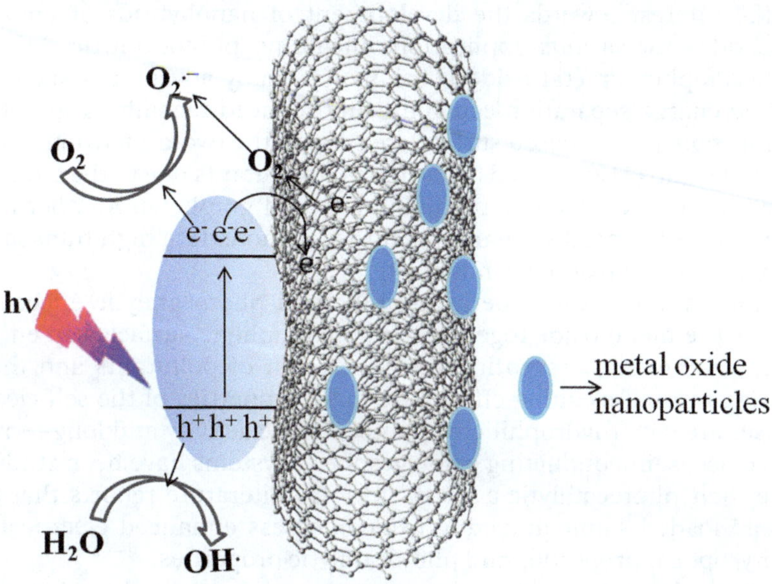

Figure 8.10 Schematic representation of the photocatalytic activity of a CNT–semiconductor oxide.

under UV and visible light using model dyes.[127,128] Excess amounts of MWNTs shield the TiO_2 from UV absorption by scattering the photons and, in contrast, an excess amount of TiO_2 yields a large gap between the TiO_2 particles and CNTs, reducing the electron trapping ability of the CNTs. Therefore, an optimum percentage composition is essential to attain maximum photocatalytic activity.[129] Similarly, many research investigations have demonstrated that good photocatalytic activity is also possessed by CNT/ZnO hybrids. Experiments have demonstrated that these hybrids exhibit increased photocatalytic activity during the degradation of model dyes, such as MB, methyl orange and rhodamine,[130–133] when compared to ZnO alone. Although the excellent hydrophilicity of TiO_2/CNT hybrids has not been adequately explored, it appears to be a very promising material suitable for superhydrophilic applications.

The adsorption of pollutants is enhanced by the larger surface area and by the presence of a large number of active sites on the CNTs. In addition, the CNTs aid in electron transfer due to their high electrical conductivity and high electron storage capacity. The Schottky barrier generated between the CNTs and metal oxides facilitates the transfer of electrons through the CNT surface, leaving holes in the valance band of the metal oxides.

8.5.2 Graphene Hybrids of Metal Oxides

The high surface area, excellent conductivity, fast electron transfer capabilities and high transparency of graphene, along with its ability to accept electrons from the conduction band of metal oxides, have spurred great scientific interest towards the development of nanohybrids of graphene/metal oxides for various applications including photocatalytic self-cleaning. The graphene/metal oxide hybrids offer many advantages such as (1) improved charge separation capability and reduced recombination of electron–hole pair, (2) increased surface area due to the two-dimensional nature of graphene and (3) enhanced adsorption of pollutants due to the strong π–π interaction between the organic pollutants and graphene. Another advantage of graphene based systems is that they demonstrate high transparency and high two-dimensional conductivity.

A combination of the superhydrophilic and photocatalytic effects, possessed by the metal oxide together with the graphene surface, can enhance the antistatic effect, preventing the adsorption of pollutants and thereby improving the self-cleaning effect. The main properties of the self-cleaning processes are superhydrophilicity, photocatalytic activity and long-term stability. Hence, semiconducting oxide/graphene systems have been studied to explore their photocatalytic activity.[134–137] The literature reports that these graphene loaded semiconducting oxides possess enhanced electrical conductivity, optical properties and photocatalytic properties.[138–140]

During the last two to three years, some research studies have focused on the self-cleaning applications of these hybrids and some of these systems are discussed below. Importantly, Radich and Kamat have investigated Au NP-catalysed hydroxyl radical attack on RGO (holey graphene) which produced nanopores and wrinkles *via* oxidation of the material.[141]

8.5.3 Graphene/TiO_2 Nanohybrids

Numerous studies regarding photocatalysis have been reported, especially based on TiO_2 as the photocatalyst. The efficiency of TiO_2 is limited due to the rapid recombination of the photoexcited electron–hole pair and the lack of visible light absorption. Therefore, many studies have been performed employing TiO_2 loaded with noble metals. These materials can reduce oxygen to H_2O, peroxides and other species by increasing the recombination time;[142,143] however, they are very costly. To overcome the drawbacks of TiO_2 and noble metal loaded TiO_2, research has been recently directed towards the development of TiO_2/graphene hybrids, which utilise the high electron acceptor capabilities of graphene, so that facile electron transfer occurs through the graphene, thereby increasing the recombination time.

Anandan *et al.* have studied the self-cleaning properties of TiO_2/graphene hybrids.[144] They found that the resistivity of the TiO_2/graphene hybrid was similar to that of a graphene layer, while the resistivity of a TiO_2 film alone was very high. The photocatalytic activity of the graphene loaded TiO_2 was recorded as twofold greater than that of the analogous TiO_2 film and the photocatalytic efficiency of the material improved upon increasing the graphene concentration. These properties were attributed to the electron transfer capabilities of the graphene layer. Furthermore, the hydrophilic conversion efficiency of the hybrid was also greater than that of the TiO_2 film alone.[144]

The decomposition of surface contaminants requires a high oxidative capacity from the photogenerated holes in TiO_2 and this capacity can be increased by the presence of highly conductive graphene. Lee *et al.* reported the synthesis TiO_2 nanorod-decorated graphene sheets followed by photocatalytic studies under visible light irradiation.[145] The graphene decorated TiO_2 nanorod was prepared *via* sol–gel synthesis of TiO_2 in the presence of exfoliated graphene oxide, while the photocatalytic degradation studies, based on MB degradation, confirmed the enhanced photocatalytic activity due to the presence of graphene. It was demonstrated in this study that an optimum percentage of graphene was essential to achieve high efficiency. Employing higher percentages of graphene in the nanohybrids resulted in competition between TiO_2 and graphene for the absorption of light, leading to fewer electron–hole pairs on the TiO_2 surface. The authors reported that graphene oxide of 5 wt%, with respect to the nanohybrid, yielded maximum photocatalytic efficiency. In addition, it was clear that the photocatalytic process supported the generation of superhydrophilic surfaces and together these phenomena resulted in self-cleaning.[145] Figure 8.11 shows the charge transfer mechanism in a TiO_2/graphene film during the photocatalytic process.

Recently, Zhu and He reported the facile fabrication of a TiO_2/graphene hybrid *via* a layer-by-layer self-assembly technique, enabling an enhanced optical transparency and self-cleaning effect.[146] These hybrid films were fabricated using dip coating techniques. After illumination with a broad band spectrum of light, the films exhibited superhydrophilicity. Since TiO_2 is a

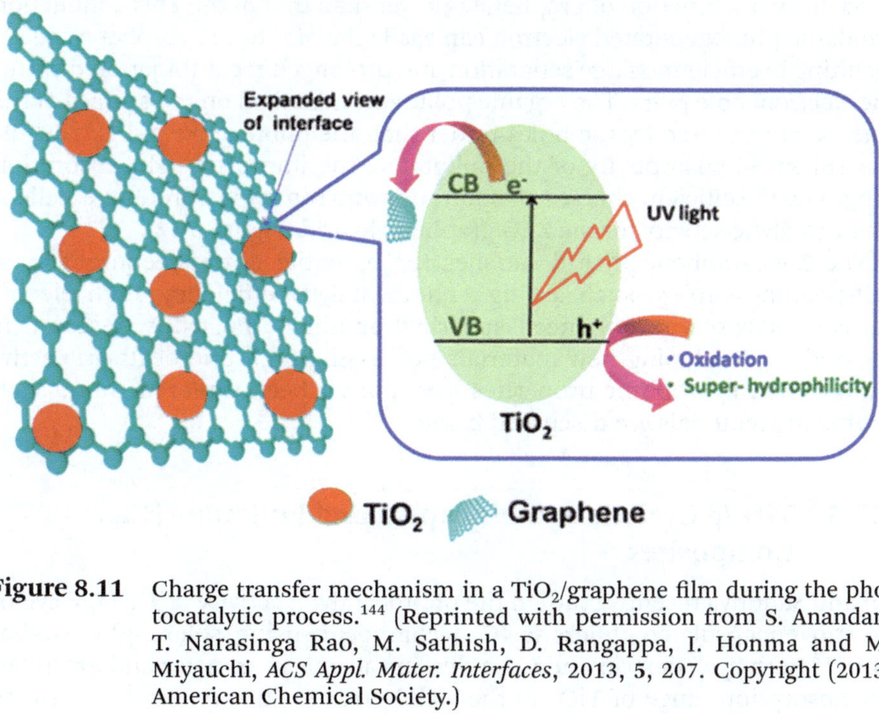

Figure 8.11 Charge transfer mechanism in a TiO_2/graphene film during the photocatalytic process.[144] (Reprinted with permission from S. Anandan, T. Narasinga Rao, M. Sathish, D. Rangappa, I. Honma and M. Miyauchi, *ACS Appl. Mater. Interfaces*, 2013, 5, 207. Copyright (2013) American Chemical Society.)

UV absorber and graphene is both a UV and visible light absorber, a greater amount of electron–hole pairs are generated. The graphene sheet acted as a supporting medium for the effective transfer of electrons, before recombination with holes in the valance band of TiO_2 could occur. The holes generated in the TiO_2 could then react with lattice oxygen leading to the adsorption of water molecules,[144] making the hybrid surface superhydrophilic. The antistatic property of this hybrid film was found to be very high and the combination of this property with the superhydrophilicity enhanced the self-cleaning property of the hybrid film.[146]

8.5.4 ZnO/Graphene Nanohybrids

Zinc oxide (ZnO) is commonly used as a heterogeneous photocatalyst due to its low cost and ease of availability.[147] The issue of fast recombination of the photogenerated electron–hole pairs can be solved by hybridising ZnO with graphene. Single layer graphene is considered a zero band gap material and hybridising it with ZnO *via* modification of graphene's chemical structure can alter the band gap.[148,149] This results in charge transfer between ZnO and graphene, enhancing the life time of photogenerated charge carriers.[150] In addition, the hybridisation of graphene can enhance the light absorption range and intensity, which will result in the visible light photocatalysis of ZnO.[151]

When a sample is irradiated with UV light, the electrons in the ZnO valence band are excited to the conduction band, leaving holes in the valence band. Since the work function of graphene is lower than that of the ZnO conduction band, the photogenerated electron can easily transfer to the graphene sheets, resulting in efficient charge separation and prolonged recombination time for the electron–hole pairs. The organic pollutants adsorbed on the surface of the catalyst are oxidised by the holes rather than the photoactive radicals. Thus, the enhanced adsorptivity of the pollutants, the improved light absorption range and the efficient charge transportation and separation produce excellent photocatalytic activity for the ZnO/graphene hybrid (Figure 8.12).

The ZnO/graphene hybrid satisfies the essential criteria required for a self-cleaning surface such as high photocatalytic efficiency, high electrical conductivity and enhanced superhydrophilicity. Presently, research is focused on developing new materials possessing high photocatalytic activity, but their hydrophilic properties have not yet been explored. Some such promising materials are discussed below.

8.5.5 TiO_2/β Cyclodextrin Encapsulated Fullerene (C_{60}) Composites

Ali and Sandhya recently studied the photocatalytic activity of a TiO_2/β cyclodextrin encapsulated fullerene (C_{60}) composite using solar light irradiation.[152] Both cyclodextrin and C_{60} aided in the charge transfer and extended the absorption range of TiO_2 to the visible range. They reported 2–5-times

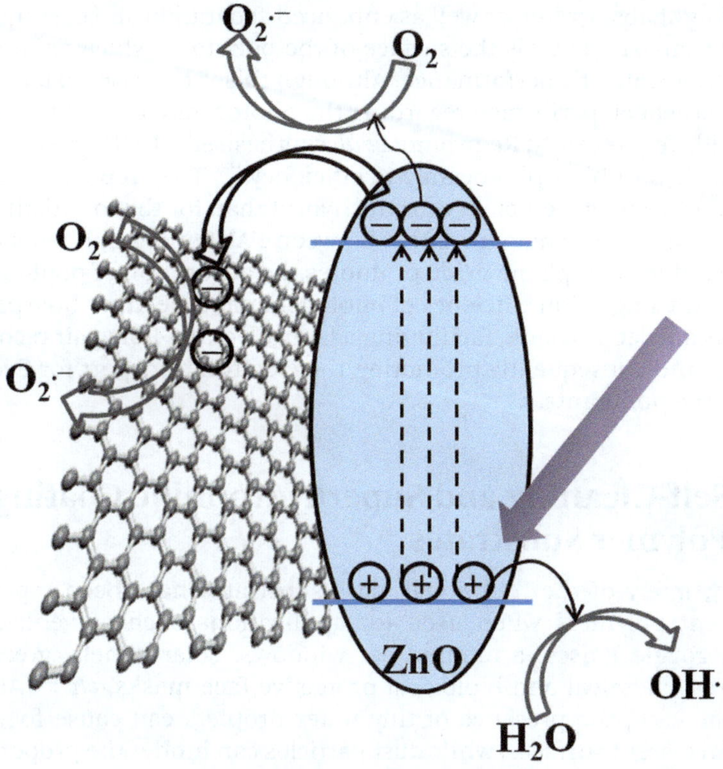

Figure 8.12 Schematic representation of photocatalytic activity in ZnO/graphene hybrids.

greater photocatalytic degradation efficiency for this composite when compared to bare TiO_2. In addition, the recombination of electron–hole pair was reduced by the presence of the cyclodextrin and C_{60} molecules.[142]

8.5.6 Conducting Polyaniline/Metal Oxide or Graphene Oxide Hybrids

Among the conducting polymers, polyaniline (PANI), a conjugated polymer with an extended π-conjugated system, is widely studied due to its ease of preparation, large doping–de-doping capability and high conductivity. PANI, with high visible light absorption capability, high charge carrier mobility and good environmental stability, is widely studied to improve the photocatalytic activity of inorganic semiconductors such as TiO_2, ZnO and CdS. This is achieved by reducing the recombination of photogenerated electron–hole pairs.[153–155]

The effective combination of graphite oxide (GO) and PANI is expected to be an ideal approach to enhance the photocatalytic properties. Graphene composites of PANI may possess high dye adsorption capability, extended range of light absorption, as well as enhanced separation and easy migration of charge carriers towards the surface of the pollutant, which enhances the overall photocatalytic performance. Although this is the case, so far, very few studies have been performed regarding the photocatalytic activity of a PANI/GO hybrid. Very recently, Remyamol *et al.* synthesised a PANI/graphene oxide hybrid with very high photocatalytic efficiency.[156] They reported ten times higher photocatalytic efficiency for the hybrid than for that of pristine PANI, possibly due to the synergetic effect between PANI and graphite oxide. The partially reduced graphene oxide promotes the adsorption of pollutants and increases the migration efficiency of photogenerated electron–hole pairs. GO can act as an electron sink, facilitating a large electron–hole pair recombination time and subsequently promoting the electron transport for decomposition of the pollutants.

8.6 Self-Cleaning and Superhydrophilic Coating on Polymer Substrates

Presently, many of the traditional glass substrates have been replaced by transparent polymers when used for applications such as vehicle windscreens, goggle lenses, armoured car windows, solar panels, greenhouse enclosures, chemical and biological protective face masks, *etc.*[157] Although this is the case, the presence of tiny water droplets can cause fogging on these transparent surfaces, while dust particles can hinder the proper utilisation of transparent polymer products. Many of the antifogging coatings that are available on the market require repeated wiping and cleaning to ensure their effectiveness. Superhydrophilic coatings can possess high antifogging and antireflective efficiency and this can ensure longer durability of the polymeric surfaces. Superhydrophilic coatings can also prevent the accumulation of dirt on the surface and are thereby considered self-cleaning.

Gemici *et al.* have developed a highly transparent, antifogging, antireflective self-cleaning coating using a layer-by-layer self-assembly technique.[157] They constructed a multilayer coating of inorganic NPs on a glass/silicon substrate using a layer-by-layer self-assembly technique while employing positively charged nano TiO_2 particles of approx. 7 nm in poly(vinyl sulfonate) and negatively charged nano SiO_2 (7–22 nm) in poly(dimethyl(diallyl)ammonium chloride). This multilayer self-cleaning coating yields high porosity and hence high superhydrophilicity. Researchers have attempted to fabricate superhydrophilic coatings on polymer substrates.[158,159] Kasanen *et al.* prepared a TiO_2 coating on a polymer surface and studied its photocatalytic effect.[160] They injected a nano TiO_2 suspension over a polyurethane (PU) surface and treated this with oxygen plasma to enhance the UV photocatalytic activity. As a measure of superhydrophilicity, they analysed the contact angle and found that the plasma treatment enhanced the photocatalytic activity and therefore in-turn the superhydrophilicity. Many research groups have attempted to prepare a photocatalytic coating on an optically transparent polycarbonate (PC) polymer.[158,161] Lam *et al.* prepared TiO_2 coated polycarbonate films and studied the influence of NaOH and UV radiation on its mechanical, superhydrophilic and self-cleaning properties.[162] They found that the NaOH treated TiO_2 coated PC films shows higher hydrophilicity, antifogging and self-cleaning properties due to the formation of greater number of –OH and –COOH functional groups. Fateh *et al.* synthesized highly hydrophilic and photocatalytic TiO_2/SiO_2 coated PC[163] and TiO_2/ZnO coated PCs[164] for self-cleaning applications.

Latthe *et al.* studied the wettability and self-cleaning property of a SiO_2/TiO_2 coating on PC using a spin-coating technique.[165] They claim that after UV irradiation the wettability of the coating has changed and the material becomes superhydrophilic. The superhydrophilicity along with nano-hardness characteristics of TiO_2 coated poly(vinyl chloride) (PVC) and polycarbonate substrates, fabricated *via* plasma enhanced chemical vapour deposition (PECVD), has been studied by Cavallin *et al.*[166] They found that the coating exhibits a contact angle of <10° which is a criterion for superhydrophilicity.

In contrast, Dong *et al.* fabricated superhydrophilic surfaces on a glass substrate using a mixture of polymers containing reactive groups such as trimethoxysilyl (TMOS).[167] They used quaternised poly[2-(dimethylamino)ethyl methacrylate] (PQDMAEMA) and poly-[3-(trimethoxysilyl)propyl methacrylate] (PTMOSPMA) polymers. SiO_2 NPs were mixed with these polymers and TMOS groups aided in the anchoring of the polymer–SiO_2 nanocomposite coatings onto glass substrates. The superhydrophilic surfaces displayed high stability, antifogging and self-cleaning characteristics. Furthermore, due to the presence of quaternary ammonium group on the polymers, the coating is very suitable for its antimicrobial activities against *E. coli*.

8.7 Commercial Materials

Self-cleaning glass has been developed by Pilkington Glass termed Activ™ along with products from other companies marketed as Radiance Ti™, Sunclean™ and Bioclean™. The Pilkington Glass product Activ™ consists of a film of TiO_2, which in the presence of UV light reduces its WCA from 67° to 0°, while possessing a low visible reflectance of approx. 7%.[168] Bioclean™, produced by Saint Gobain Glass, is constructed of a transparent coating on the glass, which can be either hydrophilic or photocatalytic. TOTO Ltd. has developed a range of products with photocatalytic technology, marketed as Hydrotect™.[20] One such product is a ceramic panel, possessing a micrometre thick layer of TiO_2 adhered to the surface. This coating is antimicrobial and has the ability to degrade pollutants. The companies Biocera and Karpery have also used their knowledge of photocatalytic materials to manufacture ceramic products with antimicrobial and deodorising properties. Similarly, the company VitrA has fabricated an antibacterial and antimicrobial tile, which can eliminate NO_x pollutants from the air. Balcony System Solutions have produced a DIY product called BalcoNano™ available in a sachet, which can be applied to windows, ceramic, porcelain and marble periodically

Figure 8.13 Representation of pitted and rough glass beneath a hydrophobic BalcoNano™ coating.[169] (Reprinted from Sol. Energy Mater. Sol. Cells, Vol. 109, K. Midtdal and B. P. Jelle, Self-cleaning glazing products: A state-of-the-art review and future research pathways, Page 126, Copyright (2013), with permission from Elsevier.)

after installation. This product is capable of inhibiting the gradual increase of material opacity and discolouration by increasing the hydrophobicity of the surface upon application (Figure 8.13).[169] Not only have TiO_2 coatings been developed for the present market but the manufacturer Gambarelli has employed the use of TiO_2 NPs in the processing of its own photocatalytic tiles.

Acknowledgements

The authors wish to acknowledge financial support under the U.S.-Ireland R&D Partnership programme from the Science Foundation Ireland (SFI-grant number 10/US/I1822(T)) and U.S. National Science Foundation-CBET (Award 1033317).

References

1. C. Han, J. Andersen, S. C. Pillai, R. Fagan, P. Falaras, J. A. Byrne, P. S. M. Dunlop, H. Choi, W. Jiang, K. O'Shea and D. D. Dionysiou, *ACS Symp. Ser.*, 2013, **1124**, 201.
2. D. Synnott, N. Nolan, D. Ryan, J. Colreavy and S. C. Pillai, in *Nanotechnology in Eco-Efficient Construction*, ed. F. Pacheco-Torgal, M. V. Diamanti, A. Nazari and C. G. Granqvist, Woodhead Publishing, Cambridge, UK, 1st edn, 2013, vol. 1, ch. 14, pp. 327–342.
3. K. Qi, W. A. Daoud, J. H. Xin, C. L. Mak, W. Tang and W. P. Cheung, *J. Mater. Chem.*, 2006, **16**, 4567.
4. W. S. Tung and W. A. Daoud, *J. Mater. Chem.*, 2011, **21**, 7858.

5. H. Yaghoubi, N. Taghavinia and E. K. Alamdari, *Surf. Coat. Technol.*, 2010, **204**, 1562.
6. W. S. Tung and W. A. Daoud, *J. Appl. Polym. Sci.*, 2009, **112**, 235.
7. F. Bondioli, R. Taurino and A. M. Ferrari, *J. Colloid Interface Sci.*, 2009, **334**, 195.
8. T.-H. Xie and J. Lin, *J. Phys. Chem. C*, 2007, **111**, 9968.
9. R. Wang, K. Hashimoto, A. Fujishima, M. Chikuni, E. Kojima, A. Kitamura, M. Shimohigoshi and T. Watanabe, *Nature*, 1997, **388**, 431.
10. I. P. Parkin and R. G. Palgrave, *J. Mater. Chem.*, 2005, **15**, 1689.
11. A. Mills, A. Lepre, N. Elliott, S. Bhopal, I. P. Parkin and S. A. O'Neill, *J. Photochem. Photobiol., A*, 2003, **160**, 213.
12. T. Watanabe, A. Nakajima, R. Wang, M. Minabe, S. Koizumi, A. Fujishima and K. Hashimoto, *Thin Solid Films*, 1999, **351**, 260.
13. R. Fateh, A. A. Ismail, R. Dillert and D. W. Bahnemann, *J. Phys. Chem. C*, 2011, **115**, 10405.
14. J. O. Carneiro, V. Teixeira, A. Portinha, A. Magalhaes, P. Coutinho, C. J. Tavares and R. Newton, *Mater. Sci. Eng., B*, 2007, **138**, 144.
15. M. Fallet, H. Mahdjoub, B. Gautier and J.-P. Bauer, *J. Non-Cryst. Solids*, 2001, **293–295**, 527.
16. A. Nazeri, P. P. Trzaskoma-Paulette and D. Bauer, *J. Sol-Gel Sci. Technol.*, 1997, **10**, 317.
17. J. C. Yu, W. Ho, J. Lin, H. Yip and P. K. Wong, *Environ. Sci. Technol.*, 2003, **37**, 2296.
18. A. Fernandez, G. Lassaletta, V. M. Jimenez, A. Justo, A. R. Gonzalez-Elipe, J. M. Herrmann, H. Tahiri and Y. Ait-Ichou, *Appl. Catal., B*, 1995, **7**, 49.
19. S. Nishimoto and B. Bhushan, *RSC Adv.*, 2013, **3**, 671.
20. L. Zhang, R. Dillert, D. Bahnemann and M. Vormoor, *Energy Environ. Sci.*, 2012, **5**, 7491.
21. A. Fujishima, T. N. Rao and D. A. Tryk, *J. Photochem. Photobiol., C*, 2000, **1**, 1.
22. R. N. Wenzel, *Ind. Eng. Chem.*, 1936, **28**, 988.
23. A. B. D. Cassie and S. Baxter, *Trans. Faraday Soc.*, 1944, **40**, 546.
24. M. Miwa, A. Nakajima, A. Fujishima, K. Hashimoto and T. Watanabe, *Langmuir*, 2000, **16**, 5754.
25. A. Borras and A. R. Gonzalez-Elipe, *Langmuir*, 2010, **26**, 15875.
26. S. Banerjee, D. D. Dionysiou and S. C. Pillai, *Appl. Catal., B*, 2015, **176**, 396.
27. T. Zubkov, D. Stahl, T. L. Thompson, D. Panayotov, O. Diwald and J. T. Yates, Jr., *J. Phys. Chem. B*, 2005, **109**, 15454.
28. N. Sakai, A. Fujishima, T. Watanabe and K. Hashimoto, *J. Phys. Chem. B*, 2001, **105**, 3023.
29. N. Sakai, A. Fujishima, T. Watanabe and K. Hashimoto, *J. Phys. Chem. B*, 2003, **107**, 1028.
30. X. Yan, R. Abe, T. Ohno, M. Toyofuku and B. Ohtani, *Thin Solid Films*, 2008, **516**, 5872.

31. M. Takeuchi, K. Sakamoto, G. Martra, S. Coluccia and M. Anpo, *J. Phys. Chem. B*, 2005, **109**, 15422.
32. M. Takeuchi, K. Sakamoto, G. Martra, S. Coluccia and M. Anpo, *J. Phys. Chem. B*, 2005, **109**, 15422.
33. M. Miyauchi, A. Nakajima, T. Watanabe and K. Hashimoto, *Chem. Mater.*, 2002, **14**, 2812.
34. A. V. Emeline, A. V. Rudakova, M. Sakai, T. Murakami and A. Fujishima, *J. Phys. Chem.C*, 2013, **117**, 12086.
35. T. K. Tseng, Y. S. Lin, Y. J. Chen and H. Chu, *Int. J. Mol. Sci.*, 2010, **11**, 2336.
36. P. Shao, J. Tian, Z. Zhao, W. Shi, S. Gao and F. Cui, *Appl. Surf. Sci.*, 2015, **324**, 35.
37. W. Raza, M. M. Haque and M. Muneer, *Appl. Surf. Sci.*, 2014, **322**, 215.
38. Z. Shen, Z. Hu, W. Wang, S.-F. Lee, D. K. L. Chan, Y. Li, T. Gu and J. C. Yu, *Nanoscale*, 2014, **6**, 14163.
39. Q. Ji, Y. Zhang, Y. Zhang and Z. Liu, *Chem. Soc. Rev.*, 2015, **44**, 2587–2602.
40. R. Quesada-Cabrera, C. Sotelo-Vazquez, J. C. Bear, J. A. Darr and I. P. Parkin, *Adv. Mater. Interfaces*, 2014, **1**, 1400069.
41. J. Liu, B. Liu, Z. Ni, Y. Deng, C. Zhong and W. Hu, *Electrochim. Acta*, 2014, **150**, 146.
42. K. R. Raksha, S. Ananda and N. M. Madegowda, *J. Mol. Catal. A: Chem.*, 2015, **396**, 319.
43. F.-X. Xiao, S.-F. Hung, H. B. Tao, J. Miao, H. B. Yang and B. Liu, *Nanoscale*, 2014, **6**, 14950.
44. H. Yang, J. Ouyang, A. Tang, Y. Xiao, X. Li, X. Dong and Y. Yu, *Mater. Res. Bull.*, 2006, **41**, 1310.
45. M. A. Mahadik, S. S. Shinde, H. M. Pathan, K. Y. Rajpure and C. H. Bhosale, *J. Photochem. Photobiol., B*, 2014, **141**, 315.
46. S. Lee, Y.-I. Kang, S.-J. Ha and J. H. Moon, *RSC Adv.*, 2014, **4**, 55371.
47. K. Ravichandran, R. Mohan, B. Sakthivel, S. Varadharajaperumal, P. Devendran, T. Alagesan and K. Pandian, *Appl. Surf. Sci.*, 2014, **321**, 310.
48. S. Yuan, Q. Zhang, B. Xu, Z. Jin, Y. Zhang, Y. Yang, M. Zhang and T. Ohno, *RSC Adv.*, 2014, **4**, 62255.
49. T. Yan, Q. Yan, X. Wang, H. Liu, M. Li, S. Lu, W. Xu and M. Sun, *Dalton Trans.*, 2015, **44**, 1601.
50. J.-C. Sin, S.-M. Lam, K.-T. Lee and A. R. Mohamed, *Mater. Lett.*, 2015, **140**, 51.
51. C. Ottone, A. Lamberti, M. Fontana and V. Cauda, *J. Electrochem. Soc.*, 2014, **161**, D484.
52. (a) L. Pinho, M. Rojas and M. J. Mosquera, *Appl. Catal., B*, 2015, **178**, 144; (b) W. Shen, C. Zhang, Q. Li, W. Zhang, L. Cao and J. Ye, *J. Cleaner Prod.*, 2015, **87**, 762.
53. D. Gregori, I. Benchenaa, F. Chaput, S. Therias, J. L. Gardette, D. Leonard, C. Guillard and S. Parola, *J. Mater. Chem. A*, 2014, **2**, 20096.

54. A. O. T. Patrocinio, L. F. Paula, R. M. Paniago, J. Freitag and D. W. Bahnemann, *ACS Appl. Mater. Interfaces*, 2014, **6**, 16859.
55. Y. Chen, C. Zhang, W. Huang, C. Yang, T. Huang, Y. Situ and H. Huang, *Surf. Coat. Technol.*, 2014, **258**, 531.
56. S. Ortelli, M. Blosi, C. Delpivo, D. Gardini, M. Dondi, I. Gualandi, D. Tonelli, V. Aina, I. Fenoglio, A. A. Gandhi, S. A. M. Tofail and A. L. Costa, *J. Photochem. Photobiol., A*, 2014, **292**, 26.
57. Q. Mao, D. Zeng, K. Xu and C. Xie, *RSC Adv.*, 2014, **4**, 58101.
58. L. Pinho and M. J. Mosquera, *J. Phys. Chem. C*, 2011, **115**, 22851.
59. S. S. Shinde, C. H. Bhosale, K. Y. Rajpure and J. H. Lee, *J. Photochem. Photobiol., B*, 2014, **141**, 210.
60. K. R. Reddy, M. Hassan and V. G. Gomes, *Appl. Catal., A*, 2015, **489**, 1.
61. Q. Wang, J. Qiao, R. Jin, X. Xu and S. Gao, *J. Power Sources*, 2015, **277**, 480.
62. J.-P. Niemela and M. Karppinen, *Dalton Trans.*, 2015, **44**, 591.
63. M. Wang, D. Zheng, M. Ye, C. Zhang, B. Xu, C. Lin, L. Sun and Z. Lin, *Small*, 2015, 1435.
64. W. S. Lee, Y.-S. Park and Y.-K. Cho, *Analyst*, 2015, **140**, 616.
65. I. Chauhan and P. Mohanty, *RSC Adv.*, 2014, **4**, 57885.
66. H. Mamane, I. Horovitz, L. Lozzi, D. D. Camillo and D. Avisar, *Chem. Eng. J.*, 2014, **257**, 159.
67. X. Zou, J. Shi and H. Zhang, *Aquat. Toxicol.*, 2014, **154**, 168.
68. S. Cao, B. Liu, L. Fan, Z. Yue, B. Liu and B. Cao, *Appl. Surf. Sci.*, 2014, **309**, 119.
69. M. Pelaez, N. T. Nolan, S. C. Pillai, M. K. Seery, P. Falaras, A. G. Kontos, P. S. M. Dunlop, J. W. J. Hamilton, J. A. Byrne, K. O'Shea, M. H. Entezari and D. D. Dionysiou, *Appl. Catal., B*, 2012, **125**, 331.
70. D. A. Keane, K. G. McGuigan, P. F. Ibanez, M. I. Polo-Lopez, J. A. Byrne, P. S. M. Dunlop, K. O'Shea, D. D. Dionysiou and S. C. Pillai, *Catal. Sci. Technol.*, 2014, **4**, 1211.
71. M. B. Fisher, D. A. Keane, P. Fernandez-Ibanez, J. Colreavy, S. J. Hinder, K. G. McGuigan and S. C. Pillai, *Appl. Catal., B*, 2013, **130–131**, 8.
72. S. Banerjee, S. C. Pillai, P. Falaras, K. E. O'Shea, J. A. Byrne and D. D. Dionysiou, *J. Phys. Chem. Lett.*, 2014, **5**, 2543.
73. Y. Sawai, S. Nishimoto, Y. Kameshima, E. Fujii and M. Miyake, *Langmuir*, 2013, **29**, 6784.
74. S. Afzal, W. A. Daoud and S. J. Langford, *ACS Appl. Mater. Interfaces*, 2013, **5**, 4753.
75. Z. Zhao, H. Tan, H. Zhao, D. Li, M. Zheng, P. Du, G. Zhang, D. Qu, Z. Sun and H. Fan, *Chem. Commun.*, 2013, 8958.
76. D. O. Scanlon, C. W. Dunnill, J. Buckeridge, S. A. Shevlin, A. J. Logsdail, S. M. Woodley, C. R. A. Catlow, M. J. Powell, R. G. Palgrave, I. P. Parkin, G. W. Watson, T. W. Keal, P. Sherwood, A. Walsh and A. A. Sokol, *Nat. Mater.*, 2013, **12**, 798.
77. L. Liu, H. Zhao, J. M. Andino and Y. Li, *ACS Catal.*, 2012, **2**, 1817.

78. T. Ohno, K. Sarukawa and M. Matsumura, *J. Phys. Chem. B*, 2001, **105**, 2417.
79. D. A. H. Hanaor and C. C. Sorrell, *J. Mater. Sci.*, 2011, **46**, 855.
80. A. Mills, J. Hepburn, D. Hazafy, C. O'Rourke, J. Krysa, M. Baudys, M. Zlamal, H. Bartkova, C. E. Hill, K. R. Winn, M. E. Simonsen, E. G. Søgaard, S. C. Pillai, N. S. Leyland, R. Fagan, F. Neumann, C. Lampe and T. Graumann, *J. Photochem. Photobiol., A*, 2013, **272**, 18.
81. A. Mills and N. Wells, *J. Photochem. Photobiol., A*, 2015, **298**, 64.
82. A. Mills, C. O'Rourke and N. Wells, *Analyst*, 2014, **139**, 5409.
83. A. Mills, J. Hepburn, D. Hazafy, C. O'Rourke, N. Wells, J. Krysa, M. Baudys, M. Zlamal, H. Bartkova, C. E. Hill, K. R. Winn, M. E. Simonsen, E. G. Søgaard, S. Banerjee, R. Fagan and S. C. Pillai, *J. Photochem. Photobiol., A*, 2014, **290**, 63.
84. L. P. Dai, H. Deng, F. Y. Mao and J. D. Zang, *J. Mater. Sci.: Mater. Electron.*, 2008, **19**, 727.
85. A. Goudarzi, G. M. Aval, R. Sahraei and H. Ahmadpoor, *Thin Solid Films*, 2008, **516**, 4953.
86. Q. Li, S.-L. Chen and W.-C. Jiang, *J. Appl. Polym. Sci.*, 2007, **103**, 412.
87. C. Liu, H. Chen, K. Dai, A. Xue, H. Chen and Q. Huang, *Mater. Res. Bull.*, 2013, **48**, 1499.
88. R. Asahi, T. Morikawa, T. Ohwaki, K. Aoki and Y. Taga, *Science*, 2001, **293**, 269.
89. N. T. Nolan, D. W. Synnott, M. K. Seery, S. J. Hinder, A. Van Wassenhoven and S. C. Pillai, *J. Hazard. Mater.*, 2012, **211–212**, 88.
90. H. Irie, Y. Watanabe and K. Hashimoto, *Chem. Lett.*, 2003, **32**, 772.
91. (a) S. Sakthivel and H. Kisch, *Angew. Chem., Int. Ed.*, 2003, **42**, 4908; (b) V. Etacheri, G. Michlits, M. K. Seery, S. J. Hinder and S. C. Pillai, *ACS Appl. Mater. Interfaces*, 2013, **5**, 1663.
92. D. Li, N. Ohashi, S. Hishita, T. Kolodiazhnyi and H. Haneda, *J. Solid State Chem.*, 2005, **178**, 3293.
93. D. B. Hamal and K. J. Klabunde, *J. Colloid Interface Sci.*, 2007, **311**, 514.
94. C. W. Dunnill, Z. Ansari, A. Kafizas, S. Perni, D. J. Morgan, M. Wilson and I. P. Parkin, *J. Mater. Chem.*, 2011, **21**, 11854.
95. E. Borgarello, J. Kiwi, M. Graetzel, E. Pelizzetti and M. Visca, *J. Am. Chem. Soc.*, 1982, **104**, 2996.
96. A. Di Paola, E. Garcia-Lopez, S. Ikeda, G. Marci, B. Ohtani and L. Palmisano, *Catal. Today*, 2002, **75**, 87.
97. M. Pelaez, B. Baruwati, R. S. Varma, R. Luque and D. D. Dionysiou, *Chem. Commun.*, 2013, 10118.
98. M. Houmard, G. Berthomé, J. C. Joud and M. Langlet, *Surf. Sci.*, 2011, **605**, 456.
99. M. Houmard, D. Riassetto, F. Roussel, A. Bourgeois, G. Berthome, J. C. Joud and M. Langlet, *Surf. Sci.*, 2008, **602**, 3364.
100. Y. Y. Liu, L. Q. Qian, C. Guo, X. Jia, J. W. Wang and W. H. Tang, *J. Alloys Compd.*, 2009, **479**, 532.

101. A. Tricoli, M. Righettoni and S. E. Pratsinis, *Langmuir*, 2009, **25**, 12578.
102. K. Guan, B. Lu and Y. Yin, *Surf. Coat. Technol.*, 2003, **173**, 219.
103. X. Zhang, F. Zhang and K.-Y. Chan, *Appl. Catal., A*, 2005, **284**, 193.
104. K. Guan, *Surf. Coat. Technol.*, 2005, **191**, 155.
105. J.-J. Wang, D.-S. Wang, J. Wang, W.-L. Zhao and C.-W. Wang, *Surf. Coat. Technol.*, 2011, **205**, 3596.
106. T. Huang, W. Huang, C. Zhou, Y. Situ and H. Huang, *Surf. Coat. Technol.*, 2012, **213**, 126.
107. E. Pakdel, W. A. Daoud and X. Wang, *Appl. Surf. Sci.*, 2013, **275**, 397.
108. E. Pakdel and W. A. Daoud, *J. Colloid Interface Sci.*, 2013, **401**, 1.
109. R. Wang, X. Wang and J. H. Xin, *ACS Appl. Mater. Interfaces*, 2009, **2**, 82.
110. P. Nostell, A. Roos and B. Karlsson, *Thin Solid Films*, 1999, **351**, 170.
111. Y.-J. Lee, D. S. Ruby, D. W. Peters, B. B. McKenzie and J. W. P. Hsu, *Nano Lett.*, 2008, **8**, 1501.
112. B. S. Richards, *Sol. Energy Mater. Sol. Cells*, 2003, **79**, 369.
113. X.-T. Zhang, O. Sato, M. Taguchi, Y. Einaga, T. Murakami and A. Fujishima, *Chem. Mater.*, 2005, **17**, 696.
114. X. Li and J. He, *ACS Appl. Mater. Interfaces*, 2013, **5**, 5282.
115. Z. Liu, X. Zhang, T. Murakami and A. Fujishima, *Sol. Energy Mater. Sol. Cells*, 2008, **92**, 1434.
116. K. Nakata, M. Sakai, T. Ochiai, T. Murakami, K. Takagi and A. Fujishima, *Langmuir*, 2011, **27**, 3275.
117. S. Guldin, P. Kohn, M. Stefik, J. Song, G. Divitini, F. Ecarla, C. Ducati, U. Wiesner and U. Steiner, *Nano Lett.*, 2013, **13**, 5329.
118. M. Faustini, L. Nicole, C. Boissière, P. Innocenzi, C. Sanchez and D. Grosso, *Chem. Mater.*, 2010, **22**, 4406.
119. A. Eshaghi and A. Eshaghi, *Appl. Surf. Sci.*, 2012, **258**, 2464.
120. A. Linsebigler, G. Lu and J. Yates, *Chem. Rev.*, 1995, **95**, 735.
121. K. Woan, G. Pyrgiotakis and W. Sigmund, *Adv. Mater.*, 2009, **21**, 2233.
122. A. Kongkanand and P. V. Kamat, *ACS Nano*, 2007, **1**, 13.
123. A. Jitianu, T. Cacciaguerra, R. Benoit, S. Delpeux, F. Beguin and S. Bonnamy, *Carbon*, 2004, **42**, 1147.
124. L. Tian, L. Ye, J. Liu and L. Zan, *Catal. Commun.*, 2012, 99.
125. K. Byrappa, A. S. Dayananda, C. P. Sajan, B. Basavalingu, M. B. Shayan, K. Soga and M. Yoshimura, *J. Mater. Sci.*, 2008, **43**, 2348.
126. T. A. Saleh and V. K. Gupta, *J. Colloid Interface Sci.*, 2011, **362**, 337.
127. Y. Cong, X. Li, Y. Qin, Z. Dong, G. Yuan, Z. Cui and X. Lai, *Appl. Catal., B*, 2011, **107**, 128.
128. T. A. Saleh and V. K. Gupta, *J. Colloid Interface Sci.*, 2012, **371**, 101.
129. Y. Li, L. Li, C. Li, W. Chen and M. Zeng, *Appl. Catal., A*, 2012, **427–428**, 1.
130. L.-P. Zhu, G.-H. Liao, W.-Y. Huang, L.-L. Ma, Y. Yang, Y. Yu and S.-Y. Fu, *Mater. Sci. Eng., B*, 2009, **163**, 194.
131. L. Jiang and L. Gao, *Mater. Chem. Phys.*, 2005, **91**, 313.
132. K. Dai, G. Dawson, S. Yang, Z. Chen and L. Lu, *Chem. Eng. J.*, 2012, **191**, 571.

133. J. Qu, C. Luo and Q. Cong, *Micro Nano Lett.*, 2012, **7**, 1064.
134. H. Zhang, X.-J. Lv, Y.-M. Li, Y. Wang and J.-H. Li, *ACS Nano*, 2010, **4**, 380.
135. J. Zhang, Z. Zhu, Y. Tang and X. Feng, *J. Mater. Chem. A*, 2013, **1**, 3752.
136. T. Xu, L. Zhang, H. Cheng and Y. Zhu, *Appl. Catal., B*, 2011, **101**, 382.
137. H. Fan, X. Zhao, J. Yang, X. Shan, L. Yang, Y. Zhang, X. Li and M. Gao, *Catal. Commun.*, 2012, **29**, 29.
138. D. Wang, D. Choi, J. Li, Z. Yang, Z. Nie, R. Kou, D. Hu, C. Wang, L. V. Saraf, J. Zhang, I. A. Aksay and J. Liu, *ACS Nano*, 2009, **3**, 907.
139. A. Cao, Z. Liu, S. Chu, M. Wu, Z. Ye, Z. Cai, Y. Chang, S. Wang, Q. Gong and Y. Liu, *Adv. Mater.*, 2010, **22**, 103.
140. Y. H. Ng, A. Iwase, A. Kudo and R. Amal, *J. Phys. Chem. Lett.*, 2010, **1**, 2607.
141. J. G. Radich and P. V. Kamat, *ACS Nano*, 2013, **7**, 5546.
142. P. D. Cozzoli, E. Fanizza, R. Comparelli, M. L. Curri, A. Agostiano and D. Laub, *J. Phys. Chem. B*, 2004, **108**, 9623.
143. Q. Lu, Z. Lu, Y. Lu, L. Lv, Y. Ning, H. Yu, Y. Hou and Y. Yin, *Nano Lett.*, 2013, **13**, 5698.
144. S. Anandan, T. Narasinga Rao, M. Sathish, D. Rangappa, I. Honma and M. Miyauchi, *ACS Appl. Mater. Interfaces*, 2013, **5**, 207.
145. E. Lee, J.-Y. Hong, H. Kang and J. Jang, *J. Hazard. Mater.*, 2012, **219–220**, 13.
146. J. Zhu and J. He, *Int. J. Nanosci.*, 2012, **11**, 1240032/1.
147. C. Tian, Q. Zhang, A. Wu, M. Jiang, Z. Liang, B. Jiang and H. Fu, *Chem. Commun.*, 2012, 2858.
148. Z. Zhan, L. Zheng, Y. Pan, G. Sun and L. Li, *J. Mater. Chem.*, 2012, **22**, 2589.
149. Q. Bao and K. P. Loh, *ACS Nano*, 2012, **6**, 3677.
150. X. Bai, L. Wang, R. Zong, Y. Lv, Y. Sun and Y. Zhu, *Langmuir*, 2013, **29**, 3097.
151. B. Li, T. Liu, Y. Wang and Z. Wang, *J. Colloid Interface Sci.*, 2012, **377**, 114.
152. M. Mukthar Ali and K. Y. Sandhya, *Carbon*, 2014, **70**, 249.
153. H. Zhang, R. Zong and Y. Zhu, *J. Phys. Chem. C*, 2009, **113**, 4605.
154. J. Li, L. Zhu, Y. Wu, Y. Harima, A. Zhang and H. Tang, *Polymer*, 2006, **47**, 7361.
155. H. Zhang and Y. Zhu, *J. Phys. Chem. C*, 2010, **114**, 5822.
156. T. Remyamol, P. Gopinath and H. John, *Mater. Res. Express*, 2014, **1**, 035051/1.
157. Z. Gemici, M. R. Rubner and R. E. Cohen, US20080038458 A1, 2008.
158. P. S. Chinthamanipeta, S. Kobukata, H. Nakata and D. A. Shipp, *Polymer*, 2008, **49**, 5636.
159. J. Yang, Z. Zhang, X. Xu, X. Zhu, X. Men and X. Zhou, *J. Mater. Chem.*, 2012, **22**, 2834.
160. J. Kasanen, M. Suvanto and T. T. Pakkanen, *J. Appl. Polym. Sci.*, 2009, **111**, 2597.
161. D. K. Hwang, J. H. Moon, Y. G. Shul, K. T. Jung, D. H. Kim and D. W. Lee, *J. Sol-Gel Sci. Technol.*, 2003, **26**, 783.

162. S. Lam, A. Soetanto and R. Amal, *J. Nanopart. Res.*, 2009, **11**, 1971.
163. R. Fateh, A. A. Ismail, R. Dillert and D. W. Bahnemann, *J. Phys. Chem. C*, 2011, **115**, 10405.
164. R. Fateh, R. Dillert and D. Bahnemann, *ACS Appl. Mater. Interfaces*, 2014, **6**, 2270.
165. S. Latthe, S. Liu, C. Terashima, K. Nakata and A. Fujishima, *Coatings*, 2014, **4**, 497.
166. T. Cavallin, N. El Habra, M. Casarin, F. Bordin, A. Sartori, M. Favaro, R. Gerbasi and G. Rossetto, *J. Nanosci. Nanotechnol.*, 2011, **11**, 8079.
167. H. Dong, P. Ye, M. Zhong, J. Pietrasik, R. Drumright and K. Matyjaszewski, *Langmuir*, 2010, **26**, 15567.
168. P. Chin and D. F. Ollis, *Catal. Today*, 2007, **123**, 177.
169. K. Midtdal and B. P. Jelle, *Sol. Energy Mater. Sol. Cells*, 2013, **109**, 126.

CHAPTER 9

Photocatalysis and Photoelectrocatalysis for Energy Generation Using PhotoFuelCells

PANAGIOTIS LIANOS*[a,b], STAVROULA SFAELOU[a], AND LUCIAN-CRISTIAN POP[a]

[a]Department of Chemical Engineering, University of Patras, University Campus, 26500 Patras, Greece; [b]FORTH/ICE-HT, P.O. Box 1414, 26504 Patras, Greece
*E-mail: lianos@upatras.gr

9.1 Introduction

Conversion of solar energy into usable forms of energy can be achieved by a few different routes. The best known is photovoltaic conversion. When two materials with different concentrations of free electrons (different Fermi levels) come into contact, in order to achieve thermodynamic equilibrium, electrons move from the material with higher to the one with lower electron concentration. This creates a potential drop across the interface of the two materials. The zone over which electrons are displaced and the potential

RSC Energy and Environment Series No. 15
Photocatalysis: Applications
Edited by Dionysios D Dionysiou, Gianluca Li Puma, Jinhua Ye, Jenny Schneider, and Detlef Bahnemann

Published by the Royal Society of Chemistry, www.rsc.org

difference is established is called the depletion zone. In the case of photovoltaic conversion, at least one of the two materials absorbs light, for example, it is a semiconductor. Absorption of photons generates electron–hole pairs. These oppositely charged carriers are separated at the interface of the two materials because of the established electric potential difference. Thus the photons are transformed into electrical energy. The separated electrons can then flow through an external circuit and produce electric current. When the electrons arrive at the material where the separated holes remain, they fill them up and thus they complete the cycle of transformation of light into electricity. The open-circuit voltage created by a photovoltaic system is theoretically equal to the difference between the Fermi levels of the two materials in contact. In classical photovoltaics, the interface may be a Schottky contact, *i.e.*, a contact between a semiconductor and a metal, or a p–n junction, *i.e.*, a contact between two semiconductors, one of n-type and the second of p-type. For example, in commercial photovoltaics, silicon is doped to become partly n-type and partly p-type, thus creating a p–n junction in the same material. In the case of the recently developed third-generation mesoscopic solar cells, the interface is formed between semiconductor nanoparticles and a liquid electrolyte. In that case, it is not proper to refer to a depletion zone but to a potential difference, which is established between the Fermi level of the semiconductor and the redox level of the electrolyte. However, mesoscopic solar cells usually involve also a sensitizer, which is usually combined with an n-type nanostructured semiconductor. The sensitizer may be a molecule or a nanoparticulate semiconductor. The sensitizer has two principal characteristics, it absorbs visible light and its conduction band or its LUMO level is more negative (higher located) than the conduction band of the majority n-type semiconductor species. When the sensitizer is excited by absorption of photons, the photogenerated electrons are injected into the conduction band of the majority species while the holes remain in the sensitizer valence band or HOMO level. This is the first stage of electron–hole separation and it constitutes the sensitization process. The injected electrons are then led through an external circuit to the counter electrode, where they are exchanged with the redox species in solution and through the redox shuttle eliminate the holes left in the sensitizer. Thus the photovoltaic cycle is again fulfilled. So-called dye-sensitized solar cells function according to this route.[1] Their name comes from the fact that they utilize a dye sensitizer. Quantum-dot sensitized solar cells utilize quantum dot semiconductor sensitizers, for example, metal sulfide nanoparticulate semiconductors.[2] In the recently developed solid state sensitized solar cells, instead of a liquid electrolyte an organic or inorganic hole conductor is used, which scavenges photogenerated holes and carries them to the counter electrode, where they are eliminated by the externally arriving electrons.[3,4] In that case, the open circuit voltage corresponds to the potential difference between the conduction band of the majority semiconductor, which is the electron accepting level, and the valence band of the hole-transporting material, which is the hole-accepting level. The recently studied perovskite solar cells function with organometal halide perovskite-type sensitizers and organic hole transporters.[5] Finally, in the case of bulk heterojunction solar cells, electron–hole separation is achieved between two materials of different LUMO levels, one acting as donor and the other as acceptor of electrons.[6] The direction of electron flow is then defined not by a depletion zone but by the electrochemical potential difference between two electrodes that collect electrons and holes. Photovoltaic conversion of solar energy can then be reached by any of the above briefly described routes.

Solar energy can be also converted into electricity or can be stored as chemical energy by photoelectrochemical processes, which are based on a different concept. Photogenerated holes can be eliminated by oxidizing a sacrificial agent. For example, an organic substance can be photocatalytically degraded to mineralization by means of a semiconductor photocatalyst. If an n-type semiconductor photocatalyst is associated with an electrode, then the holes are consumed by the degradation of the sacrificial agent while the remaining photogenerated electrons are free to move through an external circuit and produce electricity. These electrons are collected by the counter electrode where they carry out reduction reactions and are consumed. In this type of solar cell, the photogenerated charge carriers do not move in a cyclic manner but they are consumed by respective oxidation and reduction reactions. Among the most studied oxidation reactions are mineralization of organic substances, oxidation of sulfide species and oxidation of water. Most studied reduction reactions include the formation of molecular hydrogen by reduction of protons, the reduction of atmospheric oxygen, the reduction of CO_2 and the reduction of water. The very popular water-splitting processes involve either water oxidation or water reduction or both.[7,8] Production of molecular hydrogen is the best means to convert solar energy into chemical energy and if the sacrificed agent is a waste or a pollutant then this photoelectrochemical conversion provides the double benefit of renewable energy conversion and environmental remediation.

The above discussion shows that there exist several ways to convert solar energy into a usable form of energy. Photoelectrochemical conversion of solar energy dates back to 1839 and to the works of Alexander-Edmond Becquerel.[9] However, a paper published in 1972 by Fujishima and Honda[10] has set the foundations of modern photoelectrochemistry. Since then, the concept of a photoelectrochemical cell has not changed but the new materials and the new procedures of their deposition has kept the field alive and in recent years it seems to have gathered new momentum. A photoelectrochemical cell that consumes a sacrificial agent to produce electricity or hydrogen behaves like a PhotoFuelCell (PFC),[11,12] since it photocatalytically consumes a "fuel" to produce electricity or chemical energy. The present chapter studies PFCs and presents a short review of our work in this domain.

9.2 Basic Features of PhotoFuelCell Operation

A PFC, as already said, is a standard photoelectrochemical cell.[12] It consists of a photoanode, which carries an n-type semiconductor photocatalyst, a cathode (or counter electrode), which carries a reduction electrocatalyst, and an electrolyte. It is possible to utilize an electrocatalytic anode and a photocathode carrying a p-type semiconductor photocatalyst but the most efficient cells are those with standard photoanode configuration. A typical n-type semiconductor photocatalyst is mesoporous titania, the most studied semiconductor photocatalyst. Its popularity stems from its chemical stability, efficiency, low cost and the ease of its synthesis and deposition. Because titania does not absorb visible light, it is combined with a sensitizer that broadens the range of photon usage. In aqueous environment, the choice of known sensitizers is unfortunately limited to only a few metal sulfide quantum dots.[13–18] A typical reduction electrocatalyst is a mixture of platinum with carbon nanoparticles[18] but a great effort has been made to substitute Pt with less costly materials.[19–22] Finally, the electrolyte is a simple salt like Na_2SO_4 or an alkaline electrolyte like NaOH. Alkaline electrolytes facilitate the oxidation process by the intermediate of OH^- anions, which are excellent hole scavengers at the conduction band of titania and facilitate hole consumption by creation of highly oxidative $^{\bullet}OH$ radicals ($OH^- + h^+ \rightarrow {}^{\bullet}OH$). The fuel is dissolved in the electrolyte. The electrolyte may involve a redox couple but this is not necessary, because the charge carriers are consumed at the photoanode and the cathode and it is not necessary to be shuttled between electrodes. What matters is the preservation of the composition of the electrolyte, as it will be discussed below. Figure 9.1 shows a schematic representation of a PFC.

In this design, ethanol is used as a model fuel. Ethanol is indeed studied as a model fuel since it is a product of biomass and among the most efficient PFC fuels. However, other small chain alcohols, diols, organic acids and carbohydrates are equally interesting as PFC organic fuels.[13,23,24]

Table 9.1 summarizes the basic oxidation and reduction reactions taking place during PFC operation. In the absence of oxygen, ethanol is photocatalytically reformed producing H_2 according to the following overall reaction:[25–29]

$$C_2H_5OH + 3H_2O \rightarrow 2CO_2 + 6H_2 \tag{9.1}$$

Ethanol is oxidized directly or by the intermediate hydroxyl radicals and through successive steps[25–27] leading to mineralization.

During these oxidation steps, a total of 12 holes are scavenged for the mineralization of one ethanol molecule. Hydrogen ions may be formed but chances of survival are higher in acidic or moderate pH. If all hydrogen ions survive, the oxidation of one ethanol molecule would lead to the formation of 12 hydrogen ions by the simultaneous consumption of three water molecules. These hydrogen ions can be reduced at the cathode

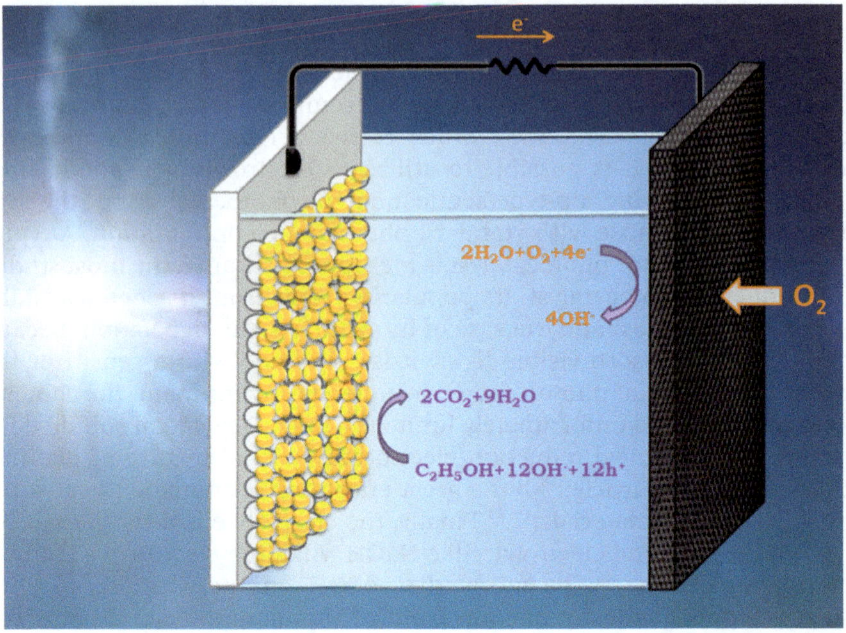

Figure 9.1 Schematic illustration of a PhotoFuelCell also showing ethanol oxidation and oxygen reduction.

(Reaction (V) of Table 9.1) to give molecular hydrogen. However, in alkaline pH, the number of hydrogen ions is very limited. In that case, hydrogen is produced by water reduction, as shown in Reaction (VI) of Table 9.1. In addition to hydrogen, hydroxyl ions are also produced thus preserving the population of hydroxyl ions in the electrolyte. Reaction (9.1) shows that, no matter what the pH value is, in the absence of oxygen, ethanol is photocatalytically reformed by consuming three water molecules and 12 photons (electron–hole pairs) to produce six hydrogen molecules for each ethanol molecule. Ethanol reforming in a PFC necessitates flow of electrons from the photoanode to the cathode electrode. The potential difference between the two electrodes under ideal conditions is defined by the level of the titania conduction band and the level of the redox reaction taking place at the cathode electrode. As seen in Figure 9.2, this difference is about 0.3 V. Due to inevitable losses, such a difference is too low to run the cell so that under real conditions it is necessary to apply an external bias in order to produce hydrogen.

In the presence of oxygen, ethanol is photocatalytically mineralized according to the following reaction:

$$C_2H_5OH + 3O_2 \rightarrow 2CO_2 + 3H_2O \tag{9.2}$$

For each ethanol molecule, 12 photons (electron–hole pairs) are consumed to perform the reaction. In an alkaline environment, reduction reaction at

Table 9.1 Typical reactions taking place in a PFC employing ethanol as model fuel.

Photoanode

(1) Reactions induced by absorption of UV photons (radiation mainly absorbed by the majority species, *i.e.* TiO_2):

$$TiO_2 \xrightarrow{h\nu} e^- + h^+ \qquad (I)$$

- Fate of the photogenerated electrons: most electrons flow in the external circuit. Some may interact with O_2, if present, producing superoxide and finally hydroxyl radicals
- Fate of the photogenerated holes:

$$C_2H_5OH + 3H_2O + 12h^+ \rightarrow 2CO_2 + 12H^+ \text{ (low pH)} \qquad (II)$$

$$OH^- + h^+ \rightarrow OH^\bullet \text{ and } C_2H_5OH + 12OH^\bullet \rightarrow 2CO_2 + 9H_2O \text{ (high pH)} \qquad (III)$$

- Intermediate steps may involve interaction of the fuel with photogenerated holes in the valence band of titania or holes injected into the valence band of the sensitizer:

$$C_2H_5OH + 2h^+ \rightarrow CH_3CHO + 2H^+ \text{ or } C_2H_5OH + h^+ \rightarrow C_2H_5O^\bullet + H^+ \qquad (IV)$$

(2) Reactions induced by absorption of visible light (exclusive excitation of the photosensitizer):

- Excited electrons are rapidly injected into the conduction band of TiO_2. Some may interact with O_2 as above. Most electrons flow through the external circuit. Holes at the valence band of the photosensitizer may go through Reactions (II)–(IV) depending on the nature of the photosensitizer
- Other schemes may also be possible

Cathode

Inert environment (no O_2 present)	Aerated electrolyte or cathode exposed to ambient air
• Low pH (potential: 0.00 V at pH 0)	• Low pH (potential: 1.23 V at pH 0)
$2H^+ + 2e^- \rightarrow H_2$ (V)	$2H^+ + 1/2O_2 + 2e^- \rightarrow H_2O$ (VII)
• Alkaline pH (potential: −0.77 V at pH 13)	• Alkaline pH (potential: 0.46 V at pH 13)
$2H_2O + 2e^- \rightarrow H_2 + 2OH^-$ (VI)	$H_2O + 1/2O_2 + 2e^- \rightarrow 2OH^-$ (VIII)

Overall cell reactions (combination of anode and cathode reactions)

- Absence of oxygen (ethanol reforming): $C_2H_5OH + 3H_2O \rightarrow 2CO_2 + 6H_2$ (IX)
- Presence of oxygen (ethanol mineralization): $C_2H_5OH + 3O_2 \rightarrow 2CO_2 + 3H_2O$ (X)

the counter electrode involves consumption of three molecules of O_2 per ethanol molecule. This reduction reaction (Reaction (VIII) of Table 9.1) takes place at a potential of +0.46 V at alkaline pH. As seen in Figure 9.2, the difference between the reduction potential and the titania conduction band is then around 1.5 V, *i.e.* it is more than enough to spontaneously run the cell and is expected to generate a high open-circuit voltage.

When titania is combined with a sensitizer, as seen in Figure 9.2, holes are accumulated in the valence band of the sensitizer and their oxidative power

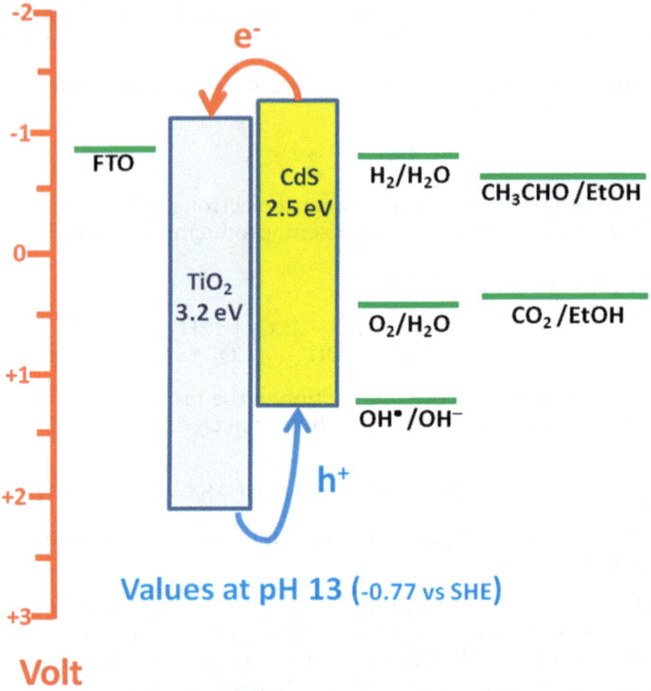

Figure 9.2 Energy levels for mesoporous titania, CdS sensitizer and a few redox reactions[18] at alkaline pH.

is decreased, since they jump to a less positive energy level. Depending on the nature of the sensitizer,[14,15,18] this modification may render certain oxidation reactions impossible. For example, CdS sensitizer is at the limit of being able to create ˙OH radicals. Other sensitizers, as it will be discussed later, are not.

Finally, it must be underlined at this point that compounds of the general chemical structure $C_xH_yO_z$ are the best fuels for PFCs. The generalized photocatalytic reforming and mineralization are given by the following reactions:

$$C_xH_yO_z + (2x - z)H_2O \rightarrow xCO_2 + (2x + y/2 - z)H_2 \quad (9.3)$$

$$C_xH_yO_z + (x + y/4 - z/2)O_2 \rightarrow xCO_2 + y/2H_2O \quad (9.4)$$

in the absence and in the presence of oxygen, respectively. The reactions at the cathode remain the same independent of the fuel. Differences between fuels demonstrate themselves by the nature of the oxidation steps that lead to mineralization.[12] The number of electrons (and holes) involved in each mineralization process is obviously equal to $2x(2x + y/2 - z) = 4x + y - 2z$.

9.3 PhotoFuelCell Configurations and Related Applications

The PFC presented in Figure 9.1 is used solely for production of electricity. The photoanode is made of an FTO (fluorine doped tin oxide) transparent electrode, which also plays the role of device window. The present research group has systematically employed nanoparticulate titania as photocatalyst and has experimented with various sensitizers, as will be discussed later. The counter electrode is usually made of carbon cloth carrying an electrocatalyst.[18] Carbon cloth is a "breathing" electrode that is permeable to air and lets oxygen diffuse between its filaments. To avoid water leakage, the electrocatalytic layer placed on the inner side of the electrode is mixed with a hydrophobic paste. A typical electrocatalyst is commercially available Pt nanoparticles mixed with carbon nanoparticles. However, various carbon based materials, for example, carbon nanotubes[19] and graphene, have been tested, in order to completely substitute Pt. As electrolyte, NaOH with added ethanol has been used. The configuration of Figure 9.1 is simple and very practical. One disadvantage is partial oxidation of ethanol at the counter electrode that decreases open-circuit voltage[15] (see also below). It is possible to divide the cell into two compartments separated by an ion transfer membrane. In that case, there are advantages and disadvantages. One advantage is that ethanol can be confined in the anode compartment. A second is that an acidic electrolyte may be used in the cathode compartment thus introducing a chemical forward bias.[12,13] The disadvantages are that the membrane introduces an internal resistance and the construction of the cell becomes then more complicated. In addition, a continuous supply of acid and base may increase the overall cell operation cost.

PFCs can be also used to produce hydrogen. If the cell is structured into two compartments, the cathode compartment can carry an acidic electrolyte, for example, H_2SO_4, and must be de-aerated. In that case, the counter electrode must be completely immersed in the electrolyte without contact with air. A characteristic design is presented in Figure 9.3. The redox level of hydrogen ion reduction (Reaction (V) of Table 9.1) in an acidic electrolyte of very low pH is close to 0 V (*vs.* NHE). At the same time, the photoanode is immersed in an alkaline electrolyte so its conduction band lies around −1.0 V (*cf.* Figure 9.2). This situation creates a strong chemical forward bias of about 1.0 V, which is more than enough to spontaneously run the cell under illumination. Hydrogen molecules are formed and can be collected by appropriate piping and storage. The device of Figure 9.3 is designed to solely produce hydrogen by short-circuiting the anode and cathode electrodes. The ion exchange membrane can be in that case a Nafion membrane, continuously activated by the acid of the cathode compartment.

To simplify cell construction and use only alkaline electrolyte, the cell design of Figure 9.4 can be employed for hydrogen production. In such a case, again photoanode and cathode were short-circuited by placing active

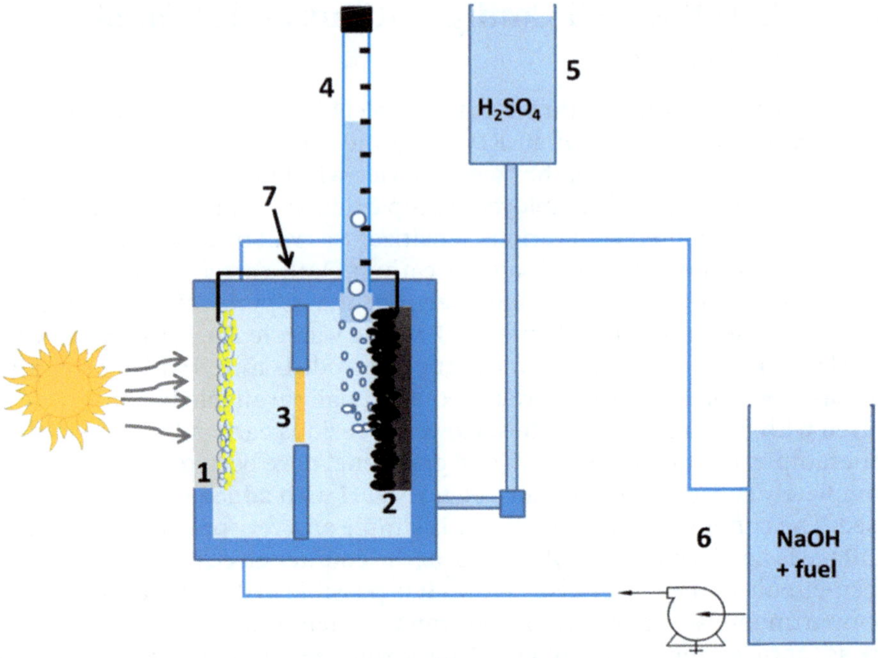

Figure 9.3 Schematic representation of a device designed for photoelectrochemical production of hydrogen: (1) FTO electrode bearing nanoparticulate titania and sensitizer playing also the role of cell window; (2) counter electrode bearing electrocatalyst; (3) Nafion membrane; (4) volumetric tube used as hydrogen trap and storage; (5) acidic electrolyte reservoir; (6) reservoir for alkaline electrolyte and fuel, circulator and piping; and (7) cable for short-circuiting anode and counter electrode (or applying electric bias if necessary).

catalysts on the same electrode. In this simplified variant the electrode is transparent and it also plays the role of device window. Since there is no more chemical or electric bias, the potential difference between photoanode and cathode is only about 0.3 V. This is not sufficient to overcome losses but, because the two electrodes are short-circuited, hydrogen is indeed produced, as it has been shown in previous publications.[30,31] The design of the electrode in Figure 9.4 places the electrocatalyst above the photocatalyst. This is not unique. Other designs might be applicable, for example alternating photocatalyst and electrocatalyst zones, as it will be shown below.

The configuration of Figure 9.4 is designed solely for hydrogen production by short-circuiting anode and cathode electrode. This design has the advantage of minimizing losses but it cannot take advantage of application of external electric bias. If the two electrodes are separated, as in Figure 9.3 or in Figure 9.5, then the device may be employed for production of hydrogen under external bias. Figure 9.5 represents a water-splitting variant where

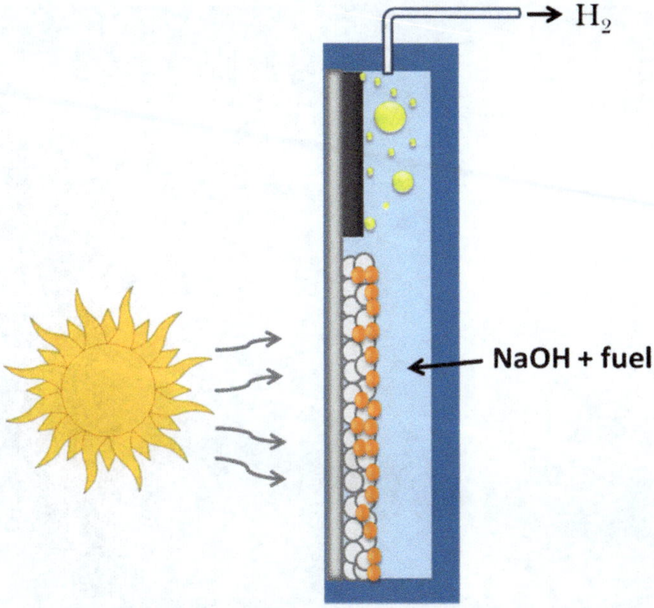

Figure 9.4 Simplified schematic representation of a device designed for photoelectrochemical production of hydrogen using one compartment cell where the electrode also plays the role of window. The lower part of the electrode bears photocatalyst and the upper part electrocatalyst.

oxygen is produced at the photoanode by water oxidation and hydrogen at the cathode by water reduction.[7,8]

The above discussion shows that a PFC may be designed in many different configurations, depending on the particular application. Two representative cases will be examined in the next section.

9.4 Selected Results and Discussion

9.4.1 PFC Used for Electricity Production Employing Ethanol as Organic Fuel

A PFC functioning with an aerated cathode can produce electricity under photoelectrochemical operation. This can be seen by the data of Figure 9.6. Curves 1 and 2 compare two- and one-compartment cell operation. The two compartments were separated by a silica frit acting as ion transfer membrane. It is observed that the open-circuit voltage (V_{oc}) and the short-circuit current density (J_{sc}) were 1.2 V and 1.0 mA cm^{-2}, respectively, in the case of the two-compartment cell (curve 1) but they became 1.0 V and 1.8 mA cm^{-2}, respectively, in the case of one-compartment cell (curve 2). In accordance with previous findings[15] and according to the above discussion, Figure 9.6 shows that when

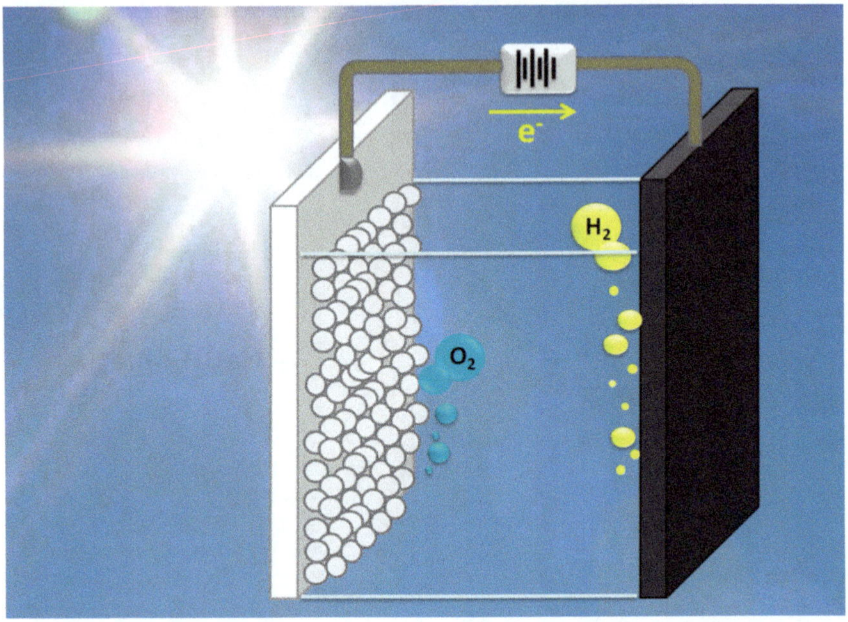

Figure 9.5 Simplified schematic representation of a device designed for photoelectrochemical production of hydrogen by water splitting and by application of an external forward electric bias. The figure presents a simplified one-compartment cell but if an ion transfer membrane is placed between the two electrodes the cell may be divided into two compartments. In a two-compartment cell electric bias may be substituted by chemical bias as in Figure 9.3.

there is no separation membrane a lower V_{oc} and a larger J_{sc} were obtained. The lower V_{oc} comes from the fact that the fuel is dispersed over the whole cell and it may be electrocatalytically oxidized on the cathode electrode, thus reversing voltage and decreasing V_{oc} by about 200 mV. The increase of the current is due to the decrease of the internal resistance of the cell by taking the membrane out. Obviously, there are advantages and disadvantages with both situations. Since the one-compartment cell is simpler to make and to operate, the rest of this subsection deals with results obtained with one-compartment cells. The current demonstrated an impressive increase when the photoanode carried visible-light-absorbing CdS quantum dot sensitizer. Indeed, the current density increased to 8.4 mA cm^{-2} in the presence of CdS (curve 3) showing the importance of broader light harvesting. Conversely, V_{oc} did not vary in the presence of CdS. This demonstrates the presence of the sensitization process. As already discussed, the V_{oc} of a PFC is related to the potential difference between the electron acceptor level and the oxygen reduction level. The electron acceptor level is the conduction band of titania (*cf.* Figure 9.2) no matter whether the electrons are generated by titania itself or by the sensitizer. For this reason, V_{oc} is the same for both curves 2 and 3.

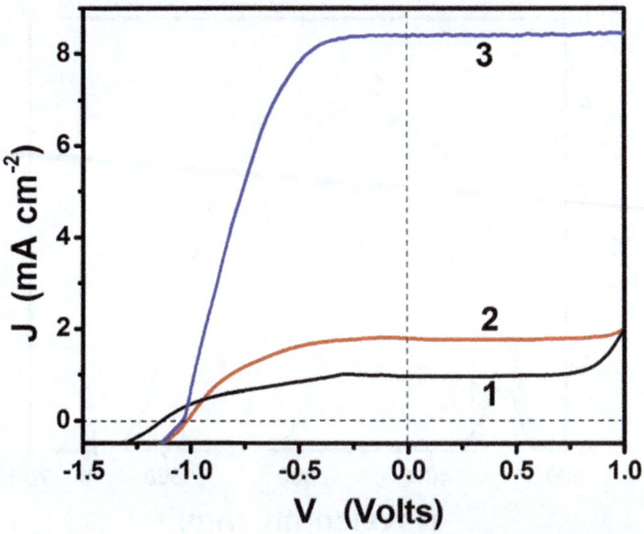

Figure 9.6 Current–voltage scans using a PFC functioning with an aerated cathode and a photoanode carrying nanoparticulate titania photocatalyst in the absence (curves 1 and 2) and in the presence (curve 3) of CdS sensitizer. The electrolyte was aqueous 0.5 mol L^{-1} NaOH containing 5 vol% ethanol. The cell was structured in two compartments separated by a silica frit (1) or it functioned as a one-compartment cell without separation membrane (2 and 3). In the first case, ethanol was introduced only in the anode compartment.

Since the purpose of using a PFC in the present case was to produce electricity, the photovoltaic efficiency of the sensitized cell has been calculated and it was found to be 4.3% (fill factor = 0.50 and simulated solar incident radiation intensity = 100 mW cm^{-2}). This efficiency is satisfactory, even though it is much lower than the one reported for other liquid-electrolyte mesoscopic solar cells, for example, dye-sensitized solar cells, where it presently amounts to 13%.[32] It must be, however, mentioned that these efficiencies are measured with tiny cells of very compact geometry. In the present case, the electrodes are relatively large with 1 cm^2 surface and with a large distance between anode and cathode electrode (a few centimeters), compared to only a few tens of micrometers in the case of dye-sensitized solar cells. It is known that when the size of the electrode increases both fill factor and current density decrease. Most important is the fact that the presently used sensitization range was limited by the absorption range of CdS (Figure 9.7).

Other sensitizers of nanoparticulate titania, which cover a larger range of the Visible, like CdSe (Figure 9.7) and PbS, have been tested in the past[15,17,18] and have been found inapplicable to PFCs. The reason, as explained in ref. 15, 17 and 18, is that the combined CdSe/TiO_2 and PbS/TiO_2 photoanode induces an oxidative state that is not powerful enough to create $^{\bullet}OH$

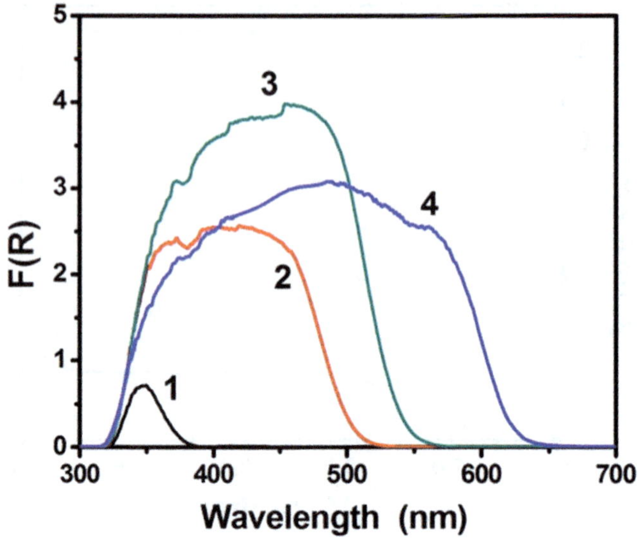

Figure 9.7 Reflection–absorption spectra of a few combined photocatalyst anode electrodes: (1) bare titania; (2) nitrate-derived SILAR deposited CdS/TiO_2; (3) acetate-derived SILAR deposited CdS/TiO_2; and (4) CdSe/CdS/TiO_2 (SILAR, successive ionic layer adsorption and reaction).

radicals. Formation of $^{\bullet}OH$ radicals competes with metal chalcogenide photocorrosion preserving their functionality. For this reason, the use of metal chalcogenide quantum dot sensitizers is unfortunately limited in the case of PFC. Even CdS made by using cadmium acetate precursor,[18,33] which leads to higher load of CdS and longer wavelength absorption threshold (Figure 9.7), is not a good sensitizer because of the decrease of the combined oxidation state. The CdS/TiO_2 combined ionization potential for cadmium-nitrate-derived CdS quantum dots is 6.7 eV and becomes 5.7 eV for acetate-derived CdS quantum dots.[18] For pure titania, it is 7.4 eV.[18] This places acetate-derived CdS/TiO_2 too high in the electrochemical scale and incapable of producing $^{\bullet}OH$ radicals. The corresponding potentials *versus* NHE for the combined photoanodes in an alkaline environment of pH 13 can be calculated by the standard formula V_{NHE} (volts) = V_a (eV) − 4.5 − 0.059 (pH), where V_{NHE} is the standard redox potential *vs.* NHE expressed in volts (V), V_a is the potential in the absolute energy scale expressed in eV and 4.5 eV is the position of the NHE. Indeed, the corresponding values were 2.13 V (bare TiO_2), 1.43 V (nitrate-derived CdS/TiO_2) and 0.43 V (acetate-derived CdS/TiO_2). The energy level corresponding to $^{\bullet}OH/OH^-$ is 1.25 V (*cf.* Figure 9.2). Clearly, only bare titania and nitrate-derived CdS/TiO_2 combined photocatalyst (absorption spectra given by curves 1 and 2 of Figure 9.7) are capable of creating $^{\bullet}OH$ radicals. Similar analysis shows that CdSe/TiO_2 and PbS/TiO_2 photoanodes are inapplicable to the presently described PFCs. This reasoning, makes the efficiency of 4.3% very satisfactory, considering the existing constraints.

Further increases of efficiency will necessitate discovery of a new class of sensitizers capable of activity in aqueous environment.

The short-circuit current density obtained with the CdS sensitized photoanode (Figure 9.6) was 8.4 mA cm^{-2}. This current is too large for CdS sensitizer. Since its band gap is 2.5 eV (496 nm, *cf.* Figure 9.2 and 9.7), the theoretical upper limit of expected current density for this band gap by assuming 100% efficiency of photon into electron conversion and for 1 sun incident light intensity[14] is around 6.3 mA cm^{-2}. The exceeding current apparently comes from "current doubling" effects.[12,34] Current doubling is attributed to the photocatalytic formation of unstable organic radicals, as in the second Reaction (IV) of Table 9.1. The radicals may interact with titania, injecting an electron into its conduction band, according to the following reactions:

$$C_2H_5OH + h^+ \rightarrow C_2H_5O^{\bullet} + H^+ \quad (9.5)$$

$$C_2H_5O^{\bullet} \rightarrow C_2H_4O + H^+ + e^- \quad (9.6)$$

Thus for each photon absorbed, there are two electrons created in the conduction band, one photogenerated and one injected by the radical. This doubles the current, hence "current doubling". This phenomenon demonstrated itself in the present work, as seen by the data of Figure 9.8. Indeed, the current in the presence of ethanol was approximately twice as large as in its absence. However, current doubling is not systematically observed[12,35] but it is frequently shadowed by the normal photocatalytic mineralization

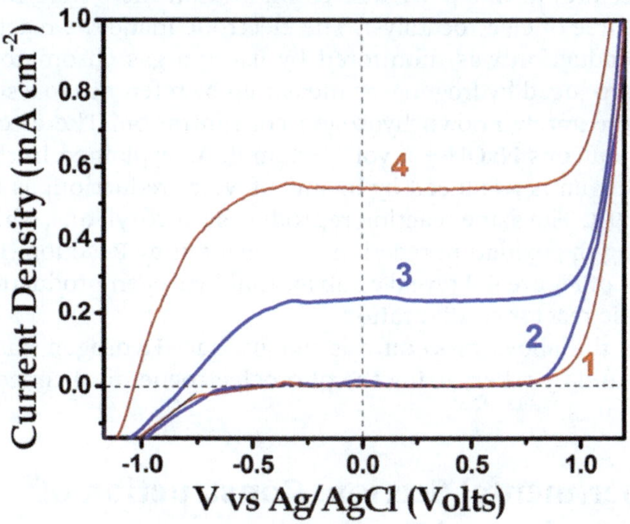

Figure 9.8 Current density–voltage curves obtained with a PFC functioning in the presence of oxygen. The aqueous electrolyte always contained 0.5 mol L^{-1} NaOH without (curves 1 and 3) or with (2 and 4) 5 vol% ethanol. Curves (1) and (2) were recorded in the dark. In all cases photoanode carried bare np-TiO_2 film. The curves were traced in a three-electrode configuration using Ag/AgCl as reference electrode.

route. If the current recorded in the absence of fuel is low, the current observed in its presence may appear several times higher and not just double.[12,35] The current observed in the absence of fuel may be small because of electron–hole recombination. In the presence of fuel, consumption of holes decreases recombination thus liberating more electrons and increasing current. Therefore, the increase of current in the presence of an organic sacrificial agent may both come from the current doubling effect and from the decrease of electron–hole recombination. Current doubling is surely responsible for unusually high currents, for example, photon into electron conversion efficiencies are sometimes higher than 100%.[14] In the present case, an unexpectedly high current density was indeed observed. Obviously, such cases demonstrate the presence of current doubling phenomena.

9.4.2 PFC Used Exclusively for Hydrogen Production Employing Ethanol as Organic Fuel

In this selected case, the reactor adopted had the configuration of Figure 9.4. The combined anode–cathode electrode was based on a FTO glass where the photocatalyst and the electrocatalyst were deposited in the form of alternating strips as in the inset of Figure 9.9. Each strip had an area of 1 × 3 cm. The black colored strips correspond to the electrocatalyst, which was a mixture of carbon paste with Pt (see Section 9.5). The photocatalyst was nanoparticulate titania. Therefore, in this particular configuration there were four strips of titania and three of electrocatalyst. The electrode made the reactor window. Hydrogen production was monitored by using a gas chromatograph. The quantity of produced hydrogen was measured by reference to a standard gas mixture of accurately known hydrogen concentration. The electrolyte was 0.5 mol L^{-1} aqueous NaOH + 5 vol% ethanol. As explained in the previous section, hydrogen is produced by means of water reduction, as in Reaction (VI) of Table 9.1. The same reaction reproduces hydroxyl ions, which are consumed during the oxidation reaction, represented by Reaction (III) of Table 9.1. The data of Figure 9.9 reveal a substantial hydrogen production capacity using a simple reactor configuration.

Of course, the above procedure is not unique. Hydrogen has been produced by us and by others using the photoelectrochemical concept in many different variants.[13,30,31,37,38]

9.5 Experimental Section: Construction of Electrodes and Devices

9.5.1 Materials

Unless otherwise indicated, reagents were obtained from Aldrich and were used as received. Commercial nanocrystalline titania Degussa P25 (specific surface area 50 m^2 g^{-1}) was used in all cell constructions and Millipore water

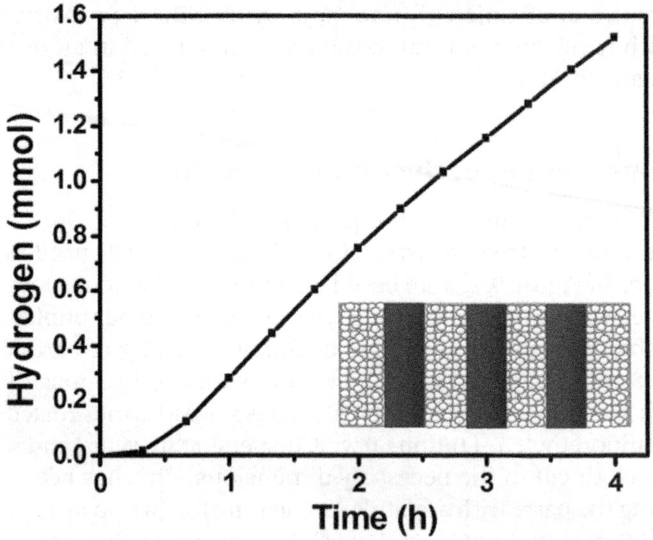

Figure 9.9 Evolution of hydrogen produced by photoelectrochemical reforming of ethanol in alkaline environment using a combined photocatalyst–electrocatalyst electrode in alternating strips. The electrolyte contained 0.5 mol L^{-1} aqueous NaOH + 0.5 vol% ethanol. The inset shows catalyst deposition pattern. Each strip had a size of 1 × 3 cm^2. The black strip represents electrocatalyst and the one composed of small circles represents nanoparticulate titania.

was used in all experiments. SnO_2:F transparent conductive electrodes (FTO, resistance 8 Ω per square) were purchased from Pilkington.

9.5.2 Preparation of TiO_2 Films and Deposition of CdS by the SILAR Method

Nanoparticulate titania films were deposited on FTO transparent electrodes by the following procedure. An FTO glass was cut in the appropriate dimensions and was carefully cleaned first with soap and then by sonication in isopropanol, water and acetone. A thin layer of compact titania was first sprayed over a patterned area by using 0.2 mol L^{-1} diisopropoxy-titanium bis(acetylacetonate) solution in ethanol and then calcined at 500 °C. Deposition of this bottom compact layer is a common practice with nanocrystalline titania photoanodes, since it enhances the attachment of the top thick film, prevents short circuits and facilitates electron flow towards the electrode. On the top of this compact film, a titania paste made of P25 nanoparticles was applied by doctor blading. The film was calcined up to 550 °C at a rate of 20 °C min^{-1}. The final thickness of the film, as measured by SEM, was approximately 10 μm. The geometrical area of the film varied according to the particular application. The CdS was deposited by successive ionic layer adsorption and reaction (SILAR).[14,18,36] Ten SILAR cycles were applied by using 0.1 mol L^{-1}

aqueous solutions of $Cd(NO_3)_2$ and Na_2S. After SILAR deposition and final washing with triply distilled water, the films were dried in an oven at 100 °C and were ready for use.

9.5.3 Construction of the Counter Electrode

Experimental data obtained for the purpose of the present work involved the following counter electrode constructions. Electrodes solely used for electricity production, as in Figure 9.1, were based on commercial carbon cloth, which was functionalized with carbon black and Pt, similarly to previous publications.[14,18,35] A hydrophobic layer was first applied as follows: 0.246 g of carbon black was mixed with 8 mL of distilled water by vigorous mixing in a mixer (about 2400 rpm) until it became a viscous paste. This paste was further mixed with 0.088 mL polytetrafluorethylene (Teflon 60 wt% dispersion in water) and then applied on a carbon cloth cut in the necessary dimensions. This has been achieved by first spreading the paste with a spatula, preheating for 30 min at 80 °C and finally heating for 30 min in an oven at 340 °C. Subsequently, the catalytic layer was prepared as follows: 1 g of Pt–carbon black electrocatalyst (30% on Vulcan XC72) was mixed with 8 g of Nafion perfluorinated resin (5 wt% solution in lower aliphatic alcohols and water) and 15 g of solution made of 7.5 g H_2O and 7.5 g isopropanol. The mixture was ultrasonically homogenized and then applied on the previously prepared carbon cloth bearing carbon black. The electrode was then heated at 80 °C for 30 min and the procedure was repeated as many times as necessary to load about 0.5 mg of Pt per cm^2. The thus prepared Pt–carbon black/carbon-cloth (Pt/CC) electrode was ready for use.

In the case of hydrogen production using alternating photocatalyst and electrocatalyst strips, titania was first deposited in four strips of 1 × 3 cm each, following the procedure described in the previous subsection. The spaces in-between were covered by tapes, which were subsequently removed. The remaining empty space was filled by the electrocatalyst, forming three strips of 1 × 3 cm each, as follows. First we applied a commercial carbon paste (Elcocarb, Solaronix). The quantity of Elcocarb in each strip was 2 mg (total of 6 mg). The strips were stabilised on the FTO glass by annealing at 500 °C followed by slow cooling. On the top of these carbon-paste strips we cast an ethanol solution of diaminedinitritoplatinum(II). The substrate was heated at 80 °C to evaporate the solvent. Then the electrode was again annealed at 500 °C, followed by slow cooling. The calculated quantity of added platinum was 0.12 mg, which represents 2 wt% with respect to the carbon paste. Slow cooling was applied to avoid film wrinkling and detachment.

9.5.4 Device (Reactor) Construction

Two reactors were used for the purposes of the present work. When PFC was employed as a device functioning in the presence of oxygen to produce electricity, it was made of a Plexiglas body with the two electrodes facing each other at a distance of 10 mm. As depicted in Figure 9.1, the photoanode electrode played the role of cell window while the opposite side was closed with the Pt/CC electrode. The cell was in one case divided into two compartments by a silica frit (ROBU, Germany, porosity SGQ 5, diameter 25 mm, thickness 2 mm) while in another case the frit was removed. The volume of the electrolyte was 10 mL and it contained 0.5 mol L^{-1} NaOH with or without 5 vol% ethanol. The active area of the photoanode was 1 × 1 cm = 1 cm^2 and that of the counter electrode 3.5 × 3.0 cm = 10.5 cm^2. Illumination was made with a 300 W Xenon lamp providing approximately 100 mW cm^{-2} of light intensity at the position of the photocatalyst.

In the case of hydrogen production, the reactor was made of Plexiglas accommodating the stripped electrode of 7 × 3 cm, which played the role of window. The aqueous electrolyte contained 0.5 mol L^{-1} NaOH and 5 vol% ethanol. Its volume was 90 mL. Illumination was made with a UV-enhanced light source (Osram Ultramed 400 W) providing intense UV radiation of approximately 50 mW cm^{-2} at the position of the photocatalyst.

9.5.5 Measurements

Current–voltage curves were obtained with the help of an Autolab PGSTAT128N potentiostat. They were all traced at a rate of 5 mV s^{-1}. Hydrogen was detected on-line by using an SRI 8610C gas chromatograph. The chromatograph signal was calibrated by comparison with a standard of 0.25% H_2 in Ar.

Acknowledgements

This project is implemented under the "ARISTEIA" Action of the "OPERATIONAL PROGRAMME EDUCATION AND LIFELONG LEARNING" and is co-funded by the European Social Fund (ESF) and National Resources (Project No. 2275).

References

1. M. Grätzel, *Acc. Chem. Res.*, 2009, **42**, 1788.
2. D. A. Hines and P. V. Kamat, *Appl. Mater. Interfaces*, 2014, **6**, 3041.
3. M. Grätzel, *MRS Bull.*, 2005, **30**, 23.
4. D. Angmo and F. C. Krebs, *J. Appl. Polym. Sci.*, 2013, **129**, 1.
5. M. A. Green, A. Ho-Baillie and H. J. Snaith, *Nat. Photonics*, 2014, **8**, 506.
6. M. C. Scharber and N. S. Sariciftci, *Prog. Polym. Sci.*, 2013, **38**, 1929.
7. P. V. Kamat, *J. Phys. Chem. C*, 2007, **111**, 2834.
8. K. Sivula, F. Le Formal and M. Grätzel, *ChemSusChem*, 2011, **4**, 432.
9. E. Becquerel, *C. R. Acad. Sci.*, 1839, **9**, 561.
10. A. Fujishima and K. Honda, *Nature*, 1972, **238**, 37.
11. M. Kaneko, J. Nemoto, H. Ueno, N. Gokan, K. Ohnuki, M. Horikawa, R. Saito and T. Shibata, *Electrochem. Commun.*, 2006, **8**, 336.

12. P. Lianos, *J. Hazard. Mater.*, 2011, **185**, 575.
13. M. Antoniadou and P. Lianos, *J. Photochem. Photobiol., A*, 2009, **204**, 69.
14. M. Antoniadou, D. I. Kondarides, D. D. Dionysiou and P. Lianos, *J. Phys. Chem. C*, 2012, **116**, 16901.
15. M. Antoniadou, S. Sfaelou and P. Lianos, *Chem. Eng. J.*, 2014, **254**, 245.
16. M. Antoniadou, H. Changseok, S. Sfaelou, M. Michailidi, D. D. Dionysiou and P. Lianos, *Sci. Adv. Mater.*, 2013, **5**, 1756.
17. S. Sfaelou, M. Antoniadou, V. Dracopoulos, K. Bourikas, D. I. Kondarides and P. Lianos, *J. Adv. Oxid. Technol.*, 2014, **17**, 59.
18. S. Sfaelou, L. Sygellou, V. Dracopoulos, A. Travlos and P. Lianos, *J. Phys. Chem. C*, 2014, **118**, 22873.
19. S. Sfaelou, M. Antoniadou, G. Trakakis, V. Dracopoulos, D. Tasis, J. Parthenios, C. Galiotis, K. Papagelis and P. Lianos, *Electrochim. Acta*, 2012, **80**, 399.
20. N. Balis, V. Dracopoulos, M. Antoniadou and P. Lianos, *Electrochim. Acta*, 2012, **70**, 338.
21. P. Trogadas, T. H. Fuller and P. Strasser, *Carbon*, 2014, **75**, 5.
22. P. Du and R. Eisenberg, *Energy Environ. Sci.*, 2012, **5**, 6012.
23. M. Antoniadou and P. Lianos, *J. Nanosci. Nanotechnol.*, 2010, **10**, 6240.
24. M. Antoniadou and P. Lianos, *Photochem. Photobiol. Sci.*, 2011, **10**, 431.
25. T. Sakada and T. Kawai, *Chem. Phys. Lett.*, 1981, **80**, 341.
26. M. Antoniadou, D. I. Kondarides and P. Lianos, *Catal. Lett.*, 2009, **129**, 344.
27. P. Panagiotopoulou, M. Antoniadou, D. I. Kondarides and P. Lianos, *Appl. Catal., B*, 2010, **100**, 124.
28. A. Patsoura, D. I. Kondarides and X. E. Verykios, *Catal. Today*, 2007, **124**, 94.
29. F. D. Foglia, G. L. Chiarello, M. V. Dozzi, P. Piseri, L. G. Bettini, S. Vinati, C. Ducati, P. Milani and E. Selli, *Int. J. Hydrogen Energy*, 2014, **39**, 13098.
30. N. Strataki, M. Antoniadou, V. Dracopoulos and P. Lianos, *Catal. Today*, 2010, **151**, 53.
31. V. M. Daskalaki, M. Antoniadou, G. Li Puma, D. I. Kondarides and P. Lianos, *Environ. Sci. Technol.*, 2010, **44**, 7200.
32. M. Mathew, A. Yella, P. Gao, R. Humphry-Baker, B. F. E. Curchod, N. Ashari-Astani, I. Tavernelli, U. Rothlisberger, Md. K. Nazeeruddin and M. Grätzel, *Nat. Chem.*, 2014, **6**, 242.
33. R. Zhou, Q. Zhang, J. Tian, D. Myers, M. Yin and G. Cao, *J. Phys. Chem. C*, 2013, **117**, 26948.
34. A. Fujishima, T. Kato, E. Maekawa and K. Honda, *Bull. Chem. Soc. Jpn.*, 1981, **54**, 1671.
35. M. Antoniadou and P. Lianos, *Appl. Catal., B*, 2010, **99**, 307.
36. Y. F. Nicolau, *Appl. Surf. Sci.*, 1985, **22/23**, 1061.
37. E. Selli, G. L. Chiarello, E. Quartarone, P. Mustarelli, I. Rossetti and L. Forni, *Chem. Commun.*, 2007, 5022.
38. Y. Horiuchi, T. Toyao, M. Takeuchi, M. Matsuoka and M. Anpo, *Phys. Chem. Chem. Phys.*, 2013, **15**, 13243.

CHAPTER 10

Photocatalytic Hydrogen Generation

XIN LI[a], JINGXIANG LOW[b], AND JIAGUO YU*[b]

[a]College of Materials and Energy, South China Agricultural University, Guangzhou 510642, China; [b]State Key Laboratory of Advanced Technology for Material Synthesis and Processing, Wuhan University of Technology, Wuhan 430070, China
*E-mail: jiaguoyu@yahoo.com

10.1 Introduction

According to a quite recent report,[1] the rate of global energy consumption in 2050 and 2100 will be two and three times higher than that in 2001 (13.5 TW), respectively. As a result of fossil fuel depletion, averting an energy crisis and disastrous environmental pollution have become two main challenges in the near future. Therefore, there is an urgent need to search for a clean, renewable, cheap and sustainable energy source as a viable alternative to fossil fuels to meet energy demand of human beings.[2] Hydrogen, a regenerative and environmentally friendly energy carrier, has attracted a great deal of attention by scientists over the past three decades.[3,4] However, at present, hydrogen is mainly produced from fossil fuels by steam reforming of natural gas or by high-energy consumption processes, which are not environmentally friendly and economic.[5] Therefore, it is expected that hydrogen will be generated using a source of renewable energy.

RSC Energy and Environment Series No. 15
Photocatalysis: Applications
Edited by Dionysios D Dionysiou, Gianluca Li Puma, Jinhua Ye, Jenny Schneider, and Detlef Bahnemann

Published by the Royal Society of Chemistry, www.rsc.org

It is widely accepted that solar energy is the most promising source of clean and renewable energy to effectively address the world's energy problems. Therefore, among various facile pathways using renewable energy sources, photocatalytic H_2 production from water splitting by means of solar energy and semiconductors has attracted a lot of attention and proven to be a promising strategy for solving both solar energy storage and green production of hydrogen fuel.[6–12] The key challenge in photocatalytic H_2 production is finding highly efficient and stable visible-light-active photocatalysts that have the unique potential to provide excited charge carries for both uphill H_2 or/and O_2 evolution reactions under visible light irradiation.

In the past ten years, heterogeneous photocatalytic hydrogen production has received increasing attention. To date, hundreds of semiconductor-based photocatalysts consisting of different metal and/or nonmetal elements have emerged and have been successfully applied in photocatalytic hydrogen production from water splitting.[2,6] Significant progress has been achieved on heterogeneous photocatalysts for hydrogen generation through water splitting, and many excellent research papers and reviews have been published.[6,11,13–15] However, most photocatalysts still have poor quantum efficiency, photostability and visible-light absorption for practical photosynthetic water splitting. Although some promising visible-light-driven photocatalysts (such as Ta_3N_5,[16,17] TaON,[18] GaN:ZnO solid solution,[19,20] sulfide,[21–23] CoO[24] and graphitic carbon nitride (g-C_3N_4)[25–27]) have also been discovered and applied in photocatalytic H_2 production, the solar-to-fuels conversion efficiency is still very low and far from the practical applications. At this point, it has been recognized that photocatalytic overall water splitting over a heterogeneous visible-light-active photocatalyst remains a great challenge. Thus, while visible-light-driven overall water splitting had once been regarded as one of the "Holy Grails" of chemistry,[28] it is well accepted that no single semiconductor material can meet all the requirements (a suitable band gap (1.8–2.4 eV) for a broad spectrum of absorption, matching the energy band to H_2 and O_2 evolution reaction, high quantum efficiency and stability) for photocatalytic application.[29] Therefore, besides continuous development of new visible-light-driven semiconductors, it is necessary and urgent to enhance the selectivity, stability and activity of existing semiconductors for water splitting under visible light irradiation through various modification strategies.[12,30] In this chapter, the fundamentals and many recent achievements in photocatalytic H_2 production over heterogeneous photocatalysts will be summarized from the viewpoint of highly efficient photocatalyst design. We will specifically focus on various modification methods and key design principles for enhanced photocatalytic H_2 production activity, including promoted charge separation and transport, accelerated H_2-evolution kinetics and increased stability of photocatalysts. This chapter offers some new insight into designing and understanding of highly effective semiconductor systems for photocatalytic H_2- and O_2-evolution half-reactions as well as overall water splitting.

10.2 Fundamentals of Photocatalytic Hydrogen Generation

10.2.1 Thermodynamics of Photocatalytic Hydrogen Generation

The overall water splitting reaction is composed of two half-reactions, the hydrogen evolution reaction (HER, eqn (10.1)) and oxygen evolution reaction (OER, eqn (10.2)). The free energy $\Delta G°$ and the standard redox potential $\Delta E°$ of the multi-electron water splitting are listed in eqn (10.3). Thermodynamically, clearly, water splitting into H_2 and O_2 (eqn (10.3)) is an uphill reaction because of a large positive change in Gibbs free energy ($\Delta G° = +237.2$ kJ mol^{-1}). This uphill reaction is more difficult to proceed than the downhill-type photodegradation reactions such as photo-oxidation of organic compounds using oxygen molecules.[11] In addition, it is also known that the OER (eqn (10.2)) is an essential component of light-driven water splitting systems, which provides protons and electrons for the production of hydrogen. Compared to the two-electron HER, the OER is more complex and difficult because this reaction requires a four-electron oxidation of two water molecules coupled to the removal of four protons to form a relatively weak oxygen–oxygen bond:[31]

$$2H^+ + 2e^- \rightarrow H_2(g)\ (\Delta E° = -0.41\ \text{V}) \tag{10.1}$$

$$2H_2O(l) \rightarrow O_2(g) + 4H^+ + 4e^-\ (\Delta E° = +0.82\ \text{V}) \tag{10.2}$$

$$H_2O(l) \rightarrow H_2(g) + 1/2O_2(g)\ (\Delta G° = +237.2\ \text{kJ mol}^{-1}, \Delta E° = 1.23\ \text{V}) \tag{10.3}$$

In 1972, Honda and Fujishima[32a] for the first time reported that the photodecomposition of water to H_2 and O_2 could be achieved on a Pt counter electrode and a single-crystal n-type TiO_2 (rutile) working electrode in a photoelectrochemical cell, respectively, under both UV light irradiation and a small applied electric voltage (as shown in Figure 10.1). Inspired by this pioneering work, Bard proposed a new concept concerning a short-circuited photoelectrochemical cell based on TiO_2 and Pt in 1979 (as shown in Figure 10.2).[32b–35] According to Bard's concept, the principles of semiconductor photoelectrochemical cells could be employed to design particulate or powder systems for heterogeneous photoreactions. Thus, a strong consistency in mechanism for the photoelectrochemical and photocatalytic systems lead to an increasing interest in the investigation of powder systems due to their low cost and relatively easy fabrication.[35]

In general, the fundamental principle of semiconductor photocatalysts for H_2 generation is shown in Figure 10.3. The electrons can be excited from the valence band (VB) to the conduction band (CB) of the semiconductors (SCs) under light irradiation of suitable wavelengths, leaving holes in the VB. These electrons and holes can then transfer to the surfaces of SC-based photocatalysts, and initiate the HER and OER, respectively. From the viewpoint of thermodynamics, the bottoms of the CB must be located at a more negative potential than the reduction potential of H^+ to H_2 (0 V *vs.* NHE at pH 0), while the tops of the valence bands must exceed the oxidation potential

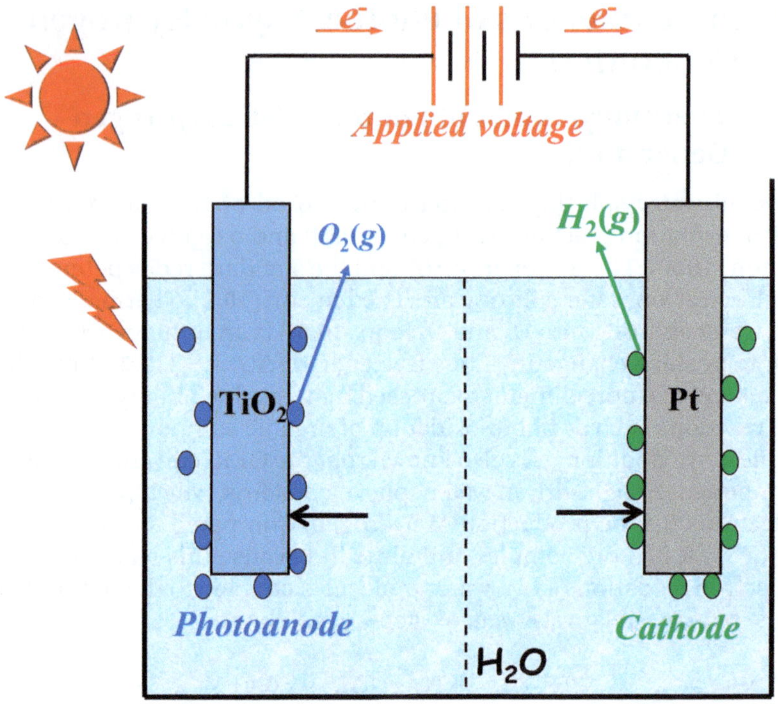

Figure 10.1 Photoelectrochemical water splitting using a TiO_2 photoanode.

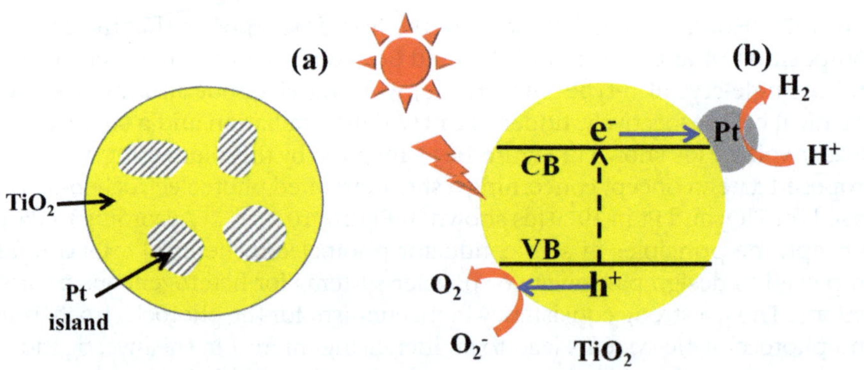

Figure 10.2 (a) Pt deposited on the surface of TiO_2 and (b) a short-circuited photoelectrochemical cell based on TiO_2 and Pt for H_2 and O_2 production.

of H_2O to O_2 (1.23 V *vs.* NHE). Thus, the band gap energy (E_g) of a SC-based photocatalyst should be >1.23 eV (<1000 nm) to convert the light energy into H_2 and O_2. To achieve photocatalytic hydrogen production under visible light irradiation, the E_g of a given photocatalyst should be <3 eV (>400 nm). Considering the energy losses, the suitable E_g of a single photocatalyst for overall water-splitting should lie in the range of 1.5–2.5 eV.[7,36]

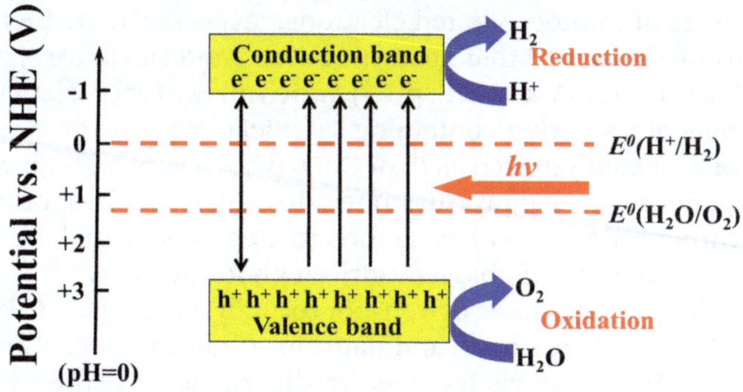

Figure 10.3 Principle of photocatalytic water splitting over semiconductor photocatalysts.

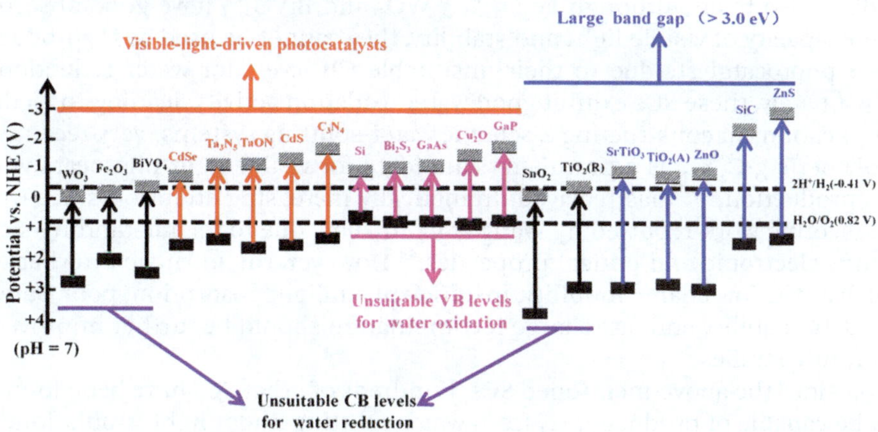

Figure 10.4 Band positions of some semiconductor photocatalysts and the redox potentials of water splitting at pH 7 in aqueous solution.

10.2.2 Materials and Systems for Photocatalytic Hydrogen Production

10.2.2.1 Materials for Photocatalytic Hydrogen Production

To date, there are many SCs with suitable CB levels for photocatalytic H_2 production. Figure 10.4 presents the band edge positions of several potential SCs. Notably, the band edge positions of SCs usually exhibit an obvious pH dependence (−0.059 V per pH unit) in an aqueous electrolyte solution.[6] According to the difference of their E_g values, these SCs could be divided into two types: visible-light-driven and wide band gap photocatalysts.

Among the SCs with a large band gap (>3.0 eV), TiO_2 is the most studied photocatalyst for photocatalytic H_2 generation due to its efficiency, low cost, non-toxicity and high stability. However, the wide band gap (~3.2 eV) and low CB level of TiO_2 (anatase) limit its utilization of the solar spectrum and the reduction power of photogenerated electrons, respectively. Furthermore, it is also noted from Figure 10.4 that ZnS with a wide band gap (3.6 eV) possesses a very high CB level (−1.85 V *vs.* NHE, pH 7), which allows for efficient generation of H_2 from aqueous solution containing sacrificial reagents even without the deposition of a co-catalyst such as Pt or RuO_2 (with a very high quantum yield of 90%).[37] Therefore, it is a powerful strategy to exploit visible-light-driven photocatalysts through doping ZnS with foreign elements or making ZnS-based solid solutions with other visible-light-driven photocatalysts.[23,38,39]

Among various visible-light-driven photocatalysts, Cu_2O,[40] CdS,[41] CoO,[24] TaON[42] and Ta_3N_5[42,43] are good candidates for photocatalytic H_2 production from water splitting under visible light irradiation because they possess narrow band gaps for visible light absorption and enough negative CB positions for water reduction. However, they are not stable in aqueous environments due to the anodic photocorrosion by photogenerated holes. Furthermore, CdS is also toxic. Although Fe_2O_3,[44–46] WO_3 and $BiVO_4$[47] have good absorption capacity of visible light and stability, they cannot be used as H_2-production photocatalysts due to their unsuitable CB levels for water reduction. Fortunately, these SCs exhibit good water oxidation activity, leading to facile applications in constructing Z-scheme water-splitting systems. Very recently, polymeric g-C_3N_4, as a promising candidate for visible-light photocatalytic H_2 production,[25,48] has received dramatically increasing attention since the pioneering work reported by Wang *et al.* in 2009, due to its outstanding stability, electronic and optical properties.[49] However, the intrinsic drawbacks such as the low charge mobility, insufficient sunlight absorption, poor water oxidation ability and fast charge recombination should be further improved in future studies.[50]

Besides the above-mentioned SCs, hundreds of other SCs have been found to be capable of producing H_2 from water splitting under light irradiation.[13] According to differences in their compositions, these photocatalysts could be divided into three types: metal oxide, non-oxide and metal-free photocatalysts, which were summarized in Figure 10.5. Multi-component metal oxides and solid-solution photocatalysts are the most promising candidates for photocatalytic H_2 production under light irradiation. Unfortunately, no one photocatalyst could meet all requirements for visible-light-driven H_2 evolution from water splitting.[11] Thus, thermodynamically speaking, it is still of importance to survey and develop new photocatalyst materials with optimized performance for photocatalytic H_2 production by combinatorial approaches[51,52] and density-functional theory (DFT) calculation.[53]

10.2.2.2 Systems for Photocatalytic Hydrogen Production

Apart from finding novel H_2-production SC-based photocatalysts, the development of various operating systems is also a significant consideration in enhancing the overall efficiency for photocatalytic hydrogen production. According to the different properties of SCs and sacrificial reagents in the systems, these water splitting systems are summarized in Figure 10.6. Generally,

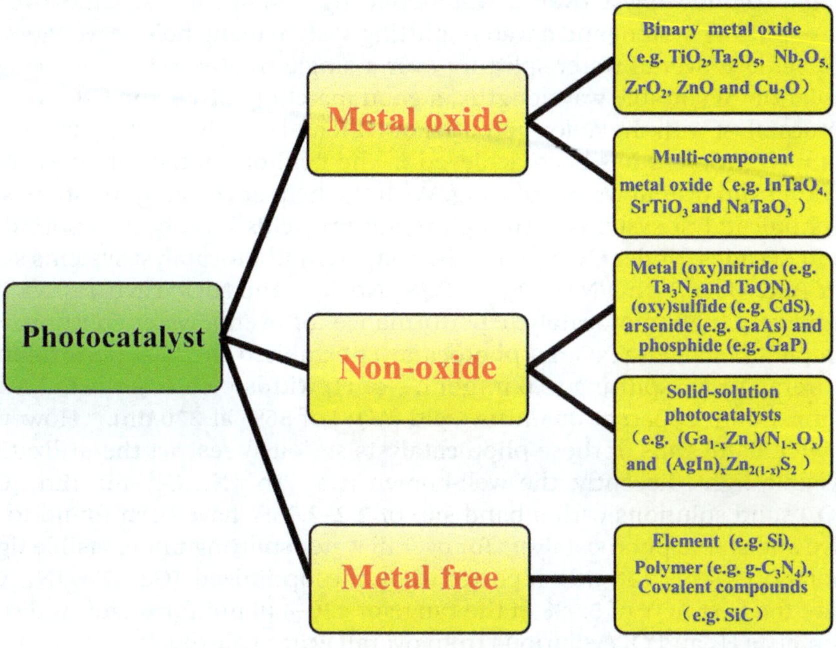

Figure 10.5 Summary of photocatalysts for photocatalytic H_2 production.

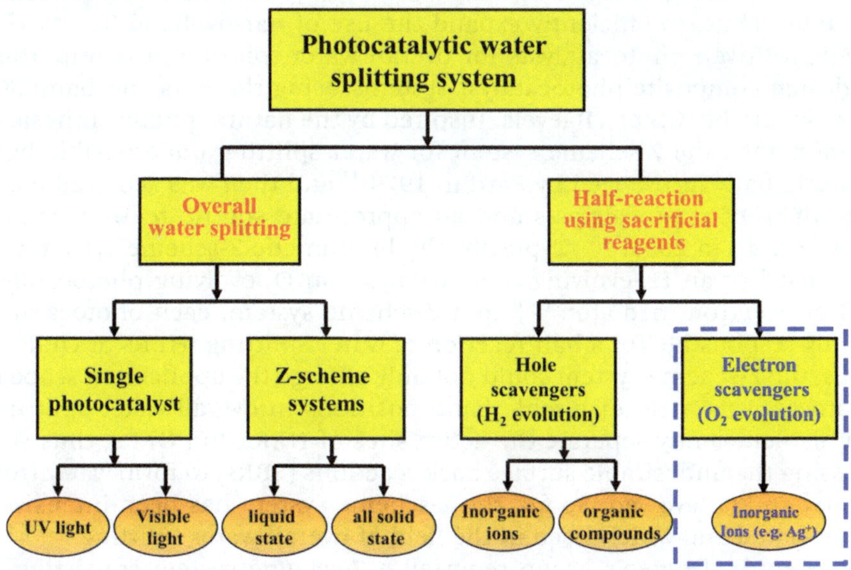

Figure 10.6 Different systems for photocatalytic water splitting.

there are three kinds of systems for photocatalytic hydrogen production: a single photocatalyst overall water-splitting system, a Z-scheme overall water-splitting system and a water-splitting system using hole scavengers.

To achieve overall water-splitting over a single photocatalyst under light irradiation of suitable wavelengths, a good matching of VB and CB levels of photocatalyst with the redox potential of HER and OER is very important. So far, much progress has been achieved in the field of photocatalytic H_2 production from overall water splitting. With the help of co-catalysts, some single-photocatalyst systems with high quantum yields have been reported in recent decades. Under UV light irradiation, many photocatalyst systems such as $Pt/RuO_2/TiO_2$,[54] $NiO/NaTaO_3$,[55,56] $Pt/Sr_2Nb_2O_7$[57] and $NiO/Sr_2Nb_xTa_{2-x}O_7$[58,59] exhibit very high photocatalytic performance for overall water splitting. For example, the $NiO/NaTaO_3$:La photocatalyst exhibited the best performance for overall water splitting (2.0 mmol h^{-1} of H_2 with stoichiometric O_2), with the maximum apparent quantum yield (AQY) of 56% at 270 nm.[56] However, the wide band gaps of these photocatalysts seriously restrict the utilization of visible light. Recently, the well-known $(Ga_{1-x}Zn_x)(N_{1-x}O_x)$ and $(Zn_{1+x}Ge)(N_2O_x)$ solid solutions with a band gap of 2.2–2.7 eV have been found to be active and stable photocatalysts for overall water splitting under visible light irradiation.[19,20,60–62] It was reported that the optimized $(Ga_{1-x}Zn_x)(N_{1-x}O_x)$ shows the best AQY of 5.9% in the range of 420–440 nm for steady and stoichiometric H_2 and O_2 evolutions from overall water splitting.[63]

Clearly, the narrow band gaps of many visible-light-driven SCs (such as WO_3, $BiVO_4$, Fe_2O_3 and Si) make them unsuitable for achieving overall water-splitting because of their unsuitable CB or VB levels for HERs or OERs. Thus, to efficiently expand the use of narrow-band SCs as visible-light-driven photocatalysts for overall water-splitting, it is important to design composite photocatalyst systems using these narrow-band SCs with unsuitable CB or VB levels. Inspired by the natural photosynthesis of green plants, the Z-scheme system for water splitting under visible light was originally introduced by Bard in 1979,[33] and then was achieved using two different photocatalysts and an appropriate shuttle redox mediator by Abe *et al.* in 2001.[64,65] Typically, the biomimetic Z-scheme system was composed of an H_2-evolving photocatalyst, an O_2-evolving photocatalyst and an electron mediator.[66–70] In a Z-scheme system, each photocatalyst is only responsible for a half-reaction of water splitting (HERs or OERs).[67] Thus, the Z-scheme system could not only enlarge the application scope of photocatalysts with unsuitable band positions in overall water splitting, but also efficiently separate the active sites of HERs and OERs, thus suppressing the undesirable surface back reactions (SBRs) to form water from the H_2 and O_2 evolved. So far, the Z-scheme system has been intensively investigated and widely used in the field of overall water splitting.[67,69,71–74] For example, Domen's group reported a Z-scheme system consisting of $Pt/ZrO_2/TaON$, Pt/WO_3 and IO_3^-/I^- pair.[71] An AQY of 6.3% at 420.5 nm was achieved, which was about six-times greater than that of a TaON analogue (Figure 10.7).

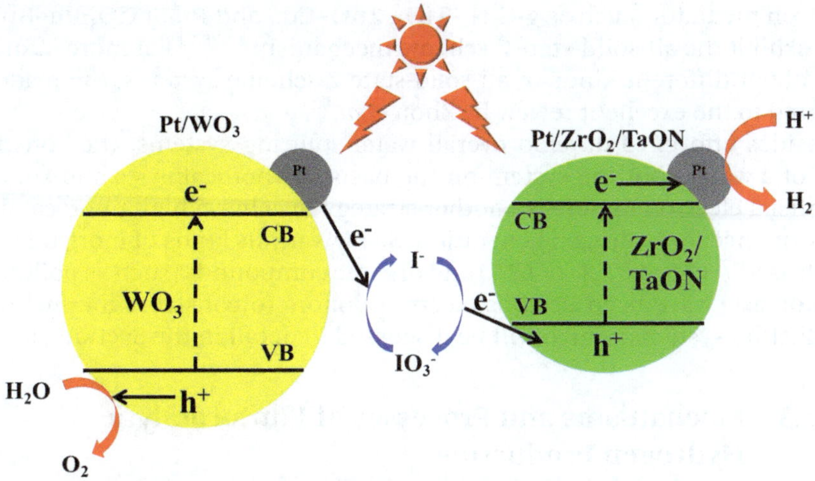

Figure 10.7 Schematic diagram of Z-scheme water splitting system containing Pt/ZrO_2/TaON and Pt/WO_3.

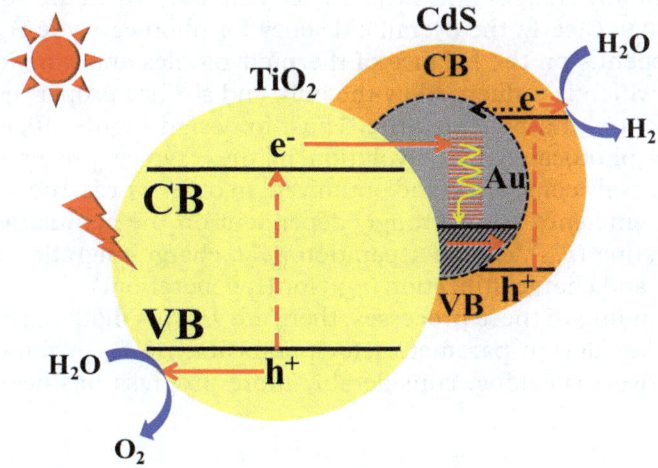

Figure 10.8 Schematic diagram of electron transfer for the all-solid-state Z-scheme CdS–Au–TiO_2 system.

Furthermore, a new and interesting concept of an all-solid-state Z-scheme has also been proposed with Tada *et al.* presenting the first three-component Z-scheme CdS/Au/TiO_2 system.[75] The proposed energy diagram is shown in Figure 10.8. Similarly, other all-solid-state Z-scheme systems such as CdS/Au/$TiO_{1.96}C_{0.04}$,[76] (Ru/$SrTiO_3$:Rh)/RGO (reduced graphene oxide)/$BiVO_4$[72] and metal sulfide/RGO/TiO_2[77] were also reported recently. In theory, any conductor (such as metal nanoparticle, carbon nanotube and nanofiber) can serve as a solid-state redox mediator to construct the all-solid-state Z-scheme system. More interestingly, some direct contact photocatalysts without an electron mediator (such as g-C_3N_4–TiO_2, ZnO–CdS and Ru/$SrTiO_3$:Rh–$BiVO_4$) also exhibit the all-solid-state Z-scheme mechanism.[68,78,79] For more information about different kinds of all-solid-state Z-scheme systems, the reader is referred to the excellent review by Zhou *et al.*[80]

Besides efforts to develop overall water-splitting systems, the construction of a water-splitting system on the basis of photocatalysts and suitable sacrificial electron donors is another strategy to achieve high photocatalytic performance for hydrogen generation. So far, various kinds of inorganic ions (such as S^{2-}/SO_3^{2-}, Fe^{2+}, I^- or CN^-) and organic compounds (such as pollutants or biomass) have been used as electron donors to construct the hydrogen production systems, which will be discussed in detail in the Section 10.5.1.

10.2.3 Mechanisms and Processes of Photocatalytic Hydrogen Production

The photocatalysts in different systems possess similar processes for photocatalytic hydrogen production. Typically, they include light harvesting, charge separation, charge migration and transport, and charge utilization, self-corrosion by charges and surface back reaction, which are summarized in Figure 10.9. Clearly, the overall efficiency for photocatalytic H_2 evolution strongly depends on the balance of thermodynamics and kinetics of these processes, which is influenced by the bulk and surface properties and electronic structure of a photocatalyst. Thus, to design highly efficient photocatalysts for photocatalytic H_2 evolution, all these typical processes must be comprehensively considered and optimized. In other words, the solar energy conversion efficiency (η) is strongly dependent on the cumulative effect of light harvesting (η_{LH}), charge separation (η_{CS}), charge migration and transport (η_{CMT}), and charge utilization (η_{CU}) for H_2 generation.[12,81]

Corresponding to these processes, there are various methods of improvement and key design parameters for photocatalytic H_2 evolution (Figure 10.9). Relatively speaking, considerably more progress has been made in

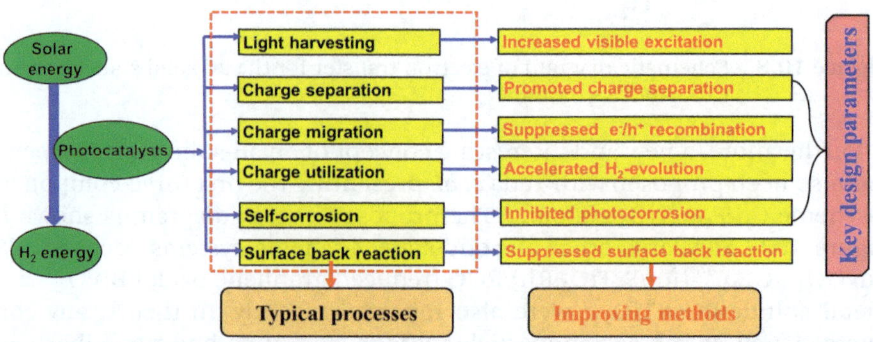

Figure 10.9 Methods of improvement and key design parameters for photocatalytic H_2 evolution.

the increased visible excitation and suppressed surface back reaction. On the one hand, more and more efficient visible-light-active photocatalysts are available.[6,11] On the other hand, suppressed SBRs could be overcome through Z-scheme photocatalyst systems or photoelectrochemical cells. Therefore, the key design parameters to enhance the overall efficiency (Figure 10.9) will be discussed in detail, including promoted charge separation and transport, accelerated H_2-evolution kinetics and increased stability of photocatalysts.

10.3 Promoted Charge Separation and Transport

As mentioned above, the charge separation and transport play crucial roles in enhancing photocatalytic efficiency for H_2 production. Therefore, much attention has been paid to promoting the charge separation and transport efficiency. Generally, the charge separation and transport are strongly dependent on several key factors: defects in the lattice, degree of crystallinity, transport distance of charge carriers to SC surface, surface state and trapper sites and interface electric fields. In the following sections, several different strategies such as improving the crystallinity, shortening charge diffusion distance, coupling SCs with nano-carbon materials and constructing the internal fields will be discussed in detail.

10.3.1 Improvement of the Crystallinity

It is widely accepted that a high degree of crystallinity is advantageous for enhancing the activity for photocatalytic H_2 production due to the higher mobility and lower recombination of excited charges within the defect-free solid.[11,82] Generally, the crystallinity could be greatly improved by calcinations at high temperature. However, the calcination process has a negative influence on the surface area and particle size, leading to a decrease in the activity for photocatalytic H_2 production.[11,83] Clearly, the final photocatalytic activity is strongly dependent on the balance of these positive and negative factors.[11] Thus, optimization of the crystallinity, surface area and crystal structure has proved to be an effective strategy to enhance photocatalytic H_2-production activity because recombination between photogenerated electrons and holes could seriously reduce the efficiency for uphill water-splitting reactions.[11] For example, it has been extensively demonstrated that TiO_2 with both good crystallinity and high surface area generally exhibits better activity for photocatalytic hydrogen production.[84–86] Bao *et al.* demonstrated that hexagonal CdS with good crystallinity and very fine particle size showed higher photocatalytic activity for hydrogen production than those samples with poor crystallinity or large particle size under visible light.[83] Recently, it was reported that the photocatalytic activity of crystalline mesoporous Ta_2O_5 for H_2 production was much higher than those of commercial Ta_2O_5 powder and amorphous mesoporous Ta_2O_5 under ultraviolet (UV) light irradiation.[87,88] In addition, crystalline g-C_3N_4 nanosheets and nanorods show much higher activity toward photocatalytic water-splitting than other pristine carbon nitrides and mesoporous C_3N_4.[89,90] Thus, fabrication of crystallized SC-based photocatalysts should be a promising way to design highly efficient photocatalysts for photocatalytic H_2 production.

10.3.2 Rational Design of Nanostructures

Another efficient strategy for accelerating the transfer and separation of the photo-induced carriers is to shorten charge diffusion distance from bulk to surface by constructing nanostructured photocatalysts. Generally, compared to their bulk materials, SC nanomaterials have several advantages for H_2 generation due to the following reasons: higher surface-to-volume ratio, stronger optical absorption, shorter charge migration length, stronger redox power and tunable electronic structure.[91] Importantly, for anisotropic nanomaterials, the photogenerated electrons and holes can also be efficiently separated due to the difference of charge diffusion distance and transfer character in different directions. The typically anisotropic nanostructures including 1D nanowires/rods, 2D nanosheets and 3D hierarchical architectures have been widely investigated, and will also be discussed in detail in this section.

So far, there have been many reports about the application of 1D nanostructures such as nanowires (NWs), nanotubes (NTs), nanofibers (NFs), nanorods (NRs) and nanobelts (NBs) in photocatalytic H_2 production, due to the shortened distance for charge diffusion to the SC/electrolyte interface and improved transport properties of charge carriers. For instance, 1D anatase-TiO_2 NFs[85,92] and Zn_2GeO_4 NRs[93] exhibited enhanced photocatalytic H_2 and O_2 evolution activity, as compared to the corresponding bulk materials. Under visible light irradiation, the activity of H_2 and O_2 evolution over C_3N_4 NRs has been improved by a factor of ~3 and ~2 over those of the bulk g-C_3N_4 due to the enlarged surface area and lowered HOMO position.[89] In particular, the CdS NRs/NWs have also been extensively investigated for photocatalytic water splitting by Yu and co-workers.[94–97] Clearly, the 1D nanostructure plays an important role in enhancing photocatalytic H_2-production activity of CdS.[97] The destruction of the NR structure can lead to a significant decrease of H_2-production activity.[94,96] In addition, a very low Pt loading on CdS NRs can greatly enhance their H_2-production activity,[98] which is strongly dependent on the sub-nanometer size and dispersion of Pt clusters.[94,99] More interestingly, Amirav *et al.* demonstrated that an AQY of 20% at 450 nm for H_2 production can be achieved by tuning the length of CdS NRs and the size of an embedded CdSe seed.[100] In an another novel system using 3-mercaptopropionic acid (MPA) as the sacrificial hole acceptor, an internal quantum efficiency (IQE) of 13.6 ± 1.0% was obtained for CdSe/CdS dot-in-rod NRs.[101] Recently, 1D core–shell nanostructures have attracted much interest due to their high solar energy conversion efficiency.[96,102–104] In addition, it is very important to precisely control the diameter, length, surface structure, crystallinity as well as the band structure of 1D nanostructures to obtain the maximum activity for photocatalytic H_2 production.

Since Lu and his co-workers pioneered the fabrication of a uniform anatase single crystal sheet with 47% (001) facets[105] there has been great interest in the development of 2D nanosheet-based SCs with dominant high-energy facets for application in photocatalytic H_2 production because of their facet-dependent properties. It is generally accepted that the (101) facets are more reductive than (001) facets as they could act as possible reservoirs of the photogenerated electrons, while (001) facets as oxidation sites play the major role in the photo-oxidative processes.[106,107] Recently, it was revealed that low index facets of anatase TiO_2 followed a photoreactivity order of (001) < (101) < (010) for photocatalytic hydrogen evolution and ˙OH radical generation.[108] Similarly, data obtained on TiO_2 nanocrystals (NCs) through a seeded growth technique also demonstrated that the (101) facets of anatase TiO_2 are more active than the (001) facets for photocatalytic water splitting.[109] More recently, based on theoretical calculations and experimental studies, it was found that the photocatalytic activity over the exposed (111) facet of anatase TiO_2 is about 5, 9, and 13 times higher than that of the TiO_2 sample exposed with dominant (010), (101), and (001) facets, respectively.[110] However, in most cases, the synergistic effect of different facets was ignored. Very recently, a new concept of surface heterojunction was first proposed in terms of the DFT calculation by Yu *et al.* (as shown in Figure 10.10).[111] Considering the overflow effects of electrons in (101) facets, the optimization of surface heterojunction formed between exposed (101) and (001) facets could greatly promote the charge separation, thereby leading to an improved photocatalytic activity for CO_2 photoreduction. Thus, the surface heterojunction may provide a new strategy for designing highly efficient photocatalysts for photocatalytic H_2 production, which deserves more attention in future studies.

Besides TiO_2, ultrathin CdS nanosheets,[112] exfoliated $CsCa_2Ta_3O_{10-x}N_y$[113] and C_3N_4 nanosheets[90,114] were also developed successfully, which showed

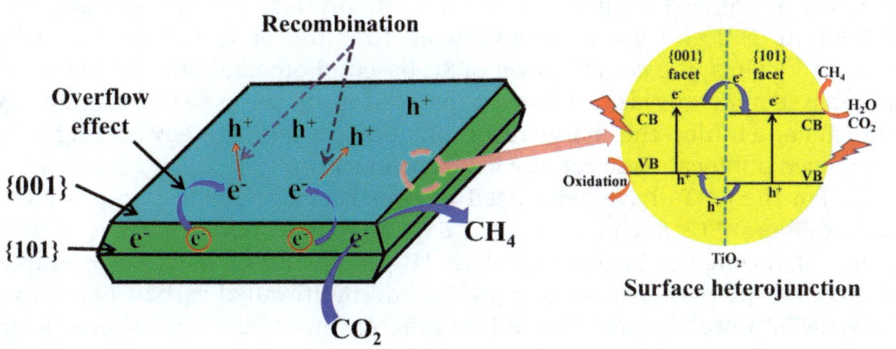

Figure 10.10 An electron overflow effect and hetero-facet junction between (101) and (001) facets in anatase.

significantly enhanced visible-light activity toward hydrogen production compared to their bulk counterpart. In future, more effort should focus on the development of feasible synthesis routes, controlling the ratio of different exposed facets, and decoration of facets with co-catalysts or other 2D nanostructured SCs to form more efficient facet-exposed SCs for extensive applications.

Recently, many studies have been directed towards more advanced three-dimensional (3D) hierarchical architectures assembled by nanoscale building blocks such as NCs, NWs, NBs and nanoplates. For example, the reported 3D-hierarchical flowerlike metal sulfide nanostructures such as $CuInS_2$,[115] ZnS,[116] CdS,[117] $ZnIn_2S_4$,[118,119] Cu_3SnS_4[120] and $AgGa_{1-x}In_xS_2$[120] exhibited enhanced H_2-generation photo-activity. Furthermore, the 3D-hierarchical architectures with heterojunctions such as core–shell carbon nanofiber@ $ZnIn_2S_4$,[121] TiO_2 nanowire arrays on RGO[122] and Z-scheme type $CdS/Au/TiO_2$[123] have been shown to achieve significantly improved photocatalytic activity for hydrogen evolution. It is generally accepted that flowerlike morphology plays an important role in enhancing photocatalytic activity. In addition, the combined effects of several factors, including large surface areas, mesoporous structures, heterojunction interfaces, suitable E_g and band positions, are also beneficial for the enhancement in photo-activity of the hierarchical photocatalysts.

10.3.3 Application of Carbon-Based Nanomaterials

In recent years, coupling SCs with carbon-based nanomaterials has proven to be another efficient strategy to promote charge separation and transport, due to the significantly increasing interest in the production and application of carbon nanomaterials, especially for carbon nanotubes (CNTs) and graphenes. The CNTs provide a large surface area (>150 $m^2\ g^{-1}$), high work function (~5 eV) and excellent conductivity and mechanical properties.[124] Furthermore, compared to the CNTs, the graphene, a sp^2-hybridized 2D carbon nanosheet, exhibits a much higher optical transmittance, conductivity (~5000 $W\ m^{-1}\ K^{-1}$), electron mobility (200 000 $cm^2\ V^{-1}\ s^{-1}$), theoretical specific surface area (~2600 $m^2\ g^{-1}$) and more suitable work function (4.42 eV) for H_2 evolution.[125–129] Thus, the combination of SC-based photocatalysts and these carbon-based materials is one of several promising strategies to favor the charge transfer and inhibit the charge recombination process, thereby leading to an enhanced photocatalytic activity for H_2 production.

So far, the CNTs have been used to construct composite photocatalysts with different SCs such as $g\text{-}C_3N_4$, CdS, Ta_2O_5 and TiO_2, leading to significant enhancements in photocatalytic H_2-production activity. For example, the composite photocatalysts of $g\text{-}C_3N_4$ and multiwalled carbon nanotubes (MWCNTs) were also synthesized by different methods and systematically investigated for H_2 evolution. Under optimal MWNTs content (~0.2–2 wt%), it was found that a 2.4- to 3.7-fold enhancement in H_2-production rate could be achieved as compared to that of pure $g\text{-}C_3N_4$,[130,131] suggesting that the enhanced activity for photocatalytic H_2 evolution could be attributed to the efficient separation of the photogenerated charge carriers because of the intimate interactions between g-C_3N_4 and MWCNT. The optimized hybrid of CNT and CdS can lead to a 17-fold enhancement in photocatalytic activity for H_2 evolution.[132] It was also revealed that the best hydrogen production rate in the binary hybrids could be obtained by using CNTs annealed at 500 °C due to improved purity and crystallinity of CNTs. In addition, the acid-treated CNT in ternary CdS/CNT/Pt hybrids could result in a larger amount of Pt adsorption on CNTs, thus significantly enhancing the amount of hydrogen production.[132] Surprisingly, a 35-fold enhancement in photocatalytic H_2-evolution activity was achieved over a nano-carbon–Ta_2O_5 hybrid photocatalysts compared to the corresponding Ta_2O_5-NTs. The remarkable enhancement is due to two effects: a single-crystalline nature of the coating, which alleviates transport of electrons to the interface, and the formation of a tight Schottky-type junction between the SC and the CNT, which facilitates charge transfer and separation at the interface.[133] Recently, a novel sandwich composite of carbon/TiO_2 nanotube/carbon nanotubes was synthesized by coating graphitic carbon laminas on the surfaces of TiO_2 nanotubes. As the result, a surprisingly high H_2-evolution rate (37.6 mmol h^{-1} g^{-1}) with an AQY of over 95% was achieved under illumination of one AM1.5G sunlight.[134] The super-high activity can be attributed to the effective separation of the photoinduced charge pairs and narrowed band gap.

To apply the RGO in photocatalytic H_2 production, different graphene-based composite systems have been constructed and applied in photocatalytic hydrogen evolution (summarized in Table 10.1). Obviously, some typical binary nanocomposite photocatalysts such as RGO/CdS,[135,136] RGO/TiO_2[137] and RGO/$Zn_{0.5}Cd_{0.5}S$[138] have been systematically studied by different groups. For example, a 41-fold enhancement in the photocatalytic H_2 evolution rate was achieved over RGO/TiO_2 as compared to that over the pure TiO_2.[137] These results further confirmed that RGO as potential co-catalyst, photosensitizer, good acceptor and reservoir of the photogenerated electrons can efficiently promote the charge transfer and separation, and increase the surface active sites for H_2 evolution.[127,128] Moreover, the photocatalytic H_2 production activity of these binary systems could be further improved by heteroatom-doped RGO or by optimizing structure/composition of SC. For example, the optimized CdS/N-doped graphene exhibits the highest H_2 production rate in Na_2S and Na_2SO_3 aqueous solution, which was 5.3-fold higher than that of CdS alone.[139] This is because N atom impurities in graphene can intrinsically tailor the electron donor or acceptor properties of graphene, thus leading to significant enhancements in electrocatalytic performance, conductivity and electron transfer efficiency of graphene.[140] Noble metal-free RGO–$Zn_xCd_{1-x}S$ composites were also constructed through a simple and economic hydrothermal method.[141] The optimized H_2-production rate of the prepared RGO–$Zn_{0.8}Cd_{0.2}S$ photocatalyst can reach up to 1824 μmol h^{-1} g^{-1} with an AQY of 23.4% at 420 nm, which was better than that of the

Table 10.1 Photocatalytic H_2-production activity of graphene-based composite photocatalysts.

Photocatalyst	Synthesis method	Co-catalyst 1	Co-catalyst 2	Light source	Sacrificial reagent	R_{H_2} (μmol g^{-1} h^{-1})	Enhanced factor by graphene	Ref.
CdS	Precipitation	5 wt% GO	—	300 W Xe-lamp, $\lambda \geq 420$ nm	0.35 M Na_2S + 0.25 M Na_2SO_3	3140	1.3	142
CdS	Solvothermal	10 wt% RGO	—	300 W Xe-lamp, $\lambda \geq 420$ nm	0.35 M Na_2S + 0.25 M Na_2SO_3	4202	1.5	143
CdS	Precipitation–hydrothermal	1 wt% RGO	—	200 W Xe-lamp, $\lambda \geq 420$ nm	0.1 M Na_2S + 0.05 M Na_2SO_3	700	4.8	136
CdS	Precipitation	2 wt% sulfonated graphene	—	Metal halide lamp, $\lambda \geq 380$ nm	0.05 M Na_2S + 0.07 M Na_2SO_3	167	—	135
CdS	Precipitation–calcination	2 wt% N-doped graphene	—	300 W Xe-lamp, $\lambda \geq 420$ nm	0.1 M Na_2S + 0.1 M Na_2SO_3	1050	5.3	144
CdS	Two-phase	333 wt% RGO	—	400 W high pressure Hg lamp	10 vol% methanol	2190	4.1	145
CdS	Solvothermal	1 wt% RGO	0.5 wt% Pt	350 W Xe-lamp, $\lambda \geq 420$ nm	10 vol% lactic acid	56 000	45.6	146
CdS	Hot injection–sonication	0.7 wt% cysteine modified RGO	2.1 wt% Pt	300 W Xe-lamp, $\lambda \geq 420$ nm	10 vol% lactic acid	29 861	—	147
CdS	Hydrothermal-mixing	0.4 wt% graphene	2 wt% MoS_2	500 W UV-vis lamp	10 vol% lactic acid	6857	71.0	148
CdS	Hydrothermal	1.33 wt% graphene	0.67 wt% MoS_2	300 W Xe-lamp, $\lambda \geq 420$ nm	20 vol% lactic acid	9000	72.0	149
CdS	Hydrothermal	33 wt% GO	1.0 wt% $Ni(OH)_2$	300 W Xe-lamp, $\lambda > 420$ nm	0.35 M Na_2S + 0.25 M Na_2SO_3	4732		150
CdS–Al_2O_3	Hydrothermal–physical mixing	1 wt% GO	—	500 W phoenix tungsten halogen lamp	0.35 M Na_2S + 0.25 M Na_2SO_3	1750	4.2	151

CdS–ZnO	Hydrothermal–physical mixing	1 wt% GO	—	500 W phoenix tungsten halogen lamp	0.35 M Na_2S + 0.25 M Na_2SO_3	3755	9.0	151
CdS–ZnO	Light irradiation-assisted precipitation	2 wt% RGO	—	300 W Xe lamp	0.1 M Na_2S + 0.1 M Na_2SO_3	5100	34.0	152
CdS–TaON	Precipitation–hydrothermal	1 wt% RGO	0.4 wt% Pt	300 W Xe-lamp, $\lambda \geq 420$ nm	0.1 M Na_2S + 0.04 M Na_2SO_3	3165	—	153
CdS QDs(3 wt%)–$ZnIn_2S_4$	Hydrothermal–precipitation–hydrothermal	1 wt% RGO	0.4 wt% Pt	300 W Xe-lamp, $\lambda \geq 420$ nm	0.1 M Na_2S + 0.04 M Na_2SO_3	27 000	—	154
TiO_2 nanosheets with exposed (001) facets	Microwave–hydrothermal	1 wt% RGO	—	50 W Xe arc lamp	25% methanol	736	41	155
g-C_3N_4	Impregnation–chemical reduction	1 wt% RGO	1.5 wt% Pt	350 W Xe-lamp, $\lambda \geq 400$ nm	25% methanol	451	3.07	156
$Zn_{0.8}Cd_{0.2}S$	Co-precipitation–hydrothermal	0.25 wt% RGO		AM1.5, 100 mW cm^{-2}, Newport 91160	0.35 M Na_2S and 0.25 M Na_2SO_3	1824	4.5	141
TiO_2	Two-step hydrothermal method	0.025 wt% RGO	0.475 wt% Pt	350 W Xe-lamp, $\lambda \geq 400$ nm, 20 mW cm^{-2}	25% ethanol	2066	5.56	157
$Zn_{0.5}Cd_{0.5}S$	One-step hydrothermal method	0.5 wt% RGO		150 W Xe-lamp, $\lambda \geq 400$ nm, 140 mW cm^{-2}	0.35 M Na_2S and 0.25 M Na_2SO_3	1060	3.2	138
TiO_2	Microwave–hydrothermal	0.1 wt% RGO	0.25 wt% Au	Four low-power 420 nm-LEDs (3 W), 80.0 mW cm^{-2}	25% methanol	296		158

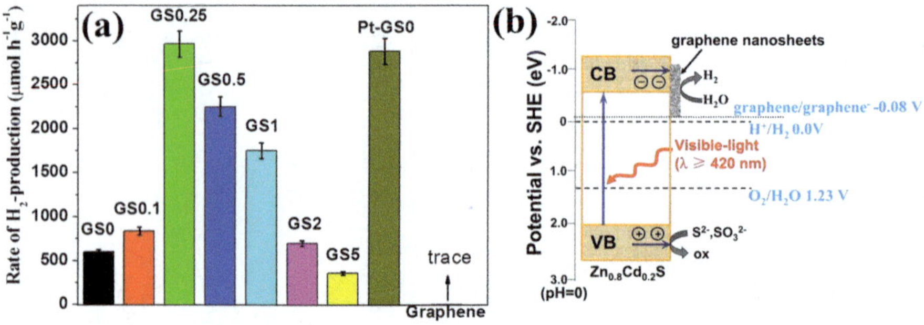

Figure 10.11 (a) Solar photocatalytic H_2-generation activity of the GS0, GS0.1, GS0.25, GS0.5, GS1, GS2, GS5, Pt-GS0 (1 wt% Pt), and RGO samples under simulated solar irradiation, and the weight ratios of GO to $Zn_{0.8}Cd_{0.2}S$ were 0, 0.1, 0.25, 0.5, 1, 2, and 5%, and the corresponding samples were labeled as GS0, GS0.1, GS0.25, GS0.5, GS1, GS2, and GS5, respectively. (b) Photocatalytic H_2-production mechanism for the RGO–$Zn_{0.8}Cd_{0.2}S$ system under solar light irradiation. (Reprinted with permission from ref. 141. Copyright 2012, American Chemical Society.)

optimized Pt–$Zn_{0.8}Cd_{0.2}S$ under the same conditions (as shown in Figure 10.11a).[141] Clearly, as a co-catalyst, RGO could promote the charge separation and accelerate water reduction kinetics, thus leading to a greatly enhanced photocatalytic activity (Figure 10.11b).

Furthermore, it was revealed that the combination of graphene and other co-catalysts could further improve the photocatalytic activity of the graphene-based photocatalysts due to the synergetic effects of dual co-catalysts.[159,160] For example, Li *et al.* reported that a CdS/RGO nanocomposite can be prepared by utilizing a one-step solvothermal method, followed by photodeposition of Pt nanoparticles (NPs) on its surface.[146] The resulting ternary CdS/Pt/graphene using 0.5 wt% Pt and 1.0 wt% RGO as dual co-catalysts exhibits a significantly enhanced photocatalytic H_2-production rate with an AQE of 22.5% at 420 nm, which was about ~45.6-times higher than that of bare CdS nanoparticles. In another report, an optimal H_2 production rate of 2.15 mmol h^{-1} ($\lambda \geq 420$ nm) with a quantum efficiency of 50.7% was accomplished over graphene nanosheets anchored by water soluble CdS quantum dots and dendritic Pt nanocrystals.[147] To replace the noble-metal co-catalyst (Pt) and increase the contact area between graphene and co-catalyst, the ternary MoS_2/CdS/RGO composite was also constructed by both Jia *et al.*[148] and Chang *et al.*[149] The optimized photocatalytic H_2 production rate of CdS–MoS_2/RGO composite was over 70-times higher than that of bare CdS. Similarly, the TiO_2/MoS_2/graphene hybrid photocatalysts using 2D graphene and layered MoS_2 as dual co-catalysts were also prepared by a two-step hydrothermal process.[157] The results revealed that the optimized TiO_2/MoS_2/graphene exhibited the highest H_2 production rate of 165.3 μmol h^{-1} with an AQE of 9.7% at 365 nm (Figure 10.12a). The significantly improved activity was attributed to the synergetic effects of graphene and earth-abundant MoS_2 co-catalysts

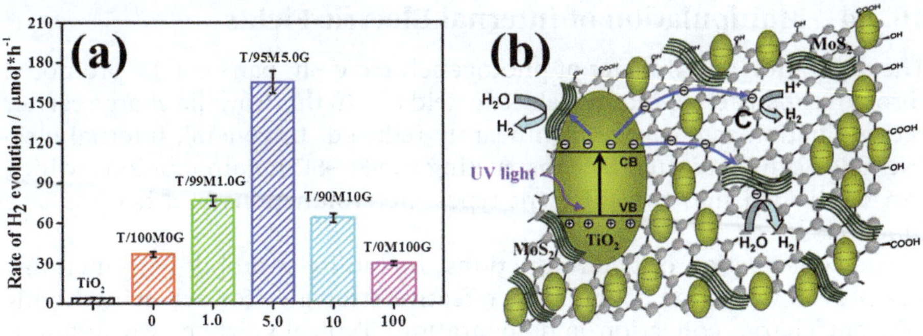

Figure 10.12 (a) Photocatalytic H_2 production of the TiO_2/MoS_2/graphene composites with different MoS_2 (M) and RGO (G) contents as co-catalyst under UV irradiation. (b) The photocatalytic H_2-production mechanism and charge separation and transfer in the TiO_2/MoS_2/graphene system. (Reprinted with permission from ref. 157. Copyright 2012, American Chemical Society.)

(Figure 10.12b). In addition, notably, an appropriate amount of graphene (≤1 wt%) plays an important role in accomplishing the highest photocatalytic H_2-production activity due to the light-shielding and active site-blocking effects of the excess graphene in composites.

In addition, hierarchical 3D macroporous graphene architectures and aerogels based on the self-assembly of 2D building units hold great potential in heterogeneous photocatalysis due to their macroscopic 3D structures, large internal surface areas and high electrical conductivities.[161,162] As very promising materials, although they have been applied in the photocatalytic CO_2 reduction[163] and degradation of pollutants,[164–171] there have been no reports about their utilization in photocatalytic H_2 evolution. Consequently, it is highly expected that these 3D graphene structures could be widely investigated in the field of photochemical H_2 evolution.

Aside from CNTs and RGO, carbon quantum dots (CQDs) were also used as co-catalysts to promote charge separation and transport. For example, it was found that CQDs and platinum NPs as dual co-catalysts could greatly enhance the photocatalytic H_2-production activity of $ZnIn_2S_4$ microspheres (ZIS MSs).[172] The highest H_2-production rate of 1032.2 μmol h^{-1} g^{-1} with an AQE of 2.2% (at 420 nm) in triethanolamine aqueous solution was achieved over optimized ZIS MSs co-loaded with CQDs and Pt under visible light irradiation. The multiple synergistic effects, including good crystallization, enhanced light absorption, high electrical conductivity of CQDs, and promoted vectorial electron transfer from ZIS to CQDs and Pt NPs, were responsible for the remarkable enhancement in photocatalytic H_2-evolution activity. Thus, it is expected that these carbon-based nanomaterials could also be applied in other systems for photocatalytic H_2 evolution in future studies. To enhance their photocatalytic activity, further optimization of these systems is also needed.

10.3.4 Manipulation of Internal Electric Fields

The separation and transfer of photogenerated e^-–h^+ pairs can be promoted through constructing internal electric fields.[173] In this way, the charge carrier recombination rate can be significantly reduced. In general, internal electric fields can be created by constructing the SC–SC heterojunctions, which can be divided into two different types: heterojunctions and heterophase junctions.[173]

The construction of heterojunctions, including p–n and n–n junctions, has proven to be one of the most effective strategies for achieving highly efficient charge collection and separation. Typically, when two different SCs were combined each other, the Fermi levels will equilibrate and internal fields will be formed, thus resulting in improved charge separation and photocatalytic activity. During the past several years, many different active heterojunction combinations have been reported. Among them, the heterojunctions between CdS and TiO_2 have attracted the most extensive research. Generally, as shown in Figure 10.13, it is believed that the internal electric fields created by the potential difference between CdS and TiO_2 could promote charge transfer and separation, thus leading to the enhanced activity for photocatalytic H_2 evolution. For example, it has been widely demonstrated that the CdS/TiO_2 composite photocatalysts exhibited enhanced H_2-production activity under visible light from water containing sulfide/sulfite or absorbed H_2S as sacrificial reagents.[174–176] The improved activity was usually attributed to efficient electron transfer from the conduction band of CdS to that of TiO_2 due to the formation of a heterojunction. In particular, an AQY of 43.4% for H_2 production under visible light irradiation (λ = 420 nm) could be achieved over CdS nanoparticles inside TiO_2 nanotubes.[177]

Meanwhile, ternary hybrid photocatalysts such as CdS/TiO_2/Pt and CdS/Au/TiO_2 have been extensively studied for photocatalytic H_2 production.[75,178–182] It was found that the deposition location of Pt co-catalyst on CdS/TiO_2 is one

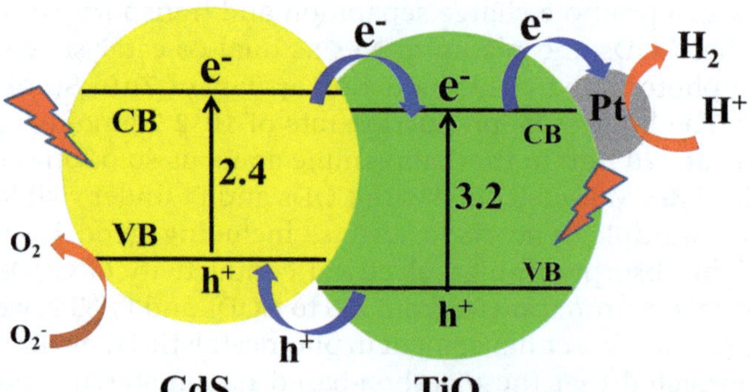

Figure 10.13 Schematic energy diagram of the heterojunction between CdS and TiO_2.

of the key factors of high-efficiency hybrid photocatalyst for hydrogen production.[178,179] In particular, Pt NPs should be deposited on the TiO_2 surface to attain a vectorial electron transfer of CdS → TiO_2 → Pt, and efficient charge transfer on the CdS/TiO_2/Pt hybrid can be achieved. Moreover, the photocatalytic H_2-production activity is closely related to the morphology of TiO_2. It was verified that the CdS-sensitized Pt/TiO_2 nanosheets with exposed (001) facets exhibited the highest photocatalytic activity as compared to that of CdS-sensitized Pt/P25 and Pt/TiO_2 nanoparticles, due to the combined effects of exposed (001) facets, surface fluorination and high specific surface area.[182] In addition, all-solid-state Z-scheme photocatalysts such as CdS/Au/TiO_2 and CdS/Au/$TiO_{1.96}C_{0.04}$ were also developed;[75,76] they exhibited an enhanced photocatalytic activity for H_2 evolution due to the efficient transfer of photogenerated electrons from the CB of TiO_2 to that of CdS.

Besides n–n junction, both p–n junction and multi-junctions have also been used to enhance the charge separation and photocatalytic activity. For example, Lee *et al.* first prepared a particulate p–n junction photocatalyst of $CaFe_2O$–$PbBi_2Nb_{1.9}W_{0.1}O_9$ according to the concept of a photochemical p–n heterojunction diode. The results showed that the efficient charge separation led to enhanced photocatalytic activity for individual H_2 and O_2 evolution from aqueous solutions containing methanol and silver nitrate, respectively.[183] Furthermore, the photo-activity could be further improved by introduction of ohmic layers between the p- and n-type SCs due to the realization of an all-solid-state Z-scheme mechanism.[184,185] For example, the optimized Cu_2O/Cu/TiO_2 nanotube ohmic heterojunction arrays showed a three-fold enhancement in the photocatalytic H_2-production activity as compared to that of the Cu_2O/TiO_2 arrays.[186]

In particular, ultrathin heterostructures or 2D layered composite photocatalysts have received growing interest in photocatalysis applications due to their significant advantages such as large specific surface areas, increased contact surface and ultrafast charge separation rates.[187] For example, ultrathin p–n junctions such as NiO/$Ca_2Nb_3O_{10}$[188] and MoS_2/N-doped RGO[189] showed significant improvement in photocatalytic activity toward H_2 generation, owing to the greatly enhanced charge generation and suppressed charge recombination. As a typical example, the MoS_2/g-C_3N_4 photocatalysts also possess a perfect 2D layered structure, which is a major reason for the enhanced photocatalytic H_2 evolution.[190,191] Thus, these 2D layered ultrathin heterostructures deserve more attention in future studies.

In addition to the heterojunctions, surface heterophase junctions can also be formed between two different phases with the same composition. It is well known that the rutile phase with a lower reduction potential could absorb visible and solar light energy more effectively and efficiently due to its much narrower band gap, whereas the anatase phase with a wider band gap is generally more photoactive due to its higher reduction potential and slower recombination rate of electrons and holes.[192,193] Thus, the combination of the anatase and rutile can lead to an enhanced photocatalytic activity. For example, the commercial Degussa P25, a well-known mixed phase TiO_2 composite, has received much attention as a benchmark photocatalyst for the investigation and comparison of photocatalytic properties due to its superior photocatalytic activity. Quite recently, Li and his co-workers demonstrated that the exposed anatase/rutile phase junctions on the TiO_2 surface can be obtained by calcination at different temperatures, which can favor the transfer of the photogenerated electron from the CB of rutile phase to the trapping sites on anatase surface, thereby improving the charge separation efficiency and enhancing the photocatalytic activity.[194] Furthermore, it was found that the optimized anatase–rutile structure of Pt/P25 photocatalyst by calcination at different temperatures could greatly promote the charge separation and adjust the surface acid/base property, thus significantly enhancing H_2-production activity and suppressing CO formation from photocatalytic biomass reforming.[195] Further results demonstrated that SO_4^{2-} as a stabilizer for the anatase phase can enhance the crystallinity of mixed-phase TiO_2 by high-temperature calcination, leading to an excellent H_2-production activity.[196]

However, the initial mechanism for this enhancement in mixed phase TiO_2 is controversial (Figure 10.14). One the one hand, the electron transfer mechanism from anatase to rutile was proposed based on different experimental results, due to the higher CB position of anatase (Figure 10.14(A)).[197,198] On the other hand, the electron transfer mechanism from rutile to the trapping sites of anatase (0.8 eV below the anatase conduction band) was also proposed by Hurum and co-workers using electron paramagnetic resonance (EPR) (Figure 10.14(B)).[198–200] More interestingly, the combination of theory

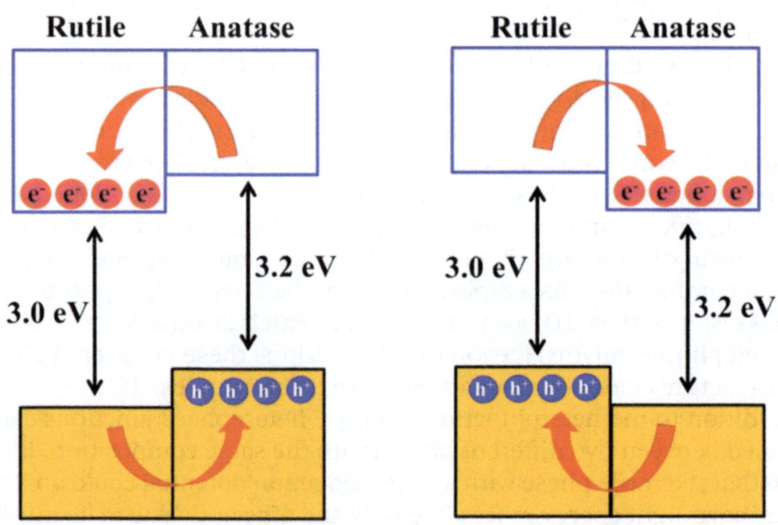

Figure 10.14 Two controversial charge transfer mechanisms between rutile and anatase: (A) from anatase to rutile; (B) from rutile to anatase. Red arrows indicate the flow of electrons (holes) in the CB (VB).[201] (Reprinted with permission from ref. 201. Copyright 2013 Nature Publishing Group.)

and experiment further verified that the higher electron affinity of anatase could also lead to the photogenerated electron transfer from rutile to anatase.[201] Although there are on-going debates on the transfer direction of photogenerated electrons, similar observations have been made in improvements in photocatalytic activity. Interestingly, surface heterophase junctions have been also reported for a Ga_2O_3-based photocatalyst. It was found that the formation of an exposed α–β phase junction of Ga_2O_3 could also significantly enhance the photocatalytic activity for overall water splitting. A detailed study demonstrated that simultaneous exposure of both phases in the outer region plays a crucial role in promoting the charge separation due to the ultrafast electron transfer process between α and β phase. All the results indicate that the construction of heterophase junctions is an efficient strategy to achieve enhanced charge separation and photocatalytic activity. In future, the design of visible-light-driven photocatalysts with heterophase[202] or twin[203] structure should be one promising direction for developing highly efficient photocatalysts.

10.4 Accelerated H_2-Evolution Kinetics

10.4.1 Increasing the Active Surface Areas

Since mesoporous TiO_2 was first synthesized by Antonelli and Ying in 1995,[204] the synthesis of SC nanoparticles with a mesoporous structural network has become a promising strategy to effectively increase the active sites, specific surface area, interfacial transport and light harvesting arising from multi-reflections in the pores.[205–207] The creation of macro/mesopores in SC-based photocatalysts has been found to improve accessibility of reactants to the catalyst surfaces and multiple scattering, and thus enhance the overall photocatalytic efficiency. All these factors can subsequently enhance the photocatalytic activity.

The higher activity of mesoporous SCs for water splitting can be verified though a series of mesoporous/nanoporous SCs such as CdS,[41] Ta_2O_5 (amorphous or crystallized)[88,208] and C_3N_4 [48,209] reported by Domen's group. All these photocatalysts synthesized by using a hard template exhibited a very large BET surface area and a very high photocatalytic activity for hydrogen production. Especially, the photocatalytic activity of the mesoporous g-C_3N_4 (mpg-C_3N_4) for hydrogen production from water can be improved ten-fold through introducing mesoporosity into polymeric C_3N_4, which can be ascribed to the increased surface area, change of SC properties and the increase of defects.[48] Meanwhile, photocatalytic H_2-evolution activity over the mesoporous anatase synthesized by using the hard template (KIT-6) method was 5.5-times higher than that of bulk anatase TiO_2 due to the larger surface areas and the enhanced textural properties.[86] Furthermore, mesoporous TiO_2 photocatalysts synthesized by a simple and facile, template-free non-hydrothermal method exhibited much higher activity for photocatalytic H_2 evolution compared to nonporous colloidal-TiO_2, commercial Degussa P25 and Hombikat UV-100.[210] The reasons for enhanced activity might be attributed to the efficient charge separation and transfer across the interfaces between the compacted TiO_2 nanoparticles. In addition, Ti-MCM-48 and activation of MCM-41 mesoporous silica by transition-metal incorporation also exhibited better activity for photocatalytic H_2 or O_2 evolution from water splitting.[211,212] In addition, the incorporation of SC nanoparticles into mesoporous substrates such as SBA-15 has also proven to be an advanced strategy for developing more efficient composite photocatalysts.[213,214]

Thus, as an important strategy, the synthesis and applications of crystalline mesoporous metal oxide have attracted extensive attention. To make full use of the unique properties of mesoporous materials and to develop practical photocatalysts, it is necessary to load suitable SCs or co-catalysts into the mesopores with a confinement effect and high crystallinity wall. Furthermore, the complicated interaction between the pore walls and loaded components in the channels also needs to be deeply studied.

10.4.2 Loading of H_2-Evolution Co-Catalysts

Since the Pt loading on the surface of TiO_2 was first introduced by Kraeutler and Bard in 1978,[32] the loading of Pt co-catalysts has been a popular strategy to improve the photocatalytic performance for hydrogen generation from water splitting because the overpotential for H_2-evolution reaction on a Pt co-catalyst is very low. It is generally believed that the co-catalysts can extract the photogenerated charges from the SC, provide reaction sites, minimize the overpotential necessary to drive the two-electron HER and decrease the activation energy for H_2 evolution.[70,215] Another function is to provide the junctions/interfaces between the co-catalysts and the SC to enhance electron–hole separation or charge transport.[215,216] To some degree, suitable H_2-evolution co-catalysts play determining roles in achieving highly efficient photocatalytic activity for water splitting, which can improve the reduction and oxidation activity of water by an order of magnitude or even more. Typically, H_2-evolution co-catalysts can be divided into noble metal-based co-catalysts, earth-abundant metal-based co-catalyst and metal free and other co-catalysts (Figure 10.15). Since metal-free co-catalysts have been introduced in Section 10.3.3, this section will focus on the other two kinds of co-catalysts: noble metal-based co-catalysts and earth-abundant metal-based co-catalysts.

Generally, the noble metals (*e.g.*, Rh, Pd, Ir and Pt) are the best known co-catalysts for H_2 evolution; they can be loaded selectively at reaction sites of SCs by an *in situ* photodeposition method. However, they could also serve as co-catalysts to promote an undesirable SBR, namely, water formation from H_2 and O_2, leading to a decrease in activity. Therefore, the development of co-catalysts with suppressed SBR has attracted increasing interest in photocatalytic H_2 evolution due to their improved charge separation and catalytic activity. For example, Domen's research group found that the Rh/Cr_2O_3 (core/shell) co-catalysts on $(Ga_{1-x}Zn_x)(N_{1-x}O_x)$ exhibit a significant enhancement in photocatalytic activity for H_2 evolution.[217] Further studies by model electrodes revealed that the Cr_2O_3 layer could suppress the SBR.[218] As illustrated in

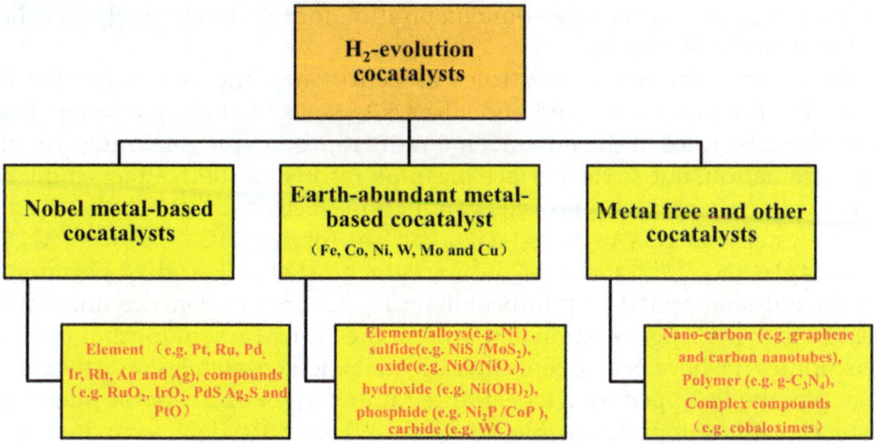

Figure 10.15 Summary of co-catalysts for photocatalytic H_2 production.

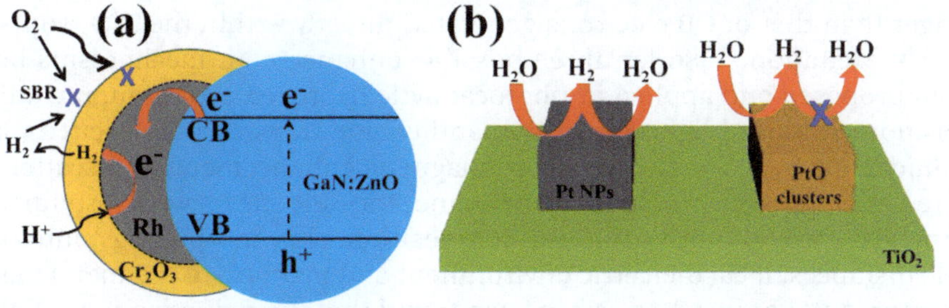

Figure 10.16 (a) Schematic illustration of H_2 evolution on core/shell structured nanoparticles as co-catalyst for photocatalytic H_2 evolution. (b) Schematic photocatalytic H_2 evolution and back-reaction processes over both PtO clusters and metallic Pt nanoparticle co-catalyst.

Figure 10.16(a), the promoted HER can occur at the interface between the Rh core and the Cr_2O_3 shell layer. In addition, it was demonstrated that the TiO_2 photocatalyst loaded by Au/Pd core–shell structured NPs yielded a remarkable H_2-evolution rate of ~0.9 mmol h^{-1} in ethanol or glycerol solutions, corresponding to an AQY of ~70%, which is superior to those of TiO_2-supported Au, Pd and their alloy NPs photocatalysts.[219] The results suggested that the reduced surface accumulation of electrons for Au/Pd core–shell structured NPs facilitated the storage and release of photo-excited electrons for catalysis. Recently, it was also found that the co-catalyst of oxidized platinum clusters in a higher oxidation state can also act as an efficient H_2 evolution site.[220] The photocatalytic activity for HER over PtO/TiO_2 is nearly four times higher than that over Pt/TiO_2 in methanol aqueous solution due to the suppression of the SBR, the oxidation of H_2 to H_2O. The possible reaction mechanism of H_2 evolution over PtO-clusters and Pt co-catalysts is illustrated in Figure 10.16(b). Controlling the valence state of platinum was expected to be a new strategy to achieve the suppression of H_2 oxidation in photocatalytic water splitting, which may provide some inspiration for the development of other high-efficient co-catalysts.

Besides the aforementioned roles as active sites and co-catalyst for H_2 evolution, the plasmonic gold and silver NPs as visible light harvesters also played crucial roles in the enhancement of H_2-production rate under visible light irradiation, due to the surface plasmon resonance (SPR) effects induced by the collective oscillations of the conduction electrons.[221–223] Obviously, the binary systems such as Au (or Ag)–TiO_2,[224–231] Au (or Ag)–ZnO,[232,233] Au–CeO_2,[234] Au (or Ag)–C_3N_4,[235–237] and Au–CdS[238] are the most widely studied plasmonic photocatalysts applied in photocatalytic H_2 generation. On the one hand, the visible-light H_2-generation activity and the enhancement mechanisms of plasmonic NPs have been deeply investigated and completely confirmed. For example, the Pt-tipped Au NRs exhibited much higher activity for photocatalytic H_2 evolution under visible/near infrared (vis/NIR) light compared with fully covered samples, indicating the visible-light H_2-generation activity of plasmonic Au nanoparticles.[239] Recently, it was also discovered that the lifetime of highly energetic "hot" electrons in Au NPs is 1–2 orders of magnitude longer than that of CB electrons generated directly within the TiO_2 support *via* UV excitation.[227] So far, three possible enhancement mechanisms have been proposed and applied in photocatalytic hydrogen evolution, including plasmonic induced light absorption and charge transfer from excited plasmonic metal to SC, near-field electromagnetic enhancement, and scattering of resonant photons.[222] On the other hand, the SPR effects were also further improved through optimizing the composition, size, morphology, interparticle distances, local dielectric environment and medium of the metal nanostructures.[240] For instance, it has been found that the reduction potential of the transferred electrons can be easily tuned by varying the size of Au NPs.[241] Recently, Seh *et al.* demonstrated that Janus Au–TiO_2 photocatalysts with 70 nm Au NPs exhibit much better activity for visible-light H_2 generation than those with 30 nm Au NPs under visible-light irradiation. This was due to the more efficiently enhanced electron–hole separation within the TiO_2 and the stronger plasmonic near-fields of the larger Au NPs.[229] Similarly, an asymmetric Au–CdS plasmonic photocatalyst exhibited a higher activity under visible light than the samples with a symmetrical shell–core structure due to improved localized SPR effects through controlling the fine structural configuration.[238] In addition to these binary systems, ternary systems such as plasmonic metal/co-catalyst/SC,[242–244] plasmonic metal/two SCs[181] and plasmonic metal/nano-carbon/SC[158] have attracted much attention in recent years. Generally, these ternary systems exhibit very high activity for photocatalytic H_2 evolution due to the multiple synergistic effects, which deserve more attention in future studies.

However, the utilization of noble metal-based co-catalysts as H_2 evolution sites is rather disadvantageous for practical applications due to the high costs. Thereby, the development and application of earth-abundant metal-based co-catalysts in photocatalytic H_2 evolution have attracted much attention recently. Table 10.2 summarizes some earth-abundant metal-based co-catalysts for photocatalytic H_2 evolution. As shown in Table 10.2, various

Table 10.2 Earth-abundant metal-based co-catalysts for photocatalytic H_2 generation.

Photocatalyst	Synthesis method	Co-catalyst	Light source	Sacrificial reagent	H_2 production (μmol h^{-1} g^{-1})	Ref.
CdS	Seeded-growth	MoS_2	Hg/Xe 50–500 W DC arc lamp (≥450 nm)	Triethanolamine	100 000	259
CdS	Impregnation	MoS_2	Xe lamp (300 W) ($\lambda > 420$ nm)	10% lactic acid (LA)	500	257
$ZnIn_2S_4$	Impregnation	MoS_2	Xe lamp (300 W) ($\lambda > 420$ nm)	0.43 M Na_2S and 0.5 M Na_2SO_3	141	258
TiO_2	Precipitation	$Ni(NO_3)_2$	Four low-power 365 nm UV-LEDs (3 W)	25% methanol	3056	245
g-C_3N_4	Precipitation	$Ni(OH)_2$	Xe lamp (300 W) ($\lambda > 400$ nm)	10% triethanolamine	152	253
g-C_3N_4	Thermal annealing	Carbon nanodots	Xe lamp (300 W) ($\lambda > 420$ nm)	Pure water	105	271
CdS	Precipitation	$Ni(OH)_2$	300 W xenon arc lamp (≥420 nm)	25% triethanolamine	5084	264
TiO_2	Light-irradiation assisted	$Ni(OH)_2$	350 W xenon arc lamp (160 mW cm^{-2})	25% methanol	2547	251
CdS	Hydrothermal	NiS	Xe lamp (300 W) ($\lambda > 420$ nm)	0.35 M Na_2S and 0.25 M Na_2SO_3	1131	95
$Zn_{0.5}Cd_{0.5}S$	Hydrothermal	NiS	Solar simulator (AM15)	0.35 M Na_2S and 0.25 M Na_2SO_3	239.5	247
CdS	Hydrothermal	NiO_x	Xe lamp (300 W) ($\lambda > 400$ nm)	30% methanol	590.8	249
CdS	Hot-injection	Ni_2P	LED light source (30 × 3 W, $\lambda \geq$ 420 nm)	Lactic acid	2 mL h^{-1}	250
CdS	Hot-injection	Nickel	Three low-power 420 nm UV-LEDs (3 W)	Lactic acid	1232 mol H_2 per mole (6 h)	246
ZnS	Hydrothermal	CuS	Xe lamp (350 W) ($\lambda > 420$ nm)	0.35 M Na_2S and 0.25 M Na_2SO_3	4147	260
TiO_2	Impregnation and calcination method	CuO	Four low-power 365 nm UV-LEDs (3 W)	0.1 M glycerol	2061	254
TiO_2	Precipitation	$Cu(OH)_2$	Four low-power 365 nm UV-LEDs (3 W)	0.09 M ethylene glycol	3418	252
CdS	Calcination	WS_2	Xe lamp (300 W) ($\lambda > 420$ nm)	10% lactic acid	420	256
CdS	Precipitation	WC	Hg-arc lamp (500 W) ($\lambda > 420$ nm)	0.1 M Na_2S and 0.02 M Na_2SO_3	140	255

earth-abundant H_2-evolution co-catalysts (*e.g.*, Ni,[245,246] NiS,[95,247] NiO,[248] NiO_x,[249] Ni_2P,[250] $Ni(OH)_2$,[251–253] CuO,[254] $Cu(OH)_2$,[252] WC,[255] WS_2,[256] MoS_2,[257,258] MoS_3[259] and CuS[260]) have been widely employed to enhance the photocatalytic H_2-production activity of different photocatalysts. In particular, in the 1980s, Domen's research group first found that the core–shell structured Ni/NiO can be used as co-catalyst to achieve the photocatalytic decomposition of H_2O into H_2 and O_2 over $SrTiO_3$ powders.[261,262] This is because the Ni metal core can improve the electron transfer between both $SrTiO_3$ and NiO, while the NiO shell layer can suppress the SBR on metal Ni. Since then, NiO_x as a popular co-catalyst for HER has been widely applied in many heterogeneous photocatalytic systems such as Ta-based photocatalysts to achieve enhanced photo-activity.[6,11,263] Besides NiO_x, $Ni(OH)_2$ and NiS as co-catalysts have also been developed and widely applied in photocatalytic H_2 generation. For example, it was found that $Ni(OH)_2$ as co-catalyst could greatly enhance the photocatalytic H_2-generation activity of g-C_3N_4, CdS and TiO_2.[251,253,264] Recently, it was also revealed that p-type NiS could promote the charge separation in CdS and $Zn_xCd_{1-x}S$ due to the formation of p–n junctions, thus leading to an enhanced activity for photocatalytic H_2 evolution.[95,247] Especially, further enhancement in photocatalytic H_2-generation activity of $Zn_xCd_{1-x}S$ could be obtained by using the dual co-catalysts of NiS and RGO.[247] Furthermore, since the hydrogen generation activity of MoS_2 with the un-coordinated S atoms at the edges was theoretically and experimentally verified,[265,266] the earth-abundant MoS_2 has also been used as an efficient co-catalyst for photocatalytic hydrogen generation. For example, Li *et al.* found that a 36-fold enhancement in activity for HER was achieved through the loading of MoS_2 onto CdS by an impregnation–sulfidation method due to the formation of heterojunction.[257] The results also verified that MoS_2 on CdS were more efficient for the H_2 evolution than Pt on CdS. Recently, Hu's research group found that amorphous MoS_x electrocatalysts in powder and films have more unsaturated S atoms and higher activity for HER than the highly crystalline molybdenum sulfide species.[267,268] Therefore, in theory, amorphous MoS_x can also be used as a co-catalyst to enhance the activity for photocatalytic hydrogen generation. The research group of Alivisatos showed that amorphous MoS_3 can be formed on the surface of CdSe-seeded CdS nanorod photoabsorbers by a microwave-deposition method. They found that the photocatalytic HER activity of initially amorphous MoS_3 was comparable to the platinum, but the activity decreased as MoS_3 was reduced to MoS_2 during the photocatalytic process.[259] In addition, earth-abundant Cu-based co-catalysts have attracted much attention in the past few years. Zhang *et al.* demonstrated that the loading of 2 mol% CuS on porous ZnS could achieve a high H_2-production rate of 4147 μmol h^{-1} g^{-1} with an AQY of 20% at 420 nm.[260] It is believed that the photoinduced interfacial charge transfer from the valence band of ZnS to CuS could partially reduce CuS to Cu_2S, thus leading to significantly enhanced H_2-production activity. More interestingly, an H_2-production rate of 3418 μmol h^{-1} g^{-1} with an AQY of 13.9% was accomplished over the 0.29 mol% $Cu(OH)_2$ loaded TiO_2, which was 205 times higher than that of pure

TiO_2.[252] The enhanced H_2-production activity is mainly attributed to the enhanced charge separation due to the presence of $Cu(OH)_2$ co-catalysts on the surface of the TiO_2. In future studies, more attention should be paid to the development and application of novel H_2-evolution electrocatalysts such as hybrid catalysts of cobalt-doped FeS_2 nanosheets–carbon nanotubes[269] and Ni_2P[270] for photocatalytic hydrogen production.

10.4.3 Elevation of Conduction Band Positions

It is known that redox reactions driven by the photogenerated electrons and holes are similar to those in electrolysis.[11] Thus, the reduction power of SC-based photocatalysts is strongly dependent on their CB positions. In general, the more negative the CB potential of a give SC is, the larger the driving force (overpotential) of photogenerated electrons on its surface for water reduction is, thus leading to more advantageous H_2 formation from water reduction. Consequently, elevating the CB positions of SCs has become another effective strategy for accelerating H_2-evolution kinetics due to increased reduction power of photogenerated electrons. However, in most cases, an increase in the CB potentials could also cause the increase of band gap, which decreases the number of photons absorbed, thus decreasing the photocatalytic activity. Therefore, the optimized photocatalytic activity is dominated by the balance of the flat band potentials (or the CB levels) and the number of photons absorbed (or the band gap). In general, there are two ways to elevate the CB positions of SCs. One is to design a solid solution through band engineering; the other is to utilize the quantum size effects of SC nanoparticles. Both will be discussed in detail in this section.

Fabricating multicomponent SCs through a solid-state reaction is a good strategy for making full use of the advantages of different components.[272] Moreover, the band gap energy and CB position of multicomponent SCs can be continuously tuned by changing the composition *via* control of constituent stoichiometries, which will be quite helpful for achieving an optimized balance between visible-light adsorption and redox potential.[20,273] To date, a great variety of multicomponent SCs, such as $La_5Ti_2MS_5O_7$ (M = Ag, Cu),[274] $(Ga_{1-x}Zn_x)(N_{1-x}O_x)$,[19,20] $BaTaO_2N$,[275] $LaTiO_2N$,[276] $(Ag_{0.75}Sr_{0.25})(Nb_{0.75}Ti_{0.25})O_3$,[273] $(Cu_xNa_{1-x})_2Ta_4O_{11}$ and $Cu_xNa_{1-x}Nb_{13}O_{33}$,[277] $(AgIn)_xZn_{2(1-x)}S_2$,[23] $ZnS–CuInS_2–AgInS_2$,[38] $Zn_xCd_{1-x}S$[138,141,247,278–283] and $ZnIn_2S_4$[22,172,284] have been available in photocatalytic hydrogen generation. For example, although ZnO and GaN are not active under visible light irradiation, their solid solutions show significant visible-light activity for water splitting due to the formation of a narrow band gap. The formation of a GaN:ZnO solid solution can be confirmed by the shifts of XRD peaks and optical absorption edges.[20] Especially, as one of the promising visible-light photocatalysts, $Zn_xCd_{1-x}S$ solid solutions have been extensively applied in photocatalytic H_2 generation. As shown in Figure 10.17(a), the calculated CB edge potential of the $Zn_{1-x}Cd_xS$ solid solution becomes more negative with increasing the content of Zn, thus leading to a stronger reduction power for the H_2 generation. However, the increasing

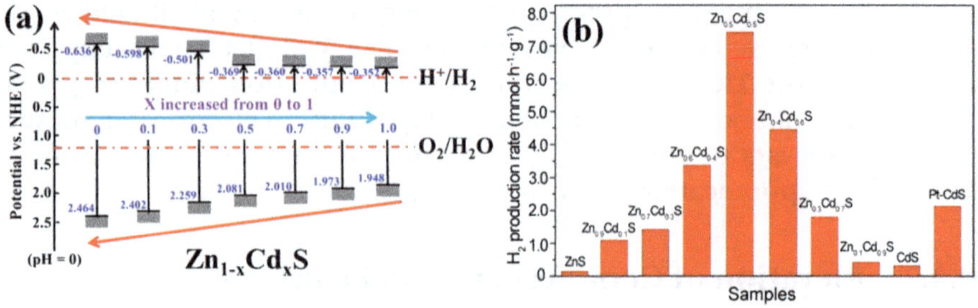

Figure 10.17 (a) CB and VB edge potential changes with composition of $Zn_{1-x}Cd_xS$; (b) comparison of photocatalytic H_2-production activity of the $Zn_{1-x}Cd_xS$ solid solution ($x = 0, 0.1, 0.3, 0.5, 0.7, 0.9$ and 1.0), ZnS, CdS and Pt–CdS samples. (Reprinted with permission from ref. 278. Copyright 2013, American Chemical Society.)

band gap has a negative effect on the photocatalytic activity.[278] Thus, the optimal H_2-production activity under visible-light irradiation was generally obtained through controlling both appropriate band gap width and suitable CB edge position by finely tuning the molar ratio of ZnS and CdS (Figure 10.17b). In addition, the photocatalytic activity of $Zn_xCd_{1-x}S$ solid solutions for H_2 generation could also be further enhanced by other efficient strategies such as Cu doping,[283,285,286] co-catalyst loading[247,287] and nano-carbon coupling.[138,141,247,282]

Furthermore, the CB positions of SCs could be elevated by using the quantum size effects. It is known that the E_g of colloidal nanocrystals along with the redox levels of CB and VB increases with decreasing particle size of a SC below a critical diameter (the Bohr radius) due to the quantum size effects.[288] It was observed that the photocatalytic H_2 production activity of CdSe nanocrystals (1.75–4.81 nm) strongly depended on the degree of quantum confinement, showing a constant decay of normalized H_2 evolution rate with increasing particle size.[289] However, a decreased activity of $SrTiO_3$ for H_2 evolution was observed with decreasing particle size, because of an increased electrochemical overpotentials for OER and a decreased light absorption due to a quantum size effect.[82] Thus, it is important to balance high surface area, crystallinity, nanosize effects and quantum size effects when optimizing photocatalytic efficiency of nanocrystals for H_2 evolution. In addition, nanocrystals should also be capped with polymers or immobilized on an inert support to prevent their agglomeration.[28]

It is also interesting to control the relative positions of CB in different SCs by combining the solid solution engineering and quantum size effects. For example, clearly, bulk CdS cannot sensitize bulk ZnS due to the positive CB position of CdS (Figure 10.18a).[281] However, the band gaps and CB position of CdS QDs increase with decreasing grain size due to quantum size effects, whereas those of $Zn_{1-x}Cd_xS$ decrease with increasing content of Cd^{2+} due to the formation of solid solution. The double effects achieve the sensitization of $Zn_{1-x}Cd_xS$ solid solutions by CdS quantum dots under visible light,

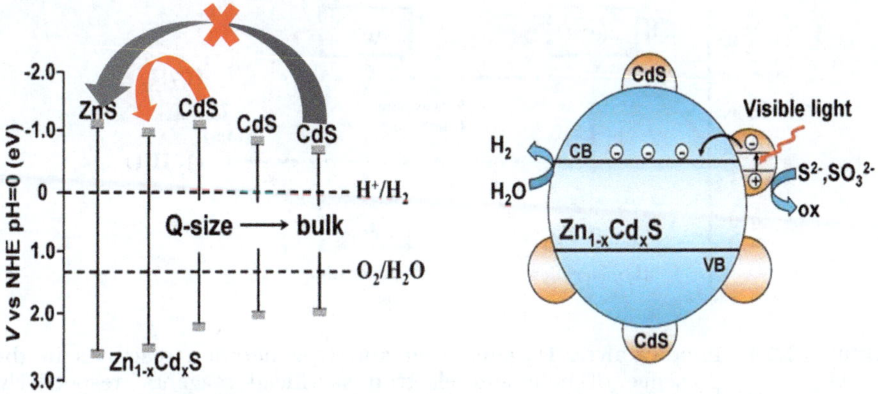

Figure 10.18 (a) Comparison of the CB and VB edge position of ZnS, $Zn_{1-x}Cd_xS$ and CdS with various size. (b) Schematic diagram for photocatalytic H_2-production mechanism in a CdS QDs-sensitized $Zn_{1-x}Cd_xS$ system under visible light. (Reprinted with permission from ref. 281. Copyright 2010, Royal Society of Chemistry.)

thus leading to improved electron transfer and enhanced photocatalytic activity (Figure 10.18b).[281] This work may offer some new ideas to optimize photocatalyst systems for maximizing the photocatalytic activity by comprehensively utilizing these two important strategies.

10.5 Increased Stability of Photocatalyst

10.5.1 Addition of Sacrificial Reagents

So far, there are few successful examples of overall water splitting systems because it is very difficult to accomplish true overall water splitting due to the uphill nature of the reaction, especially for visible-light-driven photocatalyts.[2,11] In most cases, only H_2 evolution was observed, without the O_2 evolution. Alternatively, the downhill reactions using sacrificial reagents are very easy to achieve, but are not real "overall" water splitting reactions. This is because only half-reactions of water splitting occur in these systems containing sacrificial reagents.[11] Importantly, photocatalytic hydrogen or oxygen generation rates can be greatly improved through the addition of sacrificial reagents.[290,291] Thus, the reactions utilizing sacrificial reagents are often employed to test the photocatalytic activity for water splitting.[2,11] Figure 10.19 shows the basic principle of photocatalytic HERs and OERs using sacrificial reagents. When electron donors or hole scavengers were introduced in the photocatalytic system, the photogenerated holes irreversibly oxidize the reducing electron donors instead of H_2O. Thus, the photogenerated electrons could accumulate in the CB of a given SC, leading to an enhanced H_2 evolution rate. Interestingly, both solar hydrogen production and pollutant removal by photocatalytic reactions could be simultaneously achieved if the

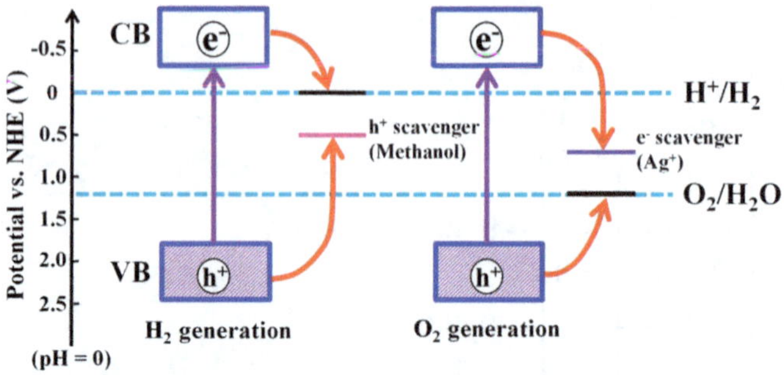

Figure 10.19 Photocatalytic H_2-generation and O_2-generation reactions in the presence of hole and electron sacrificial reagents, respectively. (Reprinted with permission from ref. 2. Copyright 2007, American Chemical Society.)

biomass and undesirable industry waste products such as alcohol and a sulfide ion can be employed as sacrificial electron donors.[11,22] Similarly, the OER could also be enhanced by fast removal of the photogenerated electrons in the CB of a given SC due to the introduction of electron acceptors or electron scavengers into the systems (Figure 10.19). So far, many efforts have been made concerning either hydrogen generation in the presence of a sacrificial electron donor or oxygen production in the presence of a sacrificial electron acceptor.

In general, hole scavengers have been widely used as sacrificial electron donors to improve the separation of electrons and holes, thus leading to an enhanced photocatalytic H_2-generation activity. According to their compositions, the sacrificial electron donors could be divided into two types: inorganic sacrificial reagents and organic sacrificial reagents.

For inorganic sacrificial reagents, the S^{2-}/SO_3^{2-} mixture, as the most widely utilized hole scavenger, has been employed to enhance the photocatalytic hydrogen generation of metal sulfide photocatalysts such as CdS, ZnS–$CuInS_2$–$AgInS_2$ and $(AgIn)_xZn_{2(1-x)}S_2$ solid solution.[22,23,38] In these systems, the sacrificial reagents (S^{2-}/SO_3^{2-}) could efficiently remove the holes on the surface of CdS, and reduce its self-photocorrosion caused by photogenerated holes, thus leading to enhanced photocatalytic activity and stability. This is because the S^{2-} and SO_3^{2-} as electron donors have higher standard reduction potentials (−0.48 and −0.93 V *vs.* normal hydrogen electrode (NHE)), leading to faster removal rates of holes.[292] Quite recently, a high AQY of about 93% at 420 nm for photocatalytic hydrogen production has been also achieved on the Pt–PdS/CdS photocatalyst, in the presence of sacrificial reagents S^{2-}/SO_3^{2-} under visible-light irradiation.[293] Furthermore, ultrafast transient absorption spectroscopy of CdS–Pt nanostructures has verified that the transfer of the holes from the VB of SC to the electron donor is an important efficiency-limiting step for the photocatalytic H_2 production and that the S^{2-}/SO_3^{2-} mixture is a better sacrificial reagent compared to methanol.[294] Moreover, H_2S can also be applied as a hole scavenger in alkaline aqueous solution for photocatalytic

H_2 generation because of the formed sulfide ions.[6] Considering the abundant sulfur compounds as electron donors in chemical industries or nature,[11] the development of metal sulfide photocatalysts for half-reactions of water splitting in the presence of sacrificial reagents is meaningful and deserves more attention in future studies.

For organic sacrificial reagents, methanol has been the most widely used hole scavenger. This is because methanol can be easily oxidized to formaldehyde (HCHO), methanoic acid (HCOOH) and even to CO_2 by the photogenerated holes. Besides methanol, other organic compounds such as alcohols (ethanol, isopropanol, *etc.*), acids (oxalic acid, formic acid, dichlorophenoxyacetic acid, chloroacetic acids, acetic acid, *etc.*), aldehydes (formaldehyde, acetaldehyde, *etc.*), organic pollutants (azo-dyes, EDTA, dithiothreitol, TEA, chlorophenol, triethanolamine, *etc.*) and biomass (starch, cellulose, glycerol, glucose, sucrose, *etc.*) have also been used as sacrificial electron donors in the construction of bifunctional photocatalytic systems. Clearly, the bifunctional photocatalytic systems could not only produce hydrogen through photoreduction of water but also achieve photocatalytic biomass reforming or decomposition of organic pollutants.[6] For example, g-C_3N_4, with a band gap of 2.7 eV, was found to be a good visible-light-driven photocatalyst for photocatalytic hydrogen evolution in an aqueous sacrificial solution containing triethanolamine.[25] A structure-controlled g-C_3N_4 photocatalyst synthesized from urea exhibits an extremely high AQY of 26.5% at 400 nm using triethanolamine as a hole scavenger. In addition, the optimization of sacrificial agent also plays an important role in enhancing the photocatalytic H_2 evolution rate. For example, CdS-based photocatalysts exhibit better activity for H_2 evolution in lactic acid solution than in Na_2S–Na_2SO_3 solution,[149,295,296] indicating that organic hole scavengers have great potential for use in photocatalytic H_2 evolution.

10.5.2 Introduction of a Protective Layer

Owing to the presence of surface trapping states, many narrow band gap SCs, such as tantalum nitride and CdS, suffer from undesirable surface recombination losses and photocorrosion.[297] Thus, passivating the SC surface with a layer of stable material not only reduces the recombination losses and photocorrosion but also serves as a hole-storage layer or an electrocatalyst to improve the poor water oxidation kinetics.[297,298]

In general, TiO_2 is regarded as an attractive protective layer of the photocatalysts for photocatalytic H_2 evolution due to its exceptional stability.[299] For example, Lewis and his co-workers reported that a Si photoanode coated with a layer of amorphous TiO_2 (4–143 nm thick) and Ni oxide electrocatalysts can exhibit high activity (>30 mA cm^{-2}) and stability (>100 h), because the transparent TiO_2 coatings with electronic defects will have a little influence on the absorption of light, but will promote hole conduction and transfer to electrocatalysts, thus leading to excellent stability against photocorrosion.[300] Furthermore, stable C_3N_4 and nano-carbon overlayers are also attractive candidates as protective shell layers.[96,301,302] In addition, for a Type I core/shell particle photocatalyst, the shell layer is always used to passivate the surface trap states of the core, thus leading to a strongly enhanced photocatalytic activity for water splitting, as well as an improved photostability. For example, an enhanced activity for HER was observed in the core–shell structured CdSe/CdS[303,304] or CdS/ZnS[305,306] system due to the passivation effects of shell layer. For this core/shell system, control of the shell thickness and porous structure is very important and deserves special attention.[306] Therefore, surface protective layers should be taken into consideration seriously in optimizing the stability and activity of photocatalysts for H_2 generation.

10.5.3 Utilization of Water Oxidation Co-Catalysts

Another effective strategy for suppressing photocorrosion is to load water oxidation co-catalysts. On the one hand, as an accompanying reaction, the water oxidation reaction is an important bottleneck for photocatalytic H_2 production because this reaction is considerably more difficult and complex. It requires a four-electron oxidation of two water molecules coupled to the removal of four protons to form a relatively weak oxygen–oxygen bond.[31,307] Therefore, to increase the photocatalytic H_2-generation activity of SCs, the difficult and slow water oxidation reaction should also be enhanced by loading water oxidation co-catalysts. On the other hand, many visible-light-driven photocatalysts suffer from serious photocorrosion due to their shallow VB levels and strong self-oxidation by photogenerated holes. Thus, loading suitable water oxidation co-catalysts could accept the holes and accelerate water oxidation kinetics, thus efficiently avoiding the self-oxidation and enhancing the stability.

Furthermore, it was verified that co-catalysts such as RuO_2 and IrO_x can be exploited to protect CdS from photocorrosion due to enhanced efficiency of water oxidation or hole transfer from the VB of CdS to the solution species.[213,308,309] Quite recently, a robust electrocatalyst for water oxidation, a cobalt phosphate system (CoPi), showed great potential to replace noble metal oxide – it was first reported by Kanan and Nocera in 2008.[31] Since then, cobalt-based electrocatalysts, such as Co-Pi, CoO_x, Co_3O_4 and $Fe_{100-y-z}CoNi_zO_x$, have attracted much attention.[10,310–314] For example, CoO_x as co-catalyst has been coupled with the SCs such as $LaTiO_2N$,[276] $Sm_2Ti_2S_2O_5$,[315] TaON,[316] Ta_3N_5[317] and $g\text{-}C_3N_4$[318] to significantly improve their stability and activity for the oxygen evolution reaction.

In addition, a few successful and reliable examples of photocatalyst systems with both H_2 and O_2 evolution co-catalysts for water splitting have been reported,[305,319–321] since the initial reports on loading both Pt and RuO_2 onto TiO_2 or CdS for water splitting by Grätzel and his co-workers.[308] Recently, Domen and co-workers demonstrated a proof-of-concept using a SiO_2–Ta_3N_5 core–shell photocatalyst loaded with Pt NPs on the inner Ta_3N_5 shell surface as an electron collector and IrO_2 or CoO_x on the outer shell surface as a hole collector. The separated co-catalysts are favorable for the separation and migration of photogenerated electrons and holes toward inner and outer surfaces, respectively, thereby suppressing their recombination and the SBR. As a consequence, significantly enhanced water reduction and oxidation performance was achieved over this core/shell photocatalyst with spatially separated co-catalysts under visible light irradiation. Tremendous research efforts are being made to obtain visible-light-driven photocatalysts with two spatially separated earth-abundant co-catalysts.

10.6 Conclusions, Perspectives and Remarks

Recent years have seen growing interest in the conversion of water and solar energy into H_2 fuels using earth-abundant materials due to the depletion of fossil fuels and serious environmental problems. Numerous visible-light-driven SCs have been developed and applied in photocatalytic hydrogen generation. To further improve the photocatalytic H_2-evolution performance of these SCs, considerable progress has also been made in constructing heterogeneous composite photocatalysts through different modification strategies. However, to date, the quantum efficiencies of visible-light-driven SCs available for photocatalytic hydrogen generation are still far from the suggested value (*ca.* 30% at 600 nm) as an initial starting point for large-scale practical applications, corresponding to a solar-to-hydrogen energy conversion efficiency of 15%.[2,11]

On the one hand, it is known that the lack of both efficient co-catalysts and visible-light-driven photocatalysts is a key factor limiting the practical applications of the SC-based photocatalytic hydrogen generation. Thus, more efforts should focus on the developments of efficient co-catalysts and visible-light-driven photocatalysts for photocatalytic hydrogen generation. On the other hand, since photocatalysts are highly functional materials, developing highly integrated SC-based photocatalysts is also of great importance to construct practical water-splitting systems. Generally, a single strategy usually only improves the photocatalytic activity for hydrogen generation to some degree. This implies that all factors in photocatalytic processes, such as charge excitation, separation and migration, and surface charge utilization should be taken simultaneously into consideration to achieve the maximum overall H_2-evolution performance. Thus, a deep understanding of the integrated effects of the different engineering strategies is highly recommended for developing better H_2-evolution SCs in future studies. More interestingly, these modification strategies appear to provide possibilities to be applied in the photocatalytic reduction of CO_2,[322,323] organic synthesis[324] and pollutant degradation.[325]

Meanwhile, future efforts need to focus on further understanding of the dynamics and kinetic details of interfacial electron and hole transfer at the fundamental level.[326–328] The manipulation of interfacial charge transfer plays a very important role in improving the photocatalytic efficiency. Importantly, a better understanding of the charge-transfer dynamics between SC and SC (or co-catalyst/sacrificial donors) will also offer some deep insights into finding a key efficiency-limiting step and designing more efficient solar-to-fuel conversion systems.[294] In addition, more attention should also be paid to photocatalytic H_2 generation based on Z-scheme or overall water splitting systems because they are favorable for in-depth investigation towards the water oxidation half-reaction or the extended applications of water-oxidation SCs in photocatalytic H_2 evolution.

From a computational viewpoint, the electronic and energy-band structures of SC could be obtained by DFT calculation, which is beneficial in understanding the relationship between structure and activity of SCs. As a useful assistant tool, the DFT calculation can even reveal the intrinsic surface reaction mechanism at the molecular level,[329] which may be helpful for designing and developing better photocatalysts and shortening the development time and process. It is highly expected in future that artificial photosynthesis devices will "see the light of day",[330] and that breakthroughs for bringing this technology to reality can occur when all these problems and engineering strategies above have been taken into consideration and the photocatalytic system has been well designed.

Acknowledgements

J. Yu would like to thank the 973 program (2013CB632402) and NSFC (51272199, 51320105001 and 21433007) and X. Li would like to thank the Industry and Research Collaborative Innovation Major Projects of Guangzhou (201508020098), NSFC (20906034), and the State Key Laboratory of Advanced Technology for Material Synthesis and Processing (Wuhan University of Technology) (2015-KF-7) for their support.

References

1. N. S. Lewis and D. G. Nocera, *Proc. Natl. Acad. Sci. U. S. A.*, 2006, **103**, 15729.
2. K. Maeda and K. Domen, *J. Phys. Chem. C*, 2007, **111**, 7851.
3. J. A. Turner, *Science*, 2004, **305**, 972.
4. L. Schlapbach and A. Zuttel, *Nature*, 2001, **414**, 353.
5. R. M. Navarro, M. A. Peña and J. L. G. Fierro, *Chem. Rev.*, 2007, **107**, 3952.
6. X. Chen, S. Shen, L. Guo and S. S. Mao, *Chem. Rev.*, 2010, **110**, 6503.
7. M. G. Walter, E. L. Warren, J. R. McKone, S. W. Boettcher, Q. Mi, E. A. Santori and N. S. Lewis, *Chem. Rev.*, 2010, **110**, 6446.
8. M. Gratzel, *Nature*, 2001, **414**, 338.
9. O. Khaselev and J. A. Turner, *Science*, 1998, **280**, 425.
10. S. Y. Reece, J. A. Hamel, K. Sung, T. D. Jarvi, A. J. Esswein, J. J. H. Pijpers and D. G. Nocera, *Science*, 2011, **334**, 645.
11. A. Kudo and Y. Miseki, *Chem. Soc. Rev.*, 2009, **38**, 253.
12. X. Li, J. Yu, J. Low, Y. Fang, J. Xiao and X. Chen, *J. Mater. Chem. A*, 2015, **3**, 2485.
13. F. E. Osterloh, *Chem. Mater.*, 2007, **20**, 35.
14. Y. Moriya, T. Takata and K. Domen, *Coord. Chem. Rev.*, 2013, **257**, 1957.
15. T. Hisatomi, J. Kubota and K. Domen, *Chem. Soc. Rev.*, 2014, **43**, 7520.

16. G. Hitoki, A. Ishikawa, T. Takata, J. N. Kondo, M. Hara and K. Domen, *Chem. Lett.*, 2002, **31**, 736.
17. W.-J. Chun, A. Ishikawa, H. Fujisawa, T. Takata, J. N. Kondo, M. Hara, M. Kawai, Y. Matsumoto and K. Domen, *J. Phys. Chem. B*, 2003, **107**, 1798.
18. G. Hitoki, T. Takata, J. N. Kondo, M. Hara, H. Kobayashi and K. Domen, *Chem. Commun.*, 2002, 1698.
19. K. Maeda, K. Teramura, D. Lu, T. Takata, N. Saito, Y. Inoue and K. Domen, *Nature*, 2006, **440**, 295.
20. K. Maeda, T. Takata, M. Hara, N. Saito, Y. Inoue, H. Kobayashi and K. Domen, *J. Am. Chem. Soc.*, 2005, **127**, 8286.
21. M. Liu, D. Jing, Z. Zhou and L. Guo, *Nat. Commun.*, 2013, **4**, 2278.
22. K. Zhang and L. Guo, *Catal. Sci. Technol.*, 2013, **3**, 1672.
23. I. Tsuji, H. Kato, H. Kobayashi and A. Kudo, *J. Am. Chem. Soc.*, 2004, **126**, 13406.
24. L. Liao, Q. Zhang, Z. Su, Z. Zhao, Y. Wang, Y. Li, X. Lu, D. Wei, G. Feng, Q. Yu, X. Cai, J. Zhao, Z. Ren, H. Fang, F. Robles-Hernandez, S. Baldelli and J. Bao, *Nat. Nanotechnol.*, 2014, **9**, 69.
25. X. Wang, K. Maeda, A. Thomas, K. Takanabe, G. Xin, J. Carlsson, K. Domen and M. Antonietti, *Nat. Mater.*, 2009, **8**, 76.
26. S. Cao, J. Low, J. Yu and M. Jaroniec, *Adv. Mater.*, 2015, **27**, 2150.
27. X. Wang, S. Blechert and M. Antonietti, *ACS Catal.*, 2012, **2**, 1596.
28. A. Bard and M. Fox, *Acc. Chem. Res.*, 1995, **28**, 141.
29. A. J. Bard, *J. Am. Chem. Soc.*, 2010, **132**, 7559.
30. A. D. Handoko, K. Li and J. Tang, *Curr. Opin. Chem. Eng.*, 2013, **2**, 200.
31. M. W. Kanan and D. G. Nocera, *Science*, 2008, **321**, 1072.
32. (a) A. Fujishima and K. Honda, *Nature*, 1972, **238**, 37; (b) B. Kraeutler and A. J. Bard, *J. Am. Chem. Soc.*, 1978, **100**, 4317.
33. A. J. Bard, *J. Photochem.*, 1979, **10**, 59.
34. A. J. Bard, *Science*, 1980, **207**, 139.
35. A. J. Bard, *J. Phys. Chem.*, 1982, **86**, 172.
36. Z. B. Chen, T. F. Jaramillo, T. G. Deutsch, A. Kleiman-Shwarsctein, A. J. Forman, N. Gaillard, R. Garland, K. Takanabe, C. Heske, M. Sunkara, E. W. McFarland, K. Domen, E. L. Miller, J. A. Turner and H. N. Dinh, *J. Mater. Res.*, 2010, **25**, 3.
37. J. F. Reber and K. Meier, *J. Phys. Chem.*, 1984, **88**, 5903.
38. I. Tsuji, H. Kato and A. Kudo, *Angew. Chem., Int. Ed.*, 2005, **44**, 3565.
39. C. Xing, Y. Zhang, W. Yan and L. Guo, *Int. J. Hydrogen Energy*, 2006, **31**, 2018.
40. A. Paracchino, V. Laporte, K. Sivula, M. Grätzel and E. Thimsen, *Nat. Mater.*, 2011, **10**, 456.
41. N. Bao, L. Shen, T. Takata and K. Domen, *Chem. Mater.*, 2007, **20**, 110.
42. M. Higashi, K. Domen and R. Abe, *Energy Environ. Sci.*, 2011, **4**, 4138.
43. Y. Li, T. Takata, D. Cha, K. Takanabe, T. Minegishi, J. Kubota and K. Domen, *Adv. Mater.*, 2012, **25**, 125.
44. M. Barroso, C. A. Mesa, S. R. Pendlebury, A. J. Cowan, T. Hisatomi, K. Sivula, M. Grätzel, D. R. Klug and J. R. Durrant, *Proc. Natl. Acad. Sci. U. S. A.*, 2012, **109**, 15640.

45. D. K. Zhong, M. Cornuz, K. Sivula, M. Gratzel and D. R. Gamelin, *Energy Environ. Sci.*, 2011, **4**, 1759.
46. F. Le Formal, S. R. Pendlebury, M. Cornuz, S. D. Tilley, M. Grätzel and J. R. Durrant, *J. Am. Chem. Soc.*, 2014, **136**, 2564.
47. A. Kudo, K. Omori and H. Kato, *J. Am. Chem. Soc.*, 1999, **121**, 11459.
48. X. C. Wang, K. Maeda, X. F. Chen, K. Takanabe, K. Domen, Y. D. Hou, X. Z. Fu and M. Antonietti, *J. Am. Chem. Soc.*, 2009, **131**, 1680.
49. S. Cao and J. Yu, *J. Phys. Chem. Lett.*, 2014, **5**, 2101.
50. J. Zhang, J. Sun, K. Maeda, K. Domen, P. Liu, M. Antonietti, X. Fu and X. Wang, *Energy Environ. Sci.*, 2011, **4**, 675.
51. J. E. Katz, T. R. Gingrich, E. A. Santori and N. S. Lewis, *Energy Environ. Sci.*, 2009, **2**, 103.
52. H. Ye, J. Lee, J. S. Jang and A. J. Bard, *J. Phys. Chem. C*, 2010, **114**, 13322.
53. Y. Wu, P. Lazic, G. Hautier, K. Persson and G. Ceder, *Energy Environ. Sci.*, 2013, **6**, 157.
54. D. Duonghong, E. Borgarello and M. Graetzel, *J. Am. Chem. Soc.*, 1981, **103**, 4685.
55. H. Kato and A. Kudo, *J. Phys. Chem. B*, 2001, **105**, 4285.
56. H. Kato, K. Asakura and A. Kudo, *J. Am. Chem. Soc.*, 2003, **125**, 3082.
57. D. Chen and J. Ye, *Chem. Mater.*, 2009, **21**, 2327.
58. M. Yoshino, M. Kakihana, W. S. Cho, H. Kato and A. Kudo, *Chem. Mater.*, 2002, **14**, 3369.
59. A. Kudo, H. Kato and S. Nakagawa, *J. Phys. Chem. B*, 2000, **104**, 571.
60. K. Maeda, K. Teramura, T. Takata, M. Hara, N. Saito, K. Toda, Y. Inoue, H. Kobayashi and K. Domen, *J. Phys. Chem. B*, 2005, **109**, 20504.
61. K. Maeda and K. Domen, *Chem. Mater.*, 2009, **22**, 612.
62. Y. Lee, K. Teramura, M. Hara and K. Domen, *Chem. Mater.*, 2007, **19**, 2120.
63. K. Maeda, K. Teramura and K. Domen, *J. Catal.*, 2008, **254**, 198.
64. K. Sayama, K. Mukasa, R. Abe, Y. Abe and H. Arakawa, *J. Photochem. Photobiol., A*, 2002, **148**, 71.
65. R. Abe, K. Sayama, K. Domen and H. Arakawa, *Chem. Phys. Lett.*, 2001, **344**, 339.
66. H. Kato, M. Hori, R. Konta, Y. Shimodaira and A. Kudo, *Chem. Lett.*, 2004, **33**, 1348.
67. A. Kudo, *MRS Bull.*, 2011, **36**, 32.
68. Y. Sasaki, H. Nemoto, K. Saito and A. Kudo, *J. Phys. Chem. C*, 2009, **113**, 17536.
69. K. Maeda, *ACS Catal.*, 2013, **3**(7), 1486, DOI: 10.1021/cs4002089.
70. K. Maeda and K. Domen, *J. Phys. Chem. Lett.*, 2010, **1**, 2655.
71. K. Maeda, M. Higashi, D. L. Lu, R. Abe and K. Domen, *J. Am. Chem. Soc.*, 2010, **132**, 5858.
72. A. Iwase, Y. H. Ng, Y. Ishiguro, A. Kudo and R. Amal, *J. Am. Chem. Soc.*, 2011, **133**, 11054.
73. D. J. Martin, P. J. T. Reardon, S. J. A. Moniz and J. Tang, *J. Am. Chem. Soc.*, 2014, **136**, 12568.
74. K. Maeda, *ACS Catal.*, 2013, **3**, 1486.

75. H. Tada, T. Mitsui, T. Kiyonaga, T. Akita and K. Tanaka, *Nat. Mater.*, 2006, **5**, 782.
76. H. J. Yun, H. Lee, N. D. Kim, D. M. Lee, S. Yu and J. Yi, *ACS Nano*, 2011, **5**, 4084.
77. K. Iwashina, A. Iwase, Y. H. Ng, R. Amal and A. Kudo, *J. Am. Chem. Soc.*, 2015, **137**, 604.
78. J. Yu, S. Wang, J. Low and W. Xiao, *Phys. Chem. Chem. Phys.*, 2013, **15**, 16883.
79. X. Wang, G. Liu, Z.-G. Chen, F. Li, L. Wang, G. Q. Lu and H.-M. Cheng, *Chem. Commun.*, 2009, 3452.
80. P. Zhou, J. Yu and M. Jaroniec, *Adv. Mater.*, 2014, **26**, 4920.
81. Y. Ma, X. Wang, Y. Jia, X. Chen, H. Han and C. Li, *Chem. Rev.*, 2014, **114**, 9987.
82. T. K. Townsend, N. D. Browning and F. E. Osterloh, *ACS Nano*, 2012, **6**, 7420.
83. N. Bao, L. Shen, T. Takata, K. Domen, A. Gupta, K. Yanagisawa and C. A. Grimes, *J. Phys. Chem. C*, 2007, **111**, 17527.
84. K. Yan, G. Wu, C. Jarvis, J. Wen and A. Chen, *Appl. Catal., B*, 2014, **148**, 281.
85. S. Chuangchote, J. Jitputti, T. Sagawa and S. Yoshikawa, *ACS Appl. Mater. Interfaces*, 2009, **1**, 1140.
86. Z. Zhang, F. Zuo and P. Feng, *J. Mater. Chem.*, 2010, **20**, 2206.
87. L. Guo, H. Hagiwara, S. Ida, T. Daio and T. Ishihara, *ACS Appl. Mater. Interfaces*, 2013, **5**, 11080.
88. Y. Noda, B. Lee, K. Domen and J. N. Kondo, *Chem. Mater.*, 2008, **20**, 5361.
89. X.-H. Li, J. Zhang, X. Chen, A. Fischer, A. Thomas, M. Antonietti and X. Wang, *Chem. Mater.*, 2011, **23**, 4344.
90. K. Schwinghammer, M. B. Mesch, V. Duppel, C. Ziegler, J. Senker and B. V. Lotsch, *J. Am. Chem. Soc.*, 2014, **136**, 1730.
91. X. Chen, C. Li, M. Gratzel, R. Kostecki and S. S. Mao, *Chem. Soc. Rev.*, 2012, **41**, 7909.
92. M.-C. Wu, J. Hiltunen, A. Sapi, A. Avila, W. Larsson, H.-C. Liao, M. Huuhtanen, G. Toth, A. Shchukarev, N. Laufer, A. Kukovecz, Z. Konya, J.-P. Mikkola, R. Keiski, W.-F. Su, Y.-F. Chen, H. Jantunen, P. M. Ajayan, R. Vajtai and K. Kordas, *ACS Nano*, 2011, **5**, 5025.
93. S. Yan, L. Wan, Z. Li and Z. Zou, *Chem. Commun.*, 2011, 5632.
94. J. Jin, J. Yu, G. Liu and P. K. Wong, *J. Mater. Chem. A*, 2013, **1**, 10927.
95. J. Zhang, S. Z. Qiao, L. Qi and J. Yu, *Phys. Chem. Chem. Phys.*, 2013, **15**, 12088.
96. J. Zhang, Y. Wang, J. Jin, J. Zhang, Z. Lin, F. Huang and J. Yu, *ACS Appl. Mater. Interfaces*, 2013, **5**, 10317.
97. J. Yu, Y. Yu, P. Zhou, W. Xiao and B. Cheng, *Appl. Catal., B*, 2014, **156–157**, 184.
98. Y. Li, Y. Hu, S. Peng, G. Lu and S. Li, *J. Phys. Chem. C*, 2009, **113**, 9352.
99. M. Berr, A. Vaneski, A. S. Susha, J. Rodríguez-Fernández, M. Döblinger, F. Jäckel, A. L. Rogach and J. Feldmann, *Appl. Phys. Lett.*, 2010, **97**, 093108.
100. L. Amirav and A. P. Alivisatos, *J. Phys. Chem. Lett.*, 2010, **1**, 1051.

101. H. Zhu, N. Song, H. Lv, C. L. Hill and T. Lian, *J. Am. Chem. Soc.*, 2012, **134**, 11701.
102. X. Wang, G. Liu, G. Q. Lu and H.-M. Cheng, *Int. J. Hydrogen Energy*, 2010, **35**, 8199.
103. G. Yang, W. Yan, Q. Zhang, S. Shen and S. Ding, *Nanoscale*, 2013, **5**, 12432.
104. J. S. Jang, H. G. Kim, U. A. Joshi, J. W. Jang and J. S. Lee, *Int. J. Hydrogen Energy*, 2008, **33**, 5975.
105. H. G. Yang, C. H. Sun, S. Z. Qiao, J. Zou, G. Liu, S. C. Smith, H. M. Cheng and G. Q. Lu, *Nature*, 2008, **453**, 638.
106. M. D'Arienzo, J. Carbajo, A. Bahamonde, M. Crippa, S. Polizzi, R. Scotti, L. Wahba and F. Morazzoni, *J. Am. Chem. Soc.*, 2011, **133**, 17652.
107. T. Tachikawa, S. Yamashita and T. Majima, *J. Am. Chem. Soc.*, 2011, **133**, 7197.
108. J. Pan, G. Liu, G. Q. Lu and H.-M. Cheng, *Angew. Chem., Int. Ed.*, 2011, **50**, 2133.
109. T. R. Gordon, M. Cargnello, T. Paik, F. Mangolini, R. T. Weber, P. Fornasiero and C. B. Murray, *J. Am. Chem. Soc.*, 2012, **134**, 6751.
110. H. Xu, P. Reunchan, S. Ouyang, H. Tong, N. Umezawa, T. Kako and J. Ye, *Chem. Mater.*, 2013, **25**, 405.
111. J. Yu, J. Low, W. Xiao, P. Zhou and M. Jaroniec, *J. Am. Chem. Soc.*, 2014, **136**, 8839.
112. Y. Xu, W. Zhao, R. Xu, Y. Shi and B. Zhang, *Chem. Commun.*, 2013, 9803.
113. S. Ida, Y. Okamoto, M. Matsuka, H. Hagiwara and T. Ishihara, *J. Am. Chem. Soc.*, 2012, **134**, 15773.
114. S. Yang, Y. Gong, J. Zhang, L. Zhan, L. Ma, Z. Fang, R. Vajtai, X. Wang and P. M. Ajayan, *Adv. Mater.*, 2013, **25**, 2452.
115. L. Zheng, Y. Xu, Y. Song, C. Wu, M. Zhang and Y. Xie, *Inorg. Chem.*, 2009, **48**, 4003.
116. J. Zhang, Y. Wang, J. Zhang, Z. Lin, F. Huang and J. Yu, *ACS Appl. Mater. Interfaces*, 2012, **5**, 1031.
117. Q. Xiang, B. Cheng and J. Yu, *Appl. Catal., B*, 2013, **138**, 299.
118. L. Shang, C. Zhou, T. Bian, H. Yu, L.-Z. Wu, C.-H. Tung and T. Zhang, *J. Mater. Chem. A*, 2013, **1**, 4552.
119. J. Shen, J. Zai, Y. Yuan and X. Qian, *Int. J. Hydrogen Energy*, 2012, **37**, 16986.
120. F. Chen, J. Zai, M. Xu and X. Qian, *J. Mater. Chem. A*, 2013, **1**, 4316.
121. Y. Chen, G. Tian, Z. Ren, K. Pan, Y. Shi, J. Wang and H. Fu, *ACS Appl. Mater. Interfaces*, 2014, **6**, 13841.
122. X. Cao, G. Tian, Y. Chen, J. Zhou, W. Zhou, C. Tian and H. Fu, *J. Mater. Chem. A*, 2014, **2**, 4366.
123. L. Ding, H. Zhou, S. Lou, J. Ding, D. Zhang, H. Zhu and T. Fan, *Int. J. Hydrogen Energy*, 2013, **38**, 8244.
124. M. Shiraishi and M. Ata, *Carbon*, 2001, **39**, 1913.
125. R. Czerw, B. Foley, D. Tekleab, A. Rubio, P. Ajayan and D. Carroll, *Phys. Rev. B*, 2002, **66**, 033408.

126. Q. J. Xiang, J. G. Yu and M. Jaroniec, *Chem. Soc. Rev.*, 2012, **41**, 782.
127. Q. Xiang and J. Yu, *J. Phys. Chem. Lett.*, 2013, **4**, 753.
128. G. Xie, K. Zhang, B. Guo, Q. Liu, L. Fang and J. R. Gong, *Adv. Mater.*, 2013, **25**, 3820.
129. F. Bonaccorso, L. Colombo, G. Yu, M. Stoller, V. Tozzini, A. C. Ferrari, R. S. Ruoff and V. Pellegrini, *Science*, 2015, 347.
130. Y. Chen, J. Li, Z. Hong, B. Shen, B. Lin and B. Gao, *Phys. Chem. Chem. Phys.*, 2014, **16**, 8106.
131. L. Ge and C. Han, *Appl. Catal., B*, 2012, **117–118**, 268.
132. Y. K. Kim and H. Park, *Energy Environ. Sci.*, 2011, **4**, 685.
133. A. S. Cherevan, P. Gebhardt, C. J. Shearer, M. Matsukawa, K. Domen and D. Eder, *Energy Environ. Sci.*, 2014, **7**, 791.
134. C. Zhao, H. Luo, F. Chen, P. Zhang, L. Yi and K. You, *Energy Environ. Sci.*, 2014, **7**, 1700.
135. X.-J. Lv, W.-F. Fu, H.-X. Chang, H. Zhang, J.-S. Cheng, G.-J. Zhang, Y. Song, C.-Y. Hu and J.-H. Li, *J. Mater. Chem.*, 2012, **22**, 1539.
136. A. Ye, W. Fan, Q. Zhang, W. Deng and Y. Wang, *Catal. Sci. Technol.*, 2012, **2**, 969.
137. S. Liu, C. Liu, W. Wang, B. Cheng and J. Yu, *Nanoscale*, 2012, **4**, 3193.
138. Q. Li, H. Meng, J. Yu, W. Xiao, Y. Zheng and J. Wang, *Chem–Eur. J.*, 2014, **20**, 1176.
139. L. Jia, D.-H. Wang, Y.-X. Huang, A.-W. Xu and H.-Q. Yu, *J. Phys. Chem. C*, 2011, **115**, 11466.
140. X. Wang, G. Sun, P. Routh, D.-H. Kim, W. Huang and P. Chen, *Chem. Soc. Rev.*, 2014, **43**, 7067.
141. J. Zhang, J. Yu, M. Jaroniec and J. R. Gong, *Nano Lett.*, 2012, **12**, 4584.
142. T. Peng, K. Li, P. Zeng, Q. Zhang and X. Zhang, *J. Phys. Chem. C*, 2012, **116**, 22720.
143. X. Fan, W. Peng, Y. Li, X. Li, S. Wang, G. Zhang and F. Zhang, *Adv. Mater.*, 2008, **20**, 4490.
144. S. Liu, Z. Chen, N. Zhang, Z.-R. Tang and Y.-J. Xu, *J. Phys. Chem. C*, 2013, **117**, 8251.
145. P. Gao, J. Liu, S. Lee, T. Zhang and D. D. Sun, *J. Mater. Chem.*, 2012, **22**, 2292.
146. Q. Li, B. Guo, J. Yu, J. Ran, B. Zhang, H. Yan and J. R. Gong, *J. Am. Chem. Soc.*, 2011, **133**, 10878.
147. Z. Fang, Y. Wang, J. Song, Y. Sun, J. Zhou, R. Xu and H. Duan, *Nanoscale*, 2013, **5**, 9830.
148. T. Jia, A. Kolpin, C. Ma, R. C.-T. Chan, W.-M. Kwok and S. E. Tsang, *Chem. Commun.*, 2014, 1185.
149. K. Chang, Z. Mei, T. Wang, Q. Kang, S. Ouyang and J. Ye, *ACS Nano*, 2014, **8**, 7078.
150. Z. Yan, X. Yu, A. Han, P. Xu and P. Du, *J. Phys. Chem. C*, 2014, **118**, 22896.
151. Z. Khan, T. R. Chetia, A. K. Vardhaman, D. Barpuzary, C. V. Sastri and M. Qureshi, *RSC Adv.*, 2012, **2**, 12122.
152. X. Wang, L. Yin and G. Liu, *Chem. Commun.*, 2014, 3460.

153. J. Hou, Z. Wang, W. Kan, S. Jiao, H. Zhu and R. Kumar, *J. Mater. Chem.*, 2012, **22**, 7291.
154. J. Hou, C. Yang, H. Cheng, Z. Wang, S. Jiao and H. Zhu, *Phys. Chem. Chem. Phys.*, 2013, **15**, 15660.
155. Q. Xiang, J. Yu and M. Jaroniec, *Nanoscale*, 2011, **3**, 3670.
156. Q. Xiang, J. Yu and M. Jaroniec, *J. Phys. Chem. C*, 2011, **115**, 7355.
157. Q. Xiang, J. Yu and M. Jaroniec, *J. Am. Chem. Soc.*, 2012, **134**, 6575.
158. Y. Wang, J. Yu, W. Xiao and Q. Li, *J. Mater. Chem. A*, 2014, **2**, 3847.
159. I. V. Lightcap, T. H. Kosel and P. V. Kamat, *Nano Lett.*, 2010, **10**, 577.
160. Q. Li, X. Li, S. Wageh, A. A. Al-Ghamdi and J. Yu, *Adv. Energy Mater.*, 2015, **5**, 1500010.
161. S. Nardecchia, D. Carriazo, M. Luisa Ferrer, M. C. Gutierrez and F. del Monte, *Chem. Soc. Rev.*, 2013, **42**, 794.
162. M. A. Worsley, P. J. Pauzauskie, T. Y. Olson, J. Biener, J. H. Satcher and T. F. Baumann, *J. Am. Chem. Soc.*, 2010, **132**, 14067.
163. W. G. Tu, Y. Zhou, Q. Liu, Z. P. Tian, J. Gao, X. Y. Chen, H. T. Zhang, J. G. Liu and Z. G. Zou, *Adv. Funct. Mater.*, 2012, **22**, 1215.
164. T. Wu, M. Chen, L. Zhang, X. Xu, Y. Liu, J. Yan, W. Wang and J. Gao, *J. Mater. Chem. A*, 2013, **1**, 7612.
165. W. Han, C. Zang, Z. Huang, H. Zhang, L. Ren, X. Qi and J. Zhong, *Int. J. Hydrogen Energy*, 2014, **39**, 19502.
166. X. Li, S. Yang, J. Sun, P. He, X. Xu and G. Ding, *Carbon*, 2014, **78**, 38.
167. B. Qiu, M. Xing and J. Zhang, *J. Am. Chem. Soc.*, 2014, **136**, 5852.
168. W. Liu, J. Cai and Z. Li, *ACS Sustainable Chem. Eng.*, 2015, **3**, 277.
169. D. A. Reddy, J. Choi, S. Lee, R. Ma and T. K. Kim, *RSC. Adv.*, 2015, **5**, 18342.
170. W. Han, L. Ren, L. Gong, X. Qi, Y. Liu, L. Yang, X. Wei and J. Zhong, *ACS Sustainable Chem. Eng.*, 2014, **2**, 741.
171. J. Du, X. Lai, N. Yang, J. Zhai, D. Kisailus, F. Su, D. Wang and L. Jiang, *ACS Nano*, 2011, **5**, 590.
172. Q. Li, C. Cui, H. Meng and J. Yu, *Chem.–Asian J.*, 2014, **9**, 1766.
173. L. Li, P. A. Salvador and G. S. Rohrer, *Nanoscale*, 2014, **6**, 24.
174. J. S. Jang, H. Gyu Kim, P. H. Borse and J. S. Lee, *Int. J. Hydrogen Energy*, 2007, **32**, 4786.
175. J. S. Jang, S. M. Ji, S. W. Bae, H. C. Son and J. S. Lee, *J. Photochem. Photobiol., A*, 2007, **188**, 112.
176. N. Serpone, E. Borgarello and M. Gratzel, *J. Chem. Soc., Chem. Commun.*, 1984, 342.
177. C. Li, J. Yuan, B. Han, L. Jiang and W. Shangguan, *Int. J. Hydrogen Energy*, 2010, **35**, 7073.
178. J. S. Jang, S. H. Choi, H. G. Kim and J. S. Lee, *J. Phys. Chem. C*, 2008, **112**, 17200.
179. H. Park, W. Choi and M. R. Hoffmann, *J. Mater. Chem.*, 2008, **18**, 2379.
180. H. Park, Y. K. Kirn and W. Choi, *J. Phys. Chem. C*, 2011, **115**, 6141.
181. J. Li, S. K. Cushing, P. Zheng, T. Senty, F. Meng, A. D. Bristow, A. Manivannan and N. Wu, *J. Am. Chem. Soc.*, 2014, **136**, 8438.

182. L. Qi, J. Yu and M. Jaroniec, *Phys. Chem. Chem. Phys.*, 2011, **13**, 8915.
183. H. G. Kim, P. H. Borse, W. Y. Choi and J. S. Lee, *Angew. Chem., Int. Ed.*, 2005, **44**, 4585.
184. Z. Liu, Z.-G. Zhao and M. Miyauchi, *J. Phys. Chem. C*, 2009, **113**, 17132.
185. H. G. Kim, E. D. Jeong, P. H. Borse, S. Jeon, K. Yong, J. S. Lee, W. Li and S. H. Oh, *Appl. Phys. Lett.*, 2006, **89**, 064103.
186. Z. Li, J. Liu, D. Wang, Y. Gao and J. Shen, *Int. J. Hydrogen Energy*, 2012, **37**, 6431.
187. J. Low, S. Cao, J. Yu and S. Wageh, *Chem. Commun.*, 2014, 10768.
188. S. Ida, A. Takashiba, S. Koga, H. Hagiwara and T. Ishihara, *J. Am. Chem. Soc.*, 2014, **136**, 1872.
189. F. Meng, J. Li, S. K. Cushing, M. Zhi and N. Wu, *J. Am. Chem. Soc.*, 2013, **135**, 10286.
190. Y. Hou, A. B. Laursen, J. Zhang, G. Zhang, Y. Zhu, X. Wang, S. Dahl and I. Chorkendorff, *Angew. Chem., Int. Ed.*, 2013, **52**, 3621.
191. L. Ge, C. Han, X. Xiao and L. Guo, *Int. J. Hydrogen Energy*, 2013, **38**, 6960.
192. M. A. Fox and M. T. Dulay, *Chem. Rev.*, 1993, **93**, 341.
193. A. Linsebigler, G. Lu and J. Yates Jr, *Chem. Rev.*, 1995, **95**, 735.
194. J. Zhang, Q. Xu, Z. Feng, M. Li and C. Li, *Angew. Chem., Int. Ed.*, 2008, **47**, 1766.
195. Q. A. Xu, Y. Ma, J. Zhang, X. L. Wang, Z. C. Feng and C. Li, *J. Catal.*, 2011, **278**, 329.
196. Y. Ma, Q. Xu, X. Zong, D. Wang, G. Wu, X. Wang and C. Li, *Energy Environ. Sci.*, 2012, **5**, 6345.
197. T. Kawahara, Y. Konishi, H. Tada, N. Tohge, J. Nishii and S. Ito, *Angew. Chem., Int. Ed.*, 2002, **41**, 2811.
198. H. Nakajima, T. Mori, Q. Shen and T. Toyoda, *Chem. Phys. Lett.*, 2005, **409**, 81.
199. D. C. Hurum, A. G. Agrios, K. A. Gray, T. Rajh and M. C. Thurnauer, *J. Phys. Chem. B*, 2003, **107**, 4545.
200. D. C. Hurum, K. A. Gray, T. Rajh and M. C. Thurnauer, *J. Phys. Chem. B*, 2004, **109**, 977.
201. D. O. Scanlon, C. W. Dunnill, J. Buckeridge, S. A. Shevlin, A. J. Logsdail, S. M. Woodley, C. R. A. Catlow, M. J. Powell, R. G. Palgrave, I. P. Parkin, G. W. Watson, T. W. Keal, P. Sherwood, A. Walsh and A. A. Sokol, *Nat. Mater.*, 2013, **12**, 798.
202. Y. Yu, G. Chen, Q. Wang and Y. Li, *Energy Environ. Sci.*, 2011, **4**, 3652.
203. M. Liu, L. Wang, G. Lu, X. Yao and L. Guo, *Energy Environ. Sci.*, 2011, **4**, 1372.
204. D. M. Antonelli and J. Y. Ying, *Angew. Chem., Int. Ed.*, 1995, **34**, 2014.
205. J. C. Yu, L. Z. Zhang and J. G. Yu, *Chem. Mater.*, 2002, **14**, 4647.
206. J. C. Yu, L. Z. Zhang, Z. Zheng and J. C. Zhao, *Chem. Mater.*, 2003, **15**, 2280.
207. X. Chen and S. S. Mao, *Chem. Rev.*, 2007, **107**, 2891.
208. Y. Takahara, J. N. Kondo, T. Takata, D. Lu and K. Domen, *Chem. Mater.*, 2001, **13**, 1194.

209. X. Chen, Y.-S. Jun, K. Takanabe, K. Maeda, K. Domen, X. Fu, M. Antonietti and X. Wang, *Chem. Mater.*, 2009, **21**, 4093.
210. N. Lakshminarasimhan, E. Bae and W. Choi, *J. Phys. Chem. C*, 2007, **111**, 15244.
211. D. Zhao, S. Budhi, A. Rodriguez and R. T. Koodali, *Int. J. Hydrogen Energy*, 2010, **35**, 5276.
212. S. Shen, J. Chen, R. T. Koodali, Y. Hu, Q. Xiao, J. Zhou, X. Wang and L. Guo, *Appl. Catal., B*, 2014, **150–151**, 138.
213. R. Peng, C.-M. Wu, J. Baltrusaitis, N. M. Dimitrijevic, T. Rajh and R. T. Koodali, *Chem. Commun.*, 2013, 3221.
214. S. A. Macías-Sánchez, R. Nava, V. Hernández-Morales, Y. J. Acosta-Silva, L. Gómez-Herrera, B. Pawelec, S. M. Al-Zahrani, R. M. Navarro and J. L. G. Fierro, *Int. J. Hydrogen Energy*, 2012, **37**, 9948.
215. F. E. Osterloh and B. A. Parkinson, *MRS Bull.*, 2011, **36**, 17.
216. J. Yang, D. Wang, H. Han and C. Li, *Acc. Chem. Res.*, 2013, **46**, 1900.
217. K. Maeda, K. Teramura, D. L. Lu, N. Saito, Y. Inoue and K. Domen, *Angew. Chem., Int. Ed.*, 2006, **45**, 7806.
218. M. Yoshida, K. Takanabe, K. Maeda, A. Ishikawa, J. Kubota, Y. Sakata, Y. Ikezawa and K. Domen, *J. Phys. Chem. C*, 2009, **113**, 10151.
219. R. Su, R. Tiruvalam, A. J. Logsdail, Q. He, C. A. Downing, M. T. Jensen, N. Dimitratos, L. Kesavan, P. P. Wells, R. Bechstein, H. H. Jensen, S. Wendt, C. R. A. Catlow, C. J. Kiely, G. J. Hutchings and F. Besenbacher, *ACS Nano*, 2014, **8**, 3490.
220. Y. H. Li, J. Xing, Z. J. Chen, Z. Li, F. Tian, L. R. Zheng, H. F. Wang, P. Hu, H. J. Zhao and H. G. Yang, *Nat. Commun.*, 2013, **4**, 2500.
221. X. Zhou, G. Liu, J. Yu and W. Fan, *J. Mater. Chem.*, 2012, **22**, 21337.
222. S. Linic, P. Christopher and D. B. Ingram, *Nat. Mater.*, 2011, **10**, 911.
223. C. Wang and D. Astruc, *Chem. Soc. Rev.*, 2014, **43**, 7188.
224. C. Gomes Silva, R. Juárez, T. Marino, R. Molinari and H. García, *J. Am. Chem. Soc.*, 2010, **133**, 595.
225. J. Fang, S.-W. Cao, Z. Wang, M. M. Shahjamali, S. C. J. Loo, J. Barber and C. Xue, *Int. J. Hydrogen Energy*, 2012, **37**, 17853.
226. A. Tanaka, S. Sakaguchi, K. Hashimoto and H. Kominami, *Catal. Sci. Technol.*, 2012, **2**, 907.
227. J. S. DuChene, B. C. Sweeny, A. C. Johnston-Peck, D. Su, E. A. Stach and W. D. Wei, *Angew. Chem., Int. Ed.*, 2014, **53**, 7887.
228. S. Pany, B. Naik, S. Martha and K. Parida, *ACS Appl. Mater. Interfaces*, 2014, **6**, 839.
229. Z. W. Seh, S. H. Liu, M. Low, S. Y. Zhang, Z. L. Liu, A. Mlayah and M. Y. Han, *Adv. Mater.*, 2012, **24**, 2310.
230. J. Long, H. Chang, Q. Gu, J. Xu, L. Fan, S. Wang, Y. Zhou, W. Wei, L. Huang, X. Wang, P. Liu and W. Huang, *Energy Environ. Sci.*, 2014, **7**, 973.
231. J.-J. Chen, J. C. S. Wu, P. C. Wu and D. P. Tsai, *J. Phys. Chem. C*, 2010, **115**, 210.
232. Q. Simon, D. Barreca, D. Bekermann, A. Gasparotto, C. Maccato, E. Comini, V. Gombac, P. Fornasiero, O. I. Lebedev, S. Turner, A. Devi, R. A. Fischer and G. Van Tendeloo, *Int. J. Hydrogen Energy*, 2011, **36**, 15527.

233. L. He, L. Li, T. T. Wang, G. Li, X. Wu, Z.-M. Su and C. Wang, *Dalton Trans.*, 2014, **43**, 16981.
234. A. Primo, T. Marino, A. Corma, R. Molinari and H. García, *J. Am. Chem. Soc.*, 2011, **133**, 6930.
235. L. Ge, C. Han, J. Liu and Y. Li, *Appl. Catal., A*, 2011, **409**, 215.
236. X. Bai, R. Zong, C. Li, D. Liu, Y. Liu and Y. Zhu, *Appl. Catal., B*, 2014, **147**, 82.
237. S. Samanta, S. Martha and K. Parida, *ChemCatChem*, 2014, **6**, 1453.
238. Q. Zhao, M. Ji, H. Qian, B. Dai, L. Weng, J. Gui, J. Zhang, M. Ouyang and H. Zhu, *Adv. Mater.*, 2014, **26**, 1387.
239. Z. Zheng, T. Tachikawa and T. Majima, *J. Am. Chem. Soc.*, 2014, **136**, 6870.
240. P. V. Kamat, *J. Phys. Chem. B*, 2002, **106**, 7729.
241. K. Qian, B. C. Sweeny, A. C. Johnston-Peck, W. Niu, J. O. Graham, J. S. DuChene, J. Qiu, Y.-C. Wang, M. H. Engelhard, D. Su, E. A. Stach and W. D. Wei, *J. Am. Chem. Soc.*, 2014, **136**, 9842.
242. A. Gallo, T. Montini, M. Marelli, A. Minguzzi, V. Gombac, R. Psaro, P. Fornasiero and V. Dal Santo, *ChemSusChem*, 2012, **5**, 1800.
243. A. Gallo, M. Marelli, R. Psaro, V. Gombac, T. Montini, P. Fornasiero, R. Pievo and V. D. Santo, *Green Chem.*, 2012, **14**, 330.
244. A. Tanaka, K. Hashimoto and H. Kominami, *J. Am. Chem. Soc.*, 2013, **136**, 586.
245. W. Wang, S. Liu, L. Nie, B. Cheng and J. Yu, *Phys. Chem. Chem. Phys.*, 2013, **15**, 12033.
246. S. Cao, C.-J. Wang, X.-J. Lv, Y. Chen and W.-F. Fu, *Appl. Catal., B*, 2015, **162**, 381.
247. J. Zhang, L. Qi, J. Ran, J. Yu and S. Z. Qiao, *Adv. Energy Mater.*, 2014, 4.
248. J. Yu, W. Wang and B. Cheng, *Chem.–Asian J.*, 2010, **5**, 2499.
249. X. Chen, W. Chen, H. Gao, Y. Yang and W. Shangguan, *Appl. Catal., B*, 2014, **152–153**, 68.
250. S. Cao, Y. Chen, C.-J. Wang, P. He and W.-F. Fu, *Chem. Commun.*, 2014, 10427.
251. J. Yu, Y. Hai and B. Cheng, *J. Phys. Chem. C*, 2011, **115**, 4953.
252. J. Yu and J. Ran, *Energy Environ. Sci.*, 2011, **4**, 1364.
253. J. Yu, S. Wang, B. Cheng, Z. Lin and F. Huang, *Catal. Sci. Technol.*, 2013, **3**, 1782.
254. J. Yu, Y. Hai and M. Jaroniec, *J. Colloid Interface Sci.*, 2011, **357**, 223.
255. J. S. Jang, D. J. Ham, N. Lakshminarasimhan, W. Y. Choi and J. S. Lee, *Appl. Catal., A*, 2008, **346**, 149.
256. X. Zong, J. Han, G. Ma, H. Yan, G. Wu and C. Li, *J. Phys. Chem. C*, 2011, **115**, 12202.
257. X. Zong, H. Yan, G. Wu, G. Ma, F. Wen, L. Wang and C. Li, *J. Am. Chem. Soc.*, 2008, **130**, 7176.
258. L. Wei, Y. Chen, Y. Lin, H. Wu, R. Yuan and Z. Li, *Appl. Catal., B*, 2014, **144**, 521.
259. M. L. Tang, D. C. Grauer, B. Lassalle-Kaiser, V. K. Yachandra, L. Amirav, J. R. Long, J. Yano and A. P. Alivisatos, *Angew. Chem., Int. Ed.*, 2011, **50**, 10203.

260. J. Zhang, J. Yu, Y. Zhang, Q. Li and J. R. Gong, *Nano Lett.*, 2011, **11**, 4774.
261. K. Domen, A. Kudo and T. Onishi, *J. Catal.*, 1986, **102**, 92.
262. K. Domen, A. Kudo, T. Onishi, N. Kosugi and H. Kuroda, *J. Phys. Chem.*, 1986, **90**, 292.
263. J. Ran, J. Zhang, J. Yu, M. Jaroniec and S. Z. Qiao, *Chem. Soc. Rev.*, 2014, **43**, 7787.
264. J. Ran, J. Yu and M. Jaroniec, *Green Chem.*, 2011, **13**, 2708.
265. B. Hinnemann, P. G. Moses, J. Bonde, K. P. Jorgensen, J. H. Nielsen, S. Horch, I. Chorkendorff and J. K. Norskov, *J. Am. Chem. Soc.*, 2005, **127**, 5308.
266. T. F. Jaramillo, K. P. Jorgensen, J. Bonde, J. H. Nielsen, S. Horch and I. Chorkendorff, *Science*, 2007, **317**, 100.
267. H. Vrubel, D. Merki and X. Hu, *Energy Environ. Sci.*, 2012, **5**, 6136.
268. D. Merki, S. Fierro, H. Vrubel and X. L. Hu, *Chem. Sci.*, 2011, **2**, 1262.
269. D.-Y. Wang, M. Gong, H.-L. Chou, C.-J. Pan, H.-A. Chen, Y. Wu, M.-C. Lin, M. Guan, J. Yang, C.-W. Chen, Y.-L. Wang, B.-J. Hwang, C.-C. Chen and H. Dai, *J. Am. Chem. Soc.*, 2015, **137**, 1587.
270. E. J. Popczun, J. R. McKone, C. G. Read, A. J. Biacchi, A. M. Wiltrout, N. S. Lewis and R. E. Schaak, *J. Am. Chem. Soc.*, 2013, **135**, 9267.
271. J. Liu, Y. Liu, N. Liu, Y. Han, X. Zhang, H. Huang, Y. Lifshitz, S.-T. Lee, J. Zhong and Z. Kang, *Science*, 2015, **347**, 970.
272. F. E. Osterloh, *Chem. Soc. Rev.*, 2013, **42**, 2294.
273. D. Wang, T. Kako and J. Ye, *J. Am. Chem. Soc.*, 2008, **130**, 2724.
274. T. Suzuki, T. Hisatomi, K. Teramura, Y. Shimodaira, H. Kobayashi and K. Domen, *Phys. Chem. Chem. Phys.*, 2012, **14**, 15475.
275. M. Higashi, K. Domen and R. Abe, *J. Am. Chem. Soc.*, 2013, **135**, 10238.
276. F. Zhang, A. Yamakata, K. Maeda, Y. Moriya, T. Takata, J. Kubota, K. Teshima, S. Oishi and K. Domen, *J. Am. Chem. Soc.*, 2012, **134**, 8348.
277. U. A. Joshi, A. Palasyuk, D. Arney and P. A. Maggard, *J. Phys. Chem. Lett.*, 2010, **1**, 2719.
278. Q. Li, H. Meng, P. Zhou, Y. Zheng, J. Wang, J. Yu and J. Gong, *ACS Catal.*, 2013, **3**, 882.
279. L. Wang, Z. Yao, F. Jia, B. Chen and Z. Jiang, *Dalton Trans.*, 2013, **42**, 9976.
280. Y. Wang, J. Wu, J. Zheng and R. Xu, *Catal. Sci. Technol.*, 2011, **1**, 940.
281. J. Yu, J. Zhang and M. Jaroniec, *Green Chem.*, 2010, **12**, 1611.
282. J. Yu, B. Yang and B. Cheng, *Nanoscale*, 2012, **4**, 2670.
283. J. Zhang, Q. Xu, S. Z. Qiao and J. Yu, *ChemSusChem*, 2013, **6**, 2009.
284. Z. Lei, W. You, M. Liu, G. Zhou, T. Takata, M. Hara, K. Domen and C. Li, *Chem. Commun.*, 2003, 2142.
285. W. Zhang, Z. Zhong, Y. Wang and R. Xu, *J. Phys. Chem. C*, 2008, **112**, 17635.
286. G. Liu, L. Zhao, L. Ma and L. Guo, *Catal. Commun.*, 2008, **9**, 126.
287. D.-H. Wang, L. Wang and A.-W. Xu, *Nanoscale*, 2012, **4**, 2046.
288. D. Beydoun, R. Amal, G. Low and S. McEvoy, *J. Nanopart. Res.*, 1999, **1**, 439.

289. M. A. Holmes, T. K. Townsend and F. E. Osterloh, *Chem. Commun.*, 2012, 371.
290. A. Galińska and J. Walendziewski, *Energy Fuels*, 2005, **19**, 1143.
291. S.-C. Moon, H. Mametsuka, S. Tabata and E. Suzuki, *Catal. Today*, 2000, **58**, 125.
292. A. J. Bard, R. Parsons and J. Jordan, *Standard potentials in aqueous solution*, CRC Press, 1985.
293. H. Yan, J. Yang, G. Ma, G. Wu, X. Zong, Z. Lei, J. Shi and C. Li, *J. Catal.*, 2009, **266**, 165.
294. K. Wu, Z. Chen, H. Lv, H. Zhu, C. L. Hill and T. Lian, *J. Am. Chem. Soc.*, 2014, **136**, 7708.
295. Z. Fang, Y. Wang, J. Song, Y. Sun, J. Zhou, R. Xu and H. Duan, *Nanoscale*, 2013, **5**, 9830.
296. T. Jia, A. Kolpin, C. Ma, R. C.-T. Chan, W.-M. Kwok and S. C. E. Tsang, *Chem. Commun.*, 2014, 1185.
297. R. Liu, Z. Zheng, J. Spurgeon, B. S. Brunschwig and X. Yang, *Energy Environ. Sci.*, 2014, **7**, 2504.
298. G. Liu, J. Shi, F. Zhang, Z. Chen, J. Han, C. Ding, S. Chen, Z. Wang, H. Han and C. Li, *Angew. Chem., Int. Ed.*, 2014, **53**, 7295.
299. B. Seger, T. Pedersen, A. B. Laursen, P. C. K. Vesborg, O. Hansen and I. Chorkendorff, *J. Am. Chem. Soc.*, 2013, **135**, 1057.
300. S. Hu, M. R. Shaner, J. A. Beardslee, M. Lichterman, B. S. Brunschwig and N. S. Lewis, *Science*, 2014, **344**, 1005.
301. Z. Zhang, R. Dua, L. Zhang, H. Zhu, H. Zhang and P. Wang, *ACS Nano*, 2013, **7**, 1709.
302. P. D. Tran, S. K. Batabyal, S. S. Pramana, J. Barber, L. H. Wong and S. C. J. Loo, *Nanoscale*, 2012, **4**, 3875.
303. A. Thibert, F. A. Frame, E. Busby, M. A. Holmes, F. E. Osterloh and D. S. Larsen, *J. Phys. Chem. Lett.*, 2011, **2**, 2688.
304. P. Tongying, F. Vietmeyer, D. Aleksiuk, G. J. Ferraudi, G. Krylova and M. Kuno, *Nanoscale*, 2014, **6**, 4117.
305. L. Huang, X. Wang, J. Yang, G. Liu, J. Han and C. Li, *J. Phys. Chem. C*, 2013, **117**, 11584.
306. Y. P. Xie, Z. B. Yu, G. Liu, X. L. Ma and H.-M. Cheng, *Energy Environ. Sci.*, 2014, **7**, 1895.
307. J. K. Hurst, *Science*, 2010, **328**, 315.
308. M. Grätzel, *Acc. Chem. Res.*, 1981, **14**, 376.
309. M. Seol, J.-W. Jang, S. Cho, J. S. Lee and K. Yong, *Chem. Mater.*, 2013, **25**, 184.
310. J. Rosen, G. S. Hutchings and F. Jiao, *J. Am. Chem. Soc.*, 2013, **135**, 4516.
311. R. D. L. Smith, M. S. Prévot, R. D. Fagan, Z. Zhang, P. A. Sedach, M. K. J. Siu, S. Trudel and C. P. Berlinguette, *Science*, 2013, **340**, 60.
312. X. L. Hu, B. M. Cossairt, B. S. Brunschwig, N. S. Lewis and J. C. Peters, *Chem. Commun.*, 2005, 4723.
313. S. Cobo, J. Heidkamp, P.-A. Jacques, J. Fize, V. Fourmond, L. Guetaz, B. Jousselme, V. Ivanova, H. Dau, S. Palacin, M. Fontecave and V. Artero, *Nat. Mater.*, 2012, **11**, 802.

314. Y. Liang, Y. Li, H. Wang, J. Zhou, J. Wang, T. Regier and H. Dai, *Nat. Mater.*, 2011, **10**, 780.
315. R. Li, Z. Chen, W. Zhao, F. Zhang, K. Maeda, B. Huang, S. Shen, K. Domen and C. Li, *J. Phys. Chem. C*, 2012, **117**, 376.
316. M. Higashi, K. Domen and R. Abe, *J. Am. Chem. Soc.*, 2012, **134**, 6968.
317. S. S. K. Ma, T. Hisatomi, K. Maeda, Y. Moriya and K. Domen, *J. Am. Chem. Soc.*, 2012, **134**, 19993.
318. J. S. Zhang, M. Grzelczak, Y. D. Hou, K. Maeda, K. Domen, X. Z. Fu, M. Antonietti and X. C. Wang, *Chem. Sci.*, 2012, **3**, 443.
319. K. Maeda, A. K. Xiong, T. Yoshinaga, T. Ikeda, N. Sakamoto, T. Hisatomi, M. Takashima, D. L. Lu, M. Kanehara, T. Setoyama, T. Teranishi and K. Domen, *Angew. Chem., Int. Ed.*, 2010, **49**, 4096.
320. S. S. K. Ma, K. Maeda, R. Abe and K. Domen, *Energy Environ. Sci.*, 2012, **5**, 8390.
321. D. Wang, T. Hisatomi, T. Takata, C. Pan, M. Katayama, J. Kubota and K. Domen, *Angew. Chem., Int. Ed.*, 2013, **52**, 11252.
322. M. Marszewski, S. Cao, J. Yu and M. Jaroniec, *Mater. Horiz.*, 2015, **2**, 261.
323. X. Li, J. Wen, J. Low, Y. Fang and J. Yu, *Sci. China Mater.*, 2014, **57**, 70.
324. X. Lang, X. Chen and J. Zhao, *Chem. Soc. Rev.*, 2014, **43**, 473.
325. H. Tong, S. X. Ouyang, Y. P. Bi, N. Umezawa, M. Oshikiri and J. H. Ye, *Adv. Mater.*, 2012, **24**, 229.
326. P. V. Kamat, *J. Phys. Chem. Lett.*, 2012, **3**, 663.
327. L. Zhang, H. H. Mohamed, R. Dillert and D. Bahnemann, *J. Photochem. Photobiol., C*, 2012, **13**, 263.
328. G. Zhang, G. Kim and W. Choi, *Energy Environ. Sci.*, 2014, **7**, 954.
329. C. Dong, X. Li, P. Jin, W. Zhao, J. Chu and J. Qi, *J. Phys. Chem. C*, 2012, **116**, 15833.
330. J. R. McKone, N. S. Lewis and H. B. Gray, *Chem. Mater.*, 2013, **26**, 407.

CHAPTER 11

New Synthetic Routes in Heterogeneous Photocatalysis

GIOVANNI PALMISANO*[a], ANGELO ALBINI*[b], GIUSEPPE MARCÌ[c], LEONARDO PALMISANO*[c], AND DAVIDE RAVELLI[b]

[a]Department of Chemical and Environmental Engineering, Institute Center for Water and Environment (iWater), Masdar Institute of Science and Technology, PO BOX 54224, Abu Dhabi, UAE; [b]PhotoGreen Lab, Department of Chemistry, University of Pavia, Viale Taramelli 12, 27100, Pavia, Italy; [c]"Schiavello-Grillone" Photocatalysis Group, DEIM Department, Università degli Studi di Palermo, Viale delle Scienze (Ed. 6), 90128 Palermo, Italy

*E-mail: gpalmisano@masdar.ac.ae, angelo.albini@unipv.it, leonardo.palmisano@unipa.it

11.1 Introduction

Heterogeneous photocatalysis is an unconventional technology that has been mainly applied to degrade organic and inorganic pollutants both in gas and liquid phases.[1] Global interest in this method has grown due to the viability of exploiting a renewable energy source such as solar light[2] and many cheap, innocuous and recyclable photocatalysts such as TiO_2, ZnO and so on. The chapter will consider only heterogeneous processes, where a solid catalyst absorbs light and promotes redox processes involving adsorbed organic species present in a fluid phase. In such cases the possible reactivity

RSC Energy and Environment Series No. 15
Photocatalysis: Applications
Edited by Dionysios D Dionysiou, Gianluca Li Puma, Jinhua Ye, Jenny Schneider, and Detlef Bahnemann

Published by the Royal Society of Chemistry, www.rsc.org

of organic species due to their ability to absorb light themselves and reacting consequently can be usually neglected, since the rate of such processes is generally lower than the heterogeneous one.[1] There are, however, cases where this is not true.[3] Operation at ambient temperature and pressure and the tunability of the oxidation/reduction power of most processes are other important advantages to consider.[4] Indeed, several works have been devoted to clarifying the mechanistic issue.[5]

Applications to synthetic chemistry routes have been studied too, but they are less common because of the frequent lack of selectivity of photocatalytic reactions.[6] Nevertheless, the first studies on the effect of heterogeneous photocatalysts on several organic reactions were carried out not much later than the birth of heterogeneous photocatalysis, although most cases cannot be referred to as actual syntheses since they were carried out on a very small (analytical) scale and the products were not always isolated but just identified and quantified.[7]

Only recently have these studies grown up due to the increasing wish to support the sustainability of chemical processes, by reducing dangerous byproducts and wastes, such as toxic organic solvents and catalysts based on hardly disposable heavy metals.[8] Thus, studies on product synthesis by means of selective oxidation[9] and reduction,[10] functionalization,[11] and C–C and C–N bond formation[12] have become more usual. These studies demonstrate that high selectivities can be obtained in photo-oxidation and photoreduction processes as compared with conventional ones.

Some key advances that can improve the feasibility of photocatalytic processes are: in the use of the cheap LED UV radiation sources instead of medium-pressure Hg UV lamps, when solar light is not available or convenient to use; the employment of highly porous nanostructured materials to support catalysts; and the coupling of photocatalysis with membrane technology to separate reagent mixture, solvent and catalyst powder from the pure product.[13]

Moreover, nowadays organic chemistry is confronting the replacement of toxic organic solvents with innocuous ones, possibly with water. Interestingly, it has recently been discovered that in many cases both water-soluble and water-insoluble molecules can react faster in water compared to traditional organic solvent.[14]

Extensive industrial application of photocatalysis in synthetic chemistry is, however, still a challenge more than a reality and much effort is needed to reach significant outcomes.

11.2 Reactions

The classification of photocatalytic reactions as oxidations, reductions and alkylations, presented in Sections 11.2.1–11.2.3, simply means that the main product(s) arise(s) from oxidation, reduction or alkylation, respectively, of the main reagent(s). Noticeably, in fact, both the oxidative and reductive pathways are present in heterogeneous photocatalysis, but sometimes the reduction pathway produces species (*e.g.* $O_2^{\bullet -}$ by trapping electrons in the conduction band of the photocatalyst) whose subsequent reaction (*e.g.* with H_2O) gives rise to hydroxyl and hydroperoxyl oxidant radicals.[15]

11.2.1 Oxidations

Concerning heterogeneous photocatalytic oxidations, it is well known that they are generally unselective; several studies and applications have been reported in the field of outdoor and indoor environment purifications when the concentration of organic and inorganic pollutants to be abated is quite low. Indeed the heterogeneous photocatalytic method appears to be efficient only in such instances because the presence of high concentrations of pollutants slows down the reaction rates and other technologies should be applied in such a case. Photo-oxidations are by far the most studied class of photocatalytic reactions, although low quantum efficiencies have often been observed.

The semiconductor photocatalyst most frequently used, both as powders (micro or nanocrystalline samples) and films, is bare polycrystalline TiO_2 in the anatase phase; some studies, though, have been carried out also with the other polymorphic rutile or brookite phases.[16] Moreover, the formation of nanorods, nanotubes and nanowires under particular experimental conditions has been reported.[17] TiO_2 has also been doped with various transition[18] or rare earth metal species[19] and non-metal species of N,[20] C,[21] S,[22] F.[23] Sensitization with (metal) phthalocyanines and porphyrins,[24] to extend the light absorption in the visible region and/or to increase the lifetime of the photo-produced electron–hole pairs, has been also proposed. Other oxides such as ZnO, WO_3, MoO_3, Fe_2O_3, ZrO_2, and sulfides such as ZnS, CdS, MoS_2[25] or polyoxometalates as the heteropolyacid $H_3PW_{12}O_{40}$,[26–28] also have been successfully tested as photocatalysts.

Some photocatalysts, for instance ZnO, are not (photo)stable in liquid–solid systems, especially in aqueous environments and the extent of this instability strongly depends on the pH conditions.

Notably, it is not possible, especially in aqueous systems, to avoid the mineralization reactions through intermediates that do not desorb into the bulk of solution, but are subjected to oxidative attack on the surface until CO_2 is formed. Indirect evidence of this behavior in many cases is the detection of CO_2 soon after the starting of irradiation. Accordingly, during oxidation, at least two parallel pathways are operating on the substrate: (1) complete mineralization yielding CO_2 and (2) formation of partially oxidized species released into the bulk of the solution. In general, one can tune the catalyst characteristics to promote one or other route.[29,30]

11.2.1.1 Oxidation of Alcohols to Aldehydes

This section reports some results on the use of heterogeneous photocatalysis for the selective oxidation of alcohols to aldehydes. The partial oxidation of alcohols to produce carbonyl compounds is of great industrial importance;

generally, it is performed in environmentally harmful organic solvents at high temperatures and pressures by using stoichiometric oxygen donors that not only are expensive and toxic but also produce large amounts of dangerous waste.[31,32]

Heterogeneous photocatalysis has been used successfully for partial oxidation of alcohols in the gas[33,34] and liquid phase, using organic solvents,[35,36] affording high yields in some cases. For instance, Palmisano *et al.* studied the influence of the features of the photocatalysts on the selective oxidation of aromatic alcohols to the corresponding aldehydes in organics-free water.[9c] The partial oxidation of benzyl alcohol (BA), 4-methoxybenzyl alcohol (MBA), 4-methylbenzyl alcohol (MeBA), and 4-nitrobenzyl alcohol (NBA) to the corresponding aldehydes was achieved with both home-prepared and commercial TiO_2 samples, and the influence of the substituent group on the reaction rate was studied.[9c]

The influence of photocatalyst crystallinity is probably the predominant feature on the photo-activity for the selective oxidation of alcohols to aldehydes. The highest rates of alcohol conversion are those of commercial catalysts (more crystalline), showing at the same time a lower selectivity for partial oxidation to the corresponding aldehydes. Moreover the least strong oxidizing catalyst and the least crystalline, among the obtained catalysts tested, is the best sample for selective oxidation. An increase in crystallinity caused a significant decrease in selectivity. The commercial catalysts, the least selective ones in the formation of aldehydes, show higher activities giving rise to many highly oxidized intermediates that result in a significant lowering of selectivity.

The detected intermediates and the values for selectivity towards aldehyde and reaction rate are very different for BA and MBA. The selectivity to aldehyde in MBA oxidation is higher than in BA oxidation. A hypothetical reaction mechanism for BA and MBA is shown in Scheme 11.1. The first step (Scheme 11.1a) is abstraction of an electron from the –OH group by a hole; subsequent transformations involve the formation of a –CHO group by means of either a positive hole or an $^{\bullet}OH$ radical. Benzaldehyde, once formed, can be subjected to further attacks by $^{\bullet}OH$ radicals on the aromatic ring (Scheme 11.1b).

11.2.1.2 *Hydroxylation of Aromatics*

The hydroxylation of aromatics is of great interest for the chemical industry. Hydroquinone, for instance, is a widely employed molecule typically produced *via* oxidation of aniline by treatment with sulfuric acid and manganese dioxide, after which a reduction step is needed by using Fe/HCl.[37]

This process generates massive quantities of waste and is not atom efficient, since only some of the atoms consumed in the process are present in the final product. On the other hand, there are efficient routes that make use of titanium silicate catalysts in hydroxylation reactions by treatment with hydrogen peroxide, in particular to produce dihydroxylated benzenes.[38]

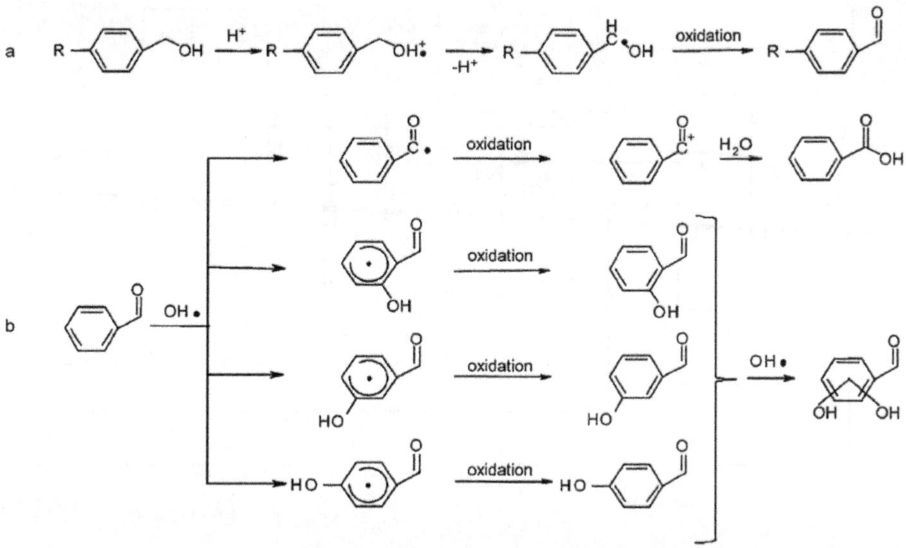

Scheme 11.1 Proposed reaction schemes for aromatic alcohol oxidation (R = H, OCH_3).[9c] (Reproduced with permission from L. Palmisano, V. Augugliaro, M. Bellardita, A. Di Paola, E. García-López, V. Loddo, G. Marcì, G. Palmisano and S. Yurdakal, *ChemSusChem*, 2011, **4**, 1431.)

The conversion of benzene into phenol is probably the most interesting reaction among photocatalytic hydroxylations: phenol is extensively employed, for instance, as a disinfectant and to synthesize resins.

An interesting example of phenol synthesis from benzene is that carried out in a photocatalytic membrane reactor, where simultaneous reaction and product separation can take place thanks to a hydrophobic polypropylene porous membrane.[39] Benzene was used as organic extractant, showing a phenol distribution coefficient of 2.1 and an extraction percentage of *ca.* 24%, thus assuring a constant feed of substrate in the reactive environment (Figure 11.1). This environment consists of a UV-irradiated batch reactor equipped with a Pyrex glass jacket to obtain a stable temperature of 25 °C. The permeation module, immersed in a thermostatic bath (Figure 11.1, 5), is made of two compartment cells separated by a flat sheet polypropylene membrane. The first compartment (Figure 11.1, 6) contains an aqueous phase which undergoes oxidation in the reactor, while the second (Figure 11.1, 7) contains an organic stripping solution (benzene).

The production and separation of phenol were investigated at different pH; better control of formation and extraction of oxidation intermediates was obtained at pH 3.1. To enhance the performance of the system, photocatalytic experiments with different metal cations (Fe^{3+}, Cu^{2+} and V^{3+}) were carried out. The results showed a positive effect in the presence of Fe(III) with a phenol flux in the organic phase almost twice that measured with no salt added.

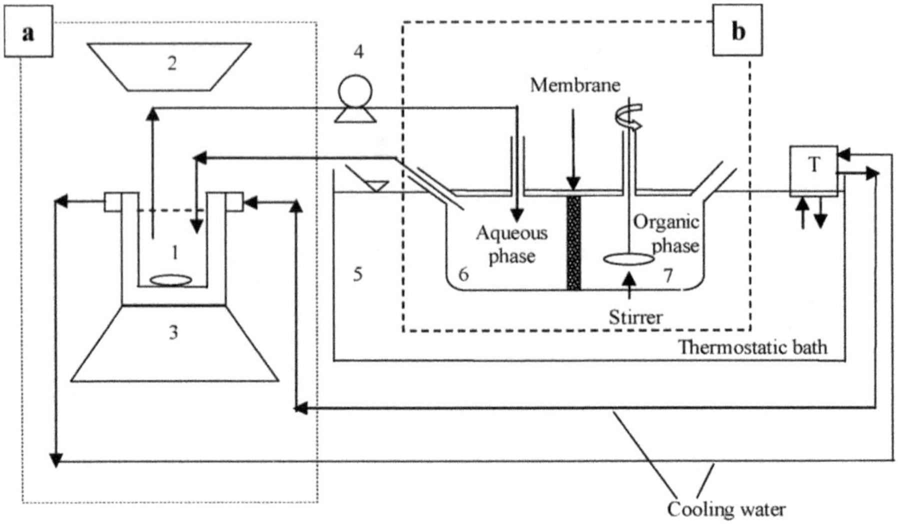

Figure 11.1 Photocatalytic membrane reactor used for benzene to phenol oxidation. (a) UV-irradiated batch reactor (1: batch reactor. 2: UV lamp, 3: magnetic stirrer, 4: peristaltic pump). (b) Permeation module with propylene membrane (5: thermostatic bath, 6: aqueous phase, 7: organic solution). For further details, see text.[39] (Reproduced with permissions from R. Molinari, A. Caruso and T. Poerio, *Catal. Today*, 2009, **144**, 81.)

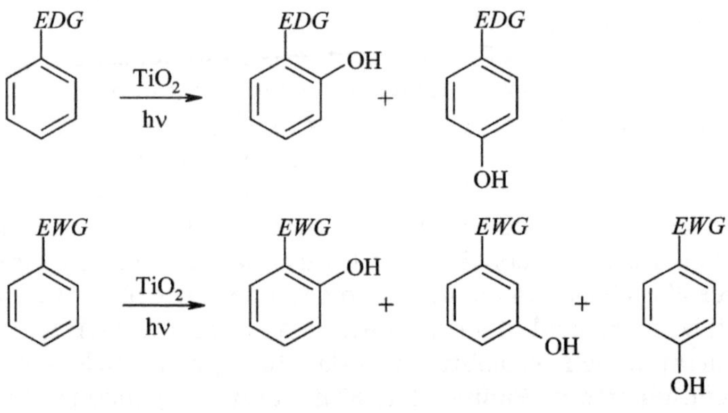

Scheme 11.2 Hydroxylation of aromatic compounds containing an electron donor (EDG) or an electron withdrawing (EWG) group.[40a] (Reproduced with permission from G. Palmisano, M. Addamo, V. Augugliaro, T. Caronna, E. García-López, V. Loddo and L. Palmisano, *Chem. Commun.*, 2006, 1012.)

One should always remember that the hydroxylation of aromatics by using TiO_2-catalyzed photocatalysis is regioselective in the presence of an electron donor group (EDG) in the aromatic ring (Scheme 11.2).[40] It is in fact possible to prepare *ortho*- and *para*-oriented hydroxy-derivatives starting from molecules such as aniline or phenol. In such cases •OH radicals attack the aromatic ring by following the selectivity rules known for homogeneous electrophilic aromatic substitution, yielding only the *ortho*- and *para*-isomers. This suggests that the photocatalytic hydroxylation could take place upon a weak photo-adsorption mode on a TiO_2 surface. In the absence of an electron donor group, on the other hand, the attack is completely unselective. Notably, the reported behavior does not depend on the irradiation source or the kind of TiO_2 used.

Besides the hydroxylation route, a competing reaction pathway to total oxidation was always observed through a total organic carbon (TOC) decrease from the beginning of irradiation: this pathway was more important for aromatics that did not contain electron donor groups. To explain this evidence the strong interaction of these molecules with the TiO_2 surface should be considered. Interestingly, when an electron donor group is present, rather high selectivities for the hydroxylated species (up to 75% with a 70% conversion) can be achieved.

The selectivity in the hydroxylation of aromatic compounds containing an EDG can be explained by examining the possible formulas of the radical intermediates (Scheme 11.3). The formula characterized by an unpaired electron on the carbon atom bonded to the EDG is possible only when the OH radical enters *ortho* and *para* positions. Conversely, formation of the hydroxylated compound is less favored when the attack occurs at the *meta* position as shown in the third row of Scheme 11.3.

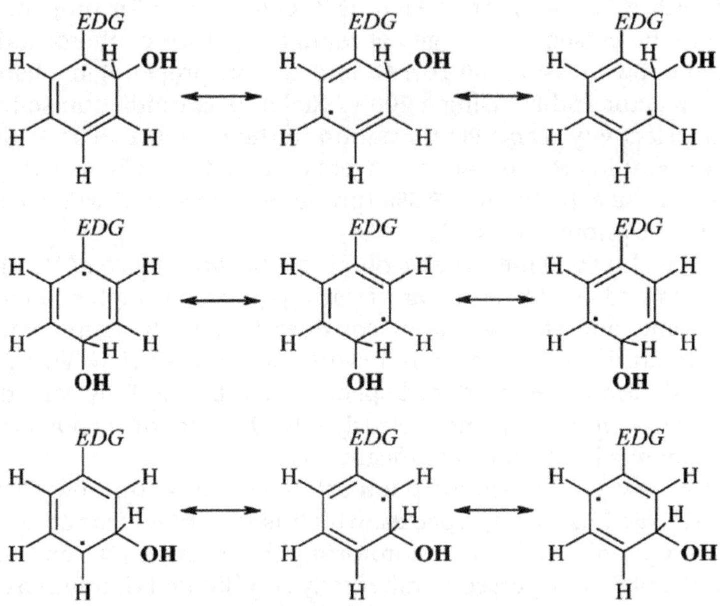

Scheme 11.3 Resonance structures of radical intermediates produced during the oxidation of a compound containing an EDG.[40a] (Reproduced with permission from G. Palmisano, M. Addamo, V. Augugliaro, T. Caronna, E. García-López, V. Loddo and L. Palmisano, *Chem. Commun.*, 2006, 1012.)

11.2.1.3 Alkenes Epoxidation

In this section we focus on the preparation of propene oxide. The partial oxidation of propene to obtain propene oxide (PO) is a reaction of paramount importance in the chemical industry. The classical methods to produce PO are based on the chlorohydrin and on the hydroperoxidation processes.[41] These processes suffer from several disadvantages; indeed, the first process is not green due to the utilization of chlorine and the production of hazardous by-products and the economy of the second one depends on the valorization of the by-products. The epoxidation process using hydrogen peroxide in the presence of methanol and titanium silicalite-1 (TS-1) catalyst[42] or by using O_2 in the presence of Au nanoparticles deposited on modified TS-1 [43] have been also reported. Compared with the classical processes, the two last methods are advantageous from the environmental point of view; however, they are often carried out at high temperature or high pressure and the development of a more environmental friendly process, such as the photocatalytic one by using only O_2, is still desirable. To the best of our knowledge few papers report the photocatalytic propene epoxidation with oxygen under mild conditions although it seems clear that metal oxide species, such as TiO_2, that are highly dispersed on silica, alumina or zeolite can exhibit photocatalytic activity for the partial oxidation of propene.[44] It is reported that silica is a good matrix that, when used as catalysts support, promotes not only catalytic thermal reactions but also photoreactions. TiO_2–SiO_2 or ZnO–SiO_2 samples prepared with a low amount of TiO_2 or ZnO are active for propene partial oxidation in the presence of oxygen at room temperature (photocatalyst bed at 50 °C) and low pressure (30 Torr O_2 and 15 Torr propene) in a batch photocatalytic reactor and by using a 200 W Xe lamp as irradiation source. The maximum selectivity *versus* PO formation in the presence of TiO_2–SiO_2 was 60%, corresponding to a propene conversion of 4.4%,[45] whereas the results were 34% (PO selectivity) and 8.6% (propene conversion) when ZnO–SiO_2 was used as the photocatalyst.[46]

Table 11.1 shows some results obtained by Murata *et al.*[45]. The conversions obtained in this study are relatively low (the highest amounting to 32.1%), and one should always consider that such figures should be increased to attain relevance from a synthetic point of view. Furthermore, the over-oxidation of the produced species must be carefully considered in all the oxidative processes, and – clearly – for higher conversions the over-oxidation extent will be more significant.

The active oxygen species for photocatalytic epoxidation over TiO_2/SiO_2 has been revealed as the O_3^- species, which is generated from the reaction between an O_2 molecule and a photoformed hole center (O^-) on the lattice oxygen of the highly dispersed tetrahedrally coordinated titanium oxide species (TiO_4).[47] It has been accepted that the isolated tetrahedrally coordinated MO_4 species are necessary for high selectivity to photocatalytic alkenes epoxidation. At the same time, however, isolated tetrahedral VO_4 species on silica are not considered to be so efficient for photocatalytic epoxidation. Indeed,

Table 11.1 Results of the photo-oxidation of propene over SiO_2, TiO_2 and TiO_2–SiO_2 samples.[45,a]

Sample	Ti content (mol%)	S.S.A. ($m^2 g^{-1}$)	Conversion[b] (%)	PO yield (%)	Selectivity[c] (%) PO	Propanal	Acetone	Acrolein	Ethanal	HC	CO_x
TiO_2–SiO_2	0.08	387	4.4	2.6	60.2	2.9	10.4	1.1	17.7	3.0	3.9
TiO_2–SiO_2	0.34	423	9.2	5.3	57.5	2.7	5.8	1.2	21.1	5.1	6.6
TiO_2–SiO_2	1.0	535	12.5	6.3	50.5	6.2	8.3	1.7	22.1	3.3	7.4
TiO_2–SiO_2	4.1	483	21.0	4.5	21.5	19.1	11.4	2.4	24.1	3.4	14.4
TiO_2–SiO_2	8.3	416	32.1	1.8	5.7	24.5	13.6	3.1	26.4	2.9	19.1
TiO_2–SiO_2[d]	0.34	423	0.2	Traces	5.5	1.6	2.9	0.0	70.0	20.0	0.0
TiO_2–SiO_2[e]	0.34	423	0.9	Traces	0.9	9.4	0.0	0.0	16.4	73.3	0.0
SiO_2		558	0.7	0.2	22.3	3.5	25.8	15.2	18.1	10.0	5.1
TiO_2[f]		34	14.1	0.0	0.0	0.0	1.3	0.0	0.8	1.7	96.2

[a]Catalyst 0.2 g, C_3H_6 100 μmol, O_2 200 μmol, reaction time 2 h.
[b]Based on propene.
[c]PO, HC and CO_x denote propene oxide, ethene + butenes, and CO + CO_2, respectively.
[d]In the dark at 330 K.
[e]In the absence of O_2.
[f]Reaction time 1 h.

in the case of propene partial photo-oxidation over V_2O_5/SiO_2 in the presence of oxygen, the formation of propene oxide was not observed.[48]

More recently, photocatalytic propene partial oxidation has been investigated by Amano *et al.* using a continuous flow reactor.[49] In this case, probably due to the short contact time, which reduced the photodecomposition of the photo-oxygenated intermediates, propene oxide formation was observed even over highly loaded V_2O_5/SiO_2. In any event, high selectivity into propene oxide formation (up to 40%) was achieved at very low loading of V_2O_5 on SiO_2. Unfortunately, the photo-oxidation mechanism over V_2O_5/SiO_2 has still not been clarified. Photocatalytic alkene epoxidation processes have been also studied, by using mesoporous Au–Ti–SBA-15, TiO_2-pillared clays, V–Ti/MCM-41 and Au/Ti based oxides semiconductors, but in these cases the process exhibits low selectivity.[50] Shiraishi *et al.* used Ti-containing silica with a hexagonal MCM-41 structure as the photocatalyst and the photoreaction was carried out in a glass tube by irradiating the system with a light with $\lambda > 280$ nm in O_2-saturated acetonitrile.[51] The selectivity, after 12 h of irradiation, of various cyclic and linear alkenes to the corresponding epoxides ranged from 11% to 99%.

11.2.1.4 Propene Hydration

The hydration of alkenes can be considered an internal dismutation as one carbon involved in the double bond is reduced and the other is oxidized. The hydration of propene at ambient conditions is a reaction of great interest. The industrial propene catalytic hydration to 2-propanol is generally carried out at moderate temperatures (*ca.* 150–200 °C) and pressure (2 MPa) in the presence of an acid catalyst such as phosphoric acid supported on silica, strong acidic resins,[52] beta-zeolite[53] and other acidic zeolites.[54]

Propene hydration is difficult to carry out because it is thermodynamically limited by the reverse dehydration reaction at high temperature. Heteropolyacids and their salts (polyoxometalates, POMs) are catalysts used industrially for the hydration of propene to 2-propanol. POMs are generally used in solution as acid and oxidation catalysts. They are very strong Brönsted acids and efficient oxidants; consequently, they can act as bifunctional catalysts.[55] The industrial hydration of propene to obtain 2-propanol has also been carried out by using aqueous solutions of $H_3PW_{12}O_{40}$. The catalytic reaction performed in the gas–solid regime has been the object of several patents[56] but only very few papers have been published.

Supported POM resulted in much more active catalysts than the corresponding bare sample. The maximum activity in hydration was measured at 130 °C, as higher temperatures led to a decrease in the rate of 2-propanol formation, due to a significant occurrence of the reverse dehydration reaction.[57]

The catalytic and catalytic photo-assisted activity of $H_3PW_{12}O_{40}$ supported on TiO_2 Evonik P25 and SiO_2 was studied and a beneficial role of the photo-catalytically active support on the reaction rate was reported.[58] The simultaneous presence of heat and UV light improved the activity of the

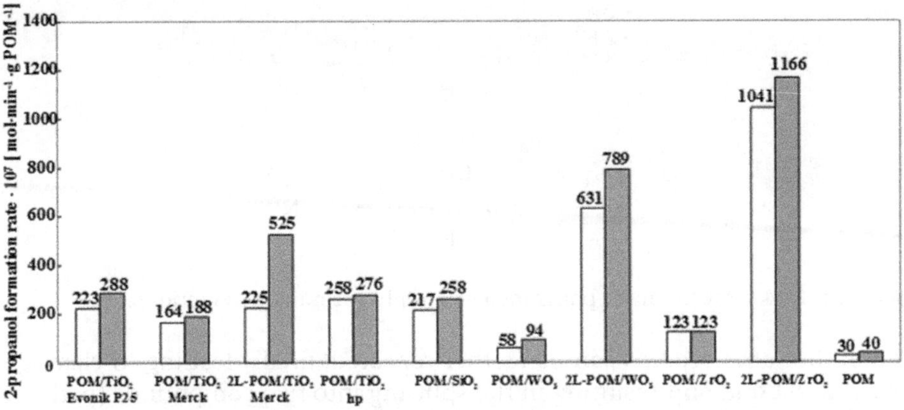

Figure 11.2 2-Propanol formation rate per gram of POM present when using various POM supported materials for the catalytic (white bars) and catalytic photo-assisted (grey bars) propene hydration reaction.[59] (Reproduced with permission from G. Marcì, E. García-López, M. Bellardita, F. Parisi, C. Colbeau-Justin, S. Sorgues, L. F. Liotta and L. Palmisano, *Phys. Chem. Chem. Phys.*, 2013, **15**, 13329.)

POM supported materials for the propene hydration reaction and the catalytic photo-assisted reaction occurred at a higher rate than the catalytic one when the support was an irradiated semiconductor. In that case, the more significant increase of reactivity was justified by considering the ability of TiO_2 to transfer electrons from the conduction band to the activated POM* species. Recently, Marcì *et al.* studied catalytic and catalytic photo-assisted activities of $H_3PW_{12}O_{40}$ supported on various acid or basic semiconductor and insulator oxides, *i.e.* TiO_2, WO_3, ZnO, SiO_2, Al_2O_3 and ZrO_2, for propene hydration.[59]

The cooperative effect of light and heat has been investigated by taking into account that the (photo)reactivity can depend on the type of support influencing the POM features.[59] Notably, the effect of local heating of the surface by photon irradiation can be excluded, due to the high quantities of cold water used as IR filter. As it can be seen in Figure 11.2, the best supports were ZrO_2 and WO_3, at least when the formation rate of 2-propanol is determined per gram of POM. The binary POM/ZnO and POM/Al_2O_3 samples, instead, were (photo) inactive. It was found, in fact, that the supported POM species were completely and partially degraded when supported on ZnO and Al_2O_3, respectively.[59]

11.2.2 Reductions

Photocatalytic reduction processes involve the consecutive transfer of electrons and protons to the reagent. The addition of two electrons and two protons to a generic A=B double bond affords the corresponding saturated derivative (path a, Scheme 11.4). Typical examples are the reduction of carbonyl derivatives (C=O)[60–62] (Section 11.2.2.1) or of (conjugated) C=C double bonds[61] (Section 11.2.2.2). Depending on the strength of the A–B sigma

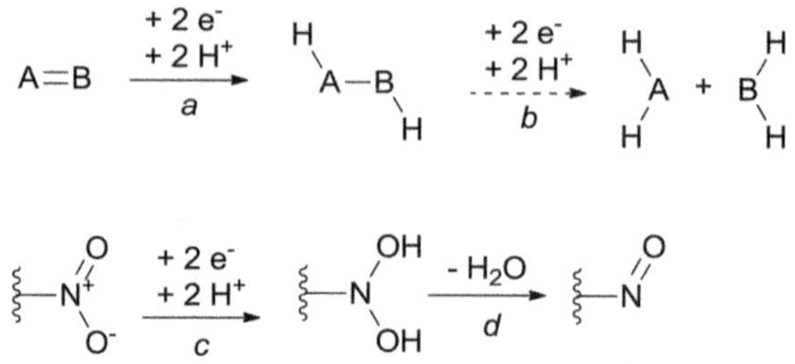

Scheme 11.4 Mechanistic paths involved in photocatalytic reductions.

bond, the same process can be further repeated, finally leading to cleavage of the A–B bond and resulting in the splitting into two compounds (path b). Azo derivatives are an example of the latter behavior, undergoing N–N bond cleavage under photocatalytic conditions.[63]

Further reactions may occur at the intermediate stage. In the case of nitro groups,[64–77] the most extensively investigated among photocatalytic reductions, the addition of two electrons and two protons occurs on the two oxygen atoms, and elimination of a water molecule leads to a nitroso derivative (paths c and d in Scheme 11.4). The nitroso derivative itself is prone to further reduction according to the scheme shown for an isolated double bond, leading to hydroxylamine or aniline (Section 11.2.2.3).

Photocatalytic reductions have a significant role and often follow a different reaction course with respect to direct irradiation, with the advantage of a better selectivity and the possibility of easily tuning the process by choosing the appropriate reaction conditions (*e.g.* different catalysts can afford completely different products).

11.2.2.1 *Carbonyl Reduction*

The photochemical reduction of benzophenone is well known to yield benzopinacol as the main product, along with several byproducts,[78] reasonably arising *via* coupling of the involved radical species (*e.g.* those arising by H-abstraction from the solvent; see Scheme 11.5a). On the other hand, benzhydrol only is produced *via* the photocatalytic route in the presence of TiO_2 under nitrogen atmosphere. Adopting the optimized conditions (2 g L^{-1} catalyst loading and a 0.5 mM initial concentration of benzophenone), in the presence of isopropanol as the electron donor, a benzhydrol yield of 70% (as from the results of HPLC analysis) was achieved (Scheme 11.5a).[60] Recently, this photocatalytic approach was extended to aromatic and heteroaromatic aldehydes, that were reduced to primary alcohols in the presence of P25 titania in acetonitrile, also yielding pinacols as by-products.[61] Indeed, it was shown that both the amount of the photocatalyst and of methanol, used as co-solvent, affected the product distribution. The best results in terms of alcohol yield were obtained upon irradiation of a 1 mg mL^{-1} solution (~0.75 equiv.) of titania P25 with UV-A light (20 h) in a MeCN/MeOH 9:1 mixture. Under the optimized conditions, several aldehydes gave a clean reaction, with high

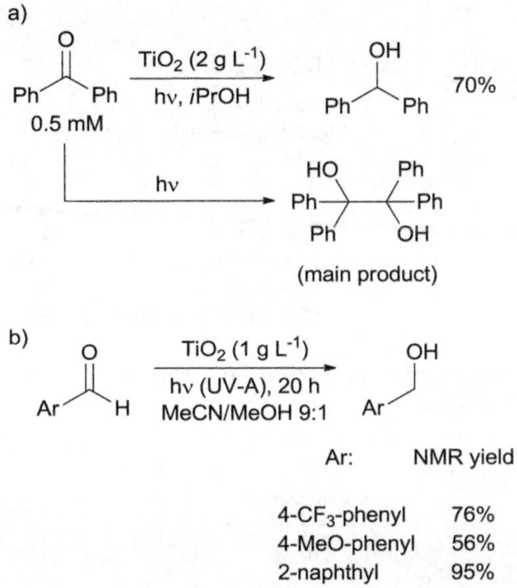

Scheme 11.5 Photocatalytic reduction of the carbonyl group in: a) benzophenone and b) (hetero)aromatic aldehydes.

yields (by NMR) of the corresponding alcohols (Scheme 11.5b).[61] A similar procedure involving benzaldehyde as the substrate was likewise performed in a micro-reactor, where titanium dioxide was immobilized. Upon irradiation with UV LEDs centered at 365 nm, the system afforded benzyl alcohol in approx. 11% yield after 60 s irradiation.[62]

11.2.2.2 C=C Double Bonds Reduction

Reduction of the conjugate C=C double bond in maleimides has been reported to take place in the presence of TiO_2 Degussa P25.[61] It was found that the reaction in neat acetonitrile gave poor performance due to the absence of an efficient hole scavenger. By contrast, when the reaction was run in neat methanol, ring-opening to the amide was observed in competition with the desired reduction. Thus, a compromise was chosen, consisting in the use of a MeCN/MeOH (9 : 1) mixture. Under the optimized conditions, several commercially available maleimides, as well as a plethora of purposely synthesized substituted maleimides, were reduced selectively to the corresponding succinimides in good to excellent isolated yields (Scheme 11.6).[61]

11.2.2.3 Reduction of N-Containing Functions

Much more extensive is the literature concerning the photocatalytic reduction of nitroaromatics,[64–77] often to the corresponding anilines. Several strategies have been developed so far, with variable results – results that also vary according to the actual experimental conditions. Different parameters can affect the final results, including the type of photocatalyst used,

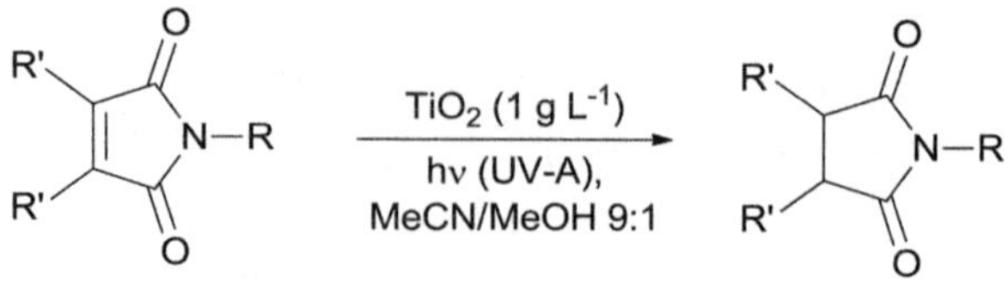

	Yield:
R' = R = H	91%
R' = H; R = Me	90%
R' = H; R = Ph	94%
R' = Me; R = 4-MeO-Ph	58%

Scheme 11.6 Photocatalytic reduction of the C=C bond in (substituted) maleimides.

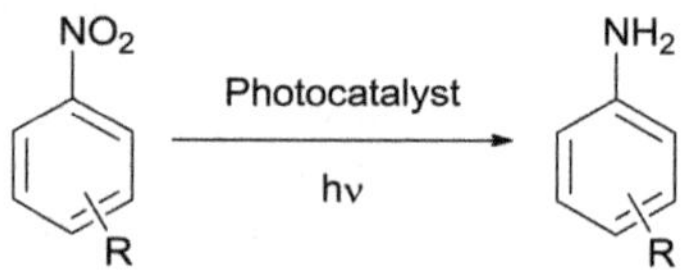

R =	*Photocatalyst*
H	TiO_2, CdS, $PbBiO_2Br$
4-NH_2	TiO_2, N-TiO_2, $SrBi_2Nb_2O_9$, $PbBi_2Nb_2O_9$
2 or 3-NH_2	TiO_2, N-TiO_2
4-OH	N-TiO_2, Arg-TiO_2, CdS
2 or 3-OH	N-TiO_2
2-OMe	CdS
3 or 4-CH_3	N-TiO_2, CdS
2-CH_3	N-TiO_2
4-Cl	TiO_2, N-TiO_2
3-Cl	CdS
2-Cl	CdS+TiO_2
4-Br	TiO_2
3-CH=CH_2	TiO_2
4-CN	CdS, $PbBiO_2Br$
4-CHO	$PbBiO_2Br$
3-$COCH_3$	CdS
3-SO_3H	TiO_2
2 or 4-NO_2	CdS

N-TiO_2 N-doped TiO_2
Arg-TiO_2 Arginine modified TiO_2

Scheme 11.7 Photocatalytic reduction of (substituted) nitrobenzenes.

the wavelength of irradiation chosen and the additives. Scheme 11.7 gathers some of the available examples, where the different substrates and the photocatalysts used are reported, to give a flavor of the scope of the method.

In the following, the main reduction processes are listed according to the photocatalyst used, starting from titania. TiO_2 is largely adopted for this purpose, either the commercially available P25-type or laboratory-made samples, usually prepared *via* hydrolysis of a titanium tetraalkoxy derivative followed by thermal decomposition and calcination.[64]

The photocatalytic reduction of nitrobenzene in the presence of laboratory-made titanium dioxide powder (80% anatase and 20% rutile mixture, with a mean grain size of 30 nm and a surface area of 50 $m^2 g^{-1}$) resulted in complete conversion of the substrate after irradiation for 6 h, with a yield of aniline of around 90% (determined by spectrophotometric analysis). Fine tuning of the reaction conditions highlighted that the best solvent was methanol, which also works as the hole scavenger, and a catalyst amount of 4.0 g L^{-1} was the optimal choice, along with an initial pH value of 4.0.[64]

In a similar approach, the photocatalytic conversion of differently substituted nitrobenzenes (*e.g.* containing a vinyl group, or a chlorine or bromine atom) in 2-propanol suspensions of TiO_2 under various conditions was examined. Anilines and acetone were simultaneously produced almost stoichiometrically and in a highly chemoselective fashion.[65]

However, since the efficiency of the process depends primarily on the efficient and irreversible hole/electron transfer, a purposely added sacrificial hole-scavenger is often required to obtain an efficient reduction process. As reported above, an alcoholic solvent can help here, but affords the corresponding oxidized (mainly carbonyls) derivatives as by-products.[65] A greener alternative would be to substitute the alcoholic solvent with water, coupled with the use of a traceless hole scavenger. Several alternatives have been proposed in this role, such as formic acid and oxalic acid, where the final by-product is gaseous carbon dioxide. Thus, the reduction of *meta*-nitrobenzenesulfonic acid in the presence of formic acid under deaerated conditions afforded *meta*-aminobenzenesulfonic acid almost quantitatively (as from HPLC analysis).[66] Apart from the role as hole scavenger, formic acid was also claimed to increase the selectivity and avoid undesired by-processes (*e.g.* re-oxidation). Thus, HCOOH had the ability to protonate the reduced amino group (to an ammonium group, $-NH_3^+$), in a way that reduced the probability of the product interacting with the (positively charged) TiO_2 surface in acidic suspensions. Moreover, the system was found to also work properly in the presence of oxygen.[66]

Another determining parameter for photocatalytic reductions is related to the surface properties of the photocatalyst. Indeed, the charge, if any, of titanium dioxide heavily affects the adsorption rate of the substrates considered, thereby influencing the associated reactivity under photocatalyzed conditions. As an example, the photocatalytic reduction of 4-nitrophenol was performed exploiting arginine-modified TiO_2 (Arg–TiO_2) nanoparticles.[67] These were prepared by simply adding an arginine solution to an aqueous TiO_2 suspension. The system afforded 4-aminophenol as the only identifiable product, as confirmed by chromatographic analysis. It was postulated that the terminal amino groups of the arginine monolayer created a positive TiO_2 surface charge over a wide range of pH values. Indeed, the reduction rate of 4-nitrophenol at pH 9 (where the Ar–OH hydroxyl group is deprotonated) proceeded very rapidly due to an improved adsorption. Furthermore, the presence of the arginine monolayer was claimed to increase the selectivity of the process.[67]

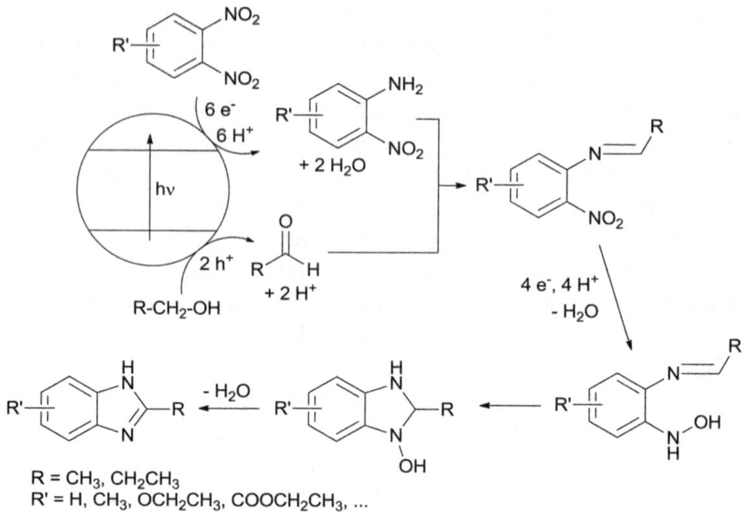

Scheme 11.8 Photocatalytic synthesis of 2-alkylbenzimidazoles from *ortho*-dinitrobenzene in alcohols.

In one instance, the reduction of *ortho*-dinitrobenzene was carried out in ethanol in the presence of TiO_2 and led to 2-methylbenzimidazole in a high isolated yield (96%).[68] This heterocyclic product was reasonably formed by reduction of one of the nitro groups to give *ortho*-nitroaniline and concomitant oxidation of the solvent to acetaldehyde (Scheme 11.8). Condensation of the amino group and the carbonyl moiety followed, then further reduction and dehydration afforded the final product, probably with the involvement of an hydroxylamine intermediate.[68] The synthetic scope of the method was investigated by introducing different substituents on the aromatic ring (*e.g.* methyl, ethoxy and ethoxycarbonyl), as well as by varying the alcoholic solvent used, always giving the expected benzimidazole in good to excellent (>70%) isolated yields (Scheme 11.8).[68]

In a slight variation of nitro group reduction, TiO_2 photocatalysis was also exploited to reduce the N=N double bond in azobenzene. Whereas under direct irradiation only photo-isomerization occurred, the TiO_2 photocatalytic reduction of azobenzene cleanly afforded hydrazobenzene.[63] Moreover, when nanometer-sized platinum particles were deposited on TiO_2, the reaction course changed dramatically, since a rate increase was observed and aniline was finally produced *via* N=N bond cleavage.[63]

One of the most important goals in photocatalysis involves the adoption of photocatalysts able to promote the desired reaction under visible light irradiation. Indeed, the reduction of nitro groups was also carried out using modified titania photocatalysts, also including N-doped TiO_2.[69,70,79]

In one instance, the N-doped TiO_2 photocatalyst was prepared adopting a sol–gel technique in the presence of urea as the nitrogen source.[69] The presence of nitrogen was demonstrated to shift the absorption properties of the material towards the visible region and N-doped TiO_2, contrary to pure TiO_2, was indeed active also under UV-filtered solar light irradiation. The role of several additives was tested. In fact, potassium iodide, working here in the role of hole scavenger, was found to increase the reaction rate and to afford excellent yields. Under optimized conditions, reduction occurred quickly (tens of minutes), with yields (as from HPLC analysis) consistently >90%, and offered a wide functional group tolerance, since nitroaromatics containing amino, hydroxy, carbonyl and carboxylic groups, as well as halogen atoms, were cleanly and selectively reduced.[69]

A different reaction course was observed in the selective reduction of aromatic nitro compounds to the corresponding symmetrically substituted azo derivatives (Scheme 11.9), again using N-doped TiO_2 nanoparticles as the photocatalyst, where urea was used during the sol–gel preparation of the material as N-source.[70] The presence of a catalytic amount of acid was mandatory for the reaction. Formic acid was found to give the best performance, but different acids, including nitric, hydrochloric and acetic acids, gave good results as well.[70]

Various azo compounds containing additional reducible substituents (*e.g.* halogen atoms, carboxylic and phenol functions) have been synthesized in a single step using the N-doped TiO_2 catalyst. The conversion occurred reasonably fast, was clean and afforded the desired products in high HPLC yield at room temperature.[70] Interestingly, the photocatalyst could be reused for at least four times with only a negligible loss of performance, with yields consistently >90%.[70]

Another widely exploited class of photocatalysts is that of CdX (with X = S, Se) derivatives, due to their capability to work under visible light irradiation.[71,72,80] When adopting blue LED irradiation (3 W), laboratory-made CdS nanoparticles were shown to perform better than commercial samples in the reduction of nitrobenzene.[71] The reaction was extended to substituted derivatives and found to be highly sensitive to the electronic properties of the aromatic ring. Nitro compounds containing electron-withdrawing groups

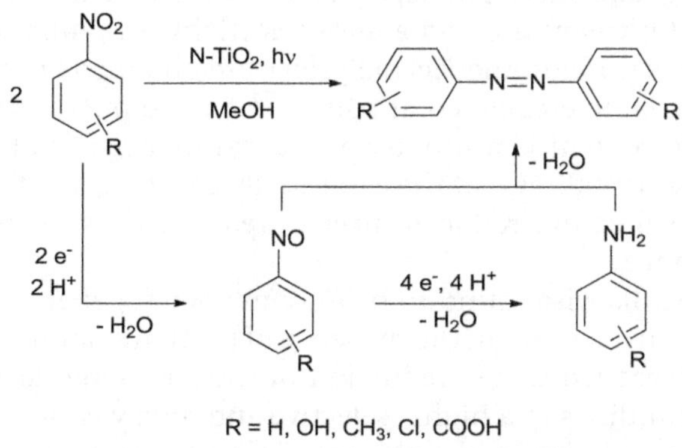

Scheme 11.9 Photocatalytic conversion of nitrobenzenes into symmetric azobenzene derivatives in the presence of N-doped TiO_2 (N–TiO_2).

(CN, COR, *etc.*) gave higher yields than those with electron-donating moieties (OMe, Me, *etc.*).[71] However, this limitation could be partially alleviated by addition of ammonium formate, in the role of hole scavenger. A further bonus of this approach was that the photocatalyst showed excellent reusability – after six runs there was no significant performance loss.[71] The particle size was found to play a key role in this kind of processes. For example, the photoreduction of 4-nitrophenol to 4-aminophenol under blue light irradiation in the presence of sodium sulfite proceeded with a high efficiency when small CdS particles (15 nm) were adopted.[72]

In a related work, CdS and CdSe nanoparticles were adopted for the highly chemoselective reduction of aromatic azides to the corresponding anilines. Noteworthy, in several cases the reaction proceeded with a quantum yield close to 0.5 (the theoretical maximum value for a two-electron process). The wide scope of the reaction was confirmed by the good performance of substrates containing either electron-withdrawing or electron-donating groups, independent of the position of the azido group. However, *meta*-substituted azides gave slightly lower yields than isomeric derivatives.[80]

Recently, a new class of visible light heterogeneous photocatalysts has been introduced. One common element of the examples reported in the literature is the presence of bismuth, often coupled with different metal elements.[73–75] A typical case is microcrystalline $SrBi_2Nb_2O_9$. This has been found to efficiently reduce 4-nitroaniline to *p*-phenylenediamine under an inert atmosphere (N_2) and UV-light irradiation (254 nm). Indeed, its photocatalytic activity was higher than that of the commercial titania P25 catalyst.[73] Further experiments indicated that ammonium oxalate (as a hole scavenger) was required for the photocatalytic process to occur. The efficient photocatalytic reduction of 4-nitroaniline over this photocatalyst has been related to the relatively high conduction band of this material.[73] Similarly, nanocrystalline $PbBi_2Nb_2O_9$ was prepared by a sol–gel method and applied in the same process as above.[74] Indeed, the resulting photocatalyst promoted the reduction of 4-nitroaniline to *p*-phenylenediamine in aqueous solution under visible light irradiation and inert atmosphere. The best results (with a complete conversion of 4-nitroaniline and a 99% selectivity for *p*-phenylenediamine, according to spectrophotometric analysis) were obtained after 150 min irradiation using a sample calcined at 650 °C.[74] This temperature played a key role, since for lower temperatures a worse performance was observed, probably due to a low crystallinity of the catalyst. By contrast, a higher calcination temperature resulted in a reduced surface area, and thus again in a poor catalytic performance.[74]

$PbBiO_2Br$ was likewise found to be a robust heterogeneous photocatalyst under blue light irradiation in the presence of triethanolamine as a hole scavenger. The process led to the reduction of nitrobenzene derivatives to the corresponding anilines in a highly selective and high yielding fashion.[75]

Some reports also deal with the use of mixed systems where two different catalysts have been coupled.[76] As an example, the cooperative action in CdS/P25 hybrid catalysts was studied under visible light irradiation. Two different samples were compared, namely, (i) P25 TiO_2 coupled with CdS quantum dots (acting as sensitizers) and (ii) a CdS–P25 hybrid prepared by a direct deposition–precipitation method in the presence of formic acid. Both materials were tested in the reduction of *ortho*-chloronitrobenzene; the former system showed an enhanced activity.[76] It was proposed that the coupled action of the two catalysts allowed an efficient spatial separation of the photogenerated charge carriers, thus preventing their recombination. Actually, the better performance offered by quantum dots was explained by the higher number of photocharges produced due to the particle size effect.[76]

The modification of nitroaromatic reduction under photocatalytic conditions in the presence of an organic additive was also considered and several photocatalytic systems were compared, including powder dispersions of TiO_2, WO_3 or CdS.[77] In the presence of cyclohexene, direct irradiation afforded a complex mixture, including azoxybenzene, azobenzene, aniline and a coupling product, namely, *N*-(2-cyclohexenyl)aniline.[77] On the other hand, the photocatalytic contribution of these semiconductors played an important role in the relative product distribution, which was observed to depend on both competitive adsorption–desorption equilibria of the involved reagents and intermediates on the solid surface and on the different reducing power of the photoexcited semiconductors.[77]

Some unconventional approaches also deserve to be reported here. In a non-synthetic application, a photosensitive plate based on sintered TiO_2 nanoparticles was used to carry out the reduction of disulfide bridges in proteins using glucose as a hole scavenger during laser desorption–ionization.[81]

Several works in the literature deal with the mechanistic issue, highlighting the full details of the reductive process. Among others, photoreduction of 4-nitroaniline to *p*-phenylenediamine over TiO_2, ZnO, $PbBi_2Nb_2O_9$ and CdS photocatalysts in the presence of methanol was investigated.[82] Furthermore, a related work analyzed the behavior of illuminated aqueous slurries of titanium dioxide (Degussa P25) in the presence of a sacrificial electron donor, 2-propanol, for the reduction of a suite of monosubstituted nitrobenzenes.[83]

11.2.3 Alkylations

Photocatalysis can be used for the mild, unusual activation of functional groups to be exploited in the formation of valuable C–C bonds. Several reviews appeared recently on this topic, but most of them deals with reactions carried out under homogeneous,[84] as opposed to heterogeneous, conditions.[85] Here we focus on (possibly recent) selected examples with the aim of giving an idea of the versatility of this approach.

The activation step under heterogeneous photocatalytic conditions involves either an oxidative (Scheme 11.10, path a) or a reductive (path b) process, and in some instances both paths are involved in the formation of the final product, with no sacrificial agent being required. Upon removal of an electron from the substrate, a radical cation is formed that can in turn

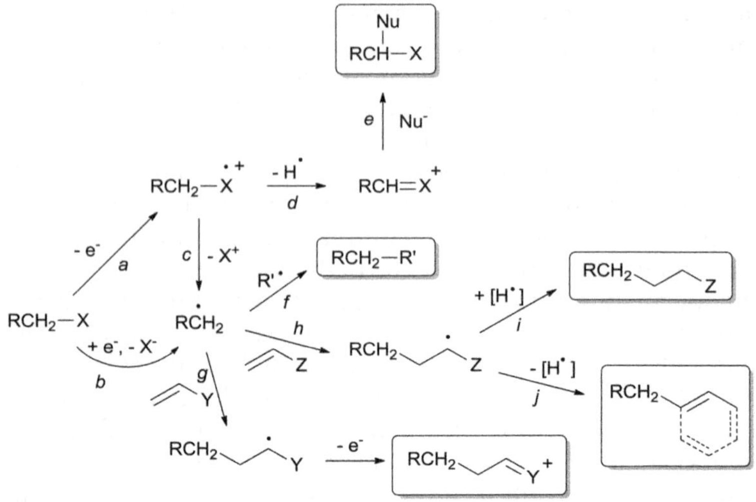

Scheme 11.10 Mechanistic paths involved in photocatalytic alkylation reactions.

undergo splitting of a suitable electrofugal group (X^+) to generate a carbon-centered radical ($RCH_2^•$) *via* path c. A second possibility is that a hydrogen atom is abstracted from the radical cation by a further intermediate involved in the reaction (usually an aggressive oxygen-based species; path d) to give a positively charged species. This cation can be conveniently trapped by a C-based nucleophile (Nu^-), leading to the formation of a new C–Nu bond (path e). The same $RCH_2^•$ radical mentioned above can also be formed by a complementary route, that is *via* mono-electronic reduction coupled to the loss of a nucleofugal group (X^-; path b). In turn, this radical can be exploited in a plethora of synthetic processes, such as the coupling with a second radical likewise formed during the process (path f). An alternative route involves the addition of the radical onto a double bond, either electron-rich (path g, followed by oxidation) or electron-poor (path h), depending on the electronic properties of the radical substituent (R). In the last case, the radical adduct formed may evolve through two different routes, that is either the formal addition of a hydrogen atom (path i) to give a saturated derivative or the loss of a hydrogen atom (path j), which is particularly favored when the original double bond is part of an aromatic system, thus allowing re-aromatization. In the following, examples will be presented according to the class of reaction involved, as previously detailed.

11.2.3.1 Conjugate Addition Reactions

The irradiation of TiO_2 in an acetonitrile suspension under inert atmosphere was carried out in the presence of several electron donors. Benzylsilanes, phenylacetic acids and alkyl aromatics were tested for performing benzylation reactions *via* mono-electronic oxidation and loss of an electrofugal group from the benzylic position at the radical-cation stage. Indeed, it was demonstrated that starting from silanes a clean benzylation of maleic acid, maleic anhydride and related nitriles was achieved, independently of the substitution pattern (the parent compound and the 4-OMe derivative were tested).[86] By contrast, phenylacetic acids were suitable reagents only provided that an electron-donating group (*e.g.* a methoxy group in the *para*- position) was present on the ring. This was because the oxidation potential of the parent derivative was too high. Finally, alkylaromatics failed to give the desired products, most probably due to a slow loss of the electrofugal group (a proton), which was completely inhibited by back electron transfer.[86] The same reaction was also carried out under solar light irradiation exploiting a TiO_2 suspension in MeCN containing maleic anhydride and (4-methoxybenzyl)trimethylsilane. Under these conditions, the expected benzylated succinic acid derivative was obtained on a gram scale (Scheme 11.11a).[87] Furthermore, adamantane was also tested as a substrate and indeed the alkylation of isopropylidenemalononitrile was obtained, though in a low yield, through a direct C–H activation process.[88]

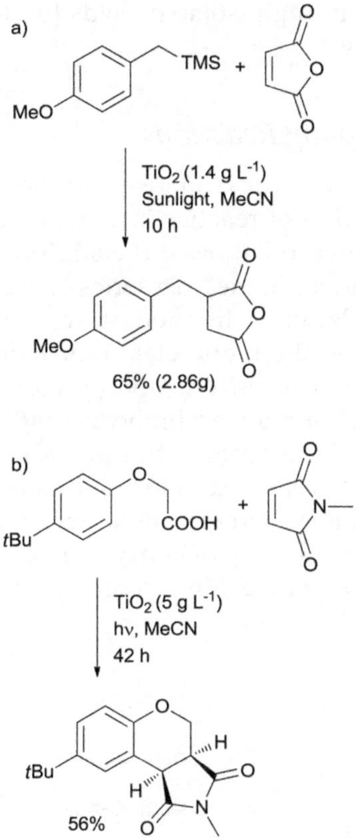

Scheme 11.11 Photocatalytic conjugate addition reactions involving: a) benzylsilanes and b) carboxylic acids.

More recently, this kind of reaction has experienced a renaissance and has been exploited in the formation of C–C bonds adopting carboxylic acids as the substrates in the presence of TiO_2. Under dry, anaerobic conditions, high isolated yields of dimers were obtained. These resulted from coupling of the radicals arising *via* decarboxylation of the reagent.[89] Upon addition of an electron-deficient alkene to the reaction mixture, the reaction course diverged completely and efficient alkylations were achieved. In particular, the reaction of phenoxyacetic acid with maleic anhydride or maleimides afforded chromene derivatives (through a cascade process) in addition to the expected addition products (Scheme 11.11b). The efficiency of alkylation varied appreciably with the substituent pattern on the carboxylic acid. The proposed mechanism, involving oxidation of the starting acid and decarboxylation, was supported by EPR spectroscopic evidence, as well as by deuterium labeling experiments.[89]

A different class of substrates is tertiary amines. These were studied as precursors of α-amino radicals and tested in the reaction with electron-deficient alkenes. TiO_2 and ZnS were used to promote the reaction, which involved a radical chain process. A menthol moiety was tethered to the alkene with the aim of directing the addition of the radical and, accordingly, the desired products were obtained in high isolated yields (up to 98%) and with a high facial diastereoselectivity.[90]

11.2.3.2 Cross-Coupling Reactions

In a different approach, both the oxidative and the reductive pathways were exploited for the generation of reactive intermediates to be incorporated in the final product. This method allowed the addition of cyclic olefins (cyclopentene, cyclohexene and α-pinene) to a series of *N*-adamantylimines affording homoallyladamantylamines in modest to excellent isolated yields.[91] The process was based on the mono-electronic oxidation and reduction of the olefin and the imine, respectively. Upon proton exchange at the radical ion stage, an allyl radical and a α-aminobenzyl radical resulted, which coupled in the final stage of the process through a C–C hetero-coupling process (Scheme 11.12). In this case, cadmium sulfide grafted onto alumina was adopted as the photocatalyst under visible light irradiation.[91] In a related application, the oxidative coupling of benzyl derivatives (benzyl alcohols and benzyl amines) was carried out, again exploiting CdS as a heterogeneous visible light photocatalyst.[92]

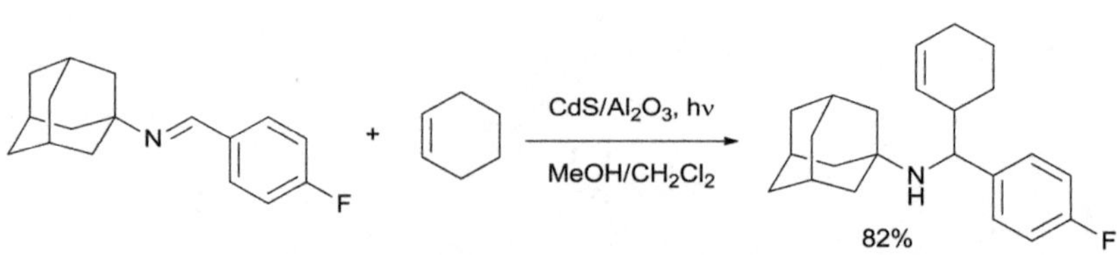

Scheme 11.12 Photocatalytic cross-coupling reaction.

11.2.3.3 Aromatic Substitution Reactions

Heterogeneous photocatalysis was also demonstrated to allow the functionalization of various heterocyclic bases in the presence of suitable radical precursors.[93–95] These included amides,[93] various oxygenated derivatives (ethers and acetals)[94] and aldehydes.[95] The reactions were carried out under sunlight irradiation in the presence of polycrystalline TiO_2 (anatase). An oxidant (either hydrogen peroxide or air) and sulfuric acid were likewise required as additives. This approach resulted in a smooth and environmentally friendly method for the photochemical functionalization of heterocyclic bases under mild conditions and with high conversions.[93–95] As an example, the reaction of quinaldine in the presence of formamide resulted in a clean and high yielding carbamoylation process, in which the formyl C–H group was selectively activated (Scheme 11.13a).[93]

A series of ethers could be likewise used to generate α-oxy radicals, including cyclic (tetrahydrofuran, tetrahydropyran, *etc.*) and acyclic (diethyl ether) derivatives.[94] Of particular interest is the reaction with 1,3,5-trioxane, since the products could be subjected to hydrolysis, affording the corresponding heterocyclic aldehydes, thus resulting in an indirect formylation procedure (Scheme 11.13b).[94] Aliphatic aldehydes were likewise adopted as radical precursors.[95] Various acyl radicals were obtained upon irradiation of TiO_2

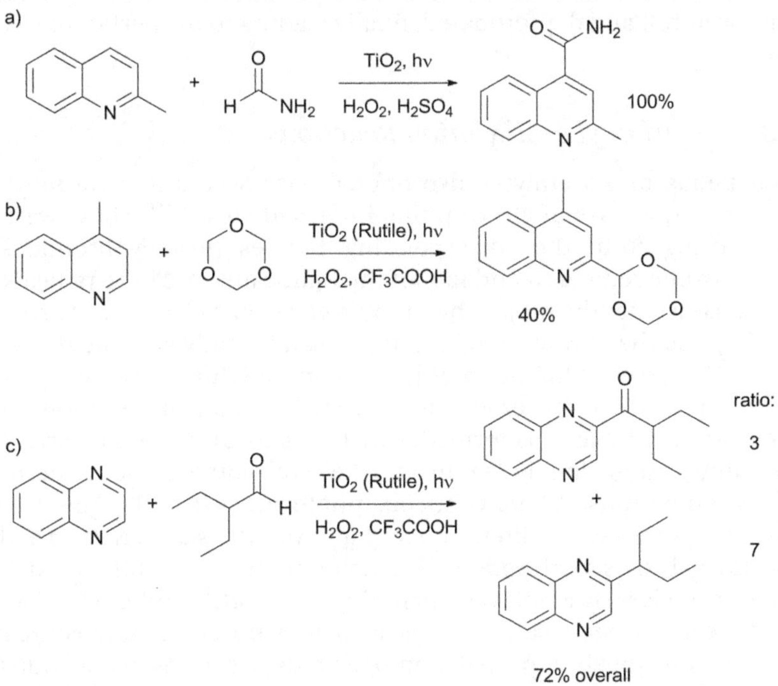

Scheme 11.13 Photocatalytic aromatic substitution reactions involving: a) amides, b) acetals and c) aldehydes.

suspensions under sunlight. The initially formed radicals could undergo decarbonylation to yield the corresponding alkyl radicals, depending on the stability of the latter species. Indeed, no decarbonylation occurred when starting from acetaldehyde, but with higher primary homologues significant amounts of alkylated derivatives were obtained, indicating that acyl radicals decarbonylated before trapping. Acylated products accounted only for a small fraction of the products (Scheme 11.13c) upon shifting to secondary aldehydes and were detected only in traces in the case of tertiary aldehydes.[95]

Other aryl–carbon bond forming reactions have been recently reported in the literature. Indeed, it was shown that the use of titanium dioxide hybridized with palladium enabled the cyanomethylation of benzene under visible light irradiation (405 nm), resulting in the clean formation of phenylacetonitrile derivatives.[96] Palladium played multiple roles here, since it acted as electron scavenger and enabled the evolution of hydrogen, formed as a by-product. Moreover, it has been proposed that it was also involved in the key C–C bond forming step.[96]

In related work, a series of perfluoroalkyl iodides were exploited as perfluoroalkyl radical precursors under irradiation conditions in the presence of TiO_2.[97] The desired radicals were formed upon mono-electronic reduction of the iodides and loss of I^- at the radical anion level. The thus formed radicals added to an aromatic derivative (benzene and some analogues, *viz.* naphthalene and benzofuran) to give a radical adduct that was oxidized by the photocatalyst and deprotonated, finally leading to the perfluoroalkylated product.[97]

11.2.3.4 Carbonyl α-Alkylation Reactions

Heterogeneous photocatalysis also offered easy access to α-carbonyl (carboxyl) radicals, a class of electrophilic intermediates.[98–100] These were generated starting from the corresponding halides (usually bromides) *via* mono-electronic reduction and loss of the halide anion. These radicals have then been conveniently trapped by suitable electron-rich olefins. Particularly interesting was the alkylation of enamines catalytically generated *in situ* by reaction of a carbonyl (often an aldehyde) and a (chiral) secondary amine, which acts here in the role of organocatalyst.[98–100] After the addition step, an α-amino radical adduct was formed, which was oxidized to the corresponding iminium cation, finally leading to the α-alkylated carbonyl derivative. Several combinations of heterogeneous photocatalysts and organocatalysts have been reported in the literature and allowed the *stereoselective* α-alkylation of aldehydes. As for the adopted semiconductors, bismuth-based derivatives have been largely employed, including $PbBiO_2Br$[98] and Bi_2O_3[99] (Scheme 11.14). Indeed, the yields and stereoselectivities observed were comparable to those of previously reported homogeneous reactions using transition metal complexes or organic dyes.[98] Moreover, it was reported that this reaction could also proceed with high efficiency when the reaction mixture was simply exposed to sunlight.[99]

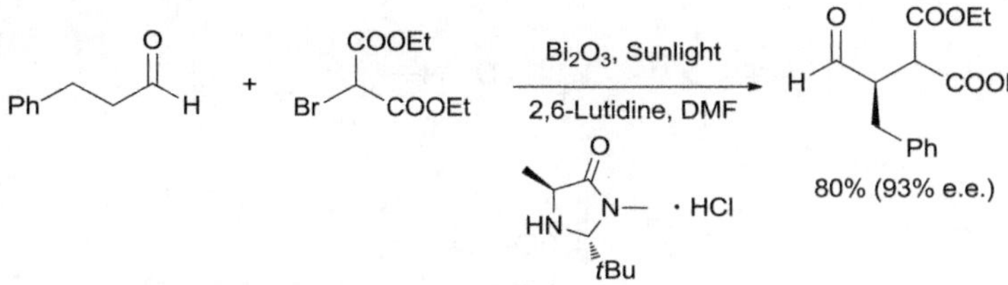

Scheme 11.14 Photocatalytic α-alkylation reaction.

In a different approach, the same reaction was carried out using chiral metal–organic frameworks (MOFs) as heterogeneous asymmetric catalysts.[100] The framework was built by incorporation of an organocatalyst (a pyrrolidin-2-yl imidazole) and a triphenylamine photoredox moiety. The resulting material acted as a cooperative catalyst able to promote the desired α-alkylation of aldehydes.[100] Indeed, it was found that the integration of both the photocatalyst and the asymmetric organocatalyst into a single MOF resulted in a higher enantioselectivity with respect to simply mixing the corresponding MOFs. The excellent performance of the photocatalyst was due to the highly ordered arrangement of the individual components within the well-defined porous, repeating structure of the MOFs.[100]

11.2.3.5 Amine α-Alkylation Reactions via Iminium Cation

Apart from radicals, heterogeneous photocatalysis can also be exploited to afford access to other kinds of intermediates, including cations. In recent years, an often-studied case was that related to the activation of *N*-aryl tetrahydroisoquinolines and related tertiary amines for the formation of iminium cations that have then been trapped in nucleophilic addition reactions.[98,101,102] The process was based on the mono-electronic oxidation of the substrate to the corresponding radical-cation. Often, the reaction was carried out in the presence of oxygen, which regenerated the reduced photocatalyst, affording the superoxide anion. This highly reactive species was then capable of abstracting a hydrogen atom from the amine radical cation to give the desired iminium ion. Trapping by a nucleophile then resulted in a cross-dehydrogenative coupling reaction. In some instances, this process has also been carried out by exploiting heterogeneous photocatalysis. Thus, different C–C (and C–P) bond forming reactions were successful, including the synthesis of nitro-amines, amino-carbonyls (carboxyls), amino-nitriles and amino-phosphonates in high chemical yield by reaction with nitronate anions, enamines (possibly organocatalytically generated from a carbonyl and a secondary amine) or enols, cyanide and phosphite derivatives, respectively.[98,101] In a related application, this strategy was exploited to gain access to various α-amino amides. The process was based on a titanium dioxide photocatalyzed, Ugi-type, three-component reaction, involving the amine, an isonitrile and water (Scheme 11.15).[102]

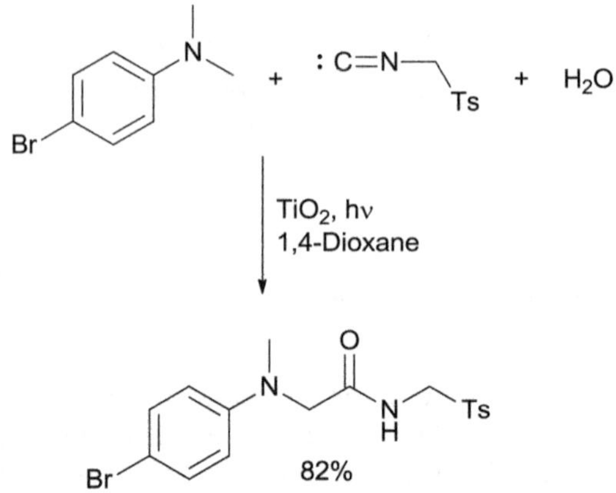

Scheme 11.15 Photocatalytic three-component (Ugi-type) reaction.

Several semiconductors were exploited for performing amine α-alkylation reactions, including metal oxides, such as TiO_2[101a,102] and ZnO,[101a] CdS,[98] mesoporous graphitic carbon nitride (mpg-C_3N_4)[101b] and iron nanoparticles supported on chemically derived graphene oxide.[101c] Furthermore, the adoption of heterogeneous conditions allowed an easy separation of the catalyst and it was demonstrated that this could be recycled and reused.[101a,c]

11.3 Influence of Catalyst Properties on Selectivity

In heterogeneous photocatalysis, several parameters affect the performance of a process since (i) the morphology, crystallinity, porosity and hydrophilicity of the solids, along with (ii) the electronic properties of both the catalyst and the reacting and formed molecules get involved; (iii) the light wavelength and intensity must be also taken into account, as the surface is modified in a different way by tuning irradiation and, as a result, different reactive sites can be activated.[103] Although many correlations have been found by researchers working in selective photocatalysis, which has become increasingly popular in the last decade, not all the obtained results can be rationalized.[104,105] Especially when more than one parameter of the above-cited ones changes, correlation of the photocatalytic activity and selectivity with the photocatalyst properties turns out to be extremely complicated.

The extent of crystallinity is a key parameter for several reasons. Semiconductors such as TiO_2 in all its polymorphic forms are known to contain a certain percentage of amorphous phase.[106] Proton magic angle spinning nuclear magnetic resonance has provided a way to study in-depth the behavior of TiO_2 precursor at the beginning of hydrolysis and the early formation of small amorphous and anatase nanoparticles. Amorphous TiO_2 is formed during crystal growth, influencing at the same time the final features of the crystals. In the case of photocatalysts synthesized at low temperature from the hydrolysis of $TiCl_4$, the formation of Ti–O–Ti bonds connecting amorphous and anatase phases is favored by the interaction between acid and basic hydroxyls

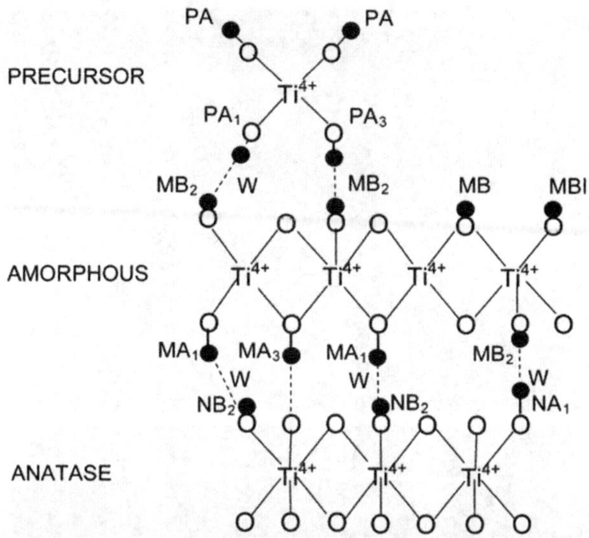

Figure 11.3 Possible interactions occurring between hydroxyls of anatase, the amorphous phase and precursor species.[106] (Reproduced with permission from J. Sanz, J. Soria, I. Sobrados, S. Yurdakal and V. Augugliaro, *J. Phys. Chem. C*, 2012, **116**, 5110.)

(Figure 11.3), with the final outcome being an increase in the agglomeration of anatase nanoparticles. Destabilization of anatase crystals, on its turn, occurs thus promoting the transformation of anatase to rutile phase, as the solution aging evolves. All these effects are thought to be due to the strong basic character of terminal hydroxyls of amorphous titania chains.[106]

A further study demonstrated how the performance of photocatalysts in the oxidation of 4-methoxybenzyl alcohol to the corresponding aldehyde is strongly dependent on the amorphous phase present, which was studied also by means of HR-TEM, revealing the presence of amorphous phase surrounding anatase (or rutile) (Figure 11.4).[107]

A significantly lower selectivity toward *para*-anisaldehyde was found for a low defective TiO_2 (BDH) sample used with respect to highly amorphous samples (6% *vs.* 36–41%). One can thus conclude that amorphous titania at the anatase surface provides the active sites where the alcohol partial oxidation takes place. At these sites 4-methoxybenzyl alcohol adsorption occurs through the alcoholic group and the strongly basic character of the terminal hydroxyls of amorphous titania, bounded to low-coordinated Ti^{4+} cations, which act as electron trapping sites in the grain boundaries. Water is then formed by reaction of alcoholic groups with those hydroxyls under irradiation; subsequently, 4-methoxybenzyl alcohol molecules are activated through abstraction of H atoms from their alcoholic group, thus facilitating the bonding of these carbon atoms with O^{2-} anions of deprotonated amorphous titania bridging hydroxyls. Since these anions are the favored sites for photogenerated hole trapping, *p*-anisaldehyde can finally form.

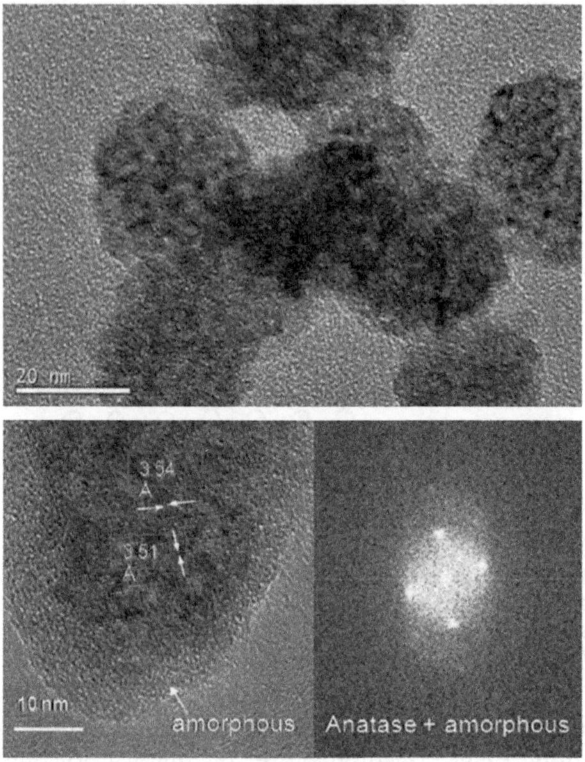

Figure 11.4 HRTEM micrographs and fast Fourier transform of a laboratory prepared TiO_2 sample containing anatase and amorphous phases.[107] (Reproduced with permission from S. Yurdakal, V. Augugliaro, J. Sanz, J. Soria, I. Sobrados and M. J. Torralvo, *J. Catal.*, 2014, **309**, 97.)

The different reactivity of TiO_2 crystal facets in several different chemical processes has been recently highlighted in a very informative report.[108] The shape and crystal forms of TiO_2 can be tuned by adjusting the environment in which they are formed and grown. The energy of the surfaces plays an important role in determining the final morphology of the crystals and the features of an adsorbate present during their preparation can deeply affect the final shape and the extent of higher or lesser reactive facets. Figure 11.5 shows that acidic conditions lead to termination, by hydrogen, that results in little change in the shape of both anatase and rutile relative to vacuum. In contrast, when surfaces are either water terminated or hydrogen-poor the resulting polymorphs are significantly elongated. Consequently, new facets (010) appear at the center of anatase particles. Predictions such as these are most important to experimentally obtain a fine tuning of morphology. Thus, for instance, an anatase phase with an excess of the highest energy (001) facets will result in a stronger reactivity of the catalyst.

These principles can be applied to several selective reactions, although reported correlations are rare so far, and more effort should be devoted to

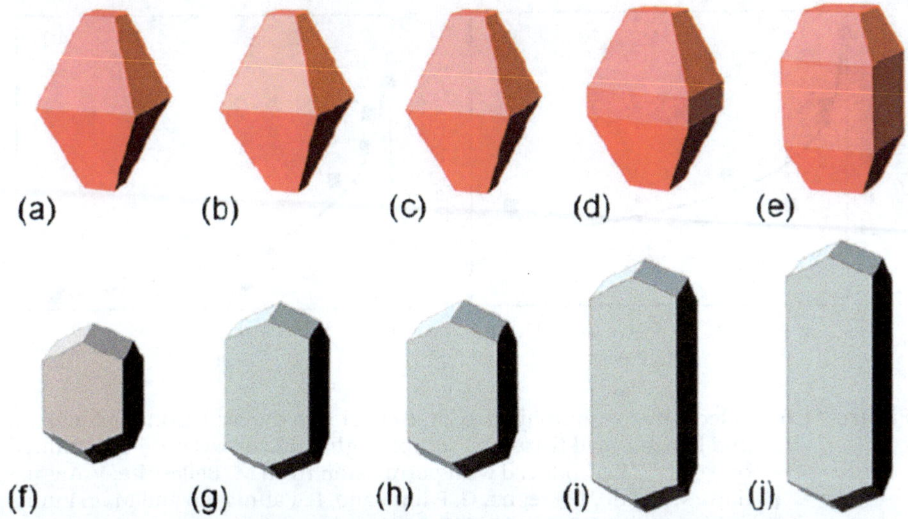

Figure 11.5 Morphology predicted for anatase (top) with (a) hydrogenated surfaces, (b) with hydrogen-rich surface adsorbates, (c) hydrated surfaces, (d) hydrogen-poor adsorbates and (e) oxygenated surfaces, and rutile (bottom) with (f) hydrogenated surfaces, (g) with hydrogen-rich surface adsorbates, (h) hydrated surfaces, (i) hydrogen-poor adsorbates and (j) oxygenated surfaces.[108] (Reproduced with permission from G. Liu, J. C. Yu, G. Q. Lu and H.-M. Cheng, *Chem. Commun.*, 2011, **47**, 6763.)

this interesting field: for example, when toluene was oxidized to benzaldehyde by using anatase TiO_2 with particle size between 23 and 94 nm and with 50–65% (001) facets selectivities comparable to those given by Degussa P25 but with triple the activity were found.[109]

In the hydroxylation of aromatic species carried out in water in the presence of different TiO_2 catalysts, the selectivity of the process was found to be strongly influenced by some peculiar physicochemical properties of the solids.[110] Phenol and benzoic acid were chosen as representatives of aromatics bearing an electron-donor (EDG) and an electron-withdrawing group (EWG), respectively. The photocatalytic results indicated that the products of the oxidation of phenol were the *ortho*- and *para*-mono-hydroxy derivatives, according to the general behavior previously found,[111] while the hydroxylation of benzoic acid took place at all ring positions. Interestingly, the *ortho*-isomer was absent or in low concentration in the presence of laboratory-prepared catalysts by oxidizing both substrates, because of the strong interaction of *ortho*-mono-hydroxy derivative with the catalyst surface. Commercial catalysts (Merck anatase) gave the highest selectivity values for both substrates, but the most relevant result is the sharp dependence of the process selectivity on the catalyst crystallinity and the total amount of OH groups present on the catalyst surface (Figure 11.6). A crystalline and poorly hydroxylated sample, such as Merck TiO_2, and HP0.5, the most hydroxylated anatase

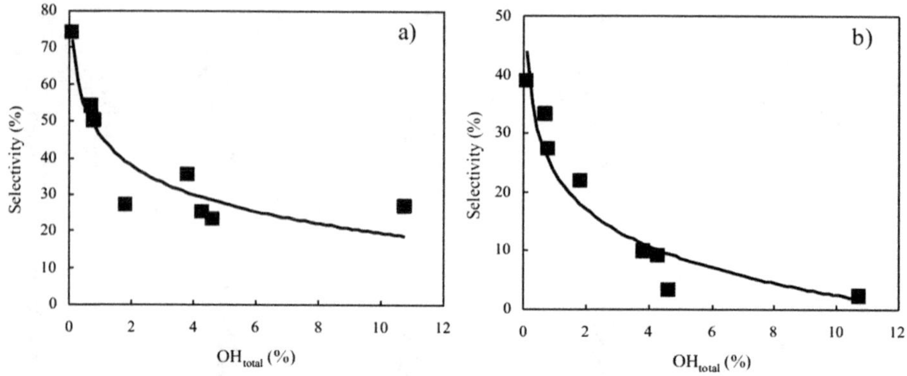

Figure 11.6 Selectivity to mono-hydroxylated species in the oxidation of phenol (a) and benzoic acid (b) *vs.* the samples' global hydroxylation determined by TGA.[110] (Reproduced with permission from M. Bellardita, V. Augugliaro, V. Loddo, B. Megna, G. Palmisano, L. Palmisano and M. A. Puma, *Appl. Catal., A*, 2012, **441–442**, 79.)

samples among those studied, were the most and the least selective catalysts, respectively.

A kinetic model taking into account both the hydroxylation and the parallel route giving rise to mineralization of the substrates showed that equilibrium adsorption constants and kinetic constants for the mineralization route were similar for both substrates, thus indicating that the nature of the substituent group does not influence the total oxidation of the species. Conversely, the partial oxidation constants were found to be very different for catalysts with different physicochemical properties.

To show how radiation at different wavelengths can activate different reactive sites, thus promoting or reducing the selectivity of a process, we can consider the example of a recently reported set of N-doped TiO_2 samples.[112] In this study, N-doped TiO_2 photocatalysts were used to selectively convert 4-methoxybenzyl alcohol into the corresponding aldehyde, under different radiation sources. The results indicated that N-doping along with solar light (instead of UV-irradiation) enhanced the selectivity of the partial oxidation process by *ca.* 20%. Poorly crystalline N-doped TiO_2 prepared at low temperature by using $TiCl_4$ as the precursor, by adding ammonium chloride as the nitrogen source, have been found to be the most selective samples, reaching a 90% selectivity under simulated solar light. When the catalysts underwent thermal treatment an improvement in crystallinity and activity was achieved, but selectivity drastically decreased, analogously to what was observed when UV light instead of solar light was used. The different preparation methods influence strongly the ability of a catalyst to absorb light of certain wavelengths, as highlighted in this study, eventually resulting in a very altered reactivity and selectivity in the studied processes. The wavelength of the light impinging onto the catalyst surface determines the activation of reaction sites, depending on the energy required relative to that reaching the surface.

Often, more than one kind of site is involved in a given reaction, each site giving different selectivities.[110,111] The energy of the light absorbed as well as the surrounding environment will then determine which sites will be predominantly involved in the given reaction. Of course, a photocatalyst with a smaller band gap energy would certainly be able to absorb radiation with a lower energy, and possibly to cause a lower oxidation extent, thus favoring the yield of partially oxidized products.

11.4 Green Organic Solvents in Photocatalysis

Soon after the early studies on heterogeneous photocatalysis carried out throughout the 1970s, investigations on selective conversions of organic molecules dissolved in various organic solvents began to appear. In one of the first comprehensive reports on such reactions,[113] the author foresaw that photocatalytic redox reactions could involve a plethora of substrates, given the compatibility between inorganic semiconductor band gaps and oxidation potentials of organic molecules. At that time, though, examples involving photocatalytic reductions were not common, so that only the first studied oxidations could be reported. Most of the reactions were carried out in acetonitrile, as in the case of the decarboxylation of acetic acid in the presence of oxygen, TiO_2 irradiated by near UV light, and tetrabutylammonium acetate. The reaction products were ethane and CO_2; the formation of methane was favored when water instead of acetonitrile was employed as the solvent.[114] The seminal papers by Bard's group made clear already at that time that the role of solvent in photocatalytic reactions is very important. Yet, the attention given to green chemistry was not as relevant as nowadays. In addition to acetonitrile, polar solvents, usually containing undesirable halogens, were also typically employed, as in the studied cases of aromatic olefin oxidative cleavage.[115]

Selective photo-oxidation of alcohols to the corresponding aldehydes and ketones was performed not only in the gas phase but also in liquid phase using acetonitrile as solvent or without solvent.[116]

More recently, the use of ionic solvents in photocatalytic oxidations increased the conversion of different aromatic alcohols into carbonyl compounds in comparison with the use of acetonitrile.[117]

However, the oxidation of aromatic alcohols can be efficiently performed in water by using poorly crystallized TiO_2 samples, as shown in 2007 and later on.[118] Certainly, water is the green solvent per excellence, being non-toxic, abundant, easily disposable after use and very cheap. Nevertheless, most organic compounds can hardly be solubilized in water and only in some cases was the employment of a surfactant successful. In recent years several scientists have tested the possibility of carrying out organic reactions in water, even in the occurrence of insoluble compounds. For many reactions, the results were surprising, since the selectivity of many syntheses was greatly changed on switching from traditional organic solvents to water. Pericyclic reactions, reactions of carbanion equivalent, reactions of carbocation equivalent, reactions of radicals and carbenes, transition-metal catalysis and oxidations–reductions, are some examples of reactions viable in aqueous solvent, as extensively reported.[119] Furthermore, aqueous organic reactions find extensive applications in biochemistry, such as in the synthesis of biological compounds from carbohydrates and in the chemical modification of biomolecules.

In heterogeneous photocatalysis, however, working in aqueous suspensions where the substrate supposed to react does not solubilize is unfeasible, because the (photo)adsorption of the reagent would be strongly prevented by its absence in the reactant solution. Therefore, other solutions have to be explored. On reviewing literature, one can find reactions carried out under solvent-free conditions, such as the selective photocatalytic oxidation of benzyl alcohol over modified TiO_2.[120] When organic solvents have been used, the activity and selectivity of many processes have been shown to change greatly when changing the properties of the solvents, for instance from a polar to a non-polar one. This can be due also to the modification of the flat-band potential (V_{fb}) of the semiconductor. For instance, Redmond and Fitzmaurice found that the V_{fb} is significantly more positive for water and nonaqueous protic solvents (MeOH and EtOH) than for nonaqueous aprotic solvents (MeCN, DMF and THF) due to the establishment of a proton adsorption–desorption equilibrium.[121]

One of the most important issues is the presence of by-products arising from the solvent oxidation, reduction or coupling reactions. Especially when carrying out reduction reactions, aliphatic alcohols are often the preferred sacrificial electron donor agents (Section 11.2.2), and when methanol for instance is used the formation of volatile formaldehyde – a very toxic, undesirable product – should be taken into account.[122]

Recent research activity has focused on applying green chemistry as one of the primary methods of pollution prevention, by following the first principle.[123] As far as organic syntheses are concerned, the solvent is surely one of the most critical choices since it often constitutes the most abundant end-of-process waste, although in (photo)catalysis one should not neglect the unwelcome stoichiometric oxidants/reductants, heavy metals and toxic reactants in general. When water is not a feasible route and solvent-free reactions are not possible, as for instance for solid species with high boiling points, the choice of an organic solvent is no longer an option. Representatively, the oxidation of polycyclic aromatic hydrocarbons depicts one of those situations. While it is possible to oxidize naphthalene in aqueous solution containing only 1% acetonitrile (notice, however, that a different selectivity from that found in pure organic solvents is obtained; see Scheme 11.16),[124] the oxidation of phenanthrene can be carried out only in organic solvents with minor amounts of water.

Interestingly, the use of dimethyl carbonate as a new solvent in photocatalysis has been very recently proposed for the oxidation of phenanthrene, whose solubility in water is extremely low (1.6 mg L^{-1}), in a fixed bed reactor where TiO_2 was immobilized on glass beads.[125] Dimethyl carbonate is an environmentally friendly organic species, which can be used as a safe

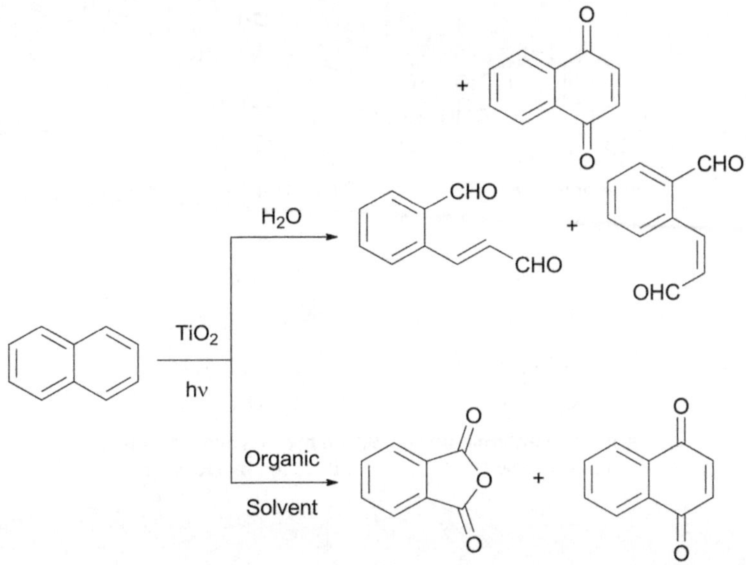

Scheme 11.16 Photocatalytic oxidation of naphthalene in aqueous and organic solvents.[124] (Reproduced with permission from F. Soana, M. Sturini, L. Cermenati and A. Albini, *J. Chem. Soc., Perkin Trans. 2*, 2000, **2**, 699.)

carboxylating or alkylating agent in place of hazardous phosgene, dimethyl sulfate and methyl halides.[126] It is biodegradable and it does not have any mutagenic or irritating effect by either contact or inhalation. Furthermore, dimethyl carbonate is widely studied also for its many potential applications, such as its possible use as a substitute for *tert*-butyl methyl ether in gasoline or diesel.[127] The decomposition of dimethyl carbonate, notably, gives rise only to CO_2 and methanol, without heavy impacts on the environment.

On the other hand, the oxidation of polycyclic aromatic hydrocarbons has been usually studied in solvents like acetonitrile,[128] mixtures of water and acetone,[129] subcritical water[130] and by employing some anionic and non-ionic surfactants.[131] However, these solutions (especially the last ones) do not ensure a high solubility of phenanthrene as is viable, for instance, in dimethyl carbonate (DMC). The latter solvent radically enhances the selectivity of supported anatase TiO_2 in the formation of high added value oxygenated products, namely 9-fluorenone and 6*H*-benzo[*c*]chromen-6-one. Performing the same reaction by using short chain alcohols (ethanol, 1- and 2-propanol) as solvents interestingly led to complex reaction mixtures. Scheme 11.17 shows the hypothesized mechanism of formation of 9-fluorenone and 6*H*-benzo[*c*]chromen-6-one. The first oxidation is carried out *via* hydroxylation of phenanthrene at positions 9 and 10 by radicals produced on the surface of TiO_2. Notably, the position of the attack is in agreement with the Dewar molecular-orbital theory.[132] Depending on the used solvent, the following pathways change significantly: a high selectivity is obtained by using

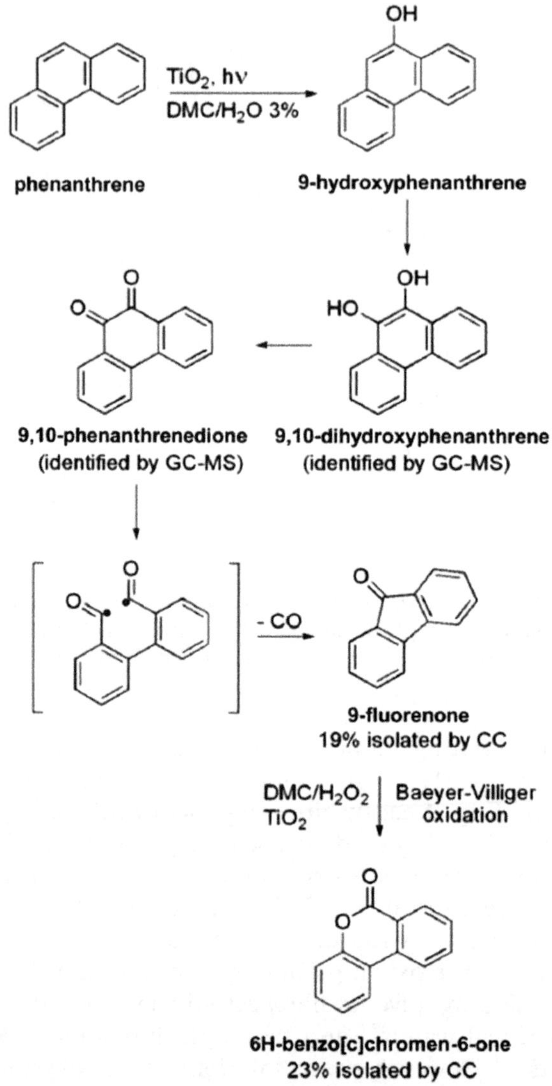

Scheme 11.17 Photocatalytic oxidation of phenanthrene to 9-fluorenone and 6*H*-benzo[*c*]chromen-6-one in dimethyl carbonate (DMC).[125] (Reproduced with permission from M. Bellardita, V. Loddo, A. Mele, W. Panzeri, F. Parrino, I. Pibiri and L. Palmisano, *RSC Adv.*, 2014, **4**, 40859.)

DMC/H_2O (3% v/v), with the following oxidation of 9,10-dihydroxyphenanthrene (formed by using both DMC and ethanol as the solvents) to the corresponding diketone. The latter undergoes homolytic rupture of the C_9–C_{10} bond, giving rise to an unstable double radical intermediate, which affords 9-fluorenone, after decarbonylation and ring closure. Finally, 9-fluorenone is converted *via* a Baeyer–Villiger reaction into 6*H*-benzo[*c*]chromen-6-one, thanks to hydroperoxide species, which are typically formed during photocatalytic reactions when H_2O and O_2 are present. Interestingly, the last step was found to take place even in the dark.

These recent results show that it is possible to find organic green solvents in which photocatalytic selective reactions can be performed. In particular, the use of dimethyl carbonate as solvent enabled the formation of 9-fluorenone and 6*H*-benzo[*c*]chromen-6-one with a selectivity of 19% and 23%, respectively, by using supported photocatalysts, thus avoiding a separation step of powder from the liquid medium. Notably, these two products were isolated before a thorough characterization. Conversely, the use of aliphatic alcohols, such as ethanol, gives rise to a plethora of products, due to the easy formation of reactive alkoxy radicals.

11.5 Conclusion

As discussed above, photocatalytic reactions of synthetic interest are still limited in number, but some of the recent examples suggest that proper attention to this aspect may increase interest to this area and allow us to exploit some of the key advantages of this method. In photocatalytic synthesis it is indeed possible to (i) use a limited amount of catalyst, which is safe, recyclable and which usually does not degrade, and (ii) work under mild temperature and pressure conditions and to use safe solvents. Accordingly, the use of heterogeneous conditions allows an easy separation of the product(s) from the reaction mixture, thus simplifying work up and purification procedures. Nonetheless, there are cases in which photocatalytic selective processes are carried out under conditions that cannot be considered as green: this happens when unsafe organic solvents are needed, or when the highly energetic UV radiation is applied.

In conclusion, it is useful to highlight some of the most challenging research areas in selective photocatalysis, which nowadays are tailoring of the structural and crystalline features of catalysts to get the desired selectivity and the replacement of unsafe organic solvents with environmentally friendly ones, when water cannot be applied. The last activity could lead to the application of photocatalytic processes as replacements for existing unsafe ones.

References

1. (a) M. R. Hoffmann, S. T. Martin, W. Choi and D. W. Bahnemann, *Chem. Rev.*, 1995, **95**, 69; (b) *Environmentally Benign Photocatalysts: Applications of Titanium Oxide-based Materials*, ed. M. Anpo and P. Kamat, Springer, U.S.A, 1st edn, 2010; (c) V. Augugliaro, V. Loddo, M. Pagliaro, G. Palmisano and L. Palmisano, *Clean by Light Irradiation - Practical Applications of Supported TiO₂*, Royal Society of Chemistry, Cambridge, U.K., 2010.

2. (a) A. Bernabeu, R. F. Vercher, L. Santos-Juanes, P. J. Simón, C. Lardín, M. A. Martínez, J. A. Vicente, R. González, C. Llosá, A. Arques and A. M. Amat, *Catal. Today*, 2011, **161**, 235; (b) S. Protti, D. Ravelli, M. Fagnoni and A. Albini, *Chem. Commun.*, 2009, 7351; (c) J. Kim and W. Choi, *Energy Environ. Sci.*, 2010, **3**, 1042; (d) C. Guarisco, G. Palmisano, G. Calogero, R. Ciriminna, G. Di Marco, V. Loddo, M. Pagliaro and F. Parrino, *Environ. Sci. Pollut. Res.*, 2014, **21**, 11135; (e) N. Zhang, Y. Zhang, M.-Q. Yang and Y.-J. Xu, *Curr. Org. Chem.*, 2013, **17**, 2503; (f) M.-Q. Yang, *Chem. Soc. Rev.*, 2014, **43**, 8240; (g) M.-Q. Yang and Y.-J. Xu, *Phys. Chem. Chem. Phys.*, 2013, **15**, 19102; (h) M.-Q. Yang, Y. Zhang, N. Zhang, Z.-R. Tang and Y.-J. Xu, *Sci. Rep.*, 2013, **3**, 3314.
3. S. Yurdakal, V. Loddo, V. Augugliaro, H. Berber, G. Palmisano and L. Palmisano, *J. Catal.*, 2007, **129**, 9.
4. (a) J. M. Herrmann, *Catal. Today*, 1999, **53**, 115; (b) T. Hirakawa, J. K. Whitesell and M. A. Fox, *J. Phys. Chem. B*, 2004, **108**, 10213; (c) B. Ohtani, *Phys. Chem. Chem. Phys.*, 2014, **16**, 1788.
5. (a) J. Schneider, M. Matsuoka, M. Takeuchi, J. Zhang, Y. Horiuchi, M. Anpo and D. W. Bahnemann, *Chem. Rev.*, 2014, **114**, 9919; (b) S. Banerjee, S. C. Pillai, P. Falaras, K. E. O'Shea, J. A. Byrne and D. D. Dionysiou, *J. Phys. Chem. Lett.*, 2014, **5**, 2543; (c) H. Kisch, *Angew. Chem., Int. Ed.*, 2013, **52**, 812.
6. (a) G. Palmisano, V. Augugliaro, M. Pagliaro and L. Palmisano, *Chem. Commun.*, 2007, 3425; (b) D. Ravelli, D. Dondi, M. Fagnoni and A. Albini, *Appl. Catal., B*, 2010, **99**, 442; (c) Y. Shiraishi and T. Hirai, *J. Photochem. Photobiol., C*, 2008, **9**, 157.
7. (a) R. I. Bickley, G. Munuera and F. S. Stone, *J. Catal.*, 1973, **31**, 398; (b) P. Pichat, H. Courbon, J. Disdier, M.-N. Mozzanega and J.-M. Herrmann, *Stud. Surf. Sci. Catal.*, 1981, **7B**, 1498; (c) M. A. Fox, *Acc. Chem. Res.*, 1983, **16**, 314; (d) T. Sakata, *J. Photochem.*, 1985, **29**, 205; (e) N. Zeug, J. Buechеler and H. Kisch, *J. Am. Chem. Soc.*, 1985, **107**, 1459.
8. (a) G. Palmisano, E. García-López, G. Marcì, V. Loddo, S. Yurdakal, V. Augugliaro and L. Palmisano, *Chem. Commun.*, 2010, 7074; (b) G. Palmisano, *Curr. Org. Chem.*, 2013, **17**, 2365; (c) X. Lang, X. Chen and J. Zhao, *Chem. Soc. Rev.*, 2014, **43**, 473; (d) M. A. Lazar and W. A. Daoud, *RSC Adv.*, 2013, **3**, 4130; (e) N. Zhang, S. Liu and Y.-J. Xu, *Nanoscale*, 2012, **4**, 2227; (f) J. C. Colmenares and R. Luque, *Chem. Soc. Rev.*, 2014, **43**, 765.
9. (a) A. Molinari, A. Maldotti and R. Amadelli, *Curr. Org. Chem.*, 2013, **17**, 2382; (b) D. Sannino, V. Vaiano and P. Ciambelli, *Curr. Org. Chem.*, 2013, **17**, 2420; (c) L. Palmisano, V. Augugliaro, M. Bellardita, A. Di Paola, E. García-López, V. Loddo, G. Marcì, G. Palmisano and S. Yurdakal, *ChemSusChem*, 2011, **4**, 1431.
10. (a) K. Mori, H. Yamashita and M. Anpo, *RSC Adv.*, 2012, **2**, 3165; (b) A. Dhakshinamoorthy, S. Navalon, A. Corma and H. Garcia, *Energy Environ. Sci*, 2012, **5**, 9217; (c) F. Zhang, R. L. Jin, J. X. Chen, C. Shao, W. L. Gao, L. Li and N. J. Guan, *J. Catal.*, 2005, **232**, 424; (d) M. Warrier, M. K. F. Lo, H. Monbouquette and M. A. Garcia-Garibay, *Photochem. Photobiol. Sci.*, 2004, **3**, 859.

11. (a) D. Ravelli, S. Protti, M. Fagnoni and A. Albini, *Curr. Org. Chem.*, 2013, **17**, 2366; (b) C. Gambarotti, L. Melone, T. Caronna and C. Punta, *Curr. Org. Chem.*, 2013, **17**, 2406; (c) C. Vila and M. Rueping, *Green Chem.*, 2013, **15**, 2056.
12. (a) M. Fagnoni, D. Dondi, D. Ravelli and A. Albini, *Chem. Rev.*, 2007, **107**, 2725; (b) A. V. Vorontsov and A. V. Arsentyev, *Curr. Org. Chem.*, 2013, **17**, 2459.
13. (a) V. Augugliaro, M. Litter, L. Palmisano and J. Soria, *J. Photochem. Photobiol., C*, 2006, **7**, 127; (b) R. Molinari, P. Argurio and C. Lavorato, *Curr. Org. Chem.*, 2013, **17**, 2516; (c) G. Camera-Roda, A. Cardillo, V. Loddo, L. Palmisano and F. Parrino, *Membranes*, 2014, **4**, 96.
14. (a) C. Li and L. Chen, *Chem. Soc. Rev.*, 2006, **35**, 68; (b) A. Chanda and V. V. Fokin, *Chem. Rev.*, 2009, **109**, 725; (c) D. Ravelli, D. Dondi, M. Fagnoni and A. Albini, *Chem. Soc. Rev.*, 2009, **38**, 1999.
15. *Heterogeneous Photocatalysis*, ed. M. Schiavello, John Wiley & Sons Ltd, Chichester, U.K., 1997, vol. 3, Series in Photoscience and Photoengineering.
16. (a) M. Yan, F. Chen, J. Zhang and M. Anpo, *J. Phys. Chem. B*, 2005, **109**, 8673; (b) K.-J. Jiang, T. Kitamura, H. Yin, S. Ito and S. Yanagida, *Chem. Lett.*, 2002, **31**, 872; (c) A. Di Paola, G. Cufalo, M. Addamo, M. Bellardita, R. Campostrini, M. Ischia, R. Ceccato and L. Palmisano, *Colloids Surf., A*, 2008, **317**, 366 and references therein; (d) M. Addamo, M. Bellardita, A. Di Paola and L. Palmisano, *Chem. Commun.*, 2006, 4943; (e) A. Di Paola, M. Addamo, M. Bellardita, E. Cazzanelli and L. Palmisano, *Thin Solid Films*, 2007, **515**, 3527; (f) M. Addamo, V. Augugliaro, A. Di Paola, E. García-López, V. Loddo, G. Marcì and L. Palmisano, *Thin Solid Films*, 2008, **516**, 3802.
17. V. Augugliaro, V. Loddo, M. Pagliaro, G. Palmisano and L. Palmisano, *Clean by Light Irradiation - Practical Applications of Supported TiO_2*, RSC Publishing, Cambridge, U.K., 2010, ch. 2.
18. (a) S. Mozia, A. Heciak and A. W. Morawski, *Catal. Today*, 2011, **161**, 189; (b) X. Z. Li, F. B. Li, C. L. Yang and W. K. Ge, *J. Photochem. Photobiol., A*, 2001, **141**, 209; (c) P. A. Kolinko and D. V. Kozlov, *Appl. Catal., B*, 2009, **90**, 126 and references therein.
19. (a) K. T. Ranjit, H. Cohen, J. Willner, S. Bossmann and A. M. Braun, *J. Mater. Sci.*, 1999, **34**, 5273; (b) F. B. Li, X. Z. Li, C. H. Ao, S. C. Lee and M. F. Hou, *Chemosphere*, 2005, **59**, 787 and references therein; (c) A. Di Paola, M. Bellardita, G. Marcì, L. Palmisano, F. Parrino and R. Amadelli, *Catal. Today*, 2011, **161**, 35; (d) M. Bellardita, A. Di Paola, L. Palmisano, F. Parrino, G. Buscarino and R. Amadelli, *Appl. Catal., B*, 2011, **104**, 291.
20. (a) R. Asahi, T. Morikawa, T. Ohwaki, K. Aoki and Y. Taga, *Science*, 2001, **293**, 269; (b) M. Bellardita, M. Addamo, A. Di Paola, L. Palmisano and A. M. Venezia, *Phys. Chem. Chem. Phys.*, 2009, **11**, 4084 and references therein.
21. (a) H. Irie, Y. Watanabe and K. Hashimoto, *Chem. Lett.*, 2003, **32**, 772; (b) J. Matos, E. García-López, L. Palmisano, A. García and G. Marcì, *Appl. Catal., B*, 2010, **99**, 70 and references therein.

22. (a) K. M. Reddy, B. Baruwati, S. V. Manorama, M. Jayalakshmi and M. M. Rao, *J. Solid State Chem.*, 2005, **178**, 3352; (b) T. Ohno, T. Tsubota, Y. Nakamura and K. Sayama, *Appl. Catal., A*, 2005, **288**, 74.
23. (a) J. C. Yu, J. Yu, W. Ho, Z. Jiang and L. Zhang, *Chem. Mater.*, 2002, **14**, 3808; (b) M. V. Dozzi, B. Ohtani and E. Selli, *Phys. Chem. Chem. Phys.*, 2011, **13**, 18217; (c) M. V. Dozzi, C. D'Andrea, B. Ohtani, G. Valentini and E. Selli, *J. Phys. Chem. C*, 2013, **117**, 25586.
24. G. Marcì, E. García-López, G. Mele, L. Palmisano, G. Dyrda and R. Słota, *Catal. Today*, 2009, **143**, 203 and references therein.
25. A. Di Paola, E. García-López, G. Marcì and L. Palmisano, *J. Hazard. Mater.*, 2012, **211–212**, 3.
26. (a) A. Hiskia, A. Mylonas and E. Papaconstantinou, *Chem. Soc. Rev.*, 2001, **30**, 62; (b) S. Yanagida, A. Nakajima, T. Sasaki, Y. Kameshima and K. Okada, *Chem. Mater.*, 2008, **20**, 3757; (c) D. Carriazo, M. Addamo, G. Marcì, C. Martín, L. Palmisano and V. Rives, *Appl. Catal., A*, 2009, **356**, 172; (d) G. Marcì, E. García-López, L. Palmisano, D. Carriazo, C. Martín and V. Rives, *Appl. Catal., B*, 2009, **90**, 497 and references therein.
27. M. Bellardita, M. Addamo, A. Di Paola, G. Marcì, L. Palmisano, L. Cassar and M. Borsa, *J. Hazard. Mater.*, 2010, **174**, 707 and references therein.
28. V. Loddo, G. Marcì, C. Martín, L. Palmisano, V. Rives and A. Sclafani, *Appl. Catal., B*, 1999, **20**, 29.
29. M. Salaices, B. Serrano and H. I. de Lasa, *Chem. Eng. Sci.*, 2004, **59**, 3.
30. G. Palmisano, V. Loddo, V. Augugliaro, L. Palmisano and S. Yurdakal, *AIChE J.*, 2007, **53**, 961.
31. D. I. Enache, J. K. Edwards, P. Landon, B. Solsona-Espriu, A. F. Carley, A. A. Herzing, M. Watanabe, C. J. Kiely, D. W. Knight and G. J. Hutchings, *Science*, 2006, **311**, 362.
32. F. A. Chavez, J. M. Rowland, M. M. Olmstead and P. K. Mascharak, *J. Am. Chem. Soc.*, 1998, **120**, 9015.
33. U. R. Pillai and E. Sahle-Demessie, *J. Catal.*, 2002, **211**, 434.
34. K. J. Green and R. Rudham, *J. Chem. Soc., Faraday Trans.*, 1993, **89**, 1867.
35. S. Farhadi, M. Afshari, M. Maleki and Z. Badazadeh, *Tetrahedron Lett.*, 2005, **46**, 8483.
36. Q. Wang, M. Zhang, C. Chen, W. Ma and J. Zhao, *Angew. Chem.*, 2010, **122**, 8148; *Angew. Chem., Int. Ed.*, 2010, **49**, 7976.
37. A. H. Reidies, Manganese compounds, in *Kirk-Othmer Encyclopedia of Chemical Technology*, John Wiley & Sons, Inc., New York, 1982, vol. 14, pp. 844–895.
38. J. S. Rafelt and J. H. Clark, *Catal. Today*, 2000, **57**, 33.
39. R. Molinari, A. Caruso and T. Poerio, *Catal. Today*, 2009, **144**, 81.
40. (a) G. Palmisano, M. Addamo, V. Augugliaro, T. Caronna, E. García-López, V. Loddo and L. Palmisano, *Chem. Commun.*, 2006, 1012; (b) G. Palmisano, M. Addamo, V. Augugliaro, T. Caronna, A. Di Paola, E. García-López, V. Loddo, G. Marcì, L. Palmisano and M. Schiavello, *Catal. Today*, 2007, **122**, 118.

41. (a) T. A. Nijhuis, M. Makkee, J. A. Moulijn and B. M. Weckhuysen, *Ind. Eng. Chem. Res.*, 2006, **45**, 3447; (b) S. B. Shin and D. Chadwick, *Ind. Eng. Chem. Res.*, 2010, **49**, 8125.
42. G. Wu, Y. Wang, L. Wang, W. Feng, H. Shi, Y. Lin, T. Zhang, X. Jin, S. Wang, X. Wu and P. Yao, *Chem. Eng. J.*, 2013, **215–216**, 306.
43. J. Huang, T. Takei, H. Ohashi and M. Haruta, *Appl. Catal., A*, 2012, **435–436**, 115.
44. A. Maldotti, A. Molinari and R. Amadelli, *Chem. Rev.*, 2002, **102**, 3811.
45. C. Murata, H. Yoshida, J. Kumagai and T. Hattori, *J. Phys. Chem. B*, 2003, **107**, 4364.
46. H. Yoshida, C. Murata and T. Hattori, *J. Catal.*, 2000, **194**, 364.
47. C. Murata, T. Hattori and H. Yoshida, *J. Catal.*, 2005, **231**, 292.
48. S. Yoshida, T. Tanaka, M. Okada and T. Funabiki, *J. Chem. Soc., Faraday Trans.*, 1984, **80**, 119.
49. F. Amano, T. Yamaguchi and T. Tanaka, *J. Phys. Chem. B*, 2006, **110**, 281.
50. (a) Y. Guo, L. Zhengwang, W. Guangjian, Y. Huang and F. Kang, *Appl. Surf. Sci.*, 2011, **258**, 1082; (b) A. Marimuthu, J. Zhang and S. Linic, *Science*, 2013, **339**, 1590; (c) S. Ouidri, C. Guillard, V. Caps and H. Khalaf, *Appl. Clay Sci.*, 2010, **48**, 431; (d) G. Wang, S. Zhang, Y. Huang, F. Kang, Z. Yang and Y. Guo, *Appl. Catal., A*, 2012, **413**, 52; (e) V. H. Nguyen, J. Wu and H. Bai, *Catal. Commun.*, 2013, **33**, 57; (f) C. Qi, J. Huang, S. Bao, H. Su, T. Akita and M. Haruta, *J. Catal.*, 2011, **281**, 12.
51. Y. Shiraishi, M. Morishita and T. Hirai, *Chem. Commun.*, 2005, 5977.
52. G. Ertl, H. Knözinger and J. Weitkamp, *Handbook of Heterogeneous Catalysis*, Wiley-VCH, Verlag GmbH, Weinheim, 2008.
53. K. Tao, W. Li, H. Li and X. Qi, *Appl. Catal., A*, 1996, **139**, 43.
54. M. H. W. Sonnemans, *Appl. Catal., A*, 1993, **94**, 215.
55. N. Mizuno and M. Misono, *Chem. Rev.*, 1998, **98**, 171.
56. (a) M. P. Atkins, US Patent 5,616,815, 1997; (b) G. J. Haining, US Patent 5,714,429, 1998.
57. A. V. Ivanov, E. Zausa, Y. Ben Taarit and N. Essayem, *Appl. Catal., A*, 2003, **256**, 225.
58. G. Marcì, E. García-López and L. Palmisano, *Appl. Catal., A*, 2012, **421–422**, 70.
59. G. Marcì, E. García-López, M. Bellardita, F. Parisi, C. Colbeau-Justin, S. Sorgues, L. F. Liotta and L. Palmisano, *Phys. Chem. Chem. Phys.*, 2013, **15**, 13329.
60. E. Albiter Escobar, M. A. Valenzuela Zapata, S. Alfaro Hernández, S. O. Flores Valle, O. Ríos Berny, V. J. González Ángeles and I. Córdova Reyes, *J. Mex. Chem. Soc.*, 2010, **54**, 133.
61. D. W. Manley, L. Buzzetti, A. MacKessack-Leitch and J. C. Walton, *Molecules*, 2014, **19**, 15324.
62. Y. Matsushita, S. Kumada, K. Wakabayashi, K. Sakeda and T. Ichimura, *Chem. Lett.*, 2006, **35**, 410.
63. H. Tada, M. Kubo, Y. Inubushi and S. Ito, *Chem. Commun.*, 2000, 977.

64. S. Chen, H. Zhang, X. Yu and W. Liu, *Chin. J. Chem.*, 2010, **28**, 21.
65. K. Imamura, T. Yoshikawa, K. Hashimoto and H. Kominami, *Appl. Catal., B*, 2013, **134–135**, 193.
66. K. Imamura, S. Iwasaki, T. Maeda, K. Hashimoto, B. Ohtani and H. Kominami, *Phys. Chem. Chem. Phys.*, 2011, **13**, 5114.
67. W. Y. Ahn, S. A. Sheeley, T. Rajh and D. M. Cropek, *Appl. Catal., B*, 2007, **74**, 103.
68. H. Wang, R. E. Partch and Y. Li, *J. Org. Chem.*, 1997, **62**, 5222.
69. H. Wang, J. Yan, W. Chang and Z. Zhang, *Catal. Commun.*, 2009, **10**, 989.
70. H. Wang, X. Yang, W. Xiong and Z. Zhang, *Res. Chem. Intermed.*, 2015, **41**, 3981.
71. P. Eskandari, F. Kazemi and Z. Zand, *J. Photochem. Photobiol., A*, 2014, **274**, 7.
72. A. Hernández-Gordillo, A. G. Romero, F. Tzompantzi, S. Oros-Ruiz and R. Gómez, *J. Photochem. Photobiol., A*, 2013, **257**, 44.
73. W. Wu, S. Liang, Y. Chen, L. Shen, H. Zheng and L. Wu, *Catal. Commun.*, 2012, **17**, 39.
74. W. Wu, G. Liu, S. Liang, Y. Chen, L. Shen, H. Zheng, R. Yuan, Y. Hou and L. Wu, *J. Catal.*, 2012, **290**, 13.
75. S. Fueldner, P. Pohla, H. Bartling, S. Dankesreiter, R. Stadler, M. Gruber, A. Pfitzner and B. Koenig, *Green Chem.*, 2011, **13**, 640.
76. P. Chang, H. Cheng, W. Li, L. Zhuo, L. He, Y. Yu and F. Zhao, *Phys. Chem. Chem. Phys.*, 2014, **16**, 16606.
77. A. Maldotti, L. Andreotti, A. Molinari, S. Tollari, A. Penoni and S. Cenini, *J. Photochem. Photobiol., A*, 2000, **133**, 129.
78. M. B. Rubin, *Tetrahedron Lett.*, 1982, **23**, 4615.
79. R. Asahi, T. Morikawa, H. Irie and T. Ohwaki, *Chem. Rev.*, 2014, **114**, 9824.
80. M. Warrier, M. K. F. Lo, H. Monbouquette and M. A. Garcia-Garibay, *Photochem. Photobiol. Sci.*, 2004, **3**, 859.
81. L. Qiao, H. Bi, J.-M. Busnel, B. Liu and H. H. Girault, *Chem. Commun.*, 2008, 6357.
82. W. Wu, L. Wen, L. Shen, R. Liang, R. Yuan and L. Wu, *Appl. Catal., B*, 2013, **130–131**, 163.
83. J. L. Ferry and W. H. Glaze, *J. Phys. Chem. B*, 1998, **102**, 2239.
84. (a) C. K. Prier, D. A. Rankic and D. W. C. MacMillan, *Chem. Rev.*, 2013, **113**, 5322; (b) J. M. R. Narayanam and C. R. J. Stephenson, *Chem. Soc. Rev.*, 2011, **40**, 102.
85. (a) M. Cherevatskaya and B. Koenig, *Russ. Chem. Rev.*, 2014, **83**, 183; (b) X. Wang, S. Blechert and M. Antonietti, *ACS Catal.*, 2012, **2**, 1596; (c) C. Gambarotti, L. Melone and C. Punta, *Artificial Photosynthesis*, ed. M. Najafpour, InTech, 2012, p. 5.
86. L. Cermenati, M. Mella and A. Albini, *Tetrahedron*, 1998, **54**, 2575.
87. L. Cermenati, C. Richter and A. Albini, *Chem. Commun.*, 1998, 805.
88. L. Cermenati, D. Dondi, M. Fagnoni and A. Albini, *Tetrahedron*, 2003, **59**, 6409.

89. (a) D. W. Manley, R. T. McBurney, P. Miller, R. F. Howe, S. Rhydderch and J. C. Walton, *J. Am. Chem. Soc.*, 2012, **134**, 13580; (b) D. W. Manley, R. T. McBurney, P. Miller and J. C. Walton, *J. Org. Chem.*, 2014, **79**, 1386.
90. S. Marinkovic and N. Hoffmann, *Int. J. Photoenergy*, 2003, **5**, 175.
91. M. Aldemir, F. W. Heinemann and H. Kisch, *Photochem. Photobiol. Sci.*, 2012, **11**, 908.
92. T. Mitkina, C. Stanglmair, W. Setzer, M. Gruber, H. Kisch and B. Koenig, *Org. Biomol. Chem.*, 2012, **10**, 3556.
93. (a) T. Caronna, C. Gambarotti, L. Palmisano, C. Punta and F. Recupero, *Chem. Commun.*, 2003, 2350; (b) T. Caronna, C. Gambarotti, A. Mele, M. Pierini, C. Punta and F. Recupero, *Res. Chem. Intermed.*, 2007, **33**, 311.
94. T. Caronna, C. Gambarotti, L. Palmisano, C. Punta and F. Recupero, *J. Photochem. Photobiol., A*, 2005, **171**, 237.
95. T. Caronna, C. Gambarotti, L. Palmisano, C. Punta, M. Pierini and F. Recupero, *J. Photochem. Photobiol., A*, 2007, **189**, 322.
96. H. Yoshida, Y. Fujimura, H. Yuzawa, J. Kumagai and T. Yoshida, *Chem. Commun.*, 2013, 3793.
97. M. Iizuka and M. Yoshida, *J. Fluorine Chem.*, 2009, **130**, 926.
98. M. Cherevatskaya, M. Neumann, S. Fueldner, C. Harlander, S. Kuemmel, S. Dankesreiter, A. Pfitzner, K. Zeitler and B. Koenig, *Angew. Chem., Int. Ed.*, 2012, **51**, 4062.
99. P. Riente, A. Matas Adams, J. Albero, E. Palomares and M. A. Pericàs, *Angew. Chem., Int. Ed.*, 2014, **53**, 9613.
100. P. Wu, C. He, J. Wang, X. Peng, X. Li, Y. An and C. Duan, *J. Am. Chem. Soc.*, 2012, **134**, 14991.
101. (a) M. Rueping, J. Zoller, D. C. Fabry, K. Poscharny, R. M. Koenigs, T. E. Weirich and J. Mayer, *Chem. Eur. J.*, 2012, **18**, 3478; (b) L. Moehlmann, M. Baar, J. Riess, M. Antonietti, X. Wang and S. Blechert, *Adv. Synth. Catal.*, 2012, **354**, 1909; (c) D. Verma, S. Verma, A. K. Sinha and S. L. Jain, *ChemPlusChem*, 2013, **78**, 860.
102. C. Vila and M. Rueping, *Green Chem.*, 2013, **15**, 2056.
103. (a) H. Xu, S. Ouyang, L. Liu, P. Reunchan, N. Umezawa and J. Ye, *J. Mater. Chem. A*, 2014, **2**, 12642; (b) N. Balázs, K. Mogyorósi, D. F. Srankó, A. Pallagi, T. Alapi, A. Oszkó, A. Dombi and P. Sipos, *Appl. Catal., B*, 2008, **84**, 356; (c) B. Ohtani, *Phys. Chem. Chem. Phys.*, 2014, **16**, 1788; (d) M. S. Hamdy, R. Amrollahi and G. Mul, *ACS Catal.*, 2012, **2**, 2641.
104. A. V. Vorontsov and A. V. Arsentyev, *Curr. Org. Chem.*, 2013, **17**, 2459.
105. X. Lang, X. Chen and J. Zhao, *Chem. Soc. Rev.*, 2014, **43**, 473.
106. J. Sanz, J. Soria, I. Sobrados, S. Yurdakal and V. Augugliaro, *J. Phys. Chem. C*, 2012, **116**, 5110.
107. S. Yurdakal, V. Augugliaro, J. Sanz, J. Soria, I. Sobrados and M. J. Torralvo, *J. Catal.*, 2014, **309**, 97.
108. G. Liu, J. C. Yu, G. Q. Lu and H.-M. Cheng, *Chem. Commun.*, 2011, 6763.
109. J. Zhu, S. H. Wang, Z. F. Bian, S. H. Xie, C. L. Cai, J. G. Wang, H. G. Yang and H. X. Li, *Cryst. Eng. Commun.*, 2010, **12**, 2219.

110. M. Bellardita, V. Augugliaro, V. Loddo, B. Megna, G. Palmisano, L. Palmisano and M. A. Puma, *Appl. Catal., A*, 2012, **441–442**, 79.
111. G. Palmisano, M. Addamo, V. Augugliaro, T. Caronna, A. Di Paola, E. García-López, V. Loddo, G. Marcì, L. Palmisano and M. Schiavello, *Catal. Today*, 2007, **122**, 118.
112. S. Yurdakal, V. Augugliaro, V. Loddo, G. Palmisano and L. Palmisano, *New J. Chem.*, 2012, **36**, 1762.
113. M. A. Fox, *Acc. Chem. Res.*, 1983, **16**, 314.
114. B. Kraeutler and A. J. Bard, *J. Am. Chem. Soc.*, 1977, **99**, 7729.
115. T. Kanno, T. Oguchi, H. Sakuragi and K. Tokumaru, *Tetrahedron Lett.*, 1980, **21**, 467.
116. O. S. Mohamed, A. E. M. Gaber and A. A. Abdel-Wahab, *J. Photochem. Photobiol., A*, 2002, **148**, 205.
117. H. Shen, H. Mao, L. Ying and Q. Xia, *J. Mol. Catal. A: Chem.*, 2007, **276**, 73.
118. G. Palmisano, S. Yurdakal, V. Augugliaro, V. Loddo and L. Palmisano, *Adv. Synth. Catal.*, 2007, **349**, 964.
119. C.-J. Li and L. Chen, *Chem. Soc. Rev.*, 2006, **35**, 68.
120. W. Feng, G. Wu, L. Li and N. Guan, *Green Chem.*, 2011, **13**, 3213.
121. G. Redmond and D. Fitzmaurice, *J. Phys. Chem.*, 1993, **97**, 1426.
122. H. Tada, T. Ishida, A. Takao and S. Ito, *Langmuir*, 2004, **20**, 7898.
123. P. Anastas and J. Warner, *Green Chemistry: Theory and Practice*, Oxford University Press, New York, 1998.
124. F. Soana, M. Sturini, L. Cermenati and A. Albini, *J. Chem. Soc., Perkin Trans. 2*, 2000, 699.
125. M. Bellardita, V. Loddo, A. Mele, W. Panzeri, F. Parrino, I. Pibiri and L. Palmisano, *RSC Adv.*, 2014, **4**, 40859.
126. P. Tundo and M. Selva, *Acc. Chem. Res.*, 2002, **35**, 706.
127. X. C. Lu, J. G. Yang, W. G. Zhang and Z. Huang, *Energy Fuels*, 2005, **19**, 1879.
128. (a) S. Kohtani, M. Tomohiro, K. Tokumura and R. Nakagaki, *Appl. Catal., B*, 2005, **58**, 265; (b) H. Jia, J. Zhao, X. Fan, K. Dilimulati and C. Wang, *Appl. Catal., B*, 2012, **123**, 43; (c) S. Higashida, A. Harada, R. Kawakatsu, N. Fujiwara and M. Matsumura, *Chem. Commun.*, 2006, 2804.
129. (a) J. Kou, Z. Li, Y. Guo, J. Gao, M. Yang and Z. Zou, *J. Mol. Catal. A: Chem.*, 2010, **325**, 48; (b) J. Kou, Z. Li, Y. Yuan, H. Zhang, Y. Wang and Z. Zou, *Environ. Sci. Technol.*, 2009, **43**, 2919; (c) A. K. Haritash and C. P. Kaushik, *J. Hazard. Mater.*, 2009, **169**, 1.
130. Y. Yang and F. Hildebrand, *Anal. Chim. Acta*, 2006, **555**, 364.
131. Y. Zhang, J. W. C. Wong, P. Liu and M. Yuan, *J. Hazard. Mater.*, 2011, **191**, 136.
132. M. J. S. Dewar, *J. Am. Chem. Soc.*, 1952, **74**, 3357.

CHAPTER 12

An Overview of the Potential Applications of TiO_2 Photocatalysis for Food Packaging, Medical Implants, and Chemical Compound Delivery

PIERRE PICHAT[a]

[a]"Photocatalyse et Environnement", CNRS/Ecole Centrale de Lyon (STMS), 69134 Ecully CEDEX, France
*E-mail: pichat@ec-lyon.fr

12.1 Introduction

The capability of photo-excited TiO_2 to degrade organic matter, including that of microorganisms, is the basis of diverse applications regarding various aspects of our environment, which covers the domain of energy production and saving. The state-of-the-art, challenges and perspectives of the most usual of these applications are presented in the other chapters of this volume. To summarize briefly, self-cleaning materials, above all glass, have been commercialised for some years.[1] Although, indoor air purifiers based,

RSC Energy and Environment Series No. 15
Photocatalysis: Applications
Edited by Dionysios D Dionysiou, Gianluca Li Puma, Jinhua Ye, Jenny Schneider, and Detlef Bahnemann

Published by the Royal Society of Chemistry, www.rsc.org

only or partially, on photocatalysis can be found on the market, their safe use is questioned because of the incomplete oxidation of the volatile organic compounds, which may result in concentrations of harmful low-molecular-mass carbonyl compounds somewhat higher at the stationary state than without air treatment.[2] The lifetime of the photocatalyst can also be a barrier to implementation, especially for centralised purifying units in buildings.[3,4] Despite extensive research,[5] photocatalytic treatment of water seems to be still limited to niche applications (see, for example, http://purifics.com) or specific applications (*e.g.*, on-site purification of agricultural waters[6]). Limitations due to low photocatalytic rates of removal of pollutants have not yet been overcome. These limitations arise not only from the recombination of photogenerated charges, as for any other application of photocatalysis, but also from the partition of hydrophilic pollutants between the semiconductor surface and the water bulk.[7] The use of solar irradiation, which could be *a priori* advantageous in some cases, is restricted by the low sensitivity in the visible region of TiO_2 despite many attempts to increase it.[8,9] Additionally, for water purification, the photocatalysts (modified TiO_2 or other semiconductors) must contain chemical elements that are both innocuous and inexpensive.[7] The uses of photocatalysis in the field of energy, especially hydrogen generation from water, and to effect carefully-selected organic syntheses through more environment-friendly routes, are considered as promising, but are still at the research stage (see other chapters in this volume).

This chapter presents three other much less-known potential applications of TiO_2 photocatalysis. These applications are related to safety, health and well-being. They correspond to important, well-identified needs for which existing or other investigated methods are not entirely satisfactory. Research regarding these applications has also been stimulated by the wish to explore possibilities for photocatalysis other than those indicated in the previous paragraph. It represents further efforts to take advantage of the properties of photo-excited TiO_2. This chapter contains three main sections, each dealing with one of the following topics in which TiO_2 photocatalysis could find its way: food packaging, medical implants, and chemical compound delivery. In addition to the aforementioned reasons, these topics, in particular the first two, were selected because they have induced a substantial number of studies. The issues remaining to be resolved and the perspectives for each of these three topics are discussed in this chapter.

12.2 Potential Advantages of TiO_2 Photocatalysis for Food Packaging

12.2.1 Generalities

In modern countries, most food is packaged for transport (over increasingly longer distances – which is indeed unfortunate for the environment), storage and sales. Therefore, food packaging is a topic of high industrial interest. Obviously, the mechanical properties of the packages are vital to both enable easy handling and protect the food from damage. The packages must also preserve the food quality – that is, freshness, taste and safety for the consumers' health – over as long a period of time as possible. Consequently, packaging must constitute a sufficient barrier against dirt, dust, moisture, dioxygen and other gases, detrimental action of sunlight, pathogenic microorganisms and any other possibly destructive or harmful substances from the exterior.[10,11]

Incorporating TiO_2 in a package material obviously allows one to protect the food from the potential oxidising effect of UV irradiation because of the TiO_2 absorption spectrum (*cf.* the use of TiO_2 in many sunscreens). Beyond that, the photocatalytic properties of TiO_2 offer several assets: (1) inactivation/degradation of the microorganisms generating spoiling, ripening and other alterations to food; (2) scavenging of O_2 capable of increasing food oxidation and inducing the growth of aerobic microorganisms; (3) oxidation of C_2H_4, a compound produced by some fruits and vegetables and which promotes their ripening.

12.2.2 Effect of TiO_2 on the Package Physical Properties

Several studies have shown good, uniform incorporation of TiO_2 in various polymers. The polymer mechanical properties were maintained or even improved.[12–14] For instance, Figure 12.1 shows the improvement of two pertinent mechanical parameters when TiO_2 particles were incorporated into polyethene; the improvement was better when the TiO_2 particles were de-agglomerated before incorporation by use of a titanium acid ester and ultrasound; moreover, it was maintained through the cycles mimicking the potential variations in temperature and relative humidity during the storage of packages. The transmission of water vapour through the polymer can be modified to some extent because of the presence of TiO_2 particles; increased transmission has been reported.[13] A possible change in O_2 transmission has also been investigated.[12] In summary, TiO_2 incorporation has been found not to alter the polymer physical properties much and can even be beneficial. This conclusion is equally valid when nanoparticles of Ag, Cu or Zn are also added to enhance the biocide properties.

12.2.3 Effect of TiO_2 on the Package Antibacterial Properties

As microorganisms consist of organic matter, they are prone to not withstand the radical chemistry generated by photo-excitation of TiO_2. This was originally shown between 1985 and 1995;[15–18] since then it has been documented in many other studies and summarised in reviews.[19–21] Most investigations have been concerned with microorganisms in liquid water. Some others have dealt with microorganisms in air, which in fact is somewhat equivalent[2] inasmuch as TiO_2 is covered by layers of adsorbed water molecules because of the humidity of ambient air. Clearly, microorganisms can be photocatalytically inhibited and degraded, although the basic mechanisms are still debated[19–21] (*cf.* Chapters 3 and 7).

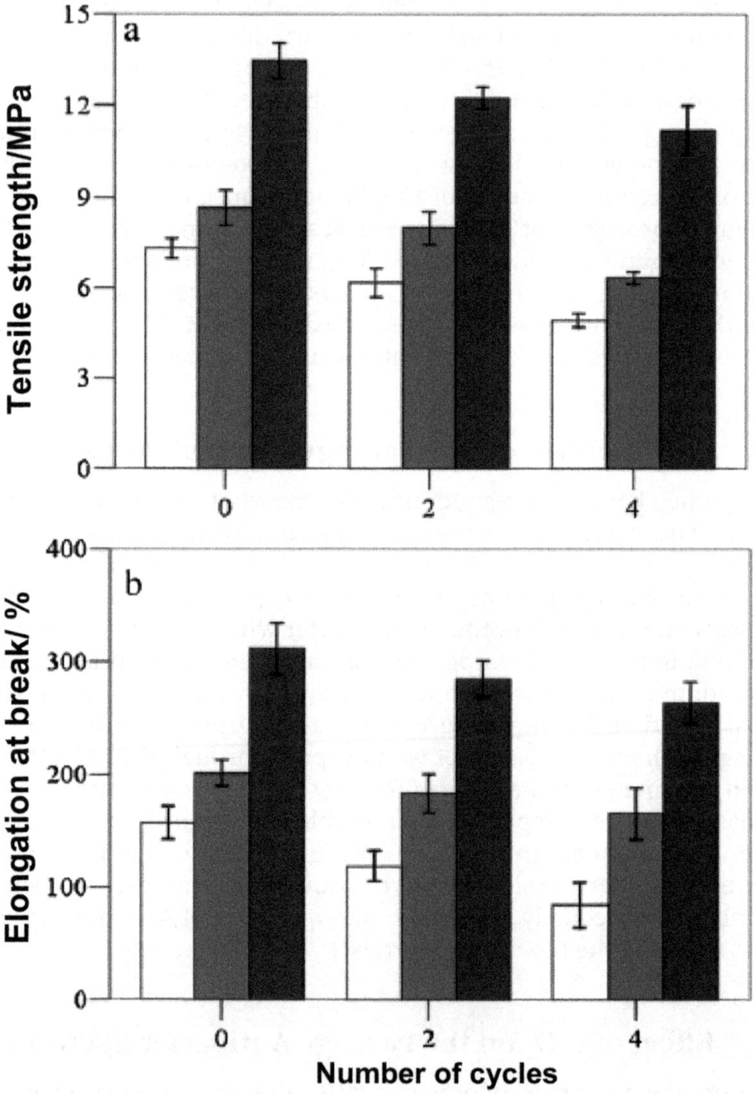

Figure 12.1 Tensile strength in MPa (a) and elongation at break in % (b) of polyethene (PE) before and after two or four 53-h-test cycles with temperature varying from −10 to 40 °C and relative humidity from 10% to 95%. In white: pristine PE; in grey: PE with unmodified TiO_2 nanoparticles; in black: PE with modified TiO_2 to increase the dispersion of the nanoparticles.[13] (Reproduced with permission from Y. Xing, X. Li, L. Zhang, Q. Xu, Z. Che, W. Li, Y. Bai and K. Li, *Prog. Org. Coat.*, 2012, 73, 219–224.)

In preliminary experiments on the use of TiO_2 in packaging for fruit and vegetables, TiO_2 powder was directly applied to the food surface to determine whether the photocatalytic properties were still observed under such conditions.[22] For example, the airborne fungus *Penicillium expansum* was inoculated into tomatoes using cuts. TiO_2 powder was then sprinkled into these cuts. The severity of the tomato rot produced by the fungus was estimated as the percentage of tomato surface area that was damaged. Figure 12.2 shows that, in the presence of UV-irradiated TiO_2, no damage occurred for 7 days and that the damage was markedly less after 14 days than with only TiO_2 or only UV-irradiation. The same team also observed the antibacterial efficacy of a polypropene film coated (using simply a suspension) with TiO_2 against *Escherichia coli* inoculated on lettuce leaves.[23]

Regarding the use of TiO_2 incorporated in various polymers packages, the aim was to determine whether the biocide properties were maintained despite the incorporation. Sometimes, Ag, Cu or Zn nanoparticles were added. Apart from *Escherichia coli*, often chosen as a kind of reference, the microorganisms used were some of those involved in the deterioration, spoilage, ripening, *etc.* of diverse types of food, as well as food-derived infection pathogens. All of the studies showed that photocatalytic inhibition of the microorganisms occurred, even though the contact between the microorganisms and TiO_2 was diminished or possibly suppressed by incorporation in the polymeric substance.

However, a dispersion of TiO_2 particles in the polymer that is as homogeneous as possible was shown to be a prerequisite. For example, in the case of a ethene–ethenol copolymer, the lower efficacy (using, separately, Gram negative *Pseudomonas aeruginosa* (Figure 12.3) and Gram positive *Enterococcus*

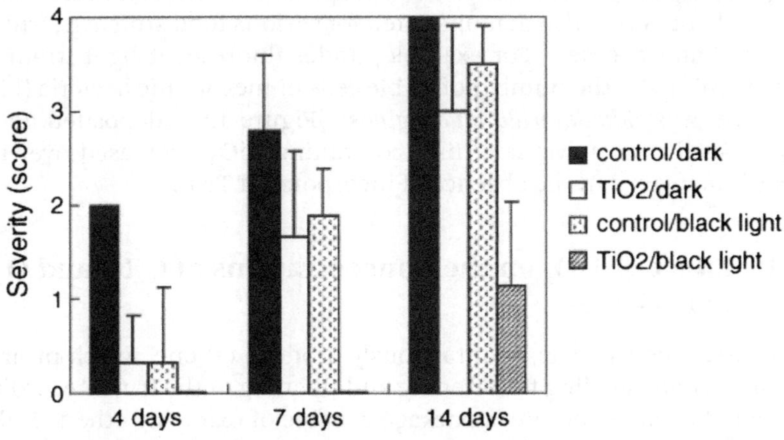

Figure 12.2 *Penicillium expansum*-generated rot on tomato during storage at 25 °C. The development of rot is expressed by a severity score: 0, none; 1, very slight; 2, less than 25% of tomato area; 3, less than 50%; 4, more than 50%. Two 20 W black light lamps provided an irradiance of 1 mW cm^{-2}.[22] (Reproduced with permission from C. Maneerat and Y. Hayata, *Int. J. Food Microbiol.*, 2006, **107**, 99.)

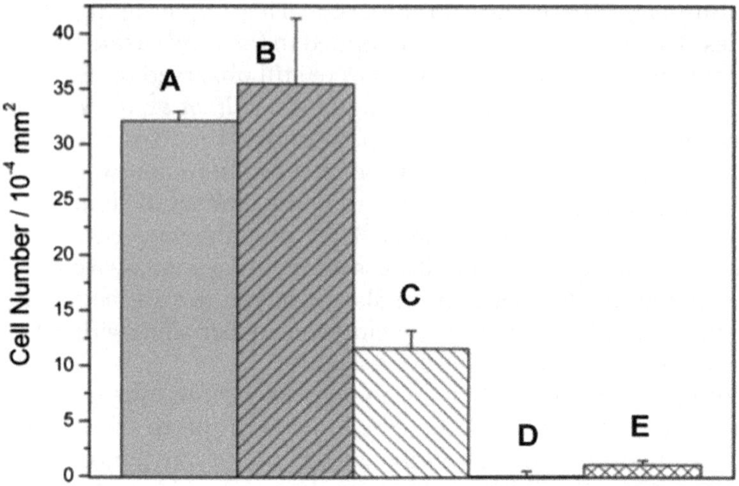

Figure 12.3 Cell number of *Pseudomonas aeruginosa* on the surface of an ethene-ethenol copolymer depending on the TiO_2 content (in wt%: A: 0; C: 0.5; D: 2; E: 5) under irradiation at 280 nm with an energy <1 kJ m^{-2}. The value for the non-irradiated copolymer containing 2 wt% of TiO_2 is also shown (B).[24] (Reproduced with permission from A. Kubacka, C. Serrano, M. Ferrer, H. Lünsdorf, P. Bielecki, M. L. Cerrada, M. Fernandez-Garcıa and M. Fernandez-Garcıa, *Nano Lett.*, 2007, 7, 2529.)

faecalis as the deposited test bacteria) at 5 wt% TiO_2 loading than at 2 wt% was attributed to the formation of some TiO_2 agglomerates within the copolymer for the highest loading according to scanning electron micrographs.[24]

The preservation of the antimicrobial properties of TiO_2 when it is incorporated in polymers was also demonstrated for various food stuff, *e.g.*, fruit,[12,25] fruit juice[26] and cheese.[27] For example, under fluorescent light irradiation (UVA < 50 μW cm^{-2}), the number of viable cells of mesophilic bacteria (Figure 12.4) or the yeast *Rhodotorula mucilaginosa* (Figure 12.5) deposited on pear fruit packaged in a polyethene film containing TiO_2 decreased over time, whereas it increased in the absence of incorporated TiO_2.[25]

12.2.4 Effect of TiO_2 on the Concentrations of C_2H_4 and O_2 in Packages

Some fruits, such as kiwis, spontaneously produce ethene, which promotes enzymatic activity leading to softening and ripening of the fruit. Accordingly, to prolong the fruit shelf-life, a package capable of oxidising ethene is desirable. In the case of kiwi fruit, with a package made of polyethene containing Ag and montmorillonite in addition to TiO_2 a marked decrease of the ethene concentration in the headspace was observed over time (Figure 12.6).[12] Concomitantly, this composite package was shown to preserve the firmness of the kiwi for a longer duration, the benefit produced by TiO_2 incorporation being especially greater during the first three weeks (Figure 12.7).

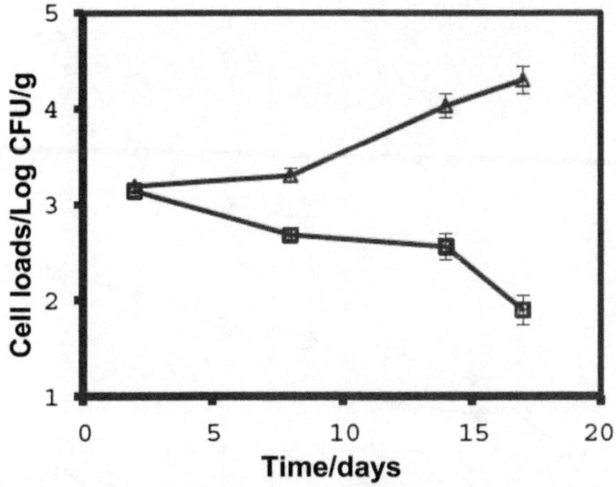

Figure 12.4 Cell loads of viable mesophilic bacteria on packaged pear fruit during storage at 5 °C under fluorescent lamps (UV A < 50 μW cm^{-2}). Polyethene package with TiO_2 (squares) or without TiO_2 (triangles).[25] (Reproduced with permission from H. Bodaghi, Y. Mostofi, A. Oromiehie, Z. Zamani, B. Ghanbarzadeh, C. Costa, A. Conte and M. A. Del Nobile, *LWT--Food Sci. Technol.*, 2013, **50**, 702.)

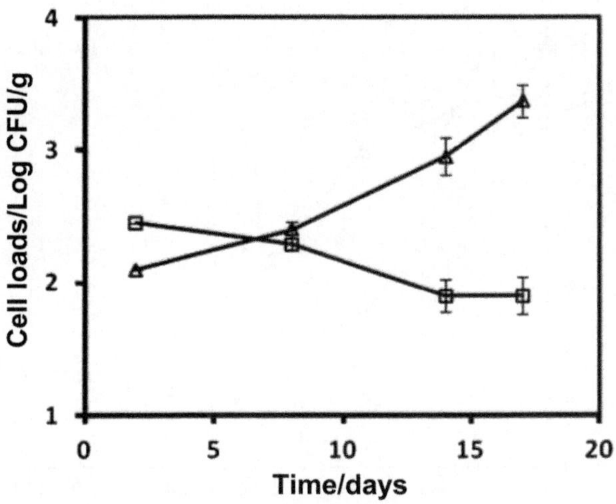

Figure 12.5 As for Figure 12.4, with the microorganism *Rhodotorula mucilaginosa* in place of mesophilic bacteria.[25] (Reproduced with permission from H. Bodaghi, Y. Mostofi, A. Oromiehie, Z. Zamani, B. Ghanbarzadeh, C. Costa, A. Conte and M. A. Del Nobile, *LWT--Food Sci. Technol.*, 2013, **50**, 702.)

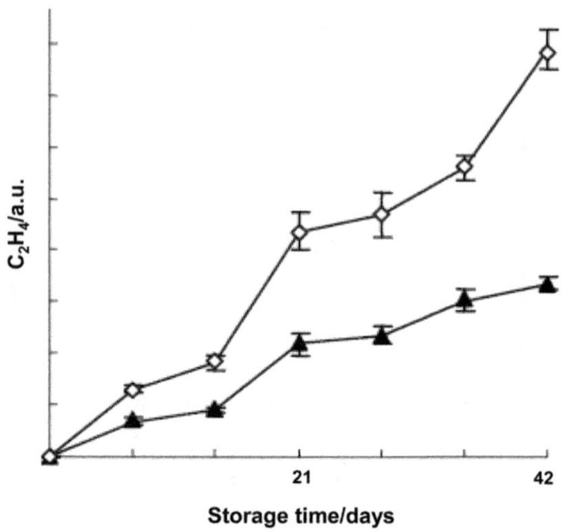

Figure 12.6 Kinetic variations in the ethene concentration in the headspace of a polyethene package containing kiwis during storage at 4 °C. Diamonds: pristine polyethene; triangles: polyethene containing TiO_2, Ag and montmorillonite.[12] (Reproduced with permission from Q. Hu, Y. Fang, Y. Yang, N. Ma and L. Zhao, *Food Res. Int.*, 2011, **44**, 1589.)

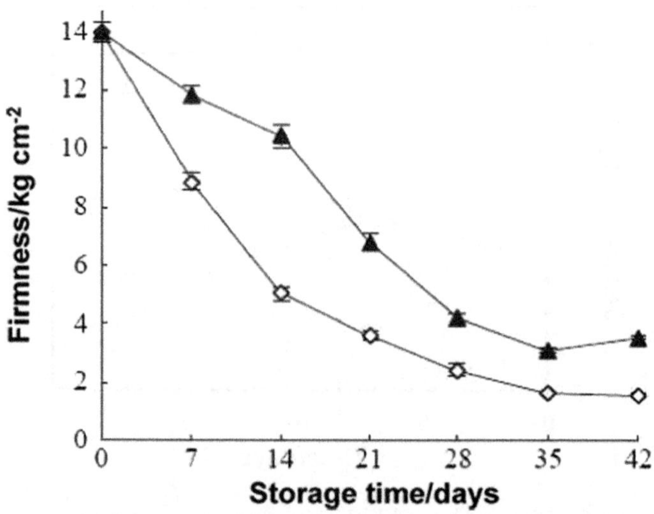

Figure 12.7 As for Figure 12.6, the kinetic variable being the kiwifruit firmness measured by use of a penetrometer.[12] (Reproduced with permission from Q. Hu, Y. Fang, Y. Yang, N. Ma and L. Zhao, *Food Res. Int.*, 2011, **44**, 1589.)

As mentioned above, O_2 can trigger or accelerate the oxidation of food components and facilitate the growth of aerobic microorganisms. Therefore, the photocatalytic scavenging of O_2 may be another reason for adding TiO_2 to the package material in order to maintain O_2 at a lower concentration. In fact, this effect can readily be assigned to the consumption of O_2 in photocatalytic reactions. Attention has been drawn to it in the case of an ethyl cellulose polymer film coated with TiO_2.[28] Note that the photocatalytic role of O_2 can be either favourable as, for instance, in the oxidation of endogenous ethene (*vide supra*) or unfavourable if compounds involved in the flavours of the packaged food are transformed.

12.2.5 Potential Barriers to the Use of TiO_2 in Packages

Despite the aforementioned potential advantages, the photocatalytic effects of TiO_2 can also be a barrier to implementing this technology in food packaging. Firstly, the package must be sufficiently resistant to photocatalytic oxidation so as not to be damaged to an unacceptable extent from the time the food is introduced into it to the time it is completely consumed by the customer. Secondly, the food in contact with the package must not be altered *via* photocatalytic oxidation. Thirdly, the intermediate products formed by photocatalytic transformation of microorganisms and gases contained in the space between the food and the package must remain at sufficiently low concentrations in order to not change the quality of the food, in particular its taste and odour. This latter condition is likely to be fulfilled insofar as detecting intermediate products of photocatalytic degradation is usually difficult because these products are more easily degraded than the initial compounds; one explanation for their faster degradation is that they are in the adsorbed state or close to the photocatalytic surface where they were formed. Indeed, to the author's knowledge, no unfavourable effect on the organoleptic quality of food has been reported. Fourthly, the possible release of micro- or nanoparticles of TiO_2 and of Ag, Cu or Zn, if these metals are also included, has also been cited in opposition to the use of this kind of packaging because of concern about the possible health effect of particles.[29,30] Regulations, though, do not seem to have been issued yet. Finally, the question of the environmental impact of the empty package when it is discarded must also be addressed.

12.3 Roles of TiO_2 and TiO_2 Photocatalysis in Medical Implants

12.3.1 Role of Non-UV-Irradiated TiO_2

Titanium is a common component of medical implants.[31] This metal is readily covered by layers of titanium oxides, among which is TiO_2. These oxides play a prominent role in facilitating the integration of the implants into the surrounding bone structure, thereby preventing implant loosening. This osseointegration (*i.e.*, bone integration) can even be improved if the future

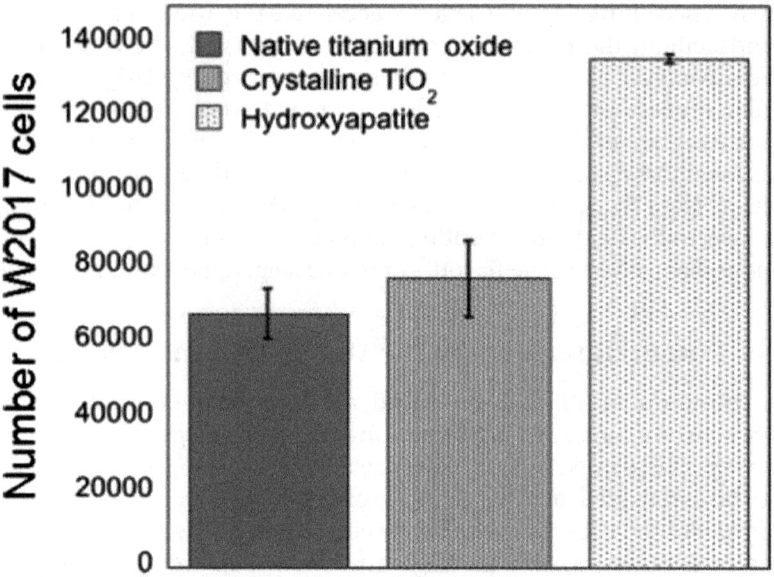

Figure 12.8 Relative viability of mesenchymal cells after 48 h on native titanium oxide (*i.e.*, spontaneously formed on Ti), crystalline TiO_2 obtained by use of physical vapour deposition, and hydroxyapatite surfaces.[33] (Reproduced with permission from S. Piskounova, J. Forsgren, U. Brohede, H. Engqvist and M. Strømme, *J. Biomed. Mater. Res., Part B*, 2009, **91B**, 780.)

implant is immersed in a solution containing calcium and phosphate ions (such as phosphate-buffered saline (PBS)), which leads to the formation of a hydroxyapatite layer ($Ca_{10}(PO_4)_6(OH)_2$). Indeed, bone contains apatite nanocrystals,[32] so that the hydroxyapatite layer at the implant surface makes the implant bioactive, that is, able to chemically bond to the bone structure. Figure 12.8 illustrates the positive effects of the deposition of both crystalline TiO_2 and hydroxyapatite on the viability of mesenchymal cells that can generate bone cells.[33] Bone formation around the implant can be further improved if bone morphogenetic proteins,[33] acting as growth agents, or osteoporosis drugs[34] are adsorbed on the hydroxyapatite layer or on the TiO_2 carrier. This functionality of the hydroxyapatite or TiO_2 surfaces can easily be obtained by simply soaking overnight the future implant in a solution of the appropriate agent for osseointegration.

12.3.2 Roles of UV-Irradiated TiO_2

Beyond the properties of titanium oxides to help promote, in the absence of UV-irradiation, bone formation around the implants, the potential effects of photo-excited TiO_2 have been explored. Indeed, UV-irradiation of the future implant has been reported to enhance osseointegration.[35–37] This effect was attributed, at least principally, to UV-induced hydrophilicity, which would be at the origin of a flattened morphology of the tissue cells in contact with the implant. Unsurprisingly, photocatalytic removal of carbon-containing species resulting from the pre-UV-irradiation was indeed shown by XPS.[37]

On the other hand, prevention of bacterial infections is obviously a major issue for implants. Therefore, the well-established capability of UV-irradiated TiO_2 to inactivate/degrade bacteria and diminish the adhesion of biofilms has been studied, perhaps particularly in the case of dental implants. The first question to be addressed has been to determine whether the incorporation of TiO_2 in the resin used as a dental adhesive is not detrimental to the implant adhesive properties. For example, 20 wt% of TiO_2 P25 in a usual dental resin (the TiO_2–resin mixture was sonicated to minimise TiO_2 aggregation) did not significantly affect the implant adhesion.[38] This amount of TiO_2 was sufficient to realize the benefit of the antibacterial effect when the resin was UV-irradiated. To wit, proportionality was found between the irradiance and the decrease in the number of viable *Staphylococcus mutans* bacteria, which were selected because of their detrimental role in the formation and growth of dental plaques (Figure 12.9).[39]

Interestingly, it was discovered that the antibacterial effect was maintained for several tens of minutes after the end of the UV-irradiation.[38] With respect to TiO_2-containing resins that had not been UV-irradiated, this effect was observed *in vitro* by use of five pathogens (*E. coli*, *Staphylococcus epidermis*, *S. pyogenes*, *S. mutans* and *E. faecalis*) which were inactivated to different extents. Unsurprisingly, the post-UV-irradiation effect did not occur in the absence of TiO_2 in the resin. It was also not found with TiO_2 particles that were not incorporated into the resin; accordingly, it was attributed to an interaction between TiO_2 and the resin.

A post-UV-irradiation effect can only arise from the persistent existence of intermediate species that are relatively stable. On the basis of the persistent blue colour of the TiO_2-containing resin over several tens of minutes, these intermediate species were proposed to be Ti^{III} ions corresponding to photogenerated electrons that had been trapped at Ti^{IV} ions.[38] Their persistence was suggested to be due to an exposure of the TiO_2 particles to O_2 much lower than in the absence of resin. These trapped electrons would intervene in the inactivation of the bacteria used to test the disinfection properties, provided the test was not carried out too long after the cessation of the UV-irradiation to avoid them being scavenged by O_2.

No change in bacterial adhesion was found for the bacterium studied (*Staphylococcus epidermis*) that was the most inactivated as a result of UV-irradiation of the TiO_2-containing resin.[38] This illustrated the stability of the bacterium structure under the irradiation used (365 nm; 10 mW cm^{-2}) for 1 h only. On the other hand, as expected, the viability of the previously adhering bacteria was decreased. Moreover, the experiment allowed one to show that the post-UV-irradiation inactivating effect decreased at times >30 min but still occurred until about 120 min (Figure 12.10). As UV-irradiation can be somewhat difficult to utilise in dentistry, this prolonged preoperative disinfection is of interest. This intriguing effect, which would leave more time for the physician to operate, has to be confirmed.

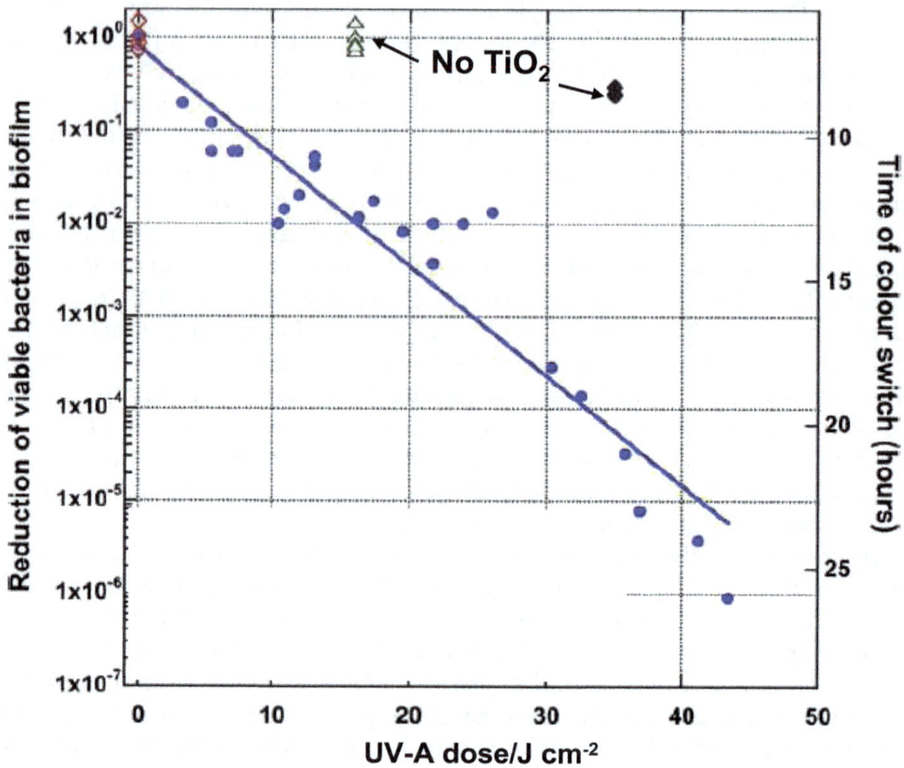

Figure 12.9 Relationship between the UV-A dose and the corresponding reduction of viable *Staphylococcus mutans* population (left-hand axis) in the biofilm. This reduction was determined using a colorimetric calibration (right-hand axis). The solid circles refer to a resin containing 20 wt% of TiO_2 P25 particles.[39] (Reproduced with permission from Y. Cai, M. Strømme, A. Melhus, H. Engqvist and K. Welch, *J. Biomed. Mater. Res., Part B*, 2014, **102B**, 62.)

In conclusion, taking into account the advantages regarding both disinfection and photo-induced cleaning related to hydrophilicity which helps integration of implants in the bone structure, the use of UV-irradiation in the field of implants has assets that can be developed.

12.4 Potential Use of TiO_2 Photocatalysis for Chemical Compound Delivery

12.4.1 Generalities

The use of photo-excited TiO_2 has been investigated for controlled delivery of chemical compounds, which is an important topic.[40] These compounds could be drugs, personal care products (*e.g.*, deodorants, fragrances), repellents, insecticides (and other pesticides), *etc.* Two delivery systems have

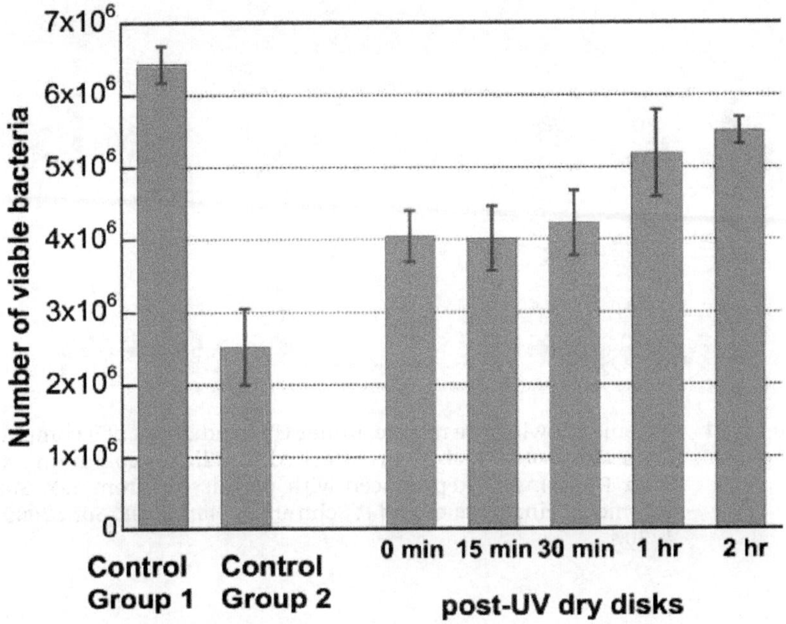

Figure 12.10 Viability of adhered *Staphylococcus epidermidis* bacteria on resin–TiO_2 disks. Post-UV dry discs were cultured following the delay periods indicated after the UV pre-treatment. Control Group 1 corresponded to resin–TiO_2 discs without UV pre-treatment. Control Group 2 corresponded to resin–TiO_2 discs also without UV pre-treatment, but irradiated during the adhesion culturing. Standard deviations were derived from three measurements.[38] (Reproduced with permission from Y. Cai, M. Strømme and K. Welch, *PLoS One*, 2013, **8**, e75929, DOI: 10.1371/journal.pone.0075929.)

been considered. Depending on its chemical structure, the compound to be delivered can be (1) directly attached to TiO_2 (as was proposed for TiO_2 nanotubes (TNT)[41,42]) *via* a coupling agent[43] designed to be broken on photo-exciting TiO_2 (Figure 12.11); (2) contained in microcapsules made of polymers easily degradable by photocatalysis[44,45] and deposited on a TiO_2 film that can be supported on various materials (Figure 12.12); alternately, TiO_2 particles can be deposited on the microcapsules. In both cases, one of the issues is to release the compound at a rate that is suitable for the application. Obviously, the greatest part of the attached or encapsulated compound must remain intact.

12.4.2 Use of TiO_2 Nanotubes Containing the Chemical to be Delivered

In the first type of systems, the internal surface of TNT was first functionalised by a silane coupling agent. This agent allowed one to bind the compound to be delivered. The underlying idea was that photo-excited TiO_2

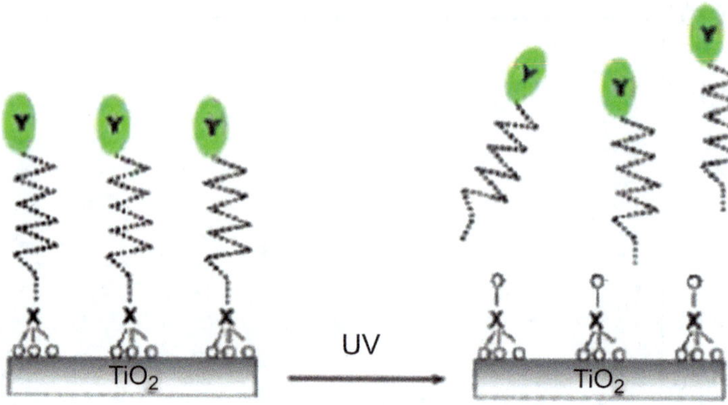

Figure 12.11 Scheme showing the release, under UV-irradiation, of a compound Y from TiO_2 onto which Y was attached by a linker containing X (*e.g.*, Si or P) atoms.[42] (Reproduced with permission from Y.-Y. Song, F. Schmidt-Stein, S. Bauer and P. Schmuki, *J. Am. Chem. Soc.*, 2009, **131**, 4230.)

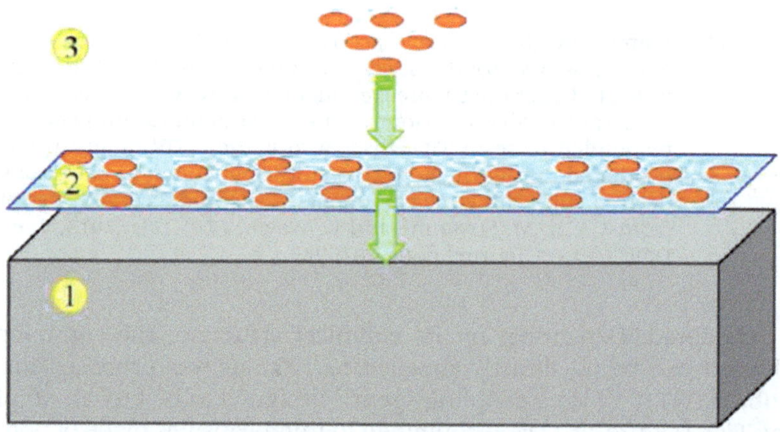

Figure 12.12 Scheme illustrating the incorporation of microcapsules containing the chemical compound to be delivered (3) into a TiO_2 film (2) supported on a substrate (1).[45] (Reproduced with permission from L. F. Oliveira, J. Marques, P. J. G. Coutinho, P. Parpot and C. J. Tavares, *J. Phys.: Conf. Ser.*, 2013, **439**, 012018, DOI: 10.1088/1742-6596/439/1/012018.)

will enable breaking of the silane link without degrading the attached compound, which thereby would be released (Figure 12.11).[41–43] In one of the studies illustrating this possibility, the compound selected, for convenient detection, was a dye fluorescing at 402 nm.[41] Accordingly, release of the dye was monitored *via* the appearance of fluorescence in the aqueous solution surrounding the TNT. The fact that the fluorescence was not altered when the dye was detached from the TNT was regarded by the authors as evidence that the dye was not significantly degraded by photocatalysis. In other words, the silane agent coordinated to TiO_2 was attacked photocatalytically before the dye was; the resulting scission of the silane chain permitted the dye to be released unaltered.

In another paper[42] from the same group, a further feature was added: the top of the TNT was capped with a hydrophobic layer of octadecylphosphonic acid. The aim was to prevent unwanted, uncontrolled release into the aqueous solution of the compound to be delivered, which was introduced into the TNT. In this case, the compound was horse radish peroxidase, a protein that can be detected by a colorimetric reaction. It was attached to TiO_2 by a silane linker using 3-(triethoxysilyl)-1-propanamine. Photo-excitation of TiO_2 was found to enable the release of horse radish peroxidase in proportion to the irradiance. Control experiments showed that both the attachment of horse radish peroxidase and the hydrophobic cap were required for a controlled release generated by UV-irradiation only (Figure 12.13). According to the authors, this procedure could be extended to the delivery of other proteins.

For medical objectives, two *a priori* interesting features were proposed. One was to coat the interior of the TNT with magnetic iron-based particles.[41] This coating will allow delivery of a drug at specific sites within the human body by use of an appropriate magnet. The other proposal was to replace

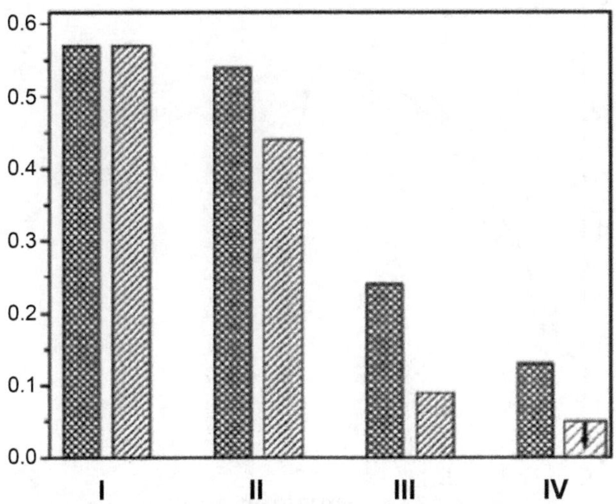

Figure 12.13 Relative fractions of horse radish peroxidase released from TNT with (clearer bars) or without (darker bars) UV-irradiation when the enzyme was initially placed in unmodified TNT (I), placed in TNT with a hydrophobic cap (II), attached in the TNT without a hydrophobic cap (III), and attached in the TNT with a hydrophobic cap (IV).[42] (Reproduced with permission from Y.-Y. Song, F. Schmidt-Stein, S. Bauer and P. Schmuki, *J. Am. Chem. Soc.*, 2009, **131**, 4230.)

UV-irradiation by X-ray irradiation.[46] Indeed, excitation of semiconductors by X-rays was reported long ago[47] as mentioned in ref. 46. X-Ray-irradiated TiO_2 was then shown to be able of oxidising and decomposing ferrocyanide anions electrochemically and the formation of electron–hole pairs was proposed to result from the relaxation process of the excited Ti cations.[48] For drug delivery application, a silane-attached Zn-porphyrin was shown to be released from the supporting TiO_2 using X-ray-irradiation (Figure 12.14).[46] The use of X-rays instead of UV radiation would permit control of drug delivery from outside the human body.

12.4.3 Use of TiO_2 Photocatalysis to Deliver Chemicals Contained in Microcapsules

Trials to demonstrate the possibility of using TiO_2 photocatalysis to release a chemical compound from microcapsules have recently been reported.[44,45] The microcapsules were made of a polyamide polymer obtained from ethane-1,2-diamine and decanedioyl dichloride (Scheme 12.1) by use of poly(ethenol) as a surfactant.

The microcapsules were spherical with diameters ranging from 20 to 100 μm (Figure 12.15a). As dodecane was added to decanedioyl dichloride during the polymerisation process, this hydrocarbon was used as the model compound

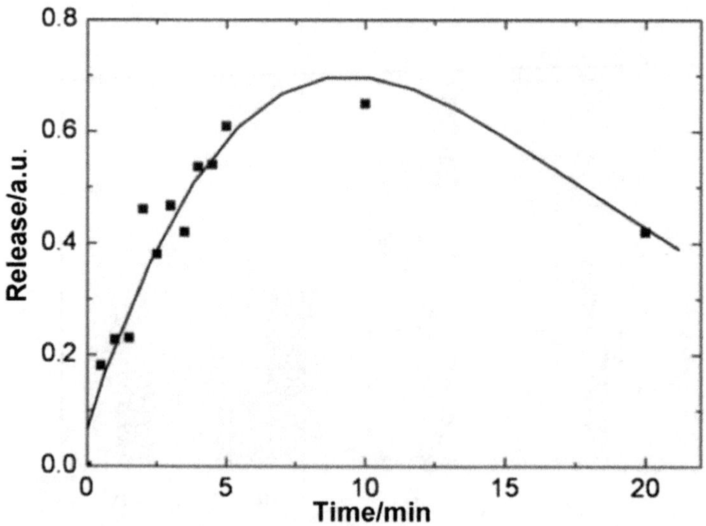

Figure 12.14 Curve showing the initial release, under X-ray irradiation, of a Zn-porphyrin attached to nanotubular TiO_2 layers. The decline indicates that the degradation rate of the released Zn-porphyrin prevailed over the release rate when most of the Zn-porphyrin was already released.[46] (Reproduced with permission from F. Schmidt-Stein, R. Hahn, J.-F. Gnichwitz, Y.-Y. Song, N. K. Shrestha, A. Hirsch and P. Schmuki, *Electrochem. Commun.*, 2009, **11**, 2077.)

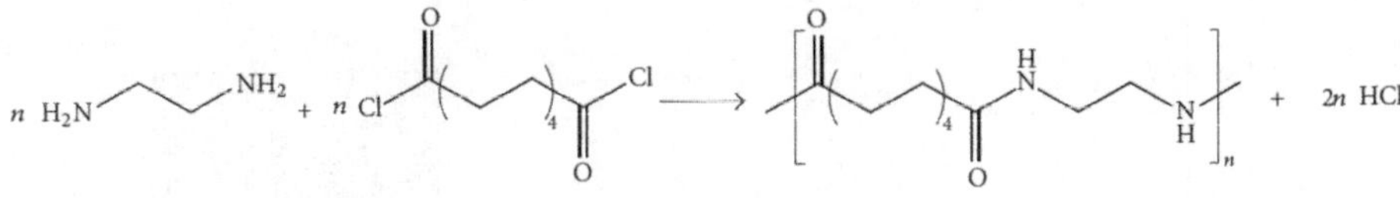

Scheme 12.1 Chemical equation corresponding to the formation of the polyamide constituting the microcapsules shown in Figure 12.15.

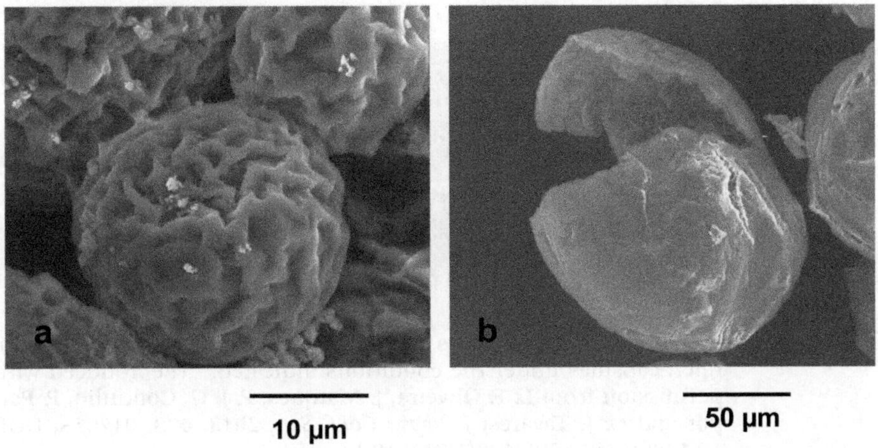

Figure 12.15 Pictures obtained by scanning electron microscopy of polyamide microcapsules loaded with TiO_2 nanoparticles before (a) and after (b) UV-A irradiation.[44] (Reproduced with permission from J. Marques, L. F. Oliveira, R. T. Pinto, P. J. G. Coutinho, P. Parpot, J. R. Góis, J. F. J. Coelho, F. D. Magalhães and C. J. Tavares, *Int. J. Photoenergy*, 2013, **2013**, 712603, DOI: 10.1155/2013/712603.)

susceptible to be released by photocatalysis. Unfortunately, dodecane was not well-contained in the microcapsules and some amount was progressively released even in the absence of UV-irradiation. This release was increased under irradiation (6.7 mW cm^{-2} with a maximum at 365 nm) in the absence of TiO_2, showing some sensitivity of the organic microcapsules to UV light (inasmuch as any increase in temperature was taken into account). By contrast, it was not significantly changed in the dark as a result of the deposition of TiO_2 P25 particles on the microcapsules. As expected, it was markedly increased when the microcapsules with deposited TiO_2 P25 were UV-irradiated. Figure 12.16 summarises these results. Opening of the organic microcapsules arising from the photoexcitation of TiO_2 was shown by scanning electron microscopy (Figure 12.15b).

To improve the delivery of compounds, such as repellents and insecticides, by sunlight in places where electricity is not available or its use is not suitable, TiO_2 samples modified by N doping/incorporation were also successfully tested for dodecane release[45] in line with the attempts, detailed elsewhere in this book, to extend TiO_2 sensitivity to longer wavelengths.[8,9]

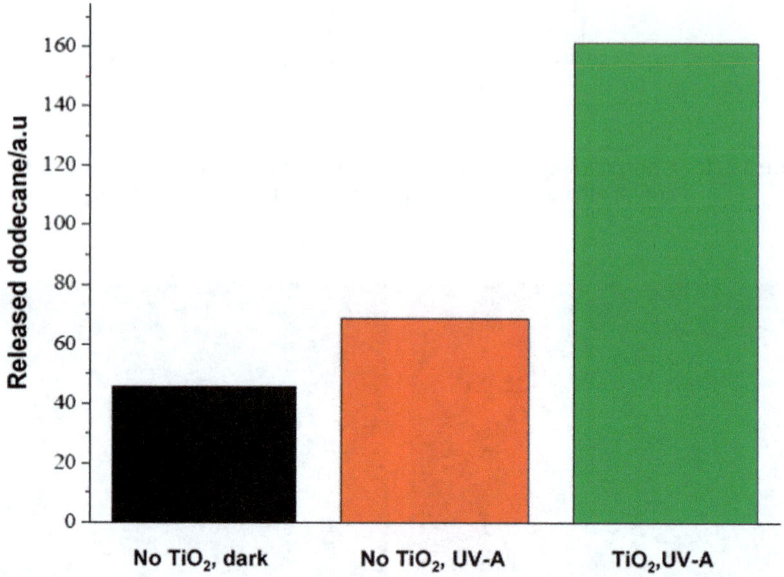

Figure 12.16 Variations in the release of dodecane contained in polyamide microcapsules under the conditions indicated.[45] (Reproduced with permission from L. F. Oliveira, J. Marques, P. J. G. Coutinho, P. Parpot and C. J. Tavares, *J. Phys.: Conf. Ser.*, 2013, **439**, 012018, DOI: 10.1088/1742-6596/439/1/012018.)

12.4.4 Issues About the Use of TiO_2 Photocatalysis for Delivery of Chemicals

In summary, the possibility to deliver a chemical compound incorporated in TNT or organic microcapsules by taking advantage of the degrading properties of photo-excited TiO_2 has been demonstrated. However, in practice, the issue of avoiding unwanted release of the active compound from the TNT or microcapsules without UV-irradiation must be solved. This involves complex engineering capable of manufacturing the materials reproducibly. The question of degradation products issued from either the microcapsules or the silane chain (or other linkers that could be used to attach the active compound) has not been addressed sufficiently, even if, for example, XPS spectra showed that the Si signal of the silane chain attaching a Zn-porphyrin was not markedly decreased upon X-ray-irradiation that caused the release of the Zn-porphyrin chosen as a model protein to be delivered (Figure 12.17).[46] In addition, more detailed studies with model compounds representative of those whose delivery would be of interest are needed to determine to what extent these compounds could be photocatalytically degraded in the process, which can potentially give rise to undesirable products. Clearly, this potential application of photocatalysis is still at the research stage.

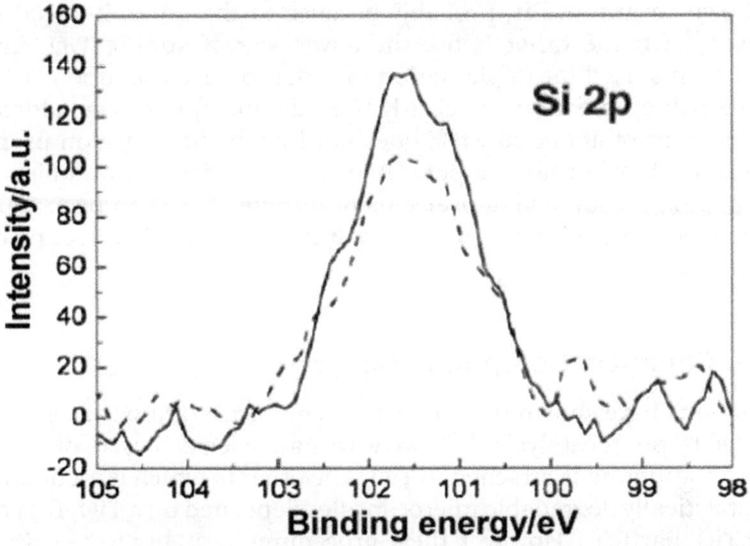

Figure 12.17 XPS spectra of Si initially (solid line) and after the complete release (dashed line), under X-ray irradiation, of a Zn-porphyrin attached to nanotubular TiO_2 layers *via* a silane chain.[46] (Reproduced with permission from F. Schmidt-Stein, R. Hahn, J.-F. Gnichwitz, Y.-Y. Song, N. K. Shrestha, A. Hirsch and P. Schmuki, *Electrochem. Commun.*, 2009, **11**, 2077.)

12.5 Conclusions

12.5.1 Food Packaging

The benefits of the use of TiO_2 are clear on the basis of various studies, including measurements dealing with not only microorganisms but also several kinds of food. Adverse effects have not been reported so far. Trials concerning the incorporation of TiO_2 into the polymeric materials commonly utilised for packaging have provided satisfactory results. Although precise information is lacking about the added cost, it should be more than counterbalanced by the increase in food shelf-life in many cases. Therefore, the main obstacle for the development of this application of photocatalysis seems to be the concern about the potential risk arising from a possible release of micro- or nanoparticles of TiO_2,[29,30] first when the package is used, then when the empty package is discarded, most likely at random by the customer (*cf.* Section 12.5.4).

12.5.2 Medical Implants

Titanium oxides spontaneously formed on Ti-implants and additional TiO_2 deposited on implants are known to facilitate integration to bone. This property does not necessitate UV-irradiation. However, pre-UV-irradiation has been shown to improve it, probably because of the photo-induced hydrophilicity.[35–37] On the other hand, the advantage of adding TiO_2 into the adhesive resins used for implantation in order to decrease, upon UV-irradiation, the risk of infection has clearly been demonstrated. The added TiO_2, up to a content of about 20 wt%, does not impair the adhesion properties of the resin.[38] Furthermore, a post-UV-irradiation disinfection effect of the TiO_2-containing resin lasting over tens of minutes has even been reported. This would be of great interest by prolonging the microbial protection during operative acts.[38]

12.5.3 Chemical Compound Delivery

Investigations have shown that delivery of chemical compounds can indeed be effected by photocatalysis.[41–46] Two ways have been explored, one in which the active compound is attached to TiO_2, the other in which it is contained in photocatalytically degradable microcapsules deposited on a TiO_2 film or supporting TiO_2 particles. However, these procedures seem not to be sufficiently mastered yet to allow applications. Further research will be needed, in particular to adjust the delivery rate to the particular application, for instance by regulating the irradiance, and to determine the fate of the microcapsules or linkers once they are degraded and/or broken. This last issue could likely limit the applications to, for instance, the delivery of repellents and insecticides, and possibly the transdermal delivery of drugs and personal care products; it could exclude the use for delivery of drugs in the body. Obviously, as for the other uses of photocatalysis, the possible release of micro- or nanoparticles of TiO_2 is also a concern.[29,30]

12.5.4 Comparisons About the Viability of These Three Applications

The viability of these applications is disparate. As mentioned above, preoperative UV-irradiation of Ti medical implants is ready to be developed through technical adjustments especially regarding coatings, and determination of good practice. By contrast, delivery of chemical compounds clearly needs further studies. Moreover, as in the case of food packaging, the present uncertainty about the risk that could arise from the possible release of micro- or nanoparticles[29,30] could be an impediment to implementation. Indeed, *in vivo* experiments have shown that fine particles inhaled or swallowed (or introduced into the body through skin lesions) can go through biological barriers and cause damage. Although the experimental methodologies can sometimes be criticised, a danger was pointed out. Therefore, to evaluate the risk, the release of particles from TiO_2-containing materials must be determined thoroughly, that is, measurement standards must be established and the results must be checked by independent agencies. Finally, the potential inhalation of TiO_2-containing particles emitted by the new materials, especially those that would be used for food packaging, will have to be compared with that of particles emitted by other sources, particularly combustions of various kinds. Extensive on-going research about other photocatalytic uses of TiO_2 particles (see the other chapters in this book) and the uses of other micro- or nanoparticles in a wide range of domains should bring more informative data on this risk and hopefully lead to proper regulations, notably relevant to thc applications considered here.

Regarding the amount of TiO_2 that would be necessary for these applications, it is obvious that food packaging would be of high interest for the manufacturers of TiO_2. Comparatively, interest in chemical compound delivery would be markedly lower. Although the amount of TiO_2 that would be used for the coating of medical implants would be extremely low, the expected profits for the manufacturers of implants should be appreciable, given that this type of application concerns a medical area of growing importance.

References

1. P. Pichat, Self-cleaning materials based on solar photocatalysis, in *New and Future Developments in Catalysis*, ed. S. L. Suib, Elsevier, 2013, vol. 7, Solar Catalysis, pp. 167–190.
2. P. Pichat, *Appl. Catal., B*, 2010, **99**, 428.
3. S. Hay, T. Obee, Z. Luo, T. Jiang, Y. Meng, J. Ye, S. Murphy and S. Suib, *Molecules*, 2014, **20**, 1319.
4. T. Ochiai, E. Ichihashi, N. Nishida, T. Machida, Y. Uchida, Y. Hayashi, Y. Morito and A. Fujishima, *Molecules*, 2014, **19**, 17424.
5. *Photocatalysis and Water Purification: From Fundamentals to Recent Applications*, ed. P. Pichat, Wiley-VCH Verlag GmbH, Weinheim, 2013.
6. P. Pichat, S. Vannier, J. Dussaud and J.-P. Rubis, *Sol. Energy*, 2004, **77**, 533.
7. P. Pichat, *Water Sci. Technol.: Water Supply*, 2015, **15**(1), 1–10, DOI: 10.2166/ws.2014.083.
8. M. A. Henderson, *Surf. Sci. Rep.*, 2011, **66**, 185.
9. M. Pelaez, N. T. Nolan, S. C. Pillai, M. K. Seery, P. Falaras, A. G. Kontos, P. S. M. Dunlop, J. W. J. Hamilton, J. A. Byrne, K. O'Shea, M. H. Entezari and D. D. Dionysiou, *Appl. Catal., B*, 2012, **125**, 331.
10. E. L. Bradley, L. Castle and Q. Chaudhry, *Trends Food Sci. Technol.*, 2011, **22**, 604.
11. A. Llorens, E. Lloret, P. A. Picouet, R. Trbojevich and A. Fernandez, *Trends Food Sci. Technol.*, 2012, **24**, 19.
12. Q. Hu, Y. Fang, Y. Yang, N. Ma and L. Zhao, *Food Res. Int.*, 2011, **44**, 1589.
13. Y. Xing, X. Li, L. Zhang, Q. Xu, Z. Che, W. Li, Y. Bai and K. Li, *Prog. Org. Coat.*, 2012, **73**, 219–224.
14. X. Wei, Z. Yang, S. L. Tay and W. Gao, *Appl. Surf. Sci.*, 2014, **290**, 274.
15. T. Matsunaga, R. Tomoda, T. Nakajima and H. Wake, *FEMS Microbiol. Lett.*, 1985, **29**, 211.
16. T. Saito, T. Iwase, J. Horie and T. Morioka, *J. Photochem. Photobiol., B*, 1992, **14**, 369.

17. J. C. Ireland, P. Klosterman, E. W. Rice and R. M. Clark, *Appl. Environ. Microbiol.*, 1993, **59**, 1668.
18. C. Wei, W. Y. Lin, Z. Zainal, N. E. Williams, K. Zhu, A. P. Kruzic, R. L. Smith and K. Rajeshwar, *Environ. Sci. Technol.*, 1994, **28**, 934.
19. O. K. Dalrymple, E. Stefanakos, M. A. Trotz and D. Y. Goswami, *Appl. Catal., B*, 2010, **98**, 27.
20. A. Markowska-Szczupak, K. Ulfig and A. W. Morawski, *Catal. Today*, 2011, **169**, 249.
21. J. A. Rengifo-Herrera, A. G. Rincón and C. Pulgarin, in *Photocatalysis and Water Purification: From Fundamentals to Recent Applications*, ed. P. Pichat, Wiley-VCH Verlag GmbH, Weinheim, 2013, pp. 295–309.
22. C. Maneerat and Y. Hayata, *Int. J. Food Microbiol.*, 2006, **107**, 99.
23. C. Chawengkijwanich and Y. Hayata, *Int. J. Food Microbiol.*, 2008, **123**, 288.
24. A. Kubacka, C. Serrano, M. Ferrer, H. Lünsdorf, P. Bielecki, M. L. Cerrada, M. Fernandez-Garcıa and M. Fernandez-Garcıa, *Nano Lett.*, 2007, **7**, 2529.
25. H. Bodaghi, Y. Mostofi, A. Oromiehie, Z. Zamani, B. Ghanbarzadeh, C. Costa, A. Conte and M. A. Del Nobile, *LWT--Food Sci. Technol.*, 2013, **50**, 702.
26. H. Li, F. Li, L. Wang, J. Sheng, Z. Xin, L. Zhao, H. Xiao, Y. Zheng and Q. Hu, *Food Chem.*, 2009, **114**, 547.
27. M. Gumiero, D. Peressini, A. Pizzariello, A. Sensidoni, L. Iacumin, G. Comi and R. Toniolo, *Food Chem.*, 2013, **138**, 1633.
28. A. Mills, G. Doyle, A. M. Peiro and J. Durrant, *J. Photochem. Photobiol., A*, 2006, **177**, 328.
29. J. L. Kang, C. Moon, H. S. Lee, H. W. Lee, E.-M. Park, H. S. Kim and V. Castranova, *J. Toxicol. Environ. Health, Part A*, 2008, **71**, 478.
30. P. Pichat, *J. Adv. Oxid. Technol.*, 2010, **13**, 238.
31. M. Textor, C. Sittig, V. Frauchiger, S. Tosatti and D. M. Brunette, in *Titanium in Medicine*, ed. D. M. Brunette, Springer, Berlin, 2001, pp. 171–230.
32. C. Rey, C. Combes, C. Drouet and M. J. Glimcher, *Osteoporosis Int.*, 2009, **20**, 1013A.
33. S. Piskounova, J. Forsgren, U. Brohede, H. Engqvist and M. Strømme, *J. Biomed. Mater. Res., Part B*, 2009, **91B**, 780.
34. N. Harmankaya, J. Karlsson, A. Palmquist, M. Halvarsson, K. Igawa, M. Andersson and P. Tengvall, *Acta Biomater.*, 2013, **9**, 7064.
35. T. Sawase, R. Jimbo, K. Baba, Y. Shibata, T. Ikeda and M. Atsuta, *Clin. Oral Implants Res.*, 2008, **19**, 491.
36. H. Aita, N. Hori, M. Takeuchi, T. Suzuki, M. Yamada and M. Anpo, *Biomater.*, 2009, **30**, 1015.
37. P. Linderbäck, N. Harmankaya, A. Askendal, S. Areva, J. Lausmaa and P. Tengvall, *Biomaterials*, 2010, **31**, 4795.
38. Y. Cai, M. Strømme and K. Welch, *PLoS One*, 2013, **8**, e75929, DOI: 10.1371/journal.pone.0075929.
39. Y. Cai, M. Strømme, A. Melhus, H. Engqvist and K. Welch, *J. Biomed. Mater. Res., Part B*, 2014, **102B**, 62.

40. K. E. Uhrich, S. M. Cannizzaro, R. S. Langer and K. M. Shakesheff, *Chem. Rev.*, 1999, **99**, 3181.
41. N. K. Shrestha, J. M. Macak, F. Schmidt-Stein, R. Hahn, C. T. Mierke, B. Fabry and P. Schmuki, *Angew. Chem.*, 2008, **120**, 1.
42. Y.-Y. Song, F. Schmidt-Stein, S. Bauer and P. Schmuki, *J. Am. Chem. Soc.*, 2009, **131**, 4230.
43. Y.-Y. Song, H. Hildebrand and P. Schmuki, *Surf. Sci.*, 2010, **604**, 346.
44. J. Marques, L. F. Oliveira, R. T. Pinto, P. J. G. Coutinho, P. Parpot, J. R. Góis, J. F. J. Coelho, F. D. Magalhães and C. J. Tavares, *Int. J. Photoenergy*, 2013, 712603, DOI: 10.1155/2013/712603.
45. L. F. Oliveira, J. Marques, P. J. G. Coutinho, P. Parpot and C. J. Tavares, *J. Phys.: Conf. Ser.*, 2013, **439**, 012018, DOI: 10.1088/1742-6596/439/1/012018.
46. F. Schmidt-Stein, R. Hahn, J.-F. Gnichwitz, Y.-Y. Song, N. K. Shrestha, A. Hirsch and P. Schmuki, *Electrochem. Commun.*, 2009, **11**, 2077.
47. K. G. MacKay, *Phys. Rev.*, 1949, **76**, 1537.
48. K. Tamura, Y. Ohko, H. Kawamura, H. Yosikawa, T. Tatsuma, A. Fujishima and J. Mizuki, *Electrochim. Acta*, 2007, **52**, 6938.

Subject Index